D0165591

SOLID STATE THEORY

SOLID STATE THEORY

Walter A. Harrison
Professor of Applied Physics
Stanford University

Dover Publications, Inc.
New York

Published in Canada by General Publishing Company, Ltd., 30 Lesmill Road, Don Mills, Toronto, Ontario.
Published in the United Kingdom by Constable and Company, Ltd., 10 Orange Street, London WC2H 7EG.

This Dover edition, first published in 1980, is an unabridged and corrected republication of the work originally published in 1970 by McGraw-Hill, Inc.

International Standard Book Number: 0-486-63948-7
Library of Congress Catalog Card Number: 79-56032

Manufactured in the United States of America
Dover Publications, Inc.
180 Varick Street
New York, N.Y. 10014

TO LUCKY AND TO MY BOYS,
Rick, John, Will and Rob

PREFACE

This manuscript has been designed to serve as a text for a one-year graduate course in solid state theory. A year of quantum mechanics is prerequisite (more advanced methods are introduced as needed) but introductory solid state is not. Thus the course attempts to span the subject matter which in many universities is covered in an introductory solid state course and in a quantum theory of solids course. As an alternative to such a two-year program, Stanford offers this one general course and a number of shorter courses on specific topics in solid state.

In this one-year program I attempt to develop a sufficient familiarity with the concepts and methods of solid state physics that a reader will feel moderately at home in any solid state seminar or when reading any article in solid state physics. I attempt also to eliminate the threshold which may prevent a specialist in another field from attacking a solid state problem which arises in the context of his own work. Both these objectives, in my view, can be accomplished in a single year of study.

In this context there are very few topics in traditional or modern solid state physics which can be omitted altogether. Yet the subject is much too broad for any kind of complete coverage in a single course. An attempt to resolve this dilemma has been made by developing the fundamental ideas and approaches in many areas but this necessitates giving only a very small number of detailed applications. The first chapter develops the fundamentals of group and representation theory and gives only three illustrative applications (though use is made of group theory at many points later in the text). In discussing band structure only one prototype in each category is considered. The transport equation is derived, but is applied to only a few properties. Included, when possible, are discussions of currently active problems such as the Mott transition, the electronic structure of disordered systems, tunneling, the Kondo effect, and fluctuations near critical points. Which topics are most active vary from year to year and few remain of central interest indefinitely. Their inclusion, however, adds an air of relevance to the subject matter and brings the reader to the frontiers of knowledge in a number of areas.

Because virtually all of the properties of solids are determined by the valence electrons, I have focused on electron states in much of the first third of the text. This is predominantly a formulation of the approximate descriptions in different systems which allow a treatment of properties rather than a discussion of the energy bands themselves; the latter provide the basis for the treatment of very few properties. The remaining portion of the text is devoted to discussions of solids and liquids using these approximate descriptions. This organization of the material allows parallel treatments of each property for metals, semiconductors, and insulators, though in some cases it separates the discussion of related properties in one material. Throughout the text emphasis is placed on self-consistent-field methods for the treatment of interaction among electrons and pseudopotential methods for the treatment of the interaction among electrons and ions.

The timing of this course in the graduate curriculum and the breadth of its coverage can make it a vehicle for a major clarification of quantum mechanics for the student. Such clarification comes in the comparison of the semiclassical (Boltzmann equation) treatment of screening and the corresponding quantum (Liouville equation) treatment. It comes in understanding of the breakdown of the Franck-Condon principle and it comes with the understanding of off-diagonal long-range order in vibrating lattices and in superconductivity.

A course at this level also provides an opportunity to present two impressive phenomenologies, the Landau theory of Fermi liquids and the Ginsburg-Landau theory of superconductivity. Both of these are presented in the spirit in which they were generated rather than as consequences of microscopic theories. This form of presentation seemed much more in tune with an effort to enable students to generate good physics rather than to simply use the physics of others.

Throughout, an attempt is made to maintain a balance between the physics and the mathematics of our understanding of solids. Where the theory provides general methods for treating solids, the attempt has been to make the derivations careful and complete. Where the physical concepts are in my belief central, the tendency is only to outline the mathematical analysis. In some cases the details are shifted to the problems following each chapter; in some cases the reader may go to the more complete discussions referred to. Sufficient completeness in most arguments is presented, however, (frequently more complete than the original source) so that the references are more likely to provide additional applications than clarification of the methods.

In total, I view this as a text in physics rather than in mathematical methods. The emphasis is on current problems and modern concepts, but the traditional background which provides the foundation for modern theory is also included. My desire is to provide a framework that the reader can fill in with his subsequent reading and work, and upon which he can build understanding where it does not exist today.

Walter A. Harrison

CONTENTS

SOLID STATE THEORY

Each of these four solid types is characterized by a rather specific crystalline structure, or arrangement of constituent atoms. The exact details of the structure appear not to be the determining factor in the properties of solids, but it will be useful for our later considerations to discuss the typical structures and to consider those aspects of the properties which are determined by structure alone.

In metals the atoms tend to be closely packed; i.e., they are packed as we might put billiard balls in a box. The three most common metallic structures are *face-centered cubic*, *body-centered cubic*, and *hexagonal close packed*. These three simple structures are illustrated in Fig. 1.1. In the face-centered cubic and body-centered cubic structures there is a framework of metallic atoms in a cubic array. If no other atoms were present these two structures would be *simple cubic* structures. In the face-centered cubic lattice, however, there are also metallic atoms in the center of each of the cube faces formed by the cubic array; in the body-centered cubic structure there is an additional atom at the center of each cube but none on the cube faces.

We should pause here to specify some of the notation that is useful in describing directions in cubic crystals. We think of three orthogonal axes, each oriented parallel to a cube edge. We then can specify the direction of a vector by giving the magnitude of its three components along the three axes. A direction is customarily specified by writing these three numbers side by side enclosed in square brackets; a negative number is usually denoted with a minus above. Thus the direction [100] represents a direction parallel to one of the cube edges; $[\bar{1}00]$ is in the opposite direction. The expression [110] represents a direction parallel to the diagonal of a cube face. The direction [111] represents the direction of a cube diagonal. The orientation of planes in crystals is similarly specified with parentheses; thus the expressions (100), (110), and (111) represent three planes perpendicular to the three directions given before. In common usage these expressions specify the orientation of the plane but not its position, although the position usually is taken through one of the atoms in the crystal. These crystallographic directions and planes are illustrated in Fig. 1.2. Similar notation has been generated for noncubic crystals, but none is universally used and the meaning of any such notation should be specified when it is used.

In the hexagonal close-packed structure there are planes of atoms closely packed, like billiard balls on a table. The next plane of atoms has the same arrangement, but is shifted so the atoms fit snugly between those in the first plane. The third plane has again the same arrangement, and each atom lies directly above one of the atoms in the first plane. The fourth plane is identical to the second and so on. Both the face-centered cubic and the hexagonal close-packed structures correspond to the densest possible

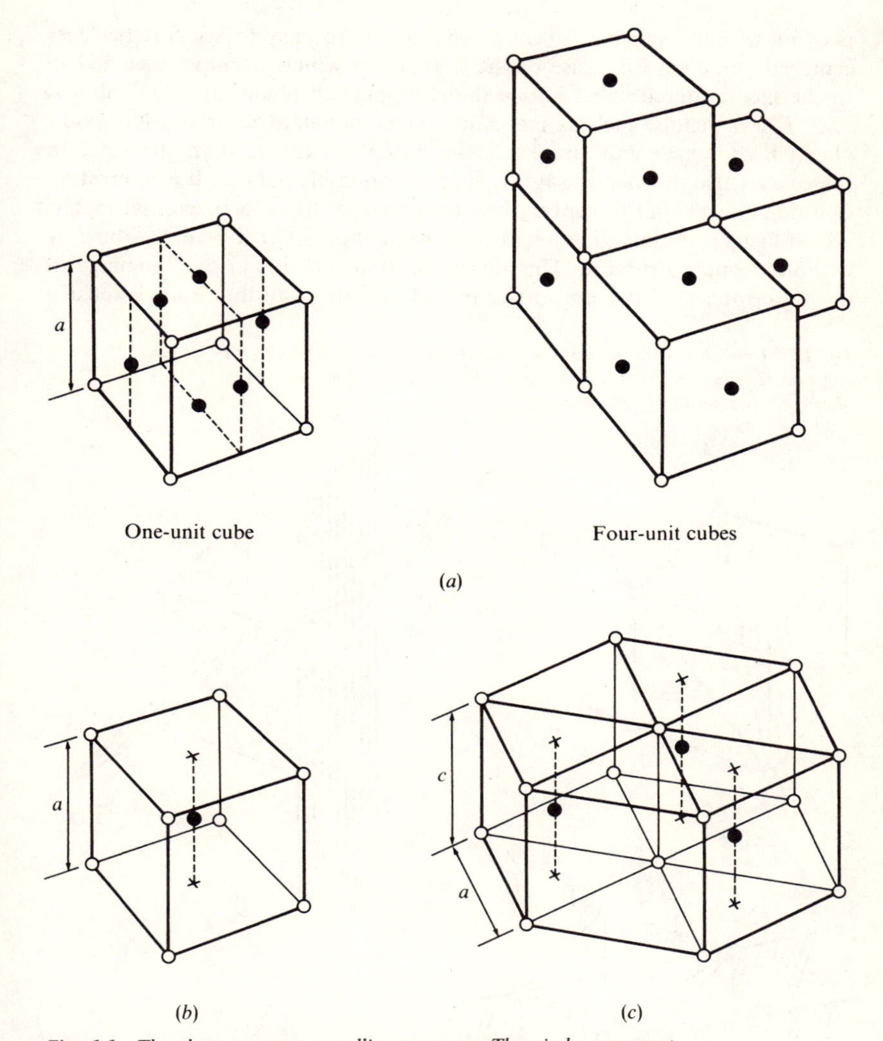

Fig. 1.1 The three common metallic structures. The circles represent atom centers. (a) In the face-centered cubic structure the atoms at face centers are shown solid to distinguish them, though in the repeated structure they may be seen to have identical environments. (b) Similarly the body-centered atom is shown solid in the body-centered cubic structure though they are equivalent positions. (c) In the hexagonal close-packed structure the environment of the solid dots differs, by a 60° rotation, from that of the empty dots.

packing of hard spheres, although this is not so easy to see for the face-centered cubic case because of the manner in which we have specified it. In the face-centered cubic lattice the close-packed planes are (111) planes.

A few metals, such as manganese and mercury, occur in more complicated structures, but those are the exception rather than the rule. In these cases also the metallic atoms are rather densely packed. It is interesting to note that only in this century has it been generally recognized that metals are in fact crystalline. They had long been thought to be amorphous, or without regular structure. This misconception remains in our language in the description of the embrittlement of metals resulting from extensive

Fig. 1.2 *Crystallographic notation in cubic crystals, here illustrated for a simple cubic structure:* (*a*) *crystallographic directions,* (*b*) *(100) and (110) planes,* (*c*) *two parallel (111) planes.*

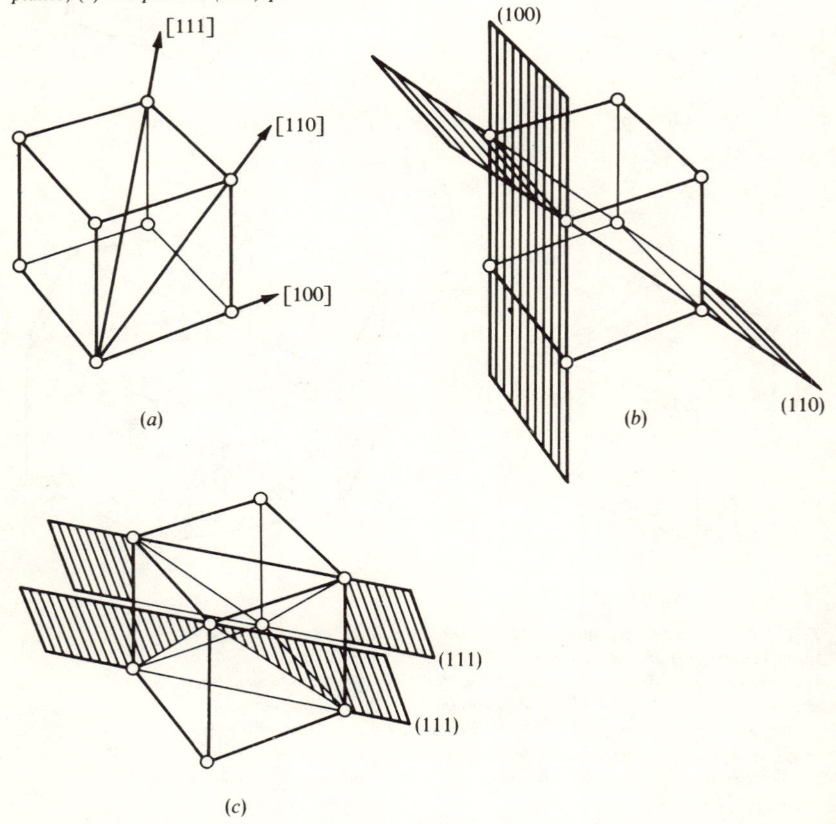

bending as "crystallization" of the metal. Although we now recognize that metals are crystalline in nature, we should perhaps emphasize that the essential feature of the metallic structure is its close packing rather than the details of its structure. If a metal is melted it loses all of its crystalline order but remains in a rather close-packed configuration. Even with this total loss of crystalline structure, however, the electrical properties remain very much the same.

A prototype insulator is the sodium chloride crystal, which is more appropriately thought of as a configuration of ions rather than of atoms. Sodium, of valence 1, gives up an electron to chlorine, of valence 7. The structure is illustrated in Fig. 1.3. Again it is based on a simple cubic array, but in this case alternate ions are sodium and chlorine. The essential feature of this insulating structure is the alternation between positive and negative ions. Sodium chloride provides perhaps the simplest structure for an insulator. Many ionic compounds form quite complex structures, but all correspond to alternate packing of positive and negative ions.

Most molecules and all rare gases form insulators when they crystallize. In cases such as ice the molecules themselves may be thought of as ionic, but in cases such as molecular hydrogen they are not. The important characteristic feature of molecular crystals is the close association of the atoms

Fig. 1.3 The sodium chloride structure. One cubic cell is shown. Note that the sodium ions form a face-centered cubic array; the chlorine ions form an interlocking face-centered cubic array.

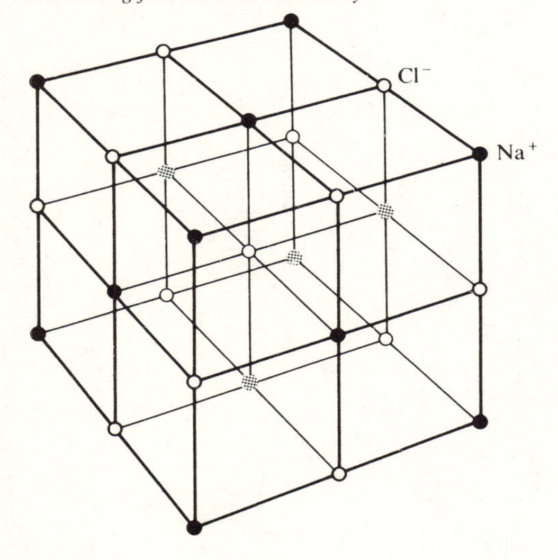

making up the molecules rather than the configuration in which the molecules are arranged.

Insulators occur also in the amorphous state, i.e., with no long-range repeating structure. Glass is the most familiar example. Thus we may think of the insulating behavior of ionic compounds as arising from the alternate arrangement of positive and negative ions rather than from the details of the crystalline structure. Similarly, when a molecular crystal is melted, the molecules generally remain intact and the insulating behavior is maintained.

Semimetallic and semiconducting behavior is more sensitive to the structure of the crystal. Bismuth is a prototype semimetal; its structure is noncubic and rather complicated. However, the packing density, viewed as the packing of spheres, is comparable to that in metals. When bismuth is melted it becomes disordered, with essentially the structure of a liquid metal, and its electrical properties become that of a metal.

Our prototype semiconductors are silicon and germanium. Both silicon and germanium have the diamond structure that is illustrated in Fig. 1.4. The diamond structure may be generated from the face-centered cubic structure by adding a second interlocking face-centered cubic structure displaced from the first one-quarter of the way along a cube diagonal [111]. This leaves any given atom surrounded by four near neighbors arranged at the corners of a regular tetrahedron centered at the given atom. The diamond structure is quite open in comparison to the metallic and semimetallic structure. If hard spheres were arranged in this configuration there would

Fig. 1.4 *The diamond structure, viewed in a direction close to the [110] direction; the overlapping circles represent atoms lying on a [110] line. Note that the full circles form the unit cube of a face-centered cubic lattice.*

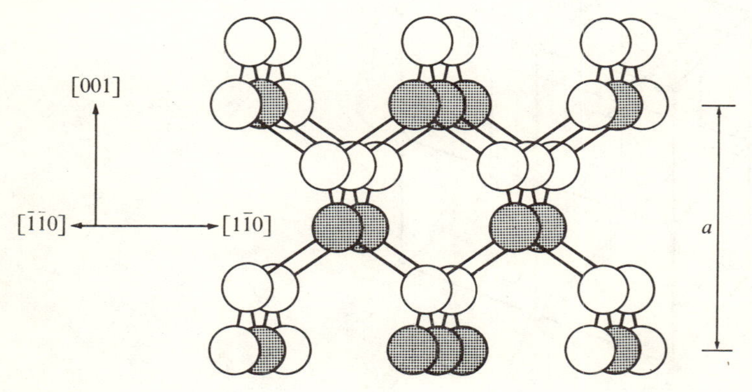

be room in the interstices to insert a second sphere for each sphere included in the original structure. Thus we might say the packing density is half that of a close-packed metal. Such open structures are typically semiconductors. They conduct only if the temperature is raised or if defects are introduced into the crystal. The semiconducting properties may be thought to be associated with the open structures. When either silicon or germanium is melted the packing density increases and they become liquid metals.

Some compounds made up of equal numbers of atoms of valence 3 and valence 5, such as indium antimonide, are also semiconductors and have structures just like the diamond structure but with the first face-centered cubic lattice being made of, say, antimony, while the second is made of indium. This is called the zincblende structure and is a structure of zinc sulfide (valence 2 and valence 6). Such two-six compounds have properties lying somewhere between those of insulators and the more familiar semiconductors.

With these differences in the arrangement of the atoms for different solids there are also significant differences in the distribution of the electrons in the crystal. In a metal, with close packing, the electrons are quite uniformly distributed through the crystal except in the immediate neighborhood of the nucleus where the density due to the core electrons is very large (and that due to the valence electrons is small). In Fig. 1.5 is a plot of the valence-electron distribution in a (110) plane in aluminum.

In ionic insulators we generally think, as we have indicated, of the outer electron on the metallic atom as having been transferred to the nonmetallic ion; however, once the ions are packed in a crystal this transfer really corresponds to only a subtle change in distribution.

In semiconductors much of the charge density due to the valence electrons ends up near the lines joining nearest neighbors. A plot of the valence-electron density in silicon is given in Fig. 1.6. This localized density forms the so-called covalent bonds in these structures. In semimetals, as in metals, the electrons are rather uniformly distributed. The importance of these differences, as well as the importance of the differences in structure, is perhaps overestimated in much thought about solids. Though the charges seem to be distributed in fundamentally different ways, they can always be rather well approximated by the superimposed electronic-charge distribution of free atoms.

A great deal can be learned about the properties of crystalline solids simply from a knowledge of their structure. Our first task will be to develop some of this knowledge. Out of this analysis will arise much of the terminology that is used to describe solids, and we will find that much of this terminology will remain appropriate when we discuss systems that do not have the simple structures.

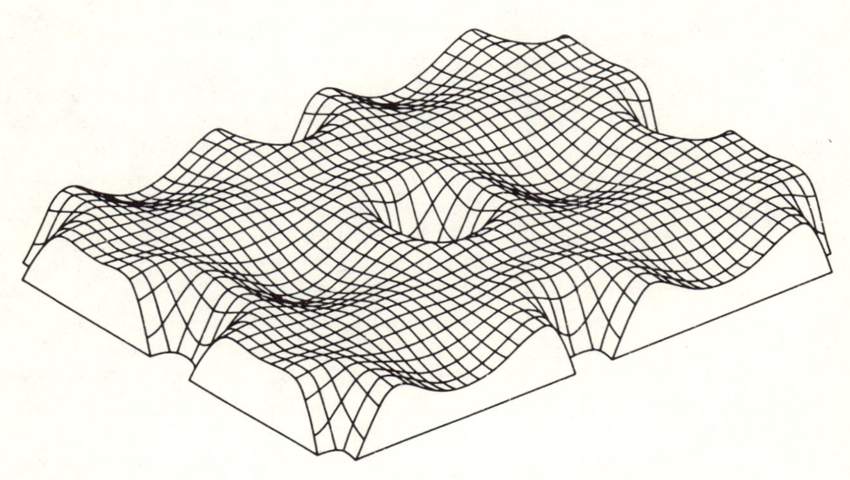

Fig. 1.5 *The valence-electron density in aluminum. The base of the figure represents a (110) plane in the crystal; the electron density is plotted above the plane. The contour lines would form a square grid if projected on the basal plane. (The calculation upon which the plot is based, as well as the plotting technique, is given by W. A. Harrison, "Pseudopotentials in the Theory of Metals," pp. 217ff., Benjamin, New York, 1965.*

Fig. 1.6 *The valence-electron density in a (110) plane in silicon. The source is the same as for Fig. 1.5.*

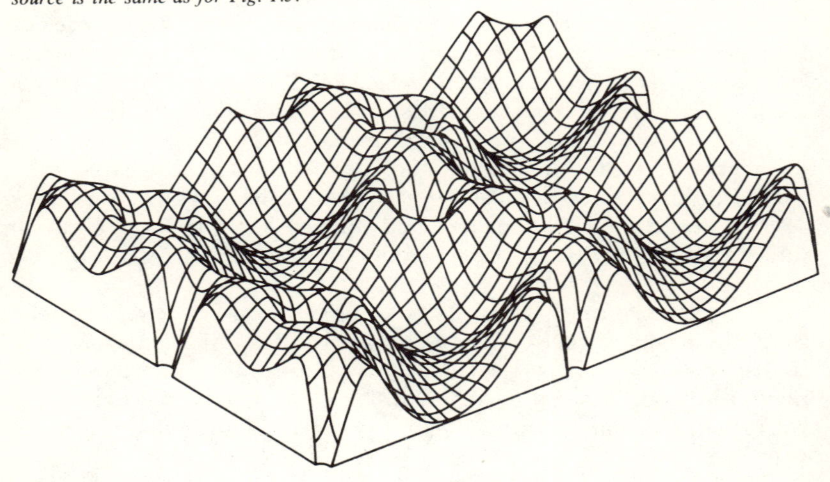

2. *Symmetry of Crystals*

The characteristic feature of all of the structures we have described is *translational invariance*. This means that there are a large number of translations,

$$\mathbf{T} = n_1\boldsymbol{\tau}_1 + n_2\boldsymbol{\tau}_2 + n_3\boldsymbol{\tau}_3 \tag{1.1}$$

which a perfect crystal may undergo and remain unchanged. This, of course, moves the boundaries, but we are interested in behavior in the interior of the crystal. The n_i are integers and any set of n_i leaves the crystal invariant.

To obtain the most complete description of the translational invariance we select the smallest $\boldsymbol{\tau}_i$, which are not coplanar, for which Eq. (1.1) is true. These then are called the *primitive lattice translations*. These are illustrated for a two-dimensional lattice in Fig. 1.7. We have let the lattice translations originate at a lattice site. The set of points shown in Fig. 1.7 and specified in Eq. (1.1) is called the *Bravais lattice*. The volume of the parallelopiped with edges, $\boldsymbol{\tau}_1$, $\boldsymbol{\tau}_2$, $\boldsymbol{\tau}_3$, is called the *primitive cell*. Clearly the crystal is made up of identical primitive cells.

Note that in the face-centered cubic structure a set of lattice translations could be taken as the edges of a cubic cell. These would not, however, be primitive lattice translations since translations to the face center are smaller yet also take the lattice into itself. By constructing a primitive cell based upon translations from a cube corner to three face centers, we obtain a cell with a volume equal to the volume per ion; i.e., a primitive cell containing only

Fig. 1.7 *A simple two-dimensional lattice showing primitive lattice translations,* $\boldsymbol{\tau}_1$ *and* $\boldsymbol{\tau}_2$, *and a primitive cell.*

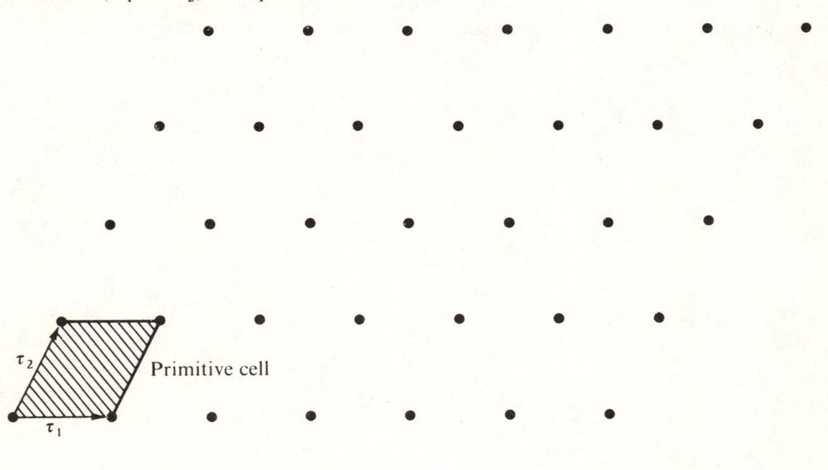

one atom. This is illustrated in Fig. 1.8. Had we used a cubic cell, we would readily see that the volume would be four times that large. The complete translational symmetry of the lattice is specified by the primitive lattice translations. A translational lattice based upon a larger cell contains only part of the symmetry information.

The specification of this translational invariance has reduced the information required to specify the structure of the crystal to the information required to specify the structure of a primitive cell.

Note that there is an atom at every point of the Bravais lattice, but there also may be other atoms within the primitive cell. The points specified, for example, might be indium ions and there also might be an antimony ion in each cell, which will be the case in indium antimonide. The points might be silicon atoms and there might be another silicon atom in each cell, yet translation of the lattice by the vector distance between the two silicon atoms will not leave the crystal invariant. This is the case for the diamond structure.

There will be other *symmetry operations* that take the crystal into itself. An inversion of the lattice through a Bravais lattice point will take the Bravais lattice into itself and it may take the crystal into itself. Various rotations, reflections, or rotary reflections may also take the crystal into itself. Clearly the combination of any two operations, each of which takes the crystal into itself, will also take the crystal into itself. The collection of all such operations, rotations, reflections, translations and combinations of these, is called the *space group* of the crystal. The space group contains all symmetry operations which take the crystal into itself.

If we take all space group operations of the crystals and omit the translation portion, the collection of all distinct rotations, reflections, and rotary reflections which remains is called the *point group*. In some crystals,

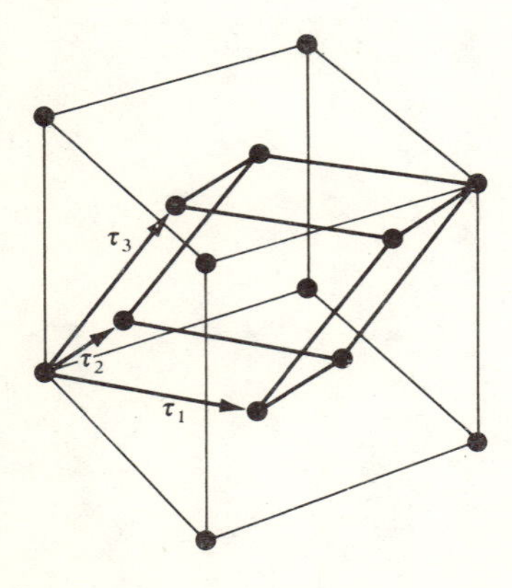

Fig. 1.8 A unit cube and the primitive cell in a face-centered cubic crystal.

certain operations of the point group will leave no atom in the crystal in the same position. Such an operation is illustrated for a two-dimensional case in Fig. 1.9. An operation combining reflection in a plane with a "partial lattice translation" (smaller than a primitive lattice translation) in the reflection plane is called *glide-plane symmetry* and the corresponding plane is called a glide plane. A second common combined operation is called a *screw axis* and consists of a rotation combined with a partial lattice translation along the rotation axis.

The symmetry operations of a crystal form a *group* (in the mathematical sense) as we shall see. The possible different point groups which are allowed in a crystal are greatly reduced from those which might be allowed in a molecule because of the translational symmetry. We may illustrate this by showing that in a crystal only rotations of 60°, 90°, or multiples of these can occur.

We construct a plane perpendicular to the axis of the rotation in question, and plot from the axis the projections on that plane of the lattice translations that take the crystal into itself. Of these projections we select the *shortest*, which will form a "star" as shown in Fig. 1.10. Note that for each projection in the star there will be a projection in the opposite direction, since for every lattice translation there is an inverse lattice translation. Let us suppose now that the rotation (counterclockwise) of an angle θ takes the crystal into itself. Then the projection **a** will be taken into the

Fig. 1.9 *Glide-plane symmetry in a two-dimensional lattice. Primitive lattice translations* τ_1 *and* τ_2 *are shown. Note that there are three atoms per primitive cell. A reflection through the line* AA *does not take the crystal into itself. However, such a reflection combined with the partial lattice translation* τ *does. The line* AA *represents a glide plane in this lattice. The space group contains this combined operation; the corresponding point group operation is not distinguished from a simple reflection.*

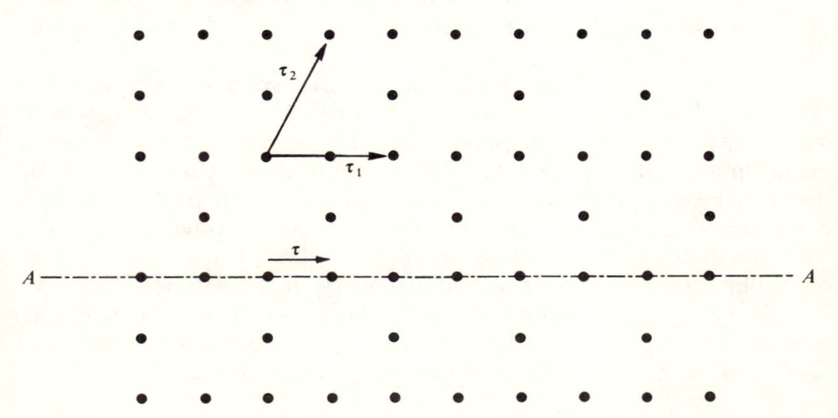

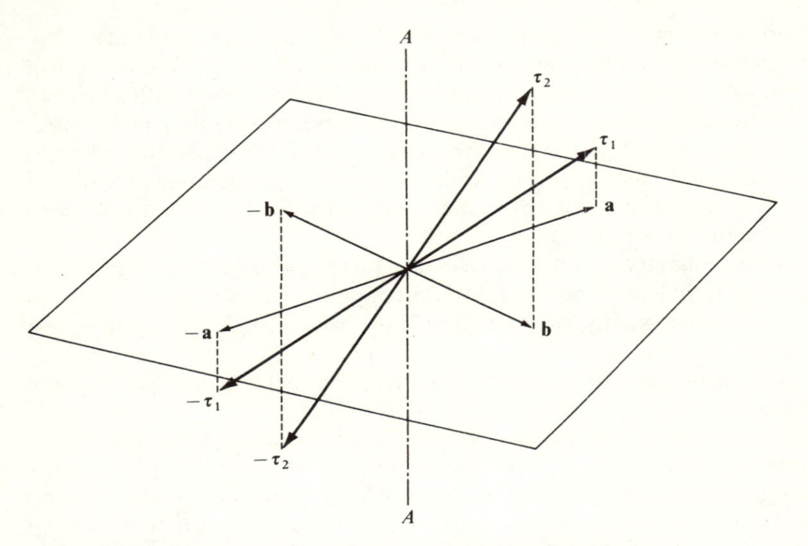

Fig. 1.10 The vectors **a**, **b**, $-$**a**, *and* $-$**b** *are of equal length and lie in a plane perpendicular to the rotation axis* AA. *They are the projections of the lattice translations* τ_1, τ_2, $-\tau_1$, *and* $-\tau_2$ *in that plane and are the shortest nonzero projections of lattice translations.*

projection **b** which must also be the projection of a lattice translation since the crystal, and therefore the lattice translations, has been supposed to be invariant under the rotation. The vector difference between **b** and **a**, however, must also be the projection of a lattice translation, since the difference between any two lattice translations is also a lattice translation. Further, this difference must not be less than the length of **b** (or **a**) since **a** and **b** have been taken to be the shortest projections. Thus, θ must be greater than or equal to 60°.

If a rotation by θ takes the lattice into itself, successive rotations by θ must also. Thus the star will be made up of projections separated by equal angles, each greater than or equal to 60°. This would allow the star to be made up of one, two, three, four, five, or six projections, but one, three, and five may be eliminated since they do not contain inverse projections. Thus the star may contain two, four, or six projections and rotations may only be through angles of 60°, 90°, or multiples of these. It is possible to have threefold rotations without sixfold symmetry. If, for example, there are translations τ_1, τ_2, and τ_3 lying above the plane in Fig. 1.10, with projections separated by 120°, then $-\tau_1$, $-\tau_2$, and $-\tau_3$ will lie below the plane and provide the negative projections completing the six-pointed star. A rotation by 60°, however, will not take the translation vectors into each other; a 120° rotation is required for that. This type of threefold symmetry arises for [111] axes in cubic crystals. The argument given, however, does rule out fivefold symmetry.

Limitations such as this reduce the possible point groups for crystals to 32 distinct groups, and all crystal structures can be classified as one of these symmetries. We can make a further classification of crystal types by noting the fuller set of operations making up the space group, and there are 230 space groups. The classification of solids in this way is part of the subject of crystallography.

A knowledge of the symmetry of the crystal tells a great deal about the possible behavior which may be found in that crystal. We may obtain such information in a systematic way by using the mathematics of groups, and we will wish to develop the rudiments of group theory for that purpose. Group theory, however, is a large subject in its own right and it will be more appropriate here to begin exploring the consequences of symmetry and to see how the concepts of group theory arise in this context. We will begin by seeing what can be learned about the physical tensors associated with crystals.

3. Physical Tensors

We will give here a very brief account of a study which has been described in detail by Nye.[1]

Many macroscopic properties of crystals can be described in terms of tensors. One such property is the conductivity, relating the electric-field vector and the current-density vector. For an isotropic material (such as a metal made of many crystalites) the conductivity is given by a scalar, or equivalently by a scalar times the unit tensor. In more complicated systems, the current will still be proportional to the field (at low fields) but may not be parallel, and the proportionality constants may depend upon the direction of the field. In such a case we may fix a coordinate system in the laboratory and orient the crystal with respect to that system. Then the components of the current are related to the components of the electric field by

$$j_i = \sum_j \sigma_{ij} \mathscr{E}_j$$

The σ_{ij} are the elements of the matrix representing the conductivity tensor of the crystal. If we know a group of symmetry operations that leave the crystal invariant, these must also leave the conductivity tensor invariant,

[1] J. F. Nye, "Physical Properties of Crystals, Their Representation by Tensors and Matrices," Clarendon, Oxford, 1957.

so we have gained information about the conductivity tensor. For this purpose all operations of the point group are applicable, including those corresponding to glide planes and screw axes.

Consider, for example, a crystal which has cubic symmetry; i.e., all of the rotations and reflections that take a cube into itself will also take this crystal into itself. We let the cube axes of the crystal lie parallel to the laboratory coordinate system. We will of course have to take our rotation axes and reflection planes with particular orientations with respect to these axes, just as we must take them with particular orientations with respect to the cube. To be specific, *we will think of a symmetry operation as a physical rotation of the crystal*, which will then rotate the conductivity tensor. This transformed conductivity tensor may be written σ'_{ij} as expressed in the same laboratory frame.

A rotation or reflection of the crystal may be represented by a matrix; i.e., any vector represented as a column vector will become a new (primed) column vector, $x' = Ux$, or more explicitly,

$$\begin{bmatrix} x'_1 \\ x'_2 \\ x'_3 \end{bmatrix} = \begin{bmatrix} U_{11} & U_{12} & U_{13} \\ U_{21} & U_{22} & U_{23} \\ U_{31} & U_{32} & U_{33} \end{bmatrix} \begin{bmatrix} x_1 \\ x_2 \\ x_3 \end{bmatrix}$$

where U is the matrix representing that rotation or reflection. A rotation of 90° around the z axis (taking a vector in the x direction into a vector in the y direction), for example, may be represented by the matrix

$$U_1 = \begin{bmatrix} 0 & -1 & 0 \\ 1 & 0 & 0 \\ 0 & 0 & 1 \end{bmatrix}.$$

A 180° rotation about the z axis is represented by

$$U_2 = \begin{bmatrix} -1 & 0 & 0 \\ 0 & -1 & 0 \\ 0 & 0 & 1 \end{bmatrix}$$

and ordinary matrix multiplication corresponds to successive operations. The corresponding rotation of a tensor is given by

$$\sigma' = U \sigma U^{-1}$$

where U^{-1} is the inverse of U. For example, $U_1{}^{-1}$ is simply

$$\begin{bmatrix} 0 & 1 & 0 \\ -1 & 0 & 0 \\ 0 & 0 & 1 \end{bmatrix}$$

The 180° rotation is a symmetry operation of the cubic group, so it is of interest to apply it to the conductivity tensor, which becomes

$$\sigma' = U_2 \begin{bmatrix} \sigma_{11} & \sigma_{12} & \sigma_{13} \\ \sigma_{21} & \sigma_{22} & \sigma_{23} \\ \sigma_{31} & \sigma_{32} & \sigma_{33} \end{bmatrix} U_2^{-1} = \begin{bmatrix} \sigma_{11} & \sigma_{12} & -\sigma_{13} \\ \sigma_{21} & \sigma_{22} & -\sigma_{23} \\ -\sigma_{31} & -\sigma_{32} & \sigma_{33} \end{bmatrix}$$

But this symmetry operation must leave the conductivity tensor unchanged, and this will only be true if σ_{13}, σ_{23}, σ_{31}, and σ_{32} all vanish. Similarly, by making a rotation of 180° around the x axis we may show that σ_{12} and σ_{21} vanish and σ is therefore diagonal. By making a 90° rotation around the z axis, we further show that $\sigma_{11} = \sigma_{22}$, and by a rotation around the x axis we show that $\sigma_{22} = \sigma_{33}$. Thus the conductivity tensor for a cubic system may be written

$$\sigma \begin{bmatrix} 1 & 0 & 0 \\ 0 & 1 & 0 \\ 0 & 0 & 1 \end{bmatrix}$$

where σ is now a scalar. The conductivity for a cubic crystal is isotropic, just as it is in an isotropic system. This result is perhaps not obvious at once. One might have thought that the conductivity measured along a cube axis in the crystal would differ from that along a cube diagonal. The assumption that current is proportional to the electric field is essential to the argument.

In crystals with lower symmetry we may be able to reduce the number of independent coefficients required to specify the conductivity, without being able to show that it is isotropic.

The piezoelectric effect in crystals is described by a tensor of third rank. This tensor specifies what electrical polarization of the crystal will occur when the crystal is strained. Note first that the state of strain of a crystal may be specified by giving the displacements $u(\mathbf{r})$ of all points $\mathbf{r}$ in the crystal. The strains ϵ_{ij} are defined by derivatives of the components of the displacements.

$$\epsilon_{ij} = \frac{\partial u_j}{\partial x_i}$$

Finally, the polarization is related to the strains by the piezoelectric tensor

$$P_k = \sum_{ij} c_{kij}\epsilon_{ij}$$

Let us consider a crystal which has a center of symmetry; i.e., a point through which the crystal may be inverted and left invariant. Inversion, however, takes the components of the piezoelectric tensor into

$$c'_{kij} = -c_{kij}$$

Thus all components of the tensor must vanish. There is no piezoelectric effect in a crystal with a center of symmetry. Note that the center of symmetry may be *between* atoms. Then the inversion leaves no atom fixed and is not a member of the point group. However, it leaves the crystal invariant and is a member of the space group. If there is no center of symmetry, there may or may not be a piezoelectric effect.

The elastic tensor of a crystal is a tensor of fourth rank. It relates the stress tensor and the strain tensor. By application of symmetry arguments to elastic tensors (which contain 81 elements) we may reduce the number of independent components. In cubic crystals, for example, there are only three independent constants.

4. Symmetry Arguments and Group Theory

We will now see how the great power of these arguments, which can be made on the basis of the symmetry of structure, can be brought to bear with the use of group theory. We will give only a brief survey of group theory and its applications to symmetry properties. A much more complete account is given by Tinkham.[1] Tinkham's text also provides proofs for theorems which we will state without proof or with plausibility arguments only.

First we will illustrate how a symmetry argument can be used to tell us about electronic eigenstates. This treatment will be very close to the more general treatment that uses group theory. We consider a structure that has a symmetry operation which we denote by R. This could be a rotation or a reflection that takes the crystal into itself. Clearly, then, R operating on the Hamiltonian will take it into itself, or equivalently, R commutes with H. We may write this statement as

$$RH = HR$$

or

$$RHR^{-1} = H$$

Thus the result of rotation or reflection of the Hamiltonian is written RHR^{-1}.

[1] Michael Tinkham, "Group Theory and Quantum Mechanics," McGraw-Hill, New York, 1964.

It is important here, and in all of our discussion of group theory, to distinguish sharply between symmetry operations and quantum-mechanical operators. The symbol H may be a quantum-mechanical operator or it may be a classical Hamiltonian. Symmetry operations are like coordinate transformations and may be applied to Hamiltonians or to wavefunctions. We will be developing the algebra of the symmetry operations; that algebra is quite distinct from the application of a Hamiltonian operator to a wavefunction.

Now let this be a one-dimensional case and let the symmetry operation be the inversion operation around the point $x = 0$; it takes x into $-x$. This particular symmetry operation is usually denoted by J. Let us now say that we have a solution of the time-independent Schroedinger equation with this Hamiltonian and we write that solution ψ_1. Then

$$H\psi_1 = E_1\psi_1$$

We may apply the inversion to this entire equation, which by our definition of symmetry operations means we flip over the Hamiltonian and the wavefunction (or equivalently invert the coordinate system) and the equation will become

$$JHJ^{-1}(J\psi_1) = E_1(J\psi_1)$$

or

$$H(J\psi_1) = E_1(J\psi_1)$$

since $JHJ^{-1} = H$. Now we see that the wavefunction $J\psi_1$ is also a solution of the same Schroedinger equation with the same energy E_1. If the energy E_1 is not degenerate, i.e., if there is only one wavefunction with this energy eigenvalue, then $J\psi_1$ must be equivalent to ψ_1. At most, $J\psi_1$ can differ from ψ_1 by a constant factor, so we may write $J\psi_1 = D_1\psi_1$. Now J^2, the repeated operation of inversion, is simply the identity (which is usually written E); thus we may write

$$J^2\psi_1 = D_1{}^2\psi_1 = \psi_1$$

It follows that $D_1 = \pm 1$; that is, ψ_1 must be even or odd around the point $x = 0$.

If the energy E_1 represents a doubly degenerate eigenvalue we may write the corresponding two states as ψ_1 and ψ_2. Then, since we can again show that $J\psi_1$ is also a solution of the Schroedinger equation with the same energy E_1, it follows that $J\psi_1$ must be a linear combination of these two states ψ_1 and ψ_2. Similarly, J operating on ψ_2 must yield a linear combination of ψ_1 and ψ_2. We write these two statements in the form

$$J\psi_1 = D_{11}\psi_1 + D_{21}\psi_2$$

$$J\psi_2 = D_{12}\psi_1 + D_{22}\psi_2$$

where the D_{ij} are coefficients which depend upon the states in question. If all of these D_{ij} are nonzero, we say that the inversion mixes the states. This could be avoided by taking new linear combinations of ψ_1 and ψ_2 which do not mix. Let these new states be related to the old by

$$\begin{aligned} \psi_1 &= U_{11}\psi'_1 + U_{21}\psi'_2 \\ \psi_2 &= U_{12}\psi'_1 + U_{22}\psi'_2 \end{aligned} \tag{1.2}$$

where the U_{ij} are chosen such that

$$\begin{aligned} J\psi'_1 &= D'_1\psi'_1 \\ J\psi'_2 &= D'_2\psi'_2 \end{aligned} \tag{1.3}$$

In matrix notation, the matrix D is diagonalized by the transformation U. As before we can show that D'_1 and D'_2 are each either $+1$ or -1. We conclude that we can always pick odd or even solutions to a Hamiltonian which has the inversion symmetry. We have assumed that these two states are degenerate and this would be called an accidental degeneracy because it is not required by symmetry. In contrast we will see that sometimes in three dimensions a degeneracy is required by symmetry. Such a symmetry degeneracy, for example, is illustrated by the three p states in a free atom which are required by the isotropy of space to be degenerate.

Group theory allows us to learn the consequences of symmetry in much more complicated systems. We do not need group theory to tell us that even potentials in one dimension will lead to even and odd solutions. However, we may state our above argument in the language of group theory and this will be directly generalizable to the more complicated case. In the language of group theory J and E form a *group* where E is the identity operation (e.g., a zero-degree rotation).

In order to define our terminology we need to represent states by vectors. A general state $\psi = a\psi_1 + b\psi_2$ we represent as a vector $\begin{pmatrix} a \\ b \end{pmatrix}$. We have found that inversion takes this state to $(aD_{11} + bD_{12})\psi_1 + (aD_{21} + bD_{22})\psi_2$. In matrix notation

$$J\begin{pmatrix} a \\ b \end{pmatrix} = D(J)\begin{pmatrix} a \\ b \end{pmatrix}$$

where $D(J)$ is the matrix

$$D(J) = \begin{bmatrix} D_{11} & D_{12} \\ D_{21} & D_{22} \end{bmatrix}$$

Similarly we define a matrix corresponding to E.

$$\mathrm{D(E)} = \begin{bmatrix} 1 & 0 \\ 0 & 1 \end{bmatrix}$$

$D(E)$ and $D(J)$ form a *representation of the group* with *basis* ψ_1 and ψ_2.

We have obtained an *equivalent representation* by taking new basis states ψ'_1 and ψ'_2 related to the old by Eq. (1.2). In the new framework, a general state given by $a'\psi'_1 + b'\psi'_2$ is represented as a vector $\begin{pmatrix} a' \\ b' \end{pmatrix}$. It is related to $\begin{pmatrix} a \\ b \end{pmatrix}$ by

$$\begin{pmatrix} a' \\ b' \end{pmatrix} = U\begin{pmatrix} a \\ b \end{pmatrix} = \begin{bmatrix} U_{11} & U_{12} \\ U_{21} & U_{22} \end{bmatrix}\begin{pmatrix} a \\ b \end{pmatrix}$$

In the new framework, inversion takes a general state into

$$J\begin{pmatrix} a' \\ b' \end{pmatrix} = D'(J)\begin{pmatrix} a' \\ b' \end{pmatrix}$$

where

$$D'(J) = U\,D(J)\,U^{-1}$$

The right-hand side is the product of the three matrices. Similarly, we find $D'(E) = U\,D(E)\,U^{-1} = D(E)$. $D'(E)$ and $D'(J)$ are an *equivalent representation* to $D(E)$ and $D(J)$; ψ'_1 and ψ'_2 form the basis of the primed representation.

From Eq. (1.3) it follows that for the example we treated,

$$D'(E) = \begin{bmatrix} 1 & 0 \\ 0 & 1 \end{bmatrix} \quad \text{and} \quad D'(J) = \begin{bmatrix} D'_1 & 0 \\ 0 & D'_2 \end{bmatrix}$$

Such a representation, which does not mix the states ψ'_1 and ψ'_2, is called a *reducible representation*. In addition, any representation equivalent to a reducible representation is called a reducible representation.

If we had more symmetry operations, R, we could similarly define $D(R)$. We might then find that this was an *irreducible* representation so that the different ψ's inevitably transform into each other and *must* be degenerate. This, for example, is true of the three p states in a free atom. This is the language that we will use in our development of group theory and its application to symmetry.

4.1 Groups The theory of groups is a very general branch of mathematics. Our interest in it here is entirely in its application to symmetry groups, so we will avoid the abstractness by illustrating every step by its application to symmetry groups. We begin by defining terms and giving elementary theorems.

A *group* is a collection of elements; e.g., a collection of symmetry operations which take the Hamiltonian into itself. For concreteness, we will continue to think of a symmetry operation on a function as a rotation, reflection, or translation of the function (*with the coordinate system remaining fixed*). This is illustrated in Fig. 1.11. The translation of a function of x (for example, $e^{-\alpha x^2}$) by a vector $\mathbf{T}$ displaces the function $Te^{-\alpha x^2} = e^{-\alpha(x-T)^2}$. This convention is called the "active point of view." In contrast, in the "passive point of view" we imagine the function remaining fixed and we rotate or translate the coordinate system. It is important to be consistent, and we will consistently take the active point of view.

For a collection of elements to form a group, it must satisfy the following conditions:

1. A product is defined. In the case of symmetry operations a product will mean the successive performance of two operations. We define the product $R_1 \cdot R_2$ as the operation R_2 performed first, then the operation R_1. We will denote a product by a dot.
2. The product must be in the group. In the case of symmetry operation, this means the product of two symmetry operations must also be a symmetry operation which is obviously true.
3. The group must contain the unit element. In the case of symmetry operations, the identity operation (i.e., leaving the system as before) is clearly a member of the group. The unit element is generally written as an E. Its defining property may be written

$$E \cdot R = R \cdot E = R$$

Fig. 1.11 *The symmetry operation T on a function $f(x)$ translates that function by the distance* $\mathbf{T}$. *This is the "active" definition of a symmetry operation.*

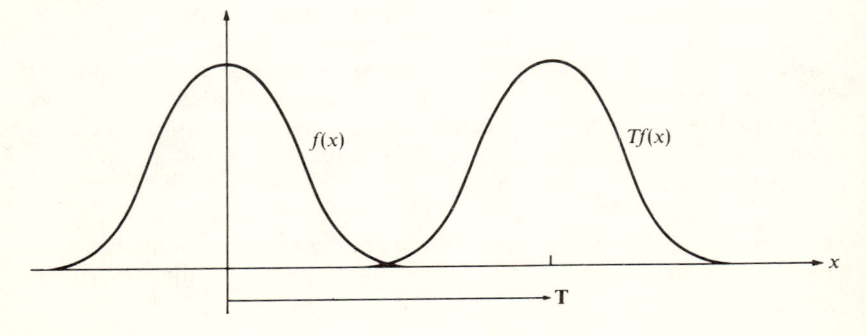

4. The multiplication is associative, which is represented by

$$R_1 \cdot (R_2 \cdot R_3) = (R_1 \cdot R_2) \cdot R_3$$

where the $(R_2 \cdot R_3)$, for example, is the element which is the product of R_2 and R_3. This relation is true for symmetry operations but is perhaps not at once obvious.
5. The inverse of every element must also be a member of the group. This is clear for symmetry operations; if a particular symmetry operation takes a Hamiltonian into itself, then clearly undoing that symmetry operation will also take that Hamiltonian into itself. The inverse of an element R is written R^{-1} and clearly $R \cdot R^{-1} = R^{-1} \cdot R = E$.

Let us illustrate a group by considering the symmetry operations of an equilateral triangle. It is not difficult to see that there are six symmetry operations which take the triangle into itself. We may see this by thinking of a triangular piece of paper. Rotating the paper by 120° clearly puts the paper into an equivalent orientation. If this were a triangular molecule such a rotation would leave the Hamiltonian unchanged. We might keep track of the symmetry operations by marking a spot on the paper, such as the E spot indicated in Fig. 1.12. The rotation takes the spot to the position marked C_1. (Rotations are conventionally written as C's.) Other symmetry operations will take the spot to the other positions marked in Fig. 1.12. Note that those spots marked with a σ involved flipping the triangle over, or reflecting it. (Reflections are conventionally written as σ's.) In this diagram each symmetry operation is represented by one of the equivalent positions into which the spot is taken under the corresponding symmetry operation.

This can of course be readily generalized to three-dimensional figures. The symmetry group of a cube may be found by again marking a spot on one of the square faces near a corner. There are six equivalent positions for that spot near the same corner. Note that half of these require an inversion of the cube. There are eight corners so altogether there are 48 symmetry operations that take the cube into itself.

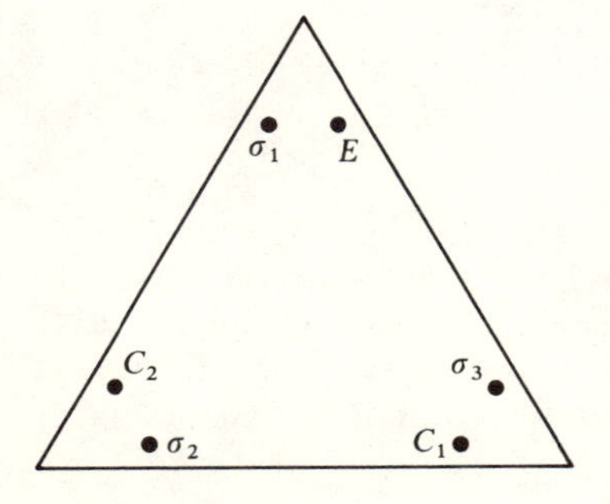

Fig. 1.12 Each symmetry operation of the group of the equilateral triangle is represented by a dot, which we imagine initially at the point E. For example, a clockwise rotation of 120° about the center is called C_1. It takes the triangle into itself and the dot to the point labeled C_1.

Let us return to the group of the triangle and obtain the multiplication table. This can be obtained from inspection of the figure. Consider for example applying the operation C_1 twice. The first time takes the spot from E to the position C_1. The second application of C_1 takes the spot to the position marked C_2. Thus we find

$$C_1 \cdot C_1 = C_2$$

Similarly, the product $C_1 \cdot \sigma_1$ takes the spot first to the position σ_1 under the operation σ_1 and then to the position σ_3 under the operation C_1. Thus we find

$$C_1 \cdot \sigma_1 = \sigma_3$$

Similarly, we can obtain the entire multiplication table which is given by

E	σ_1	σ_2	σ_3	C_1	C_2
σ_1	E	C_1	C_2	σ_2	σ_3
σ_2	C_2	E	C_1	σ_3	σ_1
σ_3	C_1	C_2	E	σ_1	σ_2
C_1	σ_3	σ_1	σ_2	C_2	E
C_2	σ_2	σ_3	σ_1	E	C_1

Here a product $R_1 \cdot R_2$ is found in the row next to R_1 and the column under R_2.

Note that the elements do not commute. For example, $\sigma_1 \cdot \sigma_2$ is not equal to $\sigma_2 \cdot \sigma_1$. In general the elements of a group do not commute. In special cases where all of the elements in a group commute, this group is called an *Abelian group.*

Two elements A and B are in the same *class* if there exists an R in the group such that

$$R \cdot A \cdot R^{-1} = B$$

Note that in an Abelian group every element is in its own class; i.e., no two different elements are in the same class. Note also that E is always in a class by itself. The group of the triangle has three classes

$$E$$

$$\sigma_1, \sigma_2, \sigma_3$$

$$C_1, C_2$$

The term *class* is a good one. Elements in the same class are very much the same operation as illustrated by the triangle. The definition of a class, $R \cdot A \cdot R^{-1} = B$, means, roughly speaking, that if we take the operation A and rotate the operation itself, it becomes the operation B. We can ordinarily divide the symmetry operations of a group into classes by inspection but the classes can also be obtained from the multiplication table.

4.2 Representations A *representation* of a group is a set of matrices which has the same multiplication table as the group. We can write a representation of the group as $D(R)$, where D is the matrix to represent symmetry operation R. The values of $D_{ij}(R)$ for various indices i and j give the various matrix elements. There may be a different matrix for each symmetry operation in the group. All of these matrices together form the representation. We stated that these must have the same multiplication table as the group. For the representation the product is ordinary matrix multiplication, which means if there are two elements R_1 and R_2 whose product is R_3, that is, $R_1 \cdot R_2 = R_3$, then it must be true that

$$D_{ij}(R_3) = \sum_l D_{il}(R_1) D_{lj}(R_2) \tag{1.4}$$

This will only make sense if for a given representation all matrices are of the same dimension, and we will always restrict our attention to unitary matrices. We will see shortly what this means.

Note that we have not required that matrices representing different symmetry elements be different. They frequently will be the same and, in fact, we note a special case in which the representation of every symmetry element is taken to be a 1; i.e., a one-dimensional matrix with its only element equal to 1. Note that this representation satisfies Eq. (1.4). This is called the unit representation and is a representation of every symmetry group.

We can already see more generally what the representation of the unit element in the symmetry group will be. We wish to require that

$$\sum_l D_{il}(E) D_{lj}(R) = D_{ij}(R)$$

We note that this will be satisfied if $D_{ij}(E)$ is equal to the unit matrix δ_{ij}. It turns out that E is always given by the unit matrix.

A unitary matrix, which we denote U, is characterized by the relation

$$U^+ = U^{-1} \tag{1.5}$$

where the superscript + (frequently a dagger is used instead) represents taking the conjugate transpose matrix (taking the complex conjugate of every element and interchanging rows with columns). $(U^+)_{ij} = U^*_{ji}$. A unitary operation on a column vector simply rotates the vector. A Hermitian matrix, in contrast, satisfies the relation

$$H^+ = H$$

Using ordinary matrix multiplication we may show that

$$(U^+U)_{ij} = \sum_l (U^+)_{il} U_{lj} = \sum_l U^*_{li} U_{lj} = E_{ij} = \delta_{ij}$$

The next to last step follows from Eq. (1.5). We see that $\sum_l U^*_{li} U_{lj} = \delta_{ij}$; that is, the columns of a unitary matrix are the components of orthogonal vectors. Similarly, from $UU^+ = E$ it follows that the rows form orthogonal vectors.

We will next define a unitary transformation on a matrix and show that it leaves the trace of a matrix unchanged. In group theory the trace (or the spur) of a matrix in a representation is called its *character*. The character of a matrix D is written

$$\chi(R) = \sum_i D_{ii}(R)$$

Let us now define a unitary transformation on the matrix D,

$$D' = UDU^+$$

or written in terms of the elements of D we have

$$D'_{ij} = \sum_{lm} U_{il} D_{lm} U^*_{jm}$$

Thus the character of D' is given by

$$\chi'(R) = \sum_{ilm} U_{il} D_{lm}(R) U^*_{im} = \sum_{lm} \delta_{lm} D_{lm}(R)$$

$$= \sum_l D_{ll}(R) = \chi(R)$$

and this proves the point. It is also readily shown that a unitary transformation on a unitary matrix leaves it unitary.

4.3 Equivalent representations Now let us consider the representation $D(R)$ of a symmetry group containing the elements R. The matrices D are to be unitary matrices. Consider another unitary matrix U of the same dimension. We may apply the same transformation $UD(R)U^+$ to all of the $D(R)$. This gives us a new set of matrices each of which is associated with one of the symmetry elements R. We may readily show that this new set of matrices also forms a representation of the group; i.e., that it has the same multiplication table. Such a new representation and the old one are called *equivalent* representations.

The fact that the trace of the matrix is unchanged by unitary transformation implies that the traces $\chi(R)$ are the same for all equivalent representations. It also means that all elements in the same class have the same trace or character for a given representation. It turns out that most of the symmetry information that we seek will be obtainable from the characters. This will mean that for many purposes we do not need to distinguish the different symmetry operations of a given class, and it will also mean that we will not have to worry about the fact that there are different equivalent representations of a given group.

If we have two representations of a group $D^1(R)$ and $D^2(R)$, which may or may not be equivalent, then clearly the set of matrices

$$D^3(R) = \begin{bmatrix} D^1(R) & 0 \\ 0 & D^2(R) \end{bmatrix}$$

is also a representation. Here the matrix D^3 has a dimension equal to the sums of the dimensions of D^1 and D^2. The matrix D^1 appears in the upper left-hand corner, the matrix D^2 appears in the lower right-hand corner, and the rest of the elements are zero. Any representation, every element of which is in this form, or any representation with an equivalent representation in this form, is called a *reducible* representation. If there is no one unitary transformation that will bring every matrix in a representation into this "diagonal-block" form, then the representation is called *irreducible*. Irreducible representations play a fundamental role in the application of group theory.

We will return to the more mathematical consequences of group theory later. For the moment it may be desirable to illustrate representations, again with the group of the triangle.

For groups with finite numbers of elements we will see that there are only a limited number of irreducible representations that are not equivalent. For the group of the triangle there are three irreducible representations, which may be written

	E	σ_1	σ_2	σ_3	C_1	C_2
Λ_1	1	1	1	1	1	1
Λ_2	1	-1	-1	-1	1	1
Λ_3	$\begin{bmatrix} 1 & 0 \\ 0 & 1 \end{bmatrix}$	$\begin{bmatrix} -1 & 0 \\ 0 & 1 \end{bmatrix}$	$\begin{bmatrix} \frac{1}{2} & -\frac{\sqrt{3}}{2} \\ -\frac{\sqrt{3}}{2} & -\frac{1}{2} \end{bmatrix}$	$\begin{bmatrix} \frac{1}{2} & \frac{\sqrt{3}}{2} \\ \frac{\sqrt{3}}{2} & -\frac{1}{2} \end{bmatrix}$	$\begin{bmatrix} -\frac{1}{2} & \frac{\sqrt{3}}{2} \\ -\frac{\sqrt{3}}{2} & -\frac{1}{2} \end{bmatrix}$	$\begin{bmatrix} -\frac{1}{2} & -\frac{\sqrt{3}}{2} \\ \frac{\sqrt{3}}{2} & -\frac{1}{2} \end{bmatrix}$

The capital lambdas are used here to denote representations. There are of course many equivalent representations to Λ_3; however, any representation of the group of the triangle must either be equivalent to one of these or it must be a reducible representation that will reduce to combinations of the Λ_1, Λ_2, and Λ_3.

4.4 Symmetry degeneracies Any complete set of orthonormal functions can be used to construct a representation of the symmetry group. We will write these functions f_i with the index i running over the set. These could be the eigenstates of some Hamiltonian, but for our argument they need not be. We would construct the representation of the group by selecting one of these functions and performing a symmetry operation on that function to obtain some new function. Since our set is complete we may write this new function as a linear combination of the old. We write this

$$Rf_i = \sum_j D_{ji}(R) f_j$$

Using this same state f_i we could generate the coefficient $D_{ji}(R)$ for each of the symmetry operations R of the group. Similarly, we could construct such coefficients for each of the functions f_i.

It could be that some set of these functions transform among themselves but not out of the set. In that case, this small set could form the basis for a representation of the group. We really only require completeness with respect to these symmetry operations in this sense. The representation of the group contains the matrices $D(R)$ for each symmetry operation R. The matrix representing R has the D_{ji} obtained above as matrix elements, the first index denoting the row and the second index denoting the column in the matrix.

In order to demonstrate that these matrices, in fact, form a representation of the group, we must show that they have the same multiplication table. Consider, for example, three symmetry operations related by $R_1 \cdot R_2 = R_3$. We will evaluate the successive operation $R_1 \cdot R_2$ on the function f_i in terms of the matrices. We will then note that the result of this operation must lead to $R_3 f_i$, which may be written in terms of the matrix for R_3. We will see that in fact the matrix representing R_3 is just the matrix product of those representing R_1 and R_2. We first write

$$R_2 f_i = \sum_j D_{ji}(R_2) f_j$$

In operating upon this equation with R_1 we will operate on each of the f_j's on the right.

$$R_1 \cdot R_2 f_i = \sum_j \sum_l D_{ji}(R_2) D_{lj}(R_1) f_l = \sum_{j,l} D_{lj}(R_1) D_{ji}(R_2) f_l$$

The left-hand side is simply $R_3 f_i$, which may be written

$$R_3 f_i = \sum_l D_{li}(R_3) f_l$$

Matching coefficients of the f_l we have

$$D_{li}(R_3) = \sum_j D_{lj}(R_1)D_{ji}(R_2)$$

This is simply matrix multiplication and indicates that the multiplication table of these matrices is indeed the same as the multiplication table of the group.

In group theory we say that the functions f_i form the *basis* of this representation of the group. These functions are sometimes called *partner functions* to the representation or it is said that they *belong* to the representation.

The representation of an operator in terms of a basis set is familiar in quantum mechanics, particularly as applied to the Hamiltonian. It will be convenient as illustration to develop a representation of the Hamiltonian with respect to the same basis f_i. This will also be useful in shedding light on symmetry degeneracies. The Hamiltonian operating upon a given function f_i will give us a new function. This again may be expanded in our complete set. Thus we may define a Hamiltonian matrix by the equation

$$\mathscr{H}f_i = \sum_j \mathscr{H}_{ji}f_j$$

Unless we have chosen the f_j judiciously this sum will run over all j.

Let us now say that this Hamiltonian is invariant under the symmetry group for which we have obtained a representation in terms of the f_i. Then if we wish to apply the Hamiltonian to some function f given by

$$f = Rf_i$$

it does not matter whether we operate with the Hamiltonian before or after the symmetry operation R; the Hamiltonian commutes with the symmetry operation. That is, $R(\mathscr{H}f_i) = \mathscr{H}(Rf_i)$ since the symmetry operation takes the Hamiltonian into itself. In terms of our representation this may be rewritten

$$\sum_{j,l} D_{lj}(R)\mathscr{H}_{ji}f_l = \sum_{l,j} \mathscr{H}_{lj}D_{ji}(R)f_l$$

Matching coefficients of the f_l we see that the representation of R commutes with the Hamiltonian matrix. This has followed from the fact that the symmetry operations take the Hamiltonian into itself.

This again is true for any choice of orthonormal f_l. If the f_l have been arbitrarily chosen many f_l will be required to generate the Hamiltonian matrix, and the corresponding representation of the symmetry group will be of very large dimension. If, on the other hand, the f_l are eigenstates of the Hamiltonian, operation with the Hamiltonian leads to a constant multiple (the energy eigenvalue) times the same function. Thus the corresponding

Hamiltonian matrix must be diagonal. Each symmetry operation must take f_l either into itself or into a degenerate eigenstate. The dimensionality of the representation arising from a given f_l must be no higher than the degeneracy of the state. Thus there is a close connection between the dimensionality of the representations of the group and the degeneracy of the states from which that representation is generated. In particular if a set of states is transformed into each other under an irreducible representation, it follows that there is no linear combination of these states (no unitary transformation) which can eliminate the transformation of these states into each other under symmetry operations. It follows then from the symmetry of the Hamiltonian that these states must be degenerate. We have arrived on rather intuitive grounds at one of the important results of group theory. The presence of symmetry groups with multidimensional irreducible representations suggests the possibility of eigenstates that must by symmetry be degenerate.

This conclusion may be formulated more precisely with the use of Schur's lemma which states:

Any matrix which commutes with all matrices in an irreducible representation must be a constant matrix (*i.e., a multiple of the unit matrix*).

This states explicitly that if we have obtained a set of functions which transform into each other under the Hamiltonian operator and which transform into each other under the symmetry operations according to an irreducible representation of the symmetry group, these functions must be degenerate eigenstates of the Hamiltonian.

The irreducible representation according to which such eigenstates transform is said to describe the *symmetry of the states*. If we find that a state transforms according to a one-dimensional representation, we might see, for example, that that state goes into its negative under an inversion or into itself under an inversion. We might similarly find that three degenerate states transform into each other under 90° rotations just as vectors transform into each other under these rotations.

It is of course possible that the Hamiltonian could have eigenstates of *different* symmetry which are also degenerate. This would depend on the detailed form of the Hamiltonian and not just the symmetry. In such a case we would say that these are *accidental* degeneracies. We may expect that accidental degeneracies will be lifted under any modification of the Hamiltonian even though it retains the initial symmetry.

Eigenstates in the free hydrogen atom represent a familiar example of symmetry and accidental degeneracies. The $2s$ and $2p$ states in the hydrogen atom are degenerate in the simplest theory. Thus these four states can form a basis for a representation of the symmetry group of the hydrogen atom. (The symmetry group of the hydrogen atom includes all rotations and all reflections.) This, however, will be a reducible representation. The ψ_j correspond to three p states and one s state. Rotations and reflections will take the p function into linear combinations of p functions, but the s state will always be taken into itself. If we were now to make a slight modification of the coulomb potential, which however retains the spherical symmetry, we will split the s state from the p states; deviations of the potential from $1/r$ within the nucleus are an example of such a change. If we know nothing of the potential except its symmetry (i.e., that it is spherically symmetric), we may conclude that the p states will be degenerate, but we cannot conclude that the s state will have the same energy.

We have seen that a knowledge of the representation to which eigenstates belong gives us information about the degeneracy of that state. Knowledge of the representations can also tell us something about the symmetry of the wavefunctions. For example, in the case of the triangular Hamiltonian we know that an eigenstate belonging to Λ_1 is transformed into itself under any of the symmetry operations of the group and that it therefore has the full symmetry of the triangle. We know that an eigenstate belonging to Λ_2 will go into itself under a rotation but will go into its negative under a reflection. Thus the wavefunction goes through zero along each of the altitudes of the triangle. Such states are illustrated in Fig. 1.13. Finally we know that eigenstates belonging to Λ_3 transform as p functions; i.e., they transform as the coordinates. In the two-dimensional case, such states occur in pairs. Their form will become clear when we discuss molecular vibrations. In the case of the group of the triangle we expect no threefold degeneracies since there are no three-dimensional irreducible representations of the group. We will see how it can be shown that there are no other irreducible representations than those we have given.

4.5 Orthogonality relation Perhaps the most important basic theorem in the theory of group representations is the *orthogonality relation*, which may be written

$$\sum_R D_{\alpha\beta}^{(i)*}(R) D_{\alpha'\beta'}^{(j)}(R) = \frac{h}{l_i} \delta_{\alpha\alpha'} \delta_{\beta\beta'} \delta_{ij}$$

where h is the number of elements in the group and l_i is the dimension of the ith irreducible representation; i and j are indices numbering the inequivalent representations. The sum is over all elements R in the group.

One way of visualizing the orthogonality relation is to construct a set of column vectors. The elements in each column vector are to be made up of particular matrix elements in the irreducible representations, e.g., the element in the upper left-hand corner of the representation Λ_3. The column vector then contains this matrix element taken from the representations of E, σ_1, σ_2, σ_3, C_1, and C_2. These are vectors with dimension equal

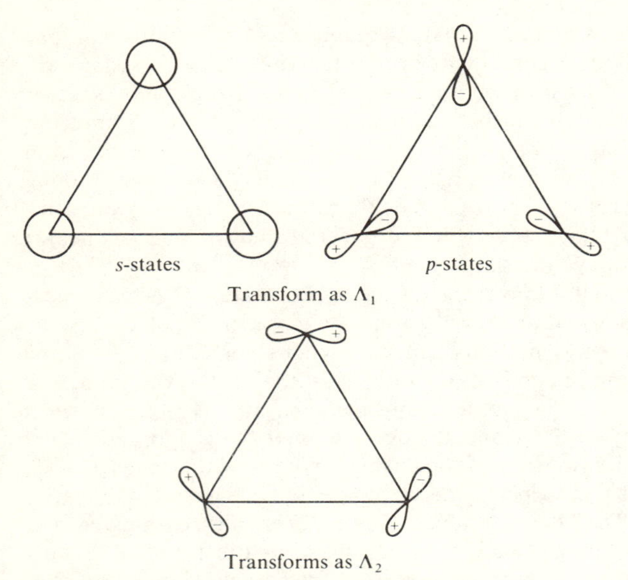

Fig. 1.13 Schematic molecular wavefunctions which transform according to the one-dimensional irreducible representations of the group of the equilateral triangle. The first is thought of as a linear combination of s states on the three atoms. The others are thought of as linear combinations of p states; the plus and minus denote the sign of the wavefunction in the corresponding lobe of the p state.

to the number of elements in the group ($h = 6$). The number of such vectors which can be constructed in this way is equal to the sum of the squares of the dimensions of the irreducible representations. The orthogonality theorem states simply that all of these vectors are orthogonal.

We may immediately draw an important conclusion. The space of these vectors is a space of h dimensions; it is of course possible to construct only h mutually orthogonal vectors in h-dimensional space. We conclude that the sum of the squares of the dimensions of the irreducible representations must be less than or equal to the number of elements in the group. It turns out to be an equality and we conclude

$$\sum_i l_i^2 = h \tag{1.6}$$

The sum is over all inequivalent irreducible representations. It follows from this that we have constructed all the irreducible representations for the triangle. The addition of another would violate Eq. (1.6) since the three we wrote exhaust the sum rule with $h = 6$.

4.6 Characters For many applications it is unnecessary to construct the representations themselves. The desired information is obtainable from the characters. Again, for each matrix, this is just the sum of the diagonal elements.

$$\chi^{(i)}(R) = \sum_{\alpha} D_{\alpha\alpha}{}^{(i)}(R)$$

We could easily write these down for each element of the group of the triangle from the irreducible representations, but it is not necessary to construct the trace for every element. We showed that the trace of a matrix is unchanged by a unitary transformation. It follows from our definition of a class that all elements in the same class will have the same characters and we need only specify for each representation the character for each class. For the case of the group of the triangle we would write the character table

	E	3σ	$2C$
Λ_1	1	1	1
Λ_2	1	-1	1
Λ_3	2	0	-1

The label 3σ indicates that there are three reflections in the class.

We note further that since the traces do not change under unitary transformation the traces (or the characters) will be the same for all equivalent representations. Thus a character table is unique and quite independent of which irreducible representation happens to have been chosen.

It will be useful to use our general orthogonality relation to obtain an orthogonality relation based upon the characters. This is readily done in two steps.

$$\chi^{(i)*}(R)\chi^{(j)}(R) = \sum_{l,m} D_{ll}{}^{(i)*}(R) D_{mm}{}^{(j)}(R)$$

Thus

$$\sum_{R} \chi^{(i)*}(R)\chi^{(j)}(R) = \sum_{l,m} \frac{h}{l_i}\,\delta_{lm}\,\delta_{lm}\,\delta_{ij} = h\,\delta_{ij}$$

We may now denumerate the classes by the index ρ; let us say that there are g_ρ elements in the ρth class. Then the above relation can be rewritten

$$\sum_{\rho} g_\rho \chi^{(i)*}(\rho)\chi^{(j)}(\rho) = h\,\delta_{ij} \tag{1.7}$$

This new orthogonality relation may also be visualized in terms of orthogonal vectors. The numbers $(\sqrt{g_\rho/h})\chi^{(i)}(\rho)$, for fixed i, may be thought of as elements of a column vector. We have a different column vector for each irreducible representation. Equation (1.7) states that these vectors are orthogonal. Each vector has a number of components equal to the number of classes. The number of mutually orthogonal vectors (irreducible representations) must be less than or equal to the number of classes. Again the equality holds and so the number of irreducible representations of a group is equal to the number of classes (three for the case of the triangle).

It is usually possible to obtain the dimensions of the irreducible representations of a group immediately from these two conditions obtained from the orthogonality relations. For example, in the group of the triangle we conclude that there are three irreducible representations and that the sum of the squares of the dimensions of these is equal to 6. The only set of three integers satisfying these conditions is 1, 1, and 2, as we have written down. We may also see immediately that an Abelian group, with every element in a class by itself, has only one-dimensional irreducible representations.

There are methods for obtaining the character table directly from a knowledge of the multiplication table for the group. In most instances, however, one can find published tables and we will not discuss further their determination.

4.7 Reduction of representations In applying group theory it will frequently happen that we know a reducible representation of the group (it may have been constructed by symmetry operations on a trial function), and we would like to know the reduction of this representation. It turns out that the character provides enough information for this reduction. Consider a representation which may be reduced into a set of representations, $D^{(i)}(R)$, each of which appears a_i times. By visualizing the corresponding matrices in reduced form, e.g.,

$$\begin{bmatrix} D_{11}{}^{(1)} & D_{12}{}^{(1)} & 0 & 0 & 0 & 0 \\ D_{21}{}^{(1)} & D_{22}{}^{(1)} & 0 & 0 & 0 & 0 \\ 0 & 0 & D^{(2)} & 0 & 0 & 0 \\ 0 & 0 & 0 & D^{(2)} & 0 & 0 \\ 0 & 0 & 0 & 0 & D_{11}{}^{(3)} & D_{12}{}^{(3)} \\ 0 & 0 & 0 & 0 & D_{21}{}^{(3)} & D_{22}{}^{(3)} \end{bmatrix}$$

it is easy to see that the character of the reducible representation is given by

$$\chi(R) = \sum_i a_i \chi^{(i)}(R)$$

We may multiply both sides by $\chi^{(j)*}(R)$ and sum over R. Applying the orthogonality theorem Eq. (1.7) for the characters, we obtain immediately

$$a_j = \frac{1}{h} \sum_R \chi^{(j)*}(R)\chi(R) = \frac{1}{h} \sum_\rho g_\rho \chi^{(j)*}(\rho)\chi(\rho)$$

thus accomplishing the reduction. This reduction is frequently written

$$D(R) = \sum_j a_j D^{(j)}(R)$$

5. *Applications of Group Theory*

We may help to clarify the abstract theory by giving some applications at this point. We will also be using the ideas and methods of group theory at various points in our later discussions.

5.1 Lowering of symmetry When the symmetry of a symmetric system is lowered by distortion of the system or by the addition of a potential, the number of symmetry operations taking the system into itself is reduced. The smaller set of symmetry operations is called a *subgroup* of the larger group. The matrices making up an irreducible representation of the larger group will provide a representation also of the subgroup (taking of course only those matrices representing elements in the subgroup), but this smaller number of matrices may correspond to a *reducible* representation of the subgroup. In this case a symmetry degeneracy of the larger group is split by the distortion. By performing the reduction we may see what the nature of the splitting will be.

As an example we may consider a triangular molecule to which we have applied a magnetic field perpendicular to the plane of the molecule. First consider the symmetry operations of the modified problem. Reflections may reverse the magnetic field; therefore, if the Hamiltonian contains the magnetic field, we may expect that a reflection will modify the Hamiltonian. On the other hand, rotations about the axis of the magnetic field do not modify the magnetic field. The subgroup, then, will be made up only of the elements E, C_1, and C_2. From our rules based upon the orthogonality relations we see that there must be 3 one-dimensional irreducible representations. (This is an Abelian group.) We may note further that all the elements in this group may be written as some power of a single element of the group C^3, C, and C^2. Such a group is called a *cyclic group*. Clearly a cyclic group is always Abelian. The character table (which in this case is also a table of the irreducible representations) is given by

	E	C_1	C_2
L_1	1	1	1
L_2	1	$e^{2\pi i/3}$	$e^{4\pi i/3}$
L_3	1	$e^{-2\pi i/3}$	$e^{-4\pi i/3}$

Cyclic groups always have irreducible representations with this periodic form. It may readily be verified that these representations have the appropriate multiplication table and therefore are correct irreducible representations of the group.

Now the representation of this subgroup, based upon the two-dimensional irreducible representation of the full group of the triangle, has characters given by

	E	C_1	C_2
Λ_3	2	-1	-1

We consider two degenerate eigenstates of the triangular molecule which belong to this irreducible representation; we may see how these are split by the magnetic field. These two wavefunctions transform among each other under the three operations of the subgroup according to the representation with character table 2, -1, -1. If by taking appropriate linear combinations we may obtain two states that do not transform into each other under the symmetry operations of the subgroup, then we need no longer expect these states to be degenerate. This linear combination of course corresponds just to making a unitary transformation of the reducible representation.

Applying our rules for the use of the character tables in the reduction of the representation, we obtain the coefficients for each of the representations.

$$a_1 = 0$$

$$a_2 = \frac{2 - e^{-2\pi i/3} - e^{-4\pi i/3}}{3} = 1$$

$$a_3 = \frac{2 - e^{2\pi i/3} - e^{4\pi i/3}}{3} = 1$$

We find of course that the degeneracy is lifted (the subgroup has no two-dimensional representation) and we find also that the states that were degenerate now have the symmetry of L_2 and L_3.

Let us see again what is being done in terms of the wavefunctions.

The representation Λ_3 describes the transformation of eigenstates without a magnetic field. In constructing the equivalent representation, we are still describing states without the field. As we apply an infinitesimal field, the symmetry group becomes smaller and each of the two states transforms only into itself under the subgroup. A degeneracy is no longer required by symmetry. An accidental degeneracy is still possible, in which case the reducible representation could describe the states. In general, however, we have no reason to believe that the states of the modified Hamiltonian will have an accidental degeneracy.

Analogous splittings occur when an atom is placed in a crystal environment. In this case a system that initially had full spherical symmetry (full rotation and inversion symmetry) is reduced to a system that has only the symmetry of the environment of the atom. For example, the fivefold degenerate *d* states of the free atom will be split into a twofold and a threefold degeneracy when the atom is put in a cubic environment. A discussion of this requires the extension of our discrete groups to a continuously infinite group, the rotation group of full spherical symmetry. For the purposes of our discussion, the inclusion of inversion symmetry will add nothing of interest, so we consider only the rotation group. The treatment parallels our brief discussion of discrete groups. We will state one or two simple results here.

This continuously infinite group may be divided into an infinite number of classes. Any rotation through the same angle is in the same class, no matter what the axis of the rotation. The irreducible representations are numbered by l. For given l the character corresponding to the class of rotations through the angle φ is given by

$$\chi^{(l)}(\varphi) = \sum_{m=-l}^{l} e^{im\varphi} \tag{1.8}$$

The character of the unit element is of course the dimension of the representation and it is immediately seen to be equal to $2l + 1$. The spherical harmonics $Y_l^m(\theta,\varphi)$ form the basis for the lth irreducible representation. There are $2l + 1$ allowed values of m for given l and the Y_l^m for these different m's transform into each other under rotations. The d states in the free atom of course belong to the $l = 2$ irreducible representation of the group and we may readily construct the characters from Eq. (1.8).

In order to describe the splitting of levels in a cubic field we will need also the character for the cubic group. It will be desirable to write this table down explicitly since the cubic group is one of the most commonly appearing symmetry groups in solid state physics.

As we indicated earlier the group of symmetry operations taking a cube into itself contains 48 elements. We consider the proper rotations (including the identity operation); there are 24 operations in this subgroup. In addition there are elements which may be constructed as the product of these same proper rotations and the inversion but we will not need to include these. The full cubic group is generally denoted by O_h. The subgroup made up of the proper rotations and the identity is generally referred to simply by O. The character table for the symmetry group O is written below.

	E	$8C_3$	$3C_2$	$6C_2$	$6C_4$
Γ_1	1	1	1	1	1
Γ_2	1	1	1	-1	-1
Γ_{12}	2	-1	2	0	0
Γ'_{15}	3	0	-1	-1	1
Γ'_{25}	3	0	-1	1	-1

All operations (except for the identity) are rotations, as indicated by the symbol C. The subscript indicates the angle of rotation; thus the operations C_3 are threefold rotations, rotations of 120°. These eight operations are rotations around cube diagonals, [111] directions. The $3C_2$'s are the 3 twofold (180°) rotations around the three [100] directions. The $6C_2$'s are the twofold rotations about the six [110] directions. The $6C_4$'s are clockwise and counterclockwise fourfold rotations about the three [100] directions. The symbol Γ is conventionally used to designate the representation of the symmetry groups O and O_h. The subscript 1 is generally used to denote the unit representation. The remaining subscripts and primes are the notation of Bouckaert, Smoluchowski, and Wigner.[1] We see that there are 2 one-dimensional and 2 three-dimensional representations and a single two-dimensional representation.

We may readily construct the character for the d states ($l = 2$) from Eq. (1.8) for the rotations which are symmetry operations of the cubic group. We obtain the characters 5, -1, $+1$, $+1$, and -1 for E, $8C_3$, $3C_2$, $6C_2$, and $6C_4$. Using our rules, then, we may readily show that this fivefold representation reduces to the representation Γ_{12} and Γ'_{25}. A d state will be split in a cubic field into twofold degenerate and threefold degenerate states. The representations Γ_{12} and Γ'_{25} are frequently referred to as d like representations of the cubic group. The threefold degeneracy of p states is not lifted in cubic symmetry, and the representation becomes Γ'_{15}. Thus the Γ'_{15} representation is frequently called a p-like representation; for similar reasons Γ_1 is called an s-like representation.

5.2 Vibrational states We will now apply the mathematics of group theory to an entirely different type of problem. This is the classical problem of vibration in a harmonic system. We will use a direct physical approach.

[1] L. P. Bouckaert, R. Smoluchowski, and E. P. Wigner, *Phys. Rev.*, **50**:58 (1936).

We consider a collection of atoms coupled to each other harmonically. The distortion of the system from its equilibrium configuration may be specified at any given time by giving a set of displacement components $x_1, x_2, \ldots, x_N$. This will ordinarily contain three displacement components for each of the atoms present. We might alternatively specify the state of the distortion by giving the value of generalized coordinates Q_i which are related to the x_i by a unitary matrix

$$Q_i = \sum_j U_{ij} x_j$$

Knowing the magnitudes of the Q_i at a given time we may of course obtain the displacements themselves by making the inverse transformation, noting that the inverse of U is simply U^+

$$x_i = \sum_j U_{ji}^* Q_j$$

A particularly convenient set of generalized coordinates are the *normal coordinates* which correspond to normal modes of vibrations of the system. The normal coordinates vary with time sinusoidally with a frequency equal to the corresponding normal mode frequency. When a single normal mode is excited, all displacements x_i vary sinusoidally with the same frequency. They are given by

$$x_i(t) = U_{ji}^* Q_j(0)(\cos \omega_j t - \tan \varphi_j \sin \omega_j t) \tag{1.9}$$

where $Q_j(0)$ is Q_j evaluated at time zero, ω_j is angular frequency of the mode, and φ_j is the phase at time zero. We may characterize the jth mode by giving U_{ji}^* for each i running from one to N.

It will be helpful to do this graphically by sketching the system in question and drawing vectors given by the components $x_i = U_{ji}^*$. Such a sketch for a particular mode in a molecule made up of four atoms is given in Fig. 1.14. Such a sketch may be thought of as a *plucking pattern*. If each atom is displaced by the vector distance indicated and released (with no velocity) at time zero, the subsequent motion will be given by Eq. (1.9) with only a single $Q_j(0)$ nonzero and with the corresponding phase $\varphi_j = 0$. This has the effect of plucking the molecule in such a way as to excite a single normal mode. The diagram displays the form of the mode in a particularly graphic way.

We may now see how the application of symmetry arguments may tell us something about the modes. If R is a symmetry operation that takes the molecule into itself, we may envisage picking up the molecule, distorted for example as shown in Fig. 1.14, and performing the operation. For example, if R were a 90° clockwise rotation, the plucking pattern would be rotated as shown below.

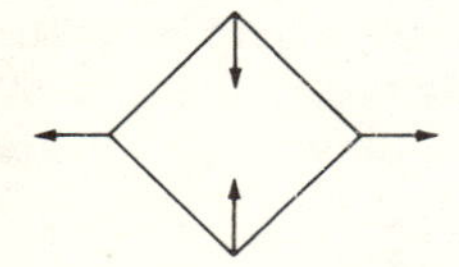

This now may be thought of as a new distortion of the system. Clearly if the system were plucked in this new way, it would oscillate with precisely the same frequency as that induced by the initial distortion. If the new mode is equivalent to the old, as it is in this case (with simply a sign change), then we have learned very little. However, if the new mode is distinct from the old, we have found a second mode with precisely the same frequency as the first. As in the quantum-mechanical treatment, we may find degeneracies required by symmetry. Thus if the normal coordinates all transform into themselves (or into equivalent coordinates) under the symmetry operations, we expect no degeneracies in the vibration spectrum. If on the other hand some of the normal coordinates transform into each other under symmetry operations, we conclude that these particular modes will be degenerate.

We may restate this point in another way. If we could make the unitary transformation from the displacement coordinates to the normal coordinates (which would require solution of the problem), the normal coordinates of

Fig. 1.14 The plucking pattern for a particular vibrational mode of a square molecule. The vectors represent the displacements of the atoms at time zero when only that mode is excited.

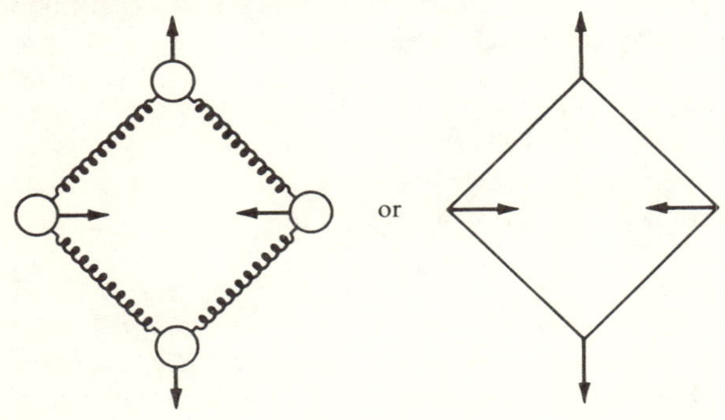

nondegenerate modes would necessarily transform only into themselves. That is, they would necessarily transform according to a one-dimensional irreducible representation. Degenerate modes, on the other hand, might transform into each other. Appropriate linear combinations could be taken (any orthogonal linear combinations of degenerate normal modes are permissible normal modes) such that the new normal coordinates transform according to irreducible representations.

This enables us to learn about the degeneracies of the modes without ever really solving for the modes themselves. We can do this because we have found that a transformation to normal coordinates will lead to irreducible representations, and group theory enables us to find the irreducible representations directly. The only uncertainty that may remain will arise if two sets of modes (degenerate or not) transform according to the same irreducible representation. Symmetry alone will not enable us to unscramble these. This is most easily seen by an example.

The procedure for reducing the coordinates is quite straightforward. We will illustrate it again for the triangular molecule in a plane. We may set up a coordinate system at each atom as shown in Fig. 1.15. The displacement of the upper atom, for example, is specified by giving the components of the displacement in the two directions indicated by x_1 and x_2. The magnitudes of these components are written x_1 and x_2. A six-dimensional column vector describes the displacements of the entire molecule. These column vectors provide the basis for a six-dimensional representation of the group. We construct this representation by envisaging a particular set of displacements for each of the atoms in the molecule. For the 120° rotation C_2, for example, we simply pick up the distorted molecule and rotate it by 120°. With respect to our old coordinate system, which has remained fixed, we now have a new set of displacements. For example, the displacement

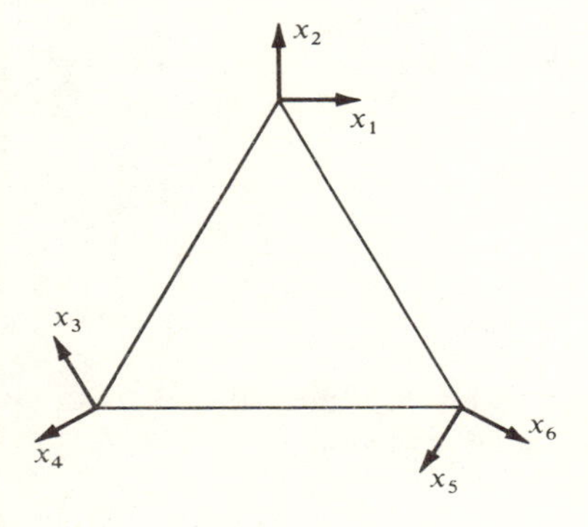

Fig. 1.15 Coordinate system for specifying the displacements of the atoms in a triangular molecule.

of the upper molecule in the x_2 direction now appears as a displacement of the lower left molecule displaced in the x_4 direction. We may write the displacements after the 120° rotation as primed x's and those before the rotation as unprimed x's. Thus we have concluded that $x_4' = x_2$. In general, then, the x' column vector is related to the x column vector by a matrix which we write $D_{ij}(C_2)$. In matrix quotation we may write

$$x_i' = \sum_j D_{ji}(C_2)x_j$$

The $D_{ij}(R)$ form a representation of the symmetry group. Transforming from the displacement column vector to normal coordinates will correspond to a unitary transformation of this representation to an equivalent one. We are interested in reducing the representation D to the irreducible representations of which it is composed. We will of course do this by first finding the characters of the initial representation D.

We may readily construct these characters by considering the diagram. Only the diagonal elements enter the determination of the character. This means that only those elements relating a displacement x_i' to a displacement x_i contribute. In the 120° rotation which we described above, all atoms are interchanged so that the primed displacements on a particular atom arise entirely from the unprimed displacement on a different atom. Thus no diagonal elements are nonvanishing and we conclude that the character is zero for all of these rotations. In a reflection, say the reflection about a vertical altitude of the triangle, the lower left and the lower right molecules are interchanged so the corresponding elements 3 through 6 do not contribute to the character. On the other hand the upper atom goes into itself so we may expect diagonal components there. In particular the displacement x_1 goes into its negative from which we conclude $D_{11}(\sigma)$ is equal to -1. On the other hand displacements x_2 go into themselves so that $D_{22}(\sigma)$ is equal to $+1$. These are the only nonvanishing diagonal matrix elements. We add them to obtain the character, and again obtain zero. In the identity operation, of course, every displacement goes into itself; all diagonal elements are unity and we have a character of 6. The character table then is given by

	E	3σ	$2C$
Λ	6	0	0

Using the character table for the irreducible representation of the triangular group we may immediately make the reduction $\Lambda = \Lambda_1 + \Lambda_2 +$

$2\Lambda_3$. These six degrees of freedom of the triangular molecule have led to two nondegenerate modes and two sets of doubly degenerate modes. Except for the possibility of accidental degeneracies this finding is unambiguous.

This is the information we sought. We will now go on to see how we can in addition find the form of these modes. We will be able to find the nondegenerate modes in detail, but some ambiguity will remain with respect to the doubly degenerate modes which transform according to the same irreducible representation. This analysis will of course require the irreducible representations themselves and we will use those which we have obtained earlier. Had we given an equivalent but different two-dimension irreducible representation to the one listed there, we would find different linear combinations of the degenerate modes from those which we will find here. They would represent an equally good, and in fact equivalent, solution of the problem. We proceed now to that problem.

Now the jth normal mode in the x representation is given simply by $x_i = U_{ji}^*$. For different values of the index i this gives the component of the displacement of the individual atoms for a unit amplitude of the mode. We again represent the normal mode by a plucking pattern; the vectors again correspond to the displacement for unit amplitude of that normal mode. The problem now is the construction of a set of such diagrams which transform according to the appropriate irreducible representation. We proceed by assuming that some particular component is nonzero and then, by making symmetry operations, deduce other components.

We look first at the irreducible representation Λ_1. We then assume that the displacement component x_1 is nonzero. We immediately see that this is inconsistent with the Λ_1 irreducible representation; if we apply a reflection about the vertical altitude, x_1 goes into its negative. That is, the arrow has reversed but the Λ_1 irreducible representation requires that the mode goes into itself. Thus we conclude immediately that x_1 is equal to zero. We next assume that x_2 is nonzero, while x_1 is zero. This arrow now goes into itself under the same reflection and is therefore consistent with Λ_1. The Λ_1 representation also tells us that a 120° clockwise rotation will take the mode into itself. After this rotation, the atom that sits in the upper position must also be displaced outward from the center. This means that the atom which has been moved to the upper position must have been displaced outward in the normal mode diagram. It also means that the atom in the position to which the upper atom has been displaced must be displaced outward. We have deduced the mode in which every atom is displaced outward by an equal amount. This is simply the breathing mode of vibration. Had we instead chosen our initial arrow to be pointing downward rather than upward, we would, of course, have obtained just the same mode but 180° out of phase. We summarize this argument diagramatically as follows:

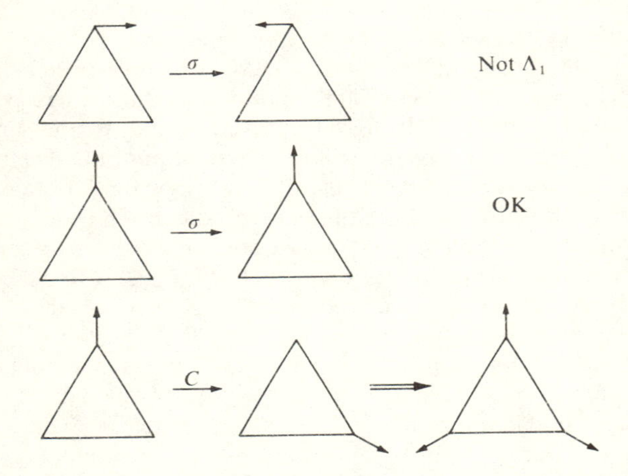

We may readily confirm by application of the other symmetry operations of the group that this mode is consistent with all of them. Also note that if we had chosen a slightly less convenient set of coordinates for the initial displacement the argument might have been more complicated but would have led to precisely the same result.

A similar argument based upon the irreducible representation Λ_2 leads directly to a normal mode in which $x_1 = x_3 = x_5$, $x_2 = x_4 = x_6 = 0$; this is a rotational mode. For a free molecule this is of course a zero-frequency mode. However, if the molecule were attached to a rigid plane by springs (which is still consistent with our symmetry requirements) then this rotational mode would have nonzero frequency.

The treatment of the Λ_3 modes is not quite so simple because we must simultaneously consider two modes. We will also see that the solution is not unique, but depends upon our choice of representation. We begin again with a discussion of the symmetry operation σ_1, the reflection around the vertical altitude. The matrix representing σ_1 in the Λ_3 representation is

$$D^{(3)}(\sigma_1) = \begin{bmatrix} -1 & 0 \\ 0 & +1 \end{bmatrix}$$

The first mode goes into its negative while the second goes into itself. This implies that the upper atom is displaced horizontally in the first mode; the upper atom is displaced vertically in the second mode. We will obtain different sets of modes depending on our choices of the sign of these arrows.

We will look first at the choice of displacements to the right and upward.

We next apply the symmetry operation C_1 which takes the displacements to those shown below.

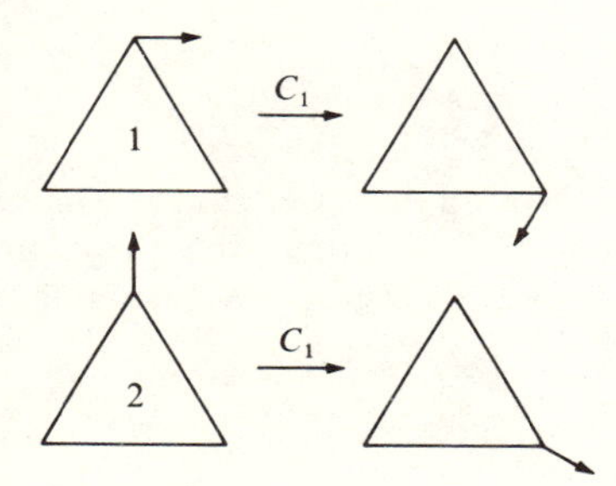

The irreducible representation corresponding to this operation is

$$D^{(3)}(C_1) = \begin{bmatrix} -\frac{1}{2} & \frac{\sqrt{3}}{2} \\ -\frac{\sqrt{3}}{2} & -\frac{1}{2} \end{bmatrix} \quad \text{or} \quad \begin{matrix} D_{11} = -\frac{1}{2} & D_{12} = \frac{\sqrt{3}}{2} \\ D_{21} = -\frac{\sqrt{3}}{2} & D_{22} = -\frac{1}{2} \end{matrix}$$

This matrix implies that the upper mode after rotation must be equal to $D_{11} = -\frac{1}{2}$ timès the old upper mode plus $D_{21} = -\sqrt{3/2}$ times the old lower mode. (Recall $Q_i' = \sum_j D_{ji} Q_j$.) Performing this combination of the upper and lower modes gives us, by virtue of $D^{(3)}(C_1)$,

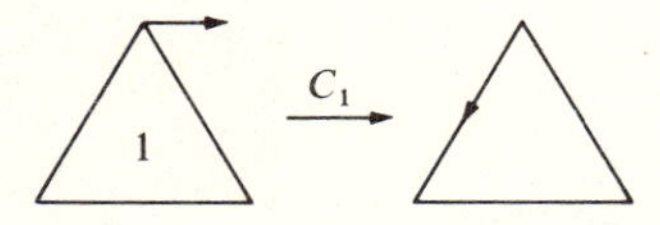

We now have two arrows on the rotated first mode.

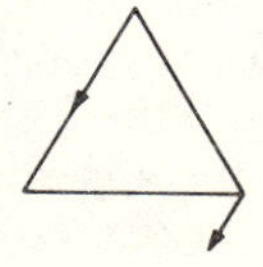

Rotating back, we have two arrows on the original first mode.

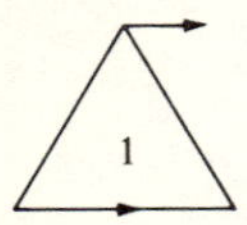

Similarly, to obtain an additional arrow on the lower mode we may add $\sqrt{3/2}$ times the arrow on the upper mode and $-\frac{1}{2}$ times the arrow on the lower mode. We draw this arrow on the rotated lower mode and rotate back to obtain

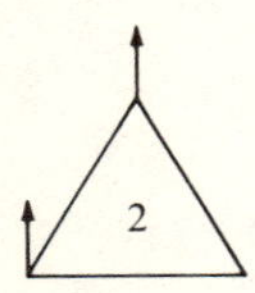

We next go through the same chain of arguments with the rotation C_2, thereby adding the other two arrows and therefore the complete form of the two modes.

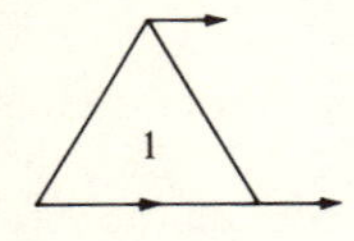

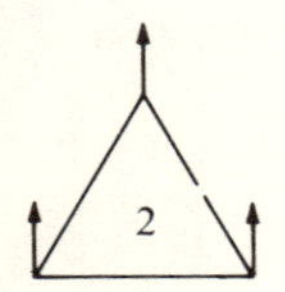

These are simply the translational modes. For the free molecule they are zero-frequency modes, but if the atoms were connected by springs to a rigid plane, the frequencies would be nonzero.

We have obtained only one of the sets of modes belonging to Λ_3. To obtain the other we start again but now choose the first mode with displacement of the upper atom to the left rather than to the right. We again choose the second mode with displacement up for the upper atom.

With precisely the same procedure we will end up with the modes shown below.

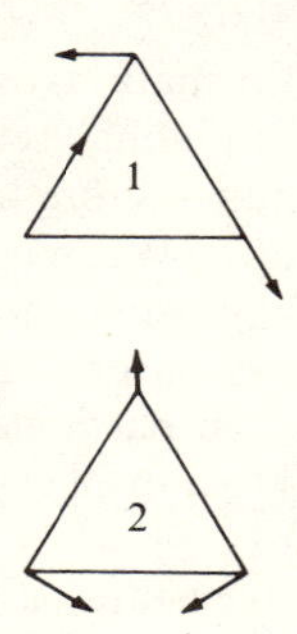

These two degenerate modes would have been difficult to pick out from the initial problem without the use of group theory. Note further that the result is independent of the magnitude of the force constants.

A more symmetric representation of this vibration could be made by taking another linear combination. We could, in particular, advance the phase of mode 2 by 90° and add it to mode 1. This gives one new mode. The second would be obtained by retarding the phase of mode 2 by 90° and adding it to mode 1. The two resulting modes correspond to the circular motion of each of the atoms, each out of phase by 120° from the rotation of its neighbor. In one mode the vibration of the atom to the right is 120° ahead in phase, in the other the motion is 120° behind in phase.

We should now note that the form of these last four modes depended upon our choice of irreducible representation. Had we used an equivalent irreducible representation, we would have mixed the vibrational and rotational modes. For the free molecule we recognize the translational modes as correct and therefore unscramble them from the vibrational modes. However, this requires our knowledge of the translational invariance of space, which was not included in the symmetry of the triangular molecule upon which our treatment is based. We could construct a system of springs—fastened to a rigid base—such that the correct solution did require mixing of the translational and vibrational modes. The unscrambling requires other information than the triangular symmetry. Similarly, a linear combination of $2p$ and $3p$ states in hydrogen will transform according to the $l = 1$ representation of the full rotation group. Knowledge of the radial dependence of the potential is required to separate them.

It is nevertheless striking that we are able to obtain so much information about these modes without ever directly utilizing the equations of motion. Of course the equations of motion would be needed if we wished to obtain the frequencies explicitly, and some further information is required for complete determination of the modes.

Since these modes correspond to a unitary transformation on the displacement coordinates, which themselves were an orthogonal set, the normal modes themselves are orthogonal; i.e., the column vectors corresponding to each of these diagrams are mutually orthogonal. We may take the dot product of two such vectors simply by superimposing the two diagrams and adding the dot products of the two vectors on each atom. By inspection we can make a quick check on our results. In particular we may note that only the two translation modes correspond to a displacement of the center of mass.

The applications of group theory to electronic eigenstates and to molecular vibrations are perhaps the two simplest nontrivial applications of symmetry arguments. We will go on now to discuss applications which are more directly related to solid state physics.

5.3 The translation group—one dimension We discussed the translational group for three-dimensional crystals in the second section of this chapter. In the first section of Chap. II we will use this three-dimensional translation group to study the structure of the electronic states in three-dimensional crystals. It may first be desirable, however, to study states in a one-dimensional crystal. All of the conceptual features are the same in one and in three dimensions, but the algebra is simpler in one dimension so that there is advantage in considering this simple system first.

We envisage a one-dimensional chain of identical atoms separated by a distance a as shown in Fig. 1.16. The symmetry translations of Eq. (1.1) become

$$T = na$$

for this one-dimensional case, where n may be a positive or negative integer. In making a connection with group theory we will need now to be more careful with the boundaries (ends) of our crystal. This is most easily achieved by considering a chain of N atoms, but now bent around in a ring. Thus the translations become circulations of the atoms around this ring, and a translation such as that given above in fact takes each atom to a position previously occupied by another.

This approach is equivalent to the application of periodic boundary conditions on a linear chain of N atoms. In terms of periodic boundary conditions we would say that any atoms translated out one end of the crystal are translated in on the opposite end. In three dimensions it will be easier to

Fig. 1.16 A chain of identical atoms in one dimension, each separated from its nearest neighbors by a.

envisage the application of periodic boundary conditions than a bending of the three-dimensional crystal into a ring in four dimensions.

In our chain there are N configurations of the atoms which can be reached by translations such as those indicated above, and we will associate with these the integers n running from zero through $N - 1$. We discard translations corresponding to other values of n since they are clearly equivalent to those we have already included. For example, a displacement corresponding to $n = N$ returns the crystal to precisely the same configuration and may therefore be considered to be the identity element, $n = 0$.

We may see immediately that the N translations defined above form a group. A product of translations is simply the subsequent performance of the two operations. Since each translation by itself takes the crystal into an indistinguishable configuration, the product also does and is therefore in the group. The unit element again is simply the translation $n = 0$. Multiplication is obviously associative and the inverse is in the group.

We further note that all elements in this group commute so that it is an Abelian group and the irreducible representations are therefore of one dimension. Furthermore, since each element may be written as the element $n = 1$ raised to some power, the group is cyclic and the irreducible representations may be immediately written down.

$$D^{(\kappa)}(T_n) = e^{-2\pi i\kappa n/N}$$

The symbol κ is an integer which indexes the representation. There are N distinguishable translations in this group and N values of κ which give distinct representations. These are then all of the irreducible representations of the group.

The index κ is conventionally taken to run from approximately $-N/2$ to $+N/2$. If N is odd we may take all positive and negative integers in this range (and zero). If N is even we take only one of the endpoints. Thus $-N/2 \leq \kappa < N/2$ is suitable for either case.

Now electron eigenstates for this system may be chosen such that each will transform according to one of the irreducible representations; i.e., the electron eigenstates form a basis for the irreducible representations. We may write these representations in a more convenient form by defining a wavenumber $k = 2\pi\kappa/Na$. We will denote an eigenstate that transforms according to the kth representation by ψ_k.

We learn about the structure of these states by seeing how they transform under the symmetry operations **T**. We distinguish the symmetry operation from the translation distance by writing the former as **T**. Thus the state ψ_k will transform according to

$$\mathbf{T}\psi_k = e^{-ikT}\psi_k$$

If we now define a new function in terms of the wavefunction, $u_k = e^{-ikx}\psi_k$, where x is a distance along the chain, we may quickly show that $\mathbf{T}u_k = u_k$. Thus u_k has the full translational symmetry of the lattice and the electron eigenstates may be written in the form $\psi_k = u_k e^{ikx}$. This is called the *Bloch form*; the u_k are often called *Bloch functions.*[1]

We may now see the significance of these manipulations. If the u_k were constant, then the eigenstates would simply be free-electron plane waves. We may also immediately see from the definition of k above that these plane waves satisfy periodic boundary conditions at the ends of the chain; i.e., if the chain were bent in a circle the wavefunctions would match smoothly at the point of joining. Even if the u_k are not constant we have been able to establish a one-to-one correspondence between states in the crystal and free-electron states with the presence of the atoms simply entering through the Bloch functions. This is a very important result; it will enable us to carry over much of the intuition and many methods from the treatment of free-electron gases to the treatment of electron states in crystals.

We must note, however, that the realm of k has been restricted to values between $-\pi/a$ and $+\pi/a$. Thus the correspondence between Bloch states and free-electron states is restricted to the small wavenumber range. We may ask then what has become of the higher wavenumber free-electron eigenstates which would be present if the potential due to the atoms in the chain were zero. Clearly any such state may be written in the form

$$(Na)^{-1/2}\, e^{ikx} = (Na)^{-1/2}\, e^{2\pi inx/a}\, e^{i(k-2\pi n/a)x}$$

where n is a positive or negative integer selected such that the *reduced wavenumber*, $k - 2\pi n/a$, lies in the realm we have chosen. We have also introduced a factor $(Na)^{-1/2}$ to normalize the plane waves. We have written this wavefunction in the Bloch form with a Bloch function given by $e^{2\pi inx/a}/\sqrt{Na}$. It may be verified immediately that this Bloch function has the full translational symmetry of the lattice as required. Each of the free-electron states is now represented by a reduced wavenumber and energies of all states may be plotted as a function of reduced wavenumber as shown in Fig. 1.17.

Note that we have drawn the eigenvalues as continuous curves, called

[1]These functions are named for F. Bloch, who first introduced them. [F. Bloch, *Z. Physik*, **52**:555 (1928).]

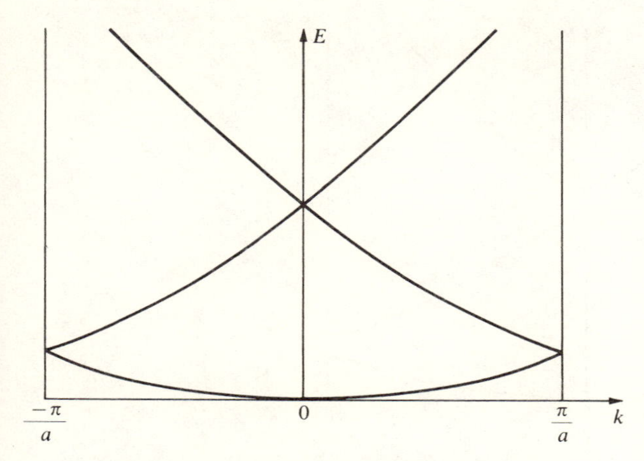

Fig. 1.17 The energy as a function of reduced wavenumber *for free electrons in one dimension.*

energy bands. This is sensible because the allowed wavenumbers are very closely spaced in a large system (either for free electrons or for states in crystals). For a chain of 10^8 atoms, a typical number of atomic distances in a macroscopic crystal, allowed wavenumbers $2\pi n/Na$ occur in the region of reduced wavenumbers for 10^8 values of n. Thus k is a quasicontinuous variable and it is sensible to consider the bands as continuous functions of k.

The free-electron eigenstates have now been represented as a series of bands, each restricted in wavenumber to the *Brillouin zone*, which is the region occupied by reduced wavenumbers in wavenumber space. The lowest-lying state at a given wavenumber is said to lie in the first band; the second-lowest-lying state is said to lie in the second band and so on. Thus any state is now specified by giving its reduced wavenumber and its *band index*. This is a complicated way of representing free-electron states but becomes quite convenient in a crystal.

We will see that if a weak potential with the periodicity of the lattice is introduced, the u_k are no longer constant and the energy of each of the states is slightly shifted. In particular it is found that the bands become separated at the points $\pm\pi/a$ and zero where they are degenerate in Fig. 1.17. The bands in this nearly-free-electron case become schematically as shown in Fig. 1.18. It turns out that in the simple metals the effect of the periodic potential *is* in fact quite weak and such a nearly-free-electron representation of the bands becomes appropriate.

We may also consider energy bands for our one-dimensional case in the other extreme limit, that in which the potential due to the ions is very

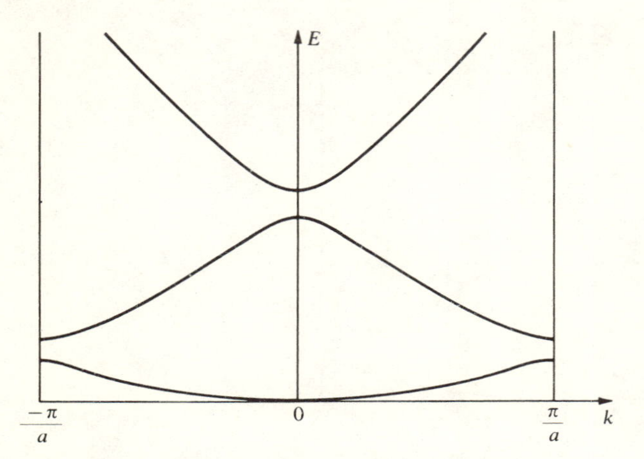

Fig. 1.18 The energy as a function of reduced wavenumber for electrons in one dimension in a weak periodic potential of period a.

large. If the potential is attractive and sufficiently large (or if the atoms are spaced widely enough), we may think of each ion as binding an atomiclike state. For an isolated atom in one dimension these atomiclike states might be written $\psi(x - na)$ for an atom at the position na. This set of N eigenstates, all with energy E_0, is a legitimate description of the states of the system. However, we may also write equivalent states in the Bloch form. Since the atomic states are degenerate we may write *Bloch sums*

$$\psi_k = N^{-1/2} \sum_n e^{ikna} \psi(x - na)$$

It is readily verified that these states are of the Bloch form and are orthonormal if the individual atomic wavefunctions do not overlap each other. This is therefore an equivalent representation of the states. The corresponding energy band is a constant energy E_0, independent of k and restricted to the Brillouin zone.

If we now weaken the potential slightly such that the atomic wavefunctions overlap each other (*the tight-binding approximation*), we will see that these Bloch functions remain approximately correct but the energy band is no longer a constant. As long as the overlap is small, however, the band will be quite narrow in its extent in energy. We will see that such narrow tight-binding bands are characteristic of some of the bands in insulators. Of course if we continued to weaken the potential these bands must ultimately be deformed into the nearly-free-electron bands which we have indicated before. When tight-binding bands are appropriate descriptions of real crystals we may associate each band with the atomic state from which it arose. Thus in the sodium chloride crystal we may think of the chlorine $3p$ bands and the sodium $3s$ bands. We will see, however, in Sec. 7 of Chap. II that when the deduced bands are very narrow, as in the case of the chlorine $3p$ band, there is good reason to believe that the band picture breaks down.

In a similar fashion we may analyze the vibrations of the crystal using symmetry arguments. We consider small displacements in the x direction of the ions (or atoms) in our linear chain. We write the displacement of the nth ion as u_n. A specification of the N displacements gives a complete specification of the state of distortion of the system.

As in the case of the vibrating molecule the normal coordinates will be linear combinations of these displacements on the individual ions and, as in the case of the molecule, the normal coordinates will transform according to the irreducible representations of the symmetry group of the system, in this case the translation group. Since the representations of this group are of one dimension and designated by the wavenumber k, we can associate with each normal mode a wavenumber k from the irreducible representation according to which it transforms.

We may begin by reducing the representation based upon the displacements just as we did in the case of the molecule. We require the characters for the representation generated by $u_n' = \sum_m D_{mn}(T)\,u_m$. The character for the unit representation ($T = 0$) is of course simply N. Every other translation moves every atom; there are no diagonal elements in the corresponding representations and the characters are zero. The reduction into irreducible representations is made according to the procedure of Sec. 4.7.

$$a_k = N^{-1}\sum_T e^{-ikT}\chi(T) = \frac{N}{N} = 1$$

There is one mode transforming according to each of the irreducible representations of the translation group.

We may now construct the form of the modes as we did for molecular vibrations. We consider a mode transforming according to the representation k. We imagine the translation as picking up the linear chain, with its associated distortion, and translating it. If the translation is by m lattice distances, the displacement of an ion at the nth position after translation is equal to the displacement u_{n-m} before displacement. Our representation tells us that the translation modifies each displacement by the factor e^{-ikam}; that is,

$$u_{n-m} = e^{-ikam}\,u_n$$

This enables us to write the displacement of each ion in terms of the displacement of the ion corresponding to $n = 0$.

$$u_m = e^{ikam} u_0$$

The form of the mode is completely specified in terms of the displacement of a single ion.

Since this is a normal coordinate the phase will change with time according to a phase factor $e^{-i\omega t}$ where ω is the angular frequency of the mode in question. Taking u_0 as the displacement of the zeroth ion at time $t = 0$, we have the displacements of every ion as a function of time given by $u_m = u_0 e^{i(kam - \omega t)}$. We have obtained propagating modes as the solutions of the problem. We can take a sum of this expression and its complex conjugate to obtain real displacements or we can work directly with the complex expression.

The physical nature of these modes is quite clear and is sketched in Fig. 1.19*a*. The modes are propagating compression waves moving down our linear chain. At long wavelengths we may expect the velocity of these modes to be a constant equal to the velocity of longitudinal sound in the chain, and therefore we expect that the frequency will be proportional to the wavenumber k. The number of modes is limited, however, by a restriction of the wavenumber to lie within the Brillouin zone. Thus there are a finite number of modes corresponding to the finite number of degrees of freedom. We will see that the frequencies at shorter wavelengths will drop below the linear curve and will flatten out such that $\partial\omega/\partial k$ is zero at $k = \pm\pi/a$, as shown in Fig. 1.19*b*. If displacements in the two other directions are included we find two transverse modes in addition to the longitudinal mode found above. We will return later to a detailed consideration of the vibration modes in three dimensions. We will have occasion in the meantime, however, to make use of the propagating form of the vibrational modes in crystals. It may therefore be desirable to generalize our findings here in anticipation of the analysis to be given in Chap. IV. Most of the details can be obtained by analogy from the structure of the electronic-energy bands which we will discuss next.

In a three-dimensional crystal modes may propagate in any direction and the wavenumber k becomes a vector. It is again limited to a Brillouin zone, which in three dimensions is a polyhedron in wavenumber space. For each wavenumber there are three modes, one with displacements approximately parallel to $\mathbf{k}$, and two with displacements approximately perpendicular to $\mathbf{k}$. (Due to elastic anisotropy the modes are not purely longitudinal nor purely transverse.) These modes give rise to three frequency bands in the Brillouin zone; these are called acoustical modes and have frequencies proportional to k at small k.

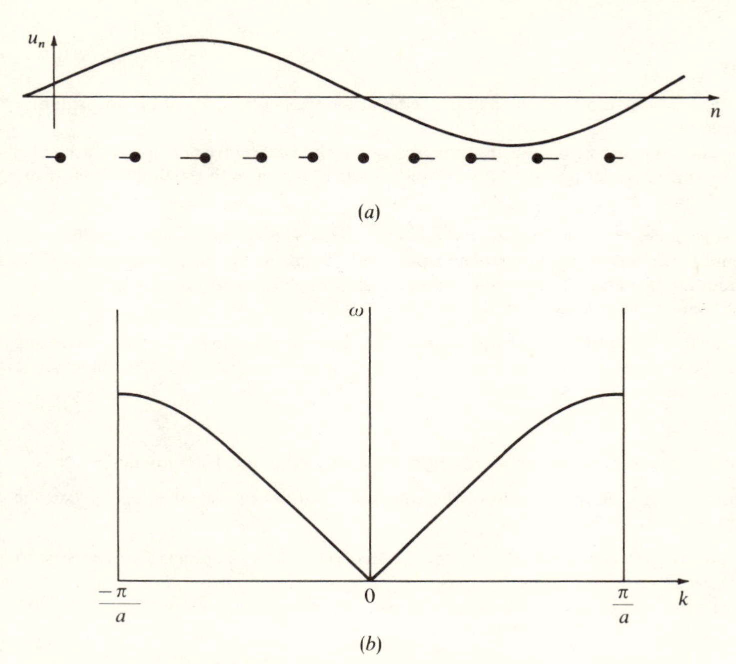

Fig. 1.19 *Vibrational modes in a one-dimensional chain. (a) A plot of the displacements for the mode with wavelength equal to the length of the chain and a sketch of the plucking pattern. (b) A sketch of the frequency as a function of wavenumber.*

If there are two (or more) atoms per primitive cell there will be additional modes, called optical modes, which may be thought of as internal vibrations within the cell, modulated by a Bloch phase factor from cell to cell. The frequencies are nonzero at all wavenumbers and lie entirely above the acoustical bands. One optical band is introduced for each additional degree of freedom in the primitive cell. This much description will suffice until we return to a detailed analysis of the lattice vibrations.

We will next formulate the translational symmetry problem for a general crystal structure, obtain some further information about the bands, using symmetry arguments, and then proceed to discuss the details of band structures in crystals. Symmetry information has given us the form of the states, but detailed calculations are necessary in order to determine the energies of individual electron states as a function of wavenumber and correspondingly to find the detailed form of the vibration frequencies as a function of wavenumber.

Problems

1.1 Consider a body-centered cubic lattice with all atoms identical. What are the primitive lattice translations?

Now add other atoms (identical to the first set) at the face-center position in each cube. What are the primitive lattice translations now? What is the primitive cell, and how many atoms does it contain?

1.2 Consider a crystal with a symmetry axis, the c axis. The symmetry group contains all rotations and reflections for the group of an equilateral triangle in a plane perpendicular to the c axis. In addition it contains a reflection of the crystal through this plane (and its product with the others). Reduce the conductivity tensor for this case.

1.3 Consider a Hamiltonian, $H(\mathbf{r}_1,\mathbf{p}_1;\mathbf{r}_2,\mathbf{p}_2)$, describing the dynamics of two interacting particles. If the particles are identical, then the Hamiltonian is unchanged by the interchange of the two particles.

$$H(\mathbf{r}_1,\mathbf{p}_1;\mathbf{r}_2,\mathbf{p}_2) = H(\mathbf{r}_2,\mathbf{p}_2;\mathbf{r}_1,\mathbf{p}_1)$$

Precisely what can you conclude about the eigenstates $\psi(\mathbf{r}_1,\mathbf{r}_2)$ of this Hamiltonian?

1.4 Construct the symmetry group of the square, its multiplication table, and divide the elements into classes.

1.5 *a.* Construct the character table for the group of the square. You may find it easiest to construct the representations first.

b. See if the twofold degeneracy is split by shearing the square (without changing lengths of edges).

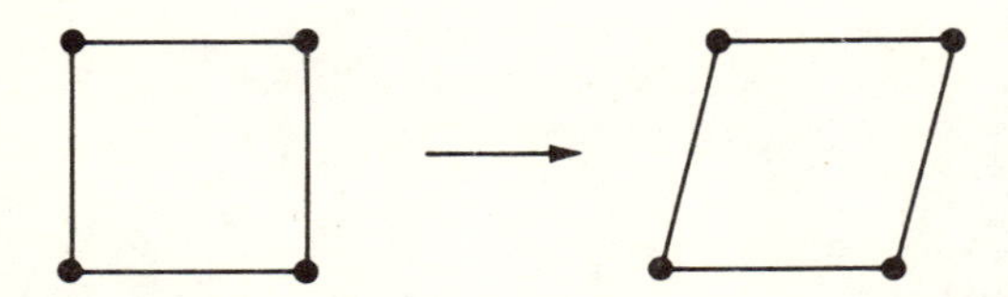

1.6 Construct the character table for the symmetry of Prob. 1.2. Find the splitting of an atomic d state in an environment of this symmetry.

1.7 Consider a symmetric molecule CO_2 with all atoms constrained to move on a line, coupled to each other by springlike bonds, but otherwise free.

a. What does the symmetry tell about the normal modes?

b. Sketch all normal modes for motion only along the line.

1.8 Find the form of the normal vibrational modes of a square molecule.

1.9 Consider electron states in an interstitial atom at a position $[110]a/4$ in a face-centered cubic crystal (see figure at the top of page 55).

a. What is the group of symmetry operations of the system leaving that point fixed; e.g., a reflection in an (001) plane?

b. Give the character table for the irreducible representations.

c. How are *s*, *p*, and *d* states split in this symmetry? (You need not give the symmetry of the states, just the splitting.)

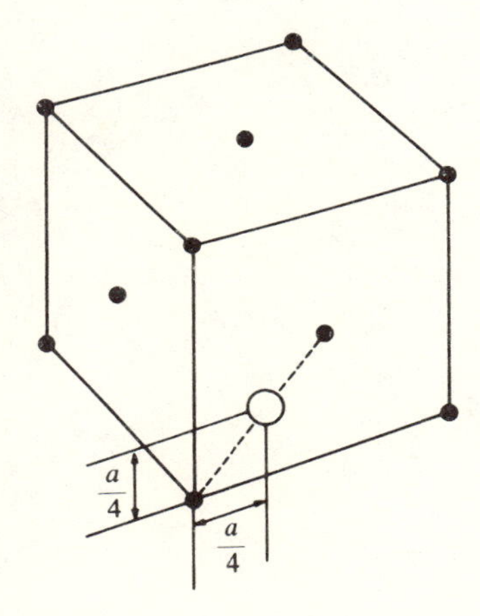

1.10 Consider a molecule with the symmetry of a regular pentagon (e.g., cyclopentane).

a. Find the group of operations in a plane which take the Hamiltonian into itself.

b. Divide them into classes.

c. Obtain part of the character table. You need only the characters for E, except for the one-dimensional representations where all characters are needed.

d. Find the form of all vibrational modes that transform according to a one-dimensional representation.

e. Consider a crystal made up of such molecules in a cubic array such that one plane looks like the accompanying figure. The plane above (and below) is the same but displaced by a in the x direction and a in the z direction. What is the translation group of the crystal?

f. What is the point group of the crystal?

g. Does the symmetry of the crystal rule out its being piezoelectric?

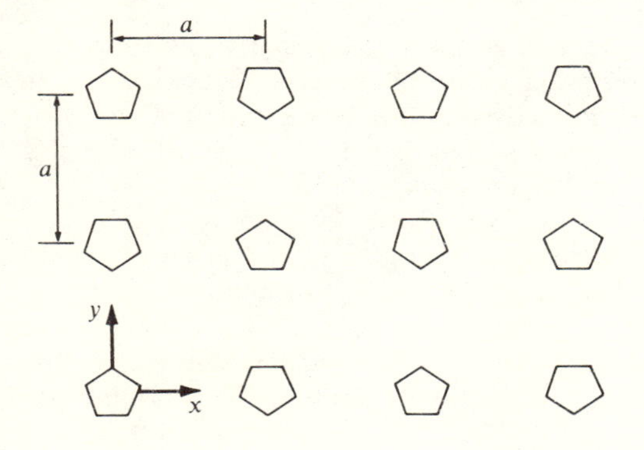

II

ELECTRON STATES

We have indicated in the preceding section how energy bands arise from the existence of the translation group in a one-dimensional crystal. We will first generalize that description to three dimensions and then look in some detail at the nature of the energy bands themselves.

1. The Structure of the Bands

We have defined for an arbitrary structure the primitive lattice translations $\boldsymbol{\tau}_1$, $\boldsymbol{\tau}_2$, and $\boldsymbol{\tau}_3$. The translation group of the crystal consists of sums of integral multiples of these three translations. For a general crystal structure these $\boldsymbol{\tau}_i$ may be of different lengths and they need not be orthogonal. However, we have required that they not be coplanar.

We will again take periodic boundary conditions as we did in the one-dimensional case. We consider a system in the shape of a parallelepiped with edges $N_1\boldsymbol{\tau}_1$, $N_2\boldsymbol{\tau}_2$, and $N_3\boldsymbol{\tau}_3$. In a translation parallel to $\boldsymbol{\tau}_1$, for example, we envisage the atoms transferred out of the boundary as reentering on the opposite face so that the crystal within our boundaries remains invariant under the translation. In addition, the periodic boundary conditions require that the slope and value of any wavefunction be the same on opposite faces of the system. The translation group contains $N_1 N_2 N_3$ distinct translations, which may be written

$$\mathbf{T}_n = n_1 \boldsymbol{\tau}_1 + n_2 \boldsymbol{\tau}_2 + n_3 \boldsymbol{\tau}_3 \tag{2.1}$$

with $0 \leq n_i < N_i$. In analogy with the one-dimensional case the irreducible representations are written

$$D^{(k)}(T_n) = e^{-ik \cdot T_n} \tag{2.2}$$

where $\mathbf{k}$ is now a vector and a dot product appears in the exponent. Each $\mathbf{k}$ corresponds to a different irreducible representation of the group. These $\mathbf{k}$'s may be written as linear combinations of *primitive lattice wavenumbers*, $\mathbf{k}_1$, $\mathbf{k}_2$, and $\mathbf{k}_3$.

$$\mathbf{k} = \frac{\kappa_1 \mathbf{k}_1}{N_1} + \frac{\kappa_2 \mathbf{k}_2}{N_2} + \frac{\kappa_3 \mathbf{k}_3}{N_3} \tag{2.3}$$

where the κ_i are integers and

$$\begin{aligned}
\mathbf{k}_1 &= \frac{2\pi \boldsymbol{\tau}_2 \times \boldsymbol{\tau}_3}{\boldsymbol{\tau}_1 \cdot (\boldsymbol{\tau}_2 \times \boldsymbol{\tau}_3)} \\
\mathbf{k}_2 &= \frac{2\pi \boldsymbol{\tau}_3 \times \boldsymbol{\tau}_1}{\boldsymbol{\tau}_2 \cdot (\boldsymbol{\tau}_3 \times \boldsymbol{\tau}_1)} \\
\mathbf{k}_3 &= \frac{2\pi \boldsymbol{\tau}_1 \times \boldsymbol{\tau}_2}{\boldsymbol{\tau}_3 \cdot (\boldsymbol{\tau}_1 \times \boldsymbol{\tau}_2)}
\end{aligned} \tag{2.4}$$

Substituting Eq. (2.1) into Eq. (2.2) and writing out $\mathbf{k}$ we may readily verify that these representations have the same multiplication table as the translation group and are therefore representations of the group. We have distinct representations for N_i consecutive integers κ_i. Thus there are as many irreducible representations given in Eq. (2.2) as there are symmetry operations in the translation group and this gives all of the irreducible representations of the translation group.

We again characterize each electron state by the wavenumber $\mathbf{k}$ corresponding to the irreducible representation according to which it transforms.

$$T_n \psi_k = e^{-ik \cdot T_n} \psi_k$$

We again define Bloch functions $u_k = e^{-ik \cdot r} \psi_k$. In terms of these the wavefunction is

$$\psi_k = u_k \, e^{ik \cdot r}$$

and u_k is seen to have the full translational symmetry of the lattice.

We may note that a representation corresponding to a wavenumber **k** is the same as a representation corresponding to

$$\mathbf{k} + m_1\mathbf{k}_1 + m_2\mathbf{k}_2 + m_3\mathbf{k}_3$$

where the m_i are integers. The addition of these extra terms does not modify the value of any of the representations of Eq. (2.2). Thus we have an arbitrariness in the choice of **k** which is to be associated with any given representation. It is desirable to prescribe a unique choice and this is customarily done by selecting for any representation the shortest wavenumber that will generate it. (This is consistent with our selection in the case of one dimension.) The resulting domain of **k** is the three-dimensional *Brillouin zone*, and the wavenumbers in the Brillouin zone are frequently called *reduced wavenumbers*.

Note that the set of wavenumbers $m_1\mathbf{k}_1 + m_2\mathbf{k}_2 + m_3\mathbf{k}_3$ forms a lattice in wavenumber space. We will call this the *wavenumber lattice* and refer to the points making up the wavenumber lattice as *lattice wavenumbers*. In diffraction theory it is more customary to discuss the reciprocal lattice which is the corresponding lattice in reciprocal space. The reciprocal lattice vectors are $(2\pi)^{-1}$ times the wavenumber lattice vectors. It is common terminology in solid state physics to call the wavenumber lattice the reciprocal lattice. However, in order to avoid the ambiguity with respect to the factor 2π, we will use the term wavenumber lattice. The symbols $\mathbf{k}_1$, $\mathbf{k}_2$, and $\mathbf{k}_3$ are primitive lattice translations in the wavenumber lattice.

Before proceeding with the formal consequences of translational symmetry, it may be helpful to give a more pictorial description of the states and see the meaning of the multiple wavenumbers which could be associated with a given state. We are assuming a potential that has the full translational symmetry of the lattice. This is illustrated for a line through a series of atoms in Fig. 2.1*a*. Because of the periodicity this potential could be expanded in a Fourier series including only plane waves with wavenumber equal to a lattice wavenumber. (This follows in detail from the Fourier integral and will be demonstrated in Sec. 4.2 of this chapter.)

We have seen that the electronic eigenstates, illustrated in Fig. 2.1*b*, can be factored into a Bloch function, u_k, and a plane wave, $e^{i\mathbf{k}\cdot\mathbf{r}}$, shown in Fig. 2.1*c* and 2.1*d*. The plane wave (as well as u_k) satisfies periodic boundary conditions. Because u_k has the full periodicity of the lattice, it could be expanded in a Fourier series including only lattice-wavenumber plane waves. It follows that the eigenstate can be expanded in a Fourier series containing plane waves of wavenumber **k** and wavenumbers differing from **k** by lattice wavenumbers; these are just the wavenumbers that generate the representation according to which ψ_k transforms.

Consequently, if we could precisely measure the true momentum

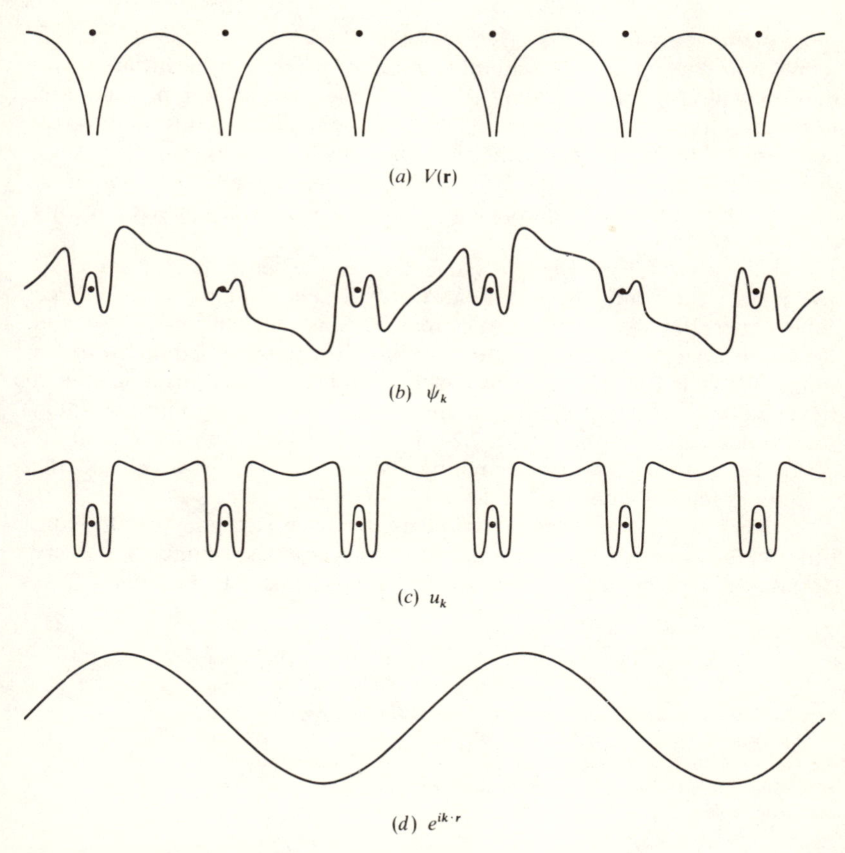

Fig. 2.1 A schematic representation of electronic eigenstates in a crystal. (a) The potential plotted along a row of atoms. (b) A sample eigenstate; the state itself is complex but only the real part is shown. This state can be factored into Bloch function (c), which has the periodicity of the lattice, and (d) a plane wave, the real part of which is shown.

of an electron in the state ψ_k, we could obtain a value corresponding to any one of these wavenumbers. The state ψ_k is not an eigenstate of true momentum since by moving in a potential field the electron is continuously exchanging momentum with the lattice, but because of the periodicity it does not contain components of all momentum values, just those corresponding to the wavenumbers which generate the appropriate representation. Such a description is one way of giving physical content to the ambiguity in the wavenumber to be associated with a state.

We have found that a given wavenumber **k** generates an irreducible representation, but any wavenumber differing from **k** by a lattice wavenumber will also lead by Eq. (2.2) to the same irreducible representation. For a given representation we have selected the smallest of this set of wavenumbers to specify that representation. In terms of our wavenumber lattice, this realm which we call the Brillouin zone contains all points closer to $\mathbf{k} = 0$ than to any other lattice wavenumber. It is bounded by planes bisecting the lattice wavenumbers.

Note that for the simple cubic lattice, the wavenumber lattice is also simple cubic and the Brillouin zone is a cube. For the face-centered cubic lattice, the wavenumber lattice may be constructed from Eq. (2.4) and is readily seen to be body-centered cubic. The corresponding Brillouin zone is shown in Fig. 2.2*a*. Similarly the body-centered cubic lattice leads to a wavenumber lattice that is face-centered cubic. The corresponding Brillouin zone in that case is shown in Fig. 2.2*b*. The hexagonal close-packed (and the hexagonal) lattice have a wavenumber lattice that is hexagonal and the corresponding Brillouin zone is a right hexagonal prism.

Every state then is characterized by a wavenumber in the Brillouin zone. We may note that as in one dimension these wavenumbers are very closely spaced within the Brillouin zone for a large crystal. Rewriting Eq. (2.3),

$$\mathbf{k} = \frac{\kappa_1 \mathbf{k}_1}{N_1} + \frac{\kappa_2 \mathbf{k}_2}{N_2} + \frac{\kappa_3 \mathbf{k}_3}{N_3}$$

Fig. 2.2 *Brillouin zones for (a) face-centered and (b) body-centered cubic crystal lattices. Each is inscribed in a cube of edge $4\pi/a$, where a is the edge of a cubic cell in the real lattice. In the face-centered cubic case the zone volume is half that of the cube; in the body-centered case, one-fourth that of the cube.*

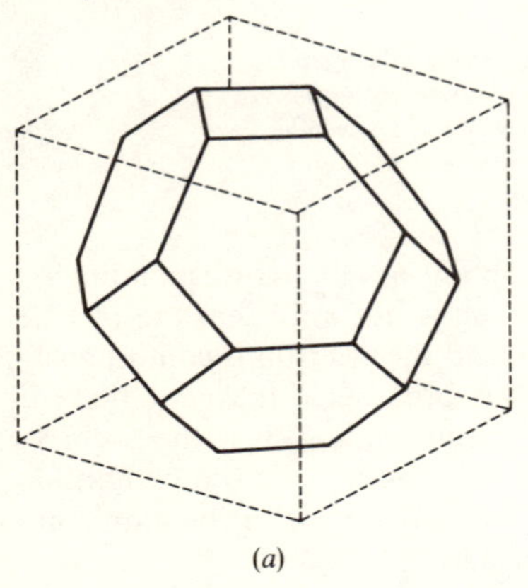

(*a*)

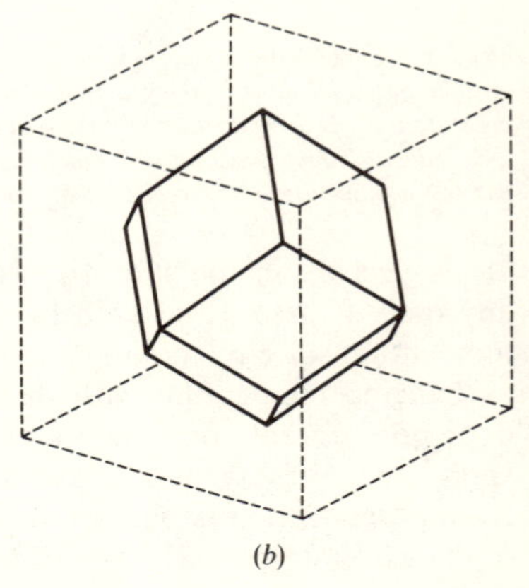

(*b*)

The N_i are large numbers, of the order of the cube root of the number of atoms present, and we have distinct states for the different integers κ_i. We may in fact readily show that there are as many **k**'s (allowed by periodic boundary conditions) in the Brillouin zone as there are primitive cells in the crystal. This is most easily seen by noting that the Brillouin zone volume is equal to the primitive cell volume in the wavenumber lattice. This follows from the fact that if a Brillouin zone is constructed around each lattice wavenumber these zones just fill space and there is one zone for each lattice wavenumber. Similarly, of course, the primitive cells in the wavenumber lattice fill space and there is one cell for each lattice wavenumber. Finally we have seen that there are $N_1 N_2 N_3$ distinct wavenumbers within the primitive cell of the wavenumber lattice, and therefore there are also that many allowed wavenumbers in the Brillouin zone. This equality between the number of states in the Brillouin zone and the number of primitive cells in the crystal will be very important when we consider the occupation of the band states by electrons in real crystals.

We note again that though there are only a finite number of wavenumbers by which to characterize states, there are an infinite number of electronic states. It follows that for each wavenumber there must be an infinite number of states. In order to specify uniquely a state, then, we specify not only its wavenumber but we number the states of each wavenumber in order of increasing energy. Just as in one dimension we call the lowest-lying state of a given wavenumber the state in the *first band*. The next lowest-lying state of the same wavenumber is said to be in the *second band*, and so forth. We uniquely specify each state by giving its wavenumber and its band index. Each state of course has a well-defined energy. We specify energies in the nth band by the function $E_n(\mathbf{k})$. It turns out that $E_n(\mathbf{k})$ is a quasicontinuous function of wavenumber for each band as it was in the one-dimensional examples. In a large crystal, with the wavenumbers allowed by periodic boundary conditions very closely spaced, the function becomes very nearly continuous. The set of functions $E_n(\mathbf{k})$ is called the energy-band structure for the crystal in question. We will see later how they are used in the calculation of properties of the solid.

We have seen examples of such bands in one dimension. It may be helpful here to see the corresponding results in two dimensions before going on to three dimensions.

We consider a two-dimensional square lattice with atomic spacing a. The primitive lattice wavenumbers may be shown to have magnitude $2\pi/a$ and to lie in directions of the primitive lattice translations. The Brillouin zone is a square and the bands give the energy as a function of the two components of **k**. Thus we may plot the energy in a third dimension as a function of the two-dimensional realm of **k**. This is illustrated in Fig. 2.3

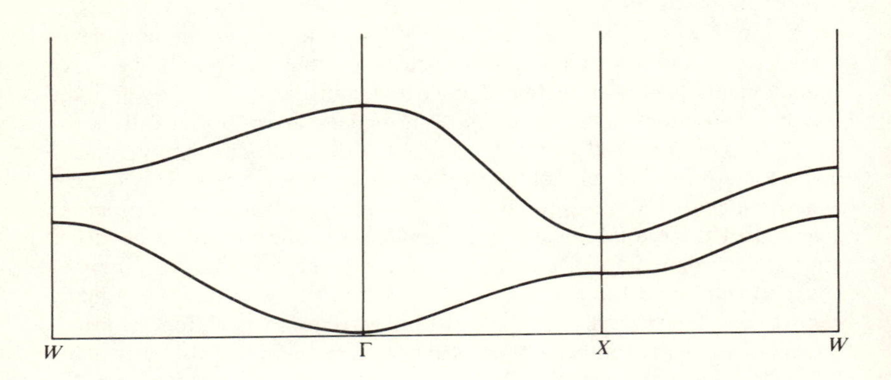

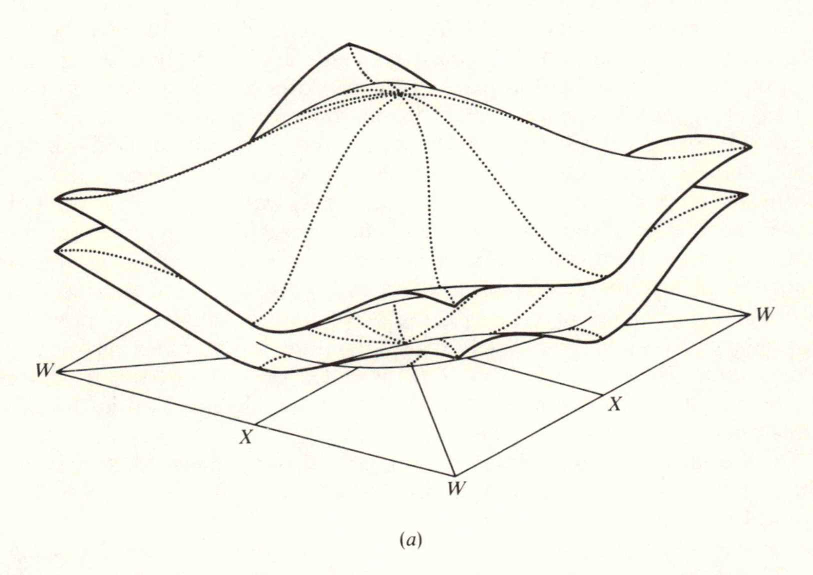

(*a*)

Fig. 2.3 *Possible energy bands for a two-dimensional square lattice. Across the top of the figure are given the first two bands along symmetry lines, shown in the conventional way. Across the bottom is a sketch of the bands throughout*

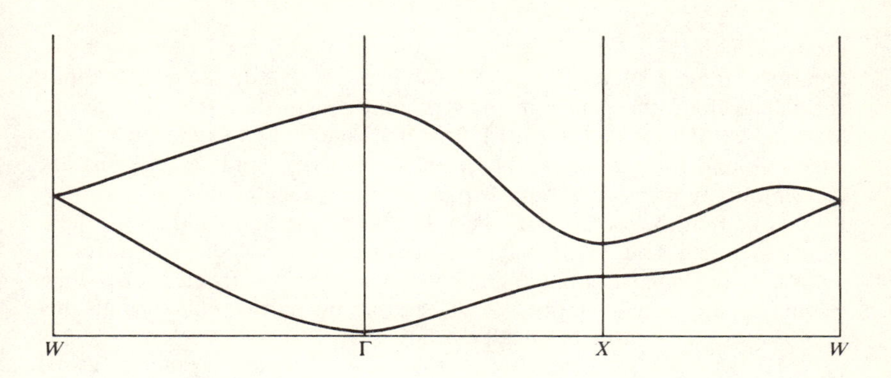

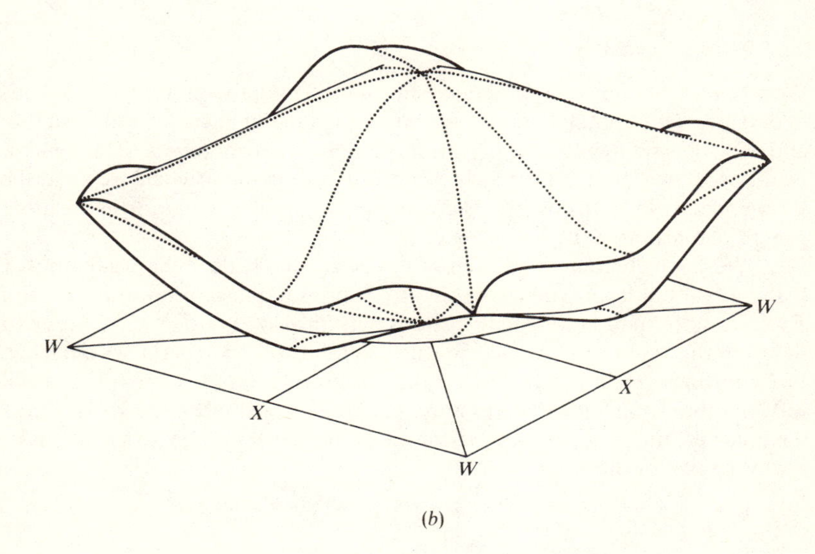

(*b*)

the Brillouin zone; energy is plotted vertically. In case (*a*) *the bands overlap, the second band being lower at X than the first band at W. In case* (*b*) *there is a degeneracy at W.*

for two situations that may arise. To the left we sketch two bands which are separated at every wavenumber. We also show at the top the energy bands calculated along three lines in the Brillouin zone, a line from the corner (denoted by W) to the center (Γ), from the center (Γ) to the center of an edge (X), and from X to W. The results of energy-band calculations are conventionally displayed by such plots along symmetry lines.

Note that the second band at X lies below the first band at W; these bands are then said to overlap. Such overlapping will be important in metals. It prevents us from filling the lowest band with electrons while leaving the upper band empty. We will see that the partially filled bands this necessitates are essential to metallic behavior.

To the right we have shown two bands that are degenerate at W, but are separated at all other points in the zone. Such "contacts" arise frequently in real crystals.

The knowledge of the translation group of the crystal has given us the structure of the energy bands, but we must of course resort to detailed calculation to find the actual $E_n(\mathbf{k})$.

2. *Electron Dynamics*

We will now obtain some general results concerning the motion of electrons in crystals. This will give more content to the idea of energy-band structure and will be useful in our description of individual materials. The general results that we obtain here will be intuitively obvious when we describe simple metals in terms of free-electron bands, but will remain true for more general situations.

We have characterized each of the electronic states by a wavenumber $\mathbf{k}$ and a band index n. The energy in any band is a quasicontinuous function $E_n(\mathbf{k})$. A state may be written in the Bloch form $\psi_k = u_k{}^n e^{ik\cdot r}$. In order to describe the motion of an electron it is helpful to construct a wavepacket. We may localize the state in one dimension, centering it on the state of wavenumber $\mathbf{k}_0$ in a particular band. Since we use a single band we suppress the index n. Summing over a series of wavenumbers $\mathbf{k}$ parallel to $\mathbf{k}_0$, we make a wavepacket of the form

$$\psi = \sum_k u_k e^{ik\cdot r} e^{-\alpha(k-k_0)^2} = e^{ik_0\cdot r} \sum_k u_k e^{i(k-k_0)\cdot r} e^{-\alpha(k-k_0)^2}$$

If α is sufficiently large, then u_k will not vary appreciably over the important terms, and this is a simple gaussian packet centered at $\mathbf{r} = 0$ and modulated by $u_{k_0} e^{ik_0\cdot r}$. We may obtain the time dependence of the wavefunction by multiplying each term by the appropriate phase factor $e^{-iE(k)t/\hbar}$. For $\mathbf{k}$ near $\mathbf{k}_0$ we may replace $E(\mathbf{k})$ by $E(\mathbf{k}_0) + (dE/d\mathbf{k})\cdot(\mathbf{k} - \mathbf{k}_0)$. The expression

$d E / d\mathbf{k}$ is of course the gradient of energy with respect to wavenumber. We may then rewrite our wavefunction in the form

$$\psi = e^{ik_0 \cdot r - iE(k_0)t/\hbar} \sum_k u_k \, e^{-\alpha(k-k_0)^2} \, e^{i(k-k_0)\cdot[r-(1/\hbar)(dE/dk)t]}$$

which is seen to be a packet of the same form as before, but now displaced in the direction of $\mathbf{k}_0$ (or $\mathbf{k} - \mathbf{k}_0$) by the corresponding component of $\mathbf{r} = \frac{1}{\hbar} \frac{dE}{d\mathbf{k}} t$. The packet moves with speed $\frac{1}{\hbar} \frac{dE}{dk}$ where the derivative is taken with $\mathbf{k}$ varying parallel to $\mathbf{k}_0$. By constructing a wavepacket localized in three dimensions we see that in general the wavepacket velocity is given by

$$\mathbf{v} = \frac{1}{\hbar} \frac{dE(\mathbf{k})}{d\mathbf{k}} \tag{2.5}$$

This is simply the counterpart of the classical result that the velocity is given by the derivative of the Hamiltonian with respect to momentum, with $\hbar\mathbf{k}$ playing the role of momentum; $\hbar\mathbf{k}$, based upon reduced wavenumber, is called *crystal momentum*. As long as the localization of the electron is in a region large compared to the interatomic distance, and therefore to the fluctuations in u_k, it is appropriate to associate this velocity with the electronic state in the crystal. For a free-electron gas this velocity is simply $\hbar\mathbf{k}/m$. For more complicated band structures the velocity will take on the more complicated form of Eq. (2.5).

We might next ask what the behavior of an electron is in the presence of some externally applied field in addition to the periodic potential of the lattice itself. If this external force is $\mathbf{F}$, then the change in energy with time of the wavepacket constructed above will be given by

$$\frac{dE}{dt} = \mathbf{F} \cdot \mathbf{v} = \frac{1}{\hbar} \mathbf{F} \cdot \frac{dE(\mathbf{k})}{d\mathbf{k}} \tag{2.6}$$

But if we are to continue to associate a wavenumber with the state, then we may write

$$\frac{dE}{dt} = \frac{d\mathbf{k}}{dt} \cdot \frac{dE(\mathbf{k})}{d\mathbf{k}} \tag{2.7}$$

Equations (2.6) and (2.7) are consistent with an equation of motion for the wavenumber given by

$$\hbar \frac{d\mathbf{k}}{dt} = \mathbf{F} \tag{2.8}$$

This is of course just Newton's equation with true momentum replaced by crystal momentum. The result, Eq. (2.8), is generally true whenever it makes sense to describe the electron in terms of wavepackets. In particular it remains true when **F** is allowed to include a magnetic force, in which case **F** becomes the Lorentz force.

$$\mathbf{F} = -e\mathscr{E} - \frac{e}{c}\mathbf{v} \times \mathbf{H}$$

Here e is the magnitude of the electronic charge, a positive number, and **v** is again the electron velocity.

We may illustrate these results in terms of a band structure such as we discussed in the last section. The electron behavior is illustrated in Fig. 2.4. We consider an electron at rest with $\mathbf{k} = 0$. If we now apply an electric field the wavenumber will change along a line parallel to the electric field with a uniform velocity in wavenumber space. The electron velocity in real space may behave in a very complicated fashion in the process as the wavenumber runs over the hills and valleys of the energy band. When the wavenumber reaches the Brillouin zone face may be represented equally well by a wavenumber on the opposite zone face, differing from the first by a lattice wavenumber. We must in fact make this switch if we are to continue to describe the electron wavenumber in the first Brillouin zone. We may then continue to follow the wavenumber through the band, again moving in a straight line parallel to the applied electric field. If our electric field had been applied along a symmetry direction, the electron would return to $\mathbf{k} = 0$ on the next pass. Thus the electron would execute a cyclic motion which we may think of as the acceleration of the electron, its diffraction by the lattice and its ultimate return to the initial state. Of course in a real crystal the electron will ordinarily be scattered by some imperfection of the lattice or by the boundaries long before it completes such a cycle.

Let us next consider the motion of an electron in a uniform magnetic field.

$$\hbar\frac{d\mathbf{k}}{dt} = -\frac{e}{c}\mathbf{v} \times \mathbf{H} \tag{2.9}$$

By Eqs. (2.5) and (2.9) we see that the motion of **k** is perpendicular to the gradient of the energy with respect to wavenumber, so the electron energy does not change with time. Thus the motion in a magnetic field only is restricted to a constant-energy surface in wavenumber space. For simplicity we consider an energy and a band structure such that this is a closed surface. We could readily extend our picture to surfaces that intersect the zone faces. We see also from Eq. (2.9) that the motion of **k** is perpendicular to the magnetic field; thus the wavenumber must move along the intersection of a plane perpendicular to **H** and the appropriate constant-energy surface.

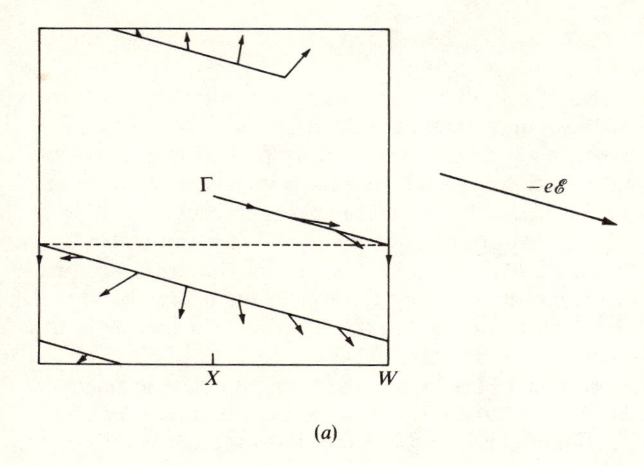

(a)

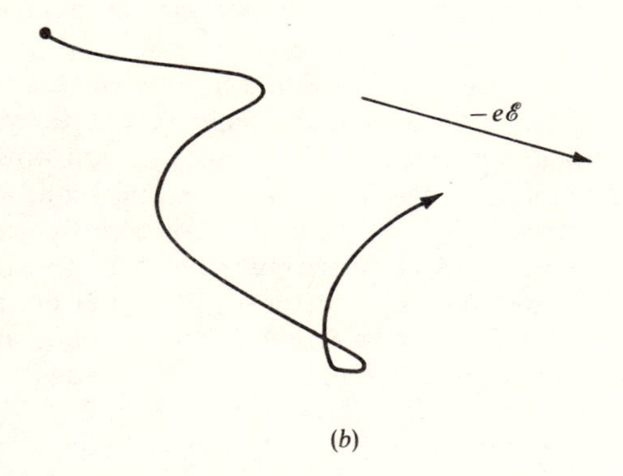

(b)

Fig. 2.4 *The motion of an electron in a two-dimensional square lattice in the presence of a uniform dc electric field* $\mathscr{E}$. (*a*) *The straight-line motion in wave-number space, beginning at* Γ. *The* $d\mathbf{k}/dt$ *is constant except for transfers across the Brillouin zone. Velocities in real space,* $\mathbf{v} = (1/\hbar)\nabla_k E$, *are shown as vectors on the trajectory, assuming an energy band such as the first band in Fig. 2.3a.* (*b*) *The shape of the corresponding electron path in real space; the scale depends on the strength of the electric field.*

Such an orbit is illustrated in Fig. 2.5*a*. The electron wavenumber moves along the intersection as drawn.

It is interesting also to ask about the electron's motion in real space. We note that in Eq. (2.9) we may write the velocity as the rate of change of position, $d\mathbf{r}/dt$. We may integrate both sides with respect to time to obtain those components of the motion perpendicular to the magnetic field. We see that the projection of the real orbit on a plane perpendicular to **H** is of precisely the same shape as the orbit in wavenumber space and scaled by a factor $\hbar c/eH$. In addition it is rotated 90° because of the cross product. Thus a knowledge of the constant-energy surfaces in the energy band will tell us precisely the shape of the orbits that the electrons will execute in the real crystal in the presence of a magnetic field.

In addition there may have been motion in the direction of the magnetic field, which did not enter Eq. (2.9). However, knowing the trajectory of the wavenumber on the constant-energy surface and knowing the band structure, we could compute the velocity at any time and obtain the complete three-dimensional orbit. With a spherical energy surface the orbit in the crystal will be of a helical shape with axis parallel to the magnetic field. For a more complicated shaped energy surface the orbit will be more complicated as illustrated in Fig. 2.5*b*, but with projection of the shape of the corresponding section of a constant-energy surface in wavenumber space.

There are one or two other aspects of the motion in a magnetic field which we might note before proceeding to detailed band structures. We have noted that if the constant-energy surface is closed, the electron wavenumber will execute a periodic motion in wavenumber space. The frequency of this motion is called the *cyclotron frequency*. We may obtain a convenient expression for the cyclotron frequency by constructing two orbits in wavenumber space corresponding to the same plane perpendicular to **H** but to slightly different energies, differing by ΔE. Two such adjacent orbits are shown in Fig. 2.6. We will let ΔE approach zero in the end. Treating ΔE as an infinitesimal we may evaluate the right-hand side of Eq. (2.9). The component of velocity perpendicular to H is simply the corresponding component $\frac{1}{\hbar}\frac{dE(\mathbf{k})}{d\mathbf{k}}$. We approximate this gradient by $\Delta E/\Delta k$ (where Δk is measured parallel to the component of $dE(\mathbf{k})/d\mathbf{k}$ in the plane) and obtain

$$\hbar\frac{d\mathbf{k}}{dt} = -\frac{eH}{\hbar c}\frac{\Delta E}{\Delta k}$$

Now $d\mathbf{k}$ is the infinitesimal change in wavenumber along the orbit in the infinitesimal time dt. We may now multiply through by dt and Δk and

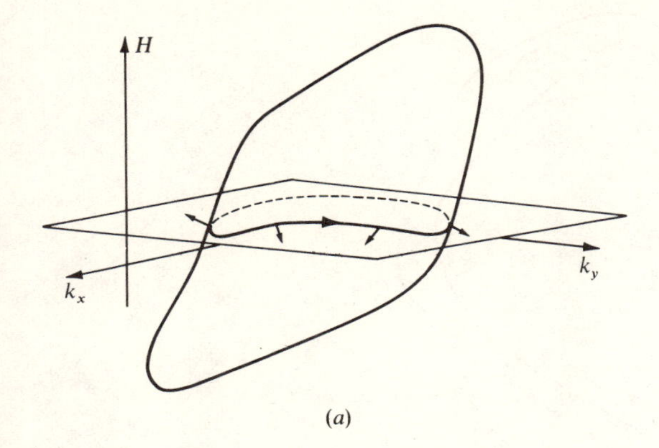

(*a*)

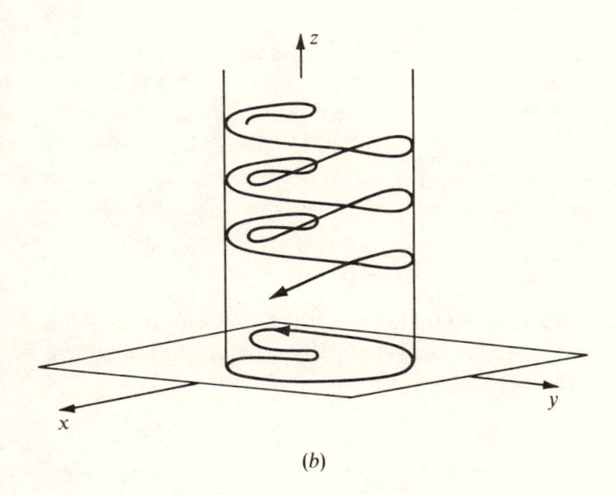

(*b*)

Fig. 2.5 *Motion of an electron in a magnetic field. In wavenumber space* (*a*) *the trajectory is the intersection of a plane perpendicular to* **H** *with a surface of constant energy. Also shown are the velocities in real space at various points on the orbit.* (*b*) *The motion in real space. The full motion is complicated, but the projection of the orbit on an x-y plane is of the same shape as the orbit in wavenumber space, but rotated 90°.*

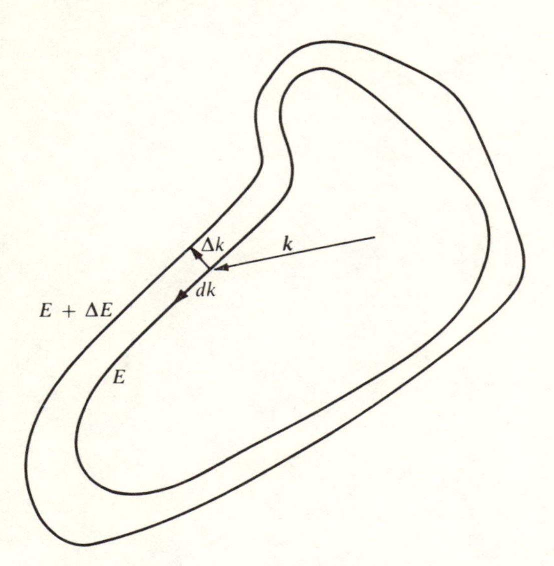

Fig. 2.6 The intersection of two constant-energy surfaces in wavenumber space with a plane perpendicular to the magnetic field **H**. *The difference in energy between them, ΔE, is small, as is the normal wavenumber distance Δk; the latter varies around the orbit. The dk is an incremental displacement along the orbit.*

integrate around the orbit. We note that the integral $\oint \Delta k \, dk$ is simply the difference in area in wavenumber space between the two orbits and of course $\oint dt$ is simply the period of the orbit, T. We have

$$\Delta A = \frac{eH}{\hbar^2 c} T \, \Delta E$$

Thus the cyclotron frequency $\omega_c = 2\pi/T$ is given by

$$\omega_c = \frac{2\pi eH}{\hbar^2 c} \left(\frac{\partial A}{\partial E} \right)^{-1} \tag{2.10}$$

where the derivative of the orbit area (in wavenumber space) with respect to energy is taken at constant component of **k** parallel to the magnetic field. For any band structure then we may compute the cyclotron frequencies for any orbits in the magnetic field. In contrast to the motion across the Brillouin zone in the presence of an electric field, it is frequently possible experimentally to have electrons completing many orbits in the magnetic field.

We can obtain the cyclotron frequency for a free-electron gas from this expression, but in fact we can obtain it much more directly by substituting the free-electron velocity in Eq. (2.9). We obtain

$$\omega_c = \frac{eH}{mc} \tag{2.11}$$

It is very common, for electrons orbiting in real band structures, to associate with any cyclotron frequency an *effective mass* or a *cyclotron mass* chosen such that if m in Eq. (2.11) is replaced by the cyclotron mass, we obtain the observed cyclotron frequency. This enables us to describe cyclotron frequencies by a dimensionless number of order unity, the ratio of the cyclotron mass to the true electronic mass.

This general approach for treating electron dynamics is the *semiclassical approach* and will be used extensively in our discussion of transport properties. The band energy which we have obtained as a function of $\mathbf{k}$ plays precisely the role of a Hamiltonian where the momentum has become $\hbar\mathbf{k}$. Thus the band calculation gives us the semiclassical Hamiltonian $\mathscr{H}(\mathbf{p})$. We may then add applied forces (which must be slowly varying on the scale of the lattice distance) simply by adding the appropriate potentials (or vector potentials for the case of a magnetic field) to obtain the semiclassical Hamiltonian $\mathscr{H}(\mathbf{p},\mathbf{r})$, and Eqs. (2.5) and (2.8) simply give us back the classical Hamilton's equations

$$\dot{\mathbf{r}} = \frac{\partial \mathscr{H}(\mathbf{p},\mathbf{r})}{\partial \mathbf{p}}$$

$$\dot{\mathbf{p}} = -\frac{\partial \mathscr{H}(\mathbf{p},\mathbf{r})}{\partial \mathbf{r}}$$

In addition when we consider many electrons at one time we will also apply the Pauli principle. The way in which this is done will be seen in our discussion of transport properties.

3. *The Self-Consistent-Field Approximation*

In discussing electronic states we have implicitly assumed that we could discuss the behavior of an individual electron moving in the presence of some given potential. Since there are always many electrons present and they interact with each other, this is necessarily an approximation. The wavefunction for the system is a function of the coordinates of all electrons present, and because of the interaction between them the Schroedinger equation is not separable. Only if we can approximate the effect of the interaction of a given electron with all others by a potential, which is then a function only of the coordinates of the given electron, does the Hamiltonian become a sum of terms, each dependent on the coordinates of a single electron; only then can the equations be separated and the electrons treated one at a time. This approximation is the *self-consistent-field approximation.* A self-consistent calculation must be made since the states themselves must be known to compute the interaction potential, which must in turn be known to compute the states.

Almost all of solid state theory is based upon such a self-consistent-field approximation. In Chapter IV, when we present second quantization, we will introduce in a direct way the same Hartree-Fock approximation developed here. We will also see in our discussion of cooperative phenomena how self-consistent fields can represent the more complicated interactions responsible for magnetism and superconductivity.

In the present section we will see, by a variational argument, how the "best" self-consistent fields are derived. In that sense, we obtain the best one-electron approximation. We will then proceed to use the one-electron approximation for a discussion of electron states in solids.

The calculation of the actual eigenstates in a many-electron system is a well-defined problem, but unfortunately it is not one that is exactly solvable. The Hamiltonian of the system contains the individual electronic kinetic energies, $-(\hbar^2/2m)\sum_i \nabla_i^2$. In addition it contains the interaction of each of these electrons with the potential arising from the nuclei. For our present purposes we regard the nuclei as classical point charges with fixed positions and write the interaction between the *ith* electron and the nuclei as $V(\mathbf{r}_i)$. Finally the Hamiltonian contains the coulomb interaction energy between all electron pairs. The wavefunction of the system will be a function of the coordinates of all N electrons so the eigenvalue equation becomes

$$\left[-\frac{\hbar^2}{2m}\sum_i \nabla_i^2 + \sum_i V(\mathbf{r}_i) + \frac{1}{2}\sum_{i,j}{}' \frac{e^2}{|\mathbf{r}_i - \mathbf{r}_j|}\right]\Psi(\mathbf{r}_1,\mathbf{r}_2,\ldots,\mathbf{r}_N)$$

$$= E\Psi(\mathbf{r}_1,\mathbf{r}_2,\ldots,\mathbf{r}_N) \quad (2.12)$$

We know in addition that the many-electron wavefunction must be antisymmetric with respect to the interchange of any two electrons. In writing this equation we have omitted relativistic effects and therefore also explicit spin effects. The symbol E is of course the total energy of the system. This equation is thought to be a correct basis for the understanding of almost all of the properties of solids. However, because of the electron-electron interaction, the equation cannot be separated to obtain independent equations in the coordinates of the individual electrons. Thus an approximation to Eq. (2.12) is required.

3.1 The Hartree approximation For the moment we will ignore the antisymmetry requirement on the wavefunction and consider the mathematical problem posed by Eq. (2.12). If the Hamiltonian could be written as the sum of individual Hamiltonians depending only upon the individual coordinates, the total wavefunction could be written as the product of one-electron wavefunctions, each depending only on the coordinates of one electron. Though this is not the case, Hartree suggested a variational calculation in which the wavefunction is approximated by such a product and the energy, $\langle\Psi|H|\Psi\rangle/\langle\Psi|\Psi\rangle$, is minimized. Even if the wavefunction itself is not so accurate we might hope that the energy would be a good approximation.

This variational procedure leads directly to the Hartree equations from which the one-electron functions that minimize the energy may be determined.[1]

$$\left[-\frac{\hbar^2}{2m}\nabla^2 + V(\mathbf{r}) + \sum_j{}' e^2 \int \frac{\psi_j^*(\mathbf{r}')\psi_j(\mathbf{r}')\,d\tau'}{|\mathbf{r}-\mathbf{r}'|}\right]\psi_i(\mathbf{r}) = \epsilon_i\psi_i(\mathbf{r})$$

The sum is over all occupied states, except the state ψ_i. The ϵ_i are variational parameters that appear in the form of one-electron energy eigenvalues. We see that in this framework the Hartree equations are very plausible. They are one-electron equations in which the potential seen by each electron is determined from the average distribution $\sum_j{}' \psi_j^*(\mathbf{r}')\psi_j(\mathbf{r}')$ of all of the other electrons. However, these ϵ_i are not truly one-electron energies. First it is never possible to uniquely specify one-particle energies in an interacting system since we can arbitrarily add energy to one and subtract it from the other without changing the total. Second, if we return to the evaluation of the total energy, we do not obtain simply a sum of the ϵ_i. We obtain

$$\frac{\langle\Psi|H|\Psi\rangle}{\langle\Psi|\Psi\rangle} = \sum_i \epsilon_i - \frac{1}{2}\sum_{ij}{}' e^2 \int \frac{\psi_j^*(\mathbf{r}')\psi_j(\mathbf{r}')\psi_i^*(\mathbf{r})\psi_i(\mathbf{r})\,d\tau\,d\tau'}{|\mathbf{r}-\mathbf{r}'|}$$

We note that in computing $\sum \epsilon_i$ we have counted the interaction between each pair of electrons twice; once when we computed the ϵ_i for the first and once when we computed the ϵ_i for the second. Therefore to correctly obtain the total energy we must subtract out a quantity equal to the interaction energy.

Note that these equations can be solved self-consistently by iteration.

[1] F. Seitz, "Modern Theory of Solids," p. 677, McGraw-Hill, New York, 1940.

We assume a particular set of approximate eigenstates, compute the potential, and recalculate the eigenstates. The improved estimates are then substituted, the potential recalculated, and the process repeated. Such a process converges and can lead to a set of states that are consistent with the potential.

We see that the only direct contact between our results and experiment is through the total energy. If we wished, for example, to compute the difference in energy between two different eigenstates (which would, for example, be equal to the energy of a photon emitted in a transition between those two eigenstates), it would in principle be necessary to redo the self-consistent calculation for the second eigenstate. This is the procedure followed in calculating the atomic term values in the free atom and such calculations have been carried out for a number of atoms and ions. We will see how this complication is customarily avoided in solid state problems after we consider the effects of the antisymmetry of the many-electron wavefunction.

3.2 The Hartree-Fock approximation The product wavefunction which we assumed in the Hartree method does not of course have the required antisymmetry. We must take linear combinations of such product wavefunctions such that the total wavefunction changes sign under the interchange of any two electrons. The corresponding antisymmetric wavefunction may be written as a Slater determinant,

$$\Psi(\mathbf{r}_1,\mathbf{r}_2,\ldots,\mathbf{r}_N)=\frac{1}{\sqrt{N!}}\begin{vmatrix}\psi_1(\mathbf{r}_1) & \psi_1(\mathbf{r}_2) & \cdots & \psi_1(\mathbf{r}_N)\\ \psi_2(\mathbf{r}_1) & \psi_2(\mathbf{r}_2) & \cdots & \cdots\\ \cdots & \cdots & \cdots & \cdots\\ \psi_N(\mathbf{r}_1) & \cdots & \cdots & \cdots\end{vmatrix} \tag{2.13}$$

We may postulate such a many-electron wavefunction and perform the variational calculation on Eq. (2.12) just as we did in the Hartree approximation. This again leads to equations for the optimized one-electron functions.[1] These are the Hartree-Fock equations.

$$\left[-\frac{\hbar^2}{2m}\nabla^2+V(\mathbf{r})+\sum_j e^2\int\frac{\psi_j^*(\mathbf{r}')\psi_j(\mathbf{r}')\,d\tau'}{|\mathbf{r}-\mathbf{r}'|}\right]\psi_i(\mathbf{r})$$
$$-\sum_j e^2\,\psi_j(\mathbf{r})\int\frac{\psi_j^*(\mathbf{r}')\psi_i(\mathbf{r}')\,d\tau'}{|\mathbf{r}-\mathbf{r}'|}=\epsilon_i\psi_i(\mathbf{r}) \tag{2.14}$$

Again these are only approximate solutions to the initial equation; an exact solution would require an infinite series of Slater determinants.

[1] *Ibid.*

We may again obtain a value for the total energy in the Hartree-Fock approximation and this will again contain corrections to the simple sum over the parameters ϵ_i.

The extra term which has appeared in the equations is called the *exchange interaction* as distinguished from the *direct interaction* which was present also in the Hartree approximation. The corresponding contribution to the total energy is called the exchange energy.

$$-\frac{e^2}{2}\sum_{ij}\int\frac{\psi_i^*(\mathbf{r})\psi_j(\mathbf{r})\psi_j^*(\mathbf{r}')\psi_i(\mathbf{r}')\,d\tau\,d\tau'}{|\mathbf{r}-\mathbf{r}'|}$$

It is purely coulombic in origin and arises from the correlated motion of the electrons due to the antisymmetry of the wavefunction. We will return to a discussion of exchange in some detail when we consider magnetism in solids.

The presence of the exchange term in the Hartree-Fock equations complicates the calculations but they remain nevertheless tractable in studies of a free atom. On the other hand it has not been possible to include exchange exactly in calculations in the solid and some approximation must always be made.

We note that because of the exchange term it is no longer necessary as it was in the Hartree approximation to omit the $i = j$ term in the summations since the corresponding terms in the Hartree and in the exchange sums cancel. We do not ordinarily worry about this distinction in the solid since the omission of one term from the potential modifies it only to one part in N.

3.3 Free-electron exchange It may be desirable in passing to indicate the general method by which exchange is most commonly approximated in solids. It is noted that the Hartree-Fock equations are exactly solvable for the case of a free-electron gas in a uniform positive compensating background. The one-electron functions become plane waves and the exchange energy may be evaluated directly. Because of the $1/r$ factor in the exchange interaction it is found that the total energy is proportional to the cube root of the total electron density ρ; i.e., proportional to $\rho^{1/3}$. It seems clear that if the electron density were varying slowly in space, the exchange energy could be calculated as a volume integral of this $\rho^{1/3}$ exchange. Slater[1] suggested that this approximation might be used even when the electron density was varying rapidly in space; i.e., varying over distances of the order of the electron wavelength.

The exchange term in the Hartree-Fock equations is replaced by a

[1] J. C. Slater, *Phys. Rev.*, **81**:385 (1951).

potential proportional to $\rho^{1/3}(\mathbf{r})$. There still remains an ambiguity with respect to the proportionality constant since the exchange term in the free-electron case varies with the wavenumber of the electron in question and a decision must be made as to which value is to be used in the equations. Slater has argued that the exchange potential should be that corresponding to an average over the occupied states in the corresponding band. Kohn and Sham[1] have used a variational argument to indicate that it should be the value for the highest-energy occupied states in the band. The coefficient of the $\rho^{1/3}$ term found by Kohn and Sham is smaller than that found by Slater by a factor of $2/3$. In either case the appropriate average is used in computing the total energy, but slightly different states are obtained in the two methods.

3.4 Koopmans' theorem In order to identify the one-particle equations in any of these approximations with the one-electron Schroedinger equation, we need a more direct relation between the parameters ϵ_i and the energies of interest in a crystal.

T. Koopmans[2] has shown that if we evaluate the total energy using a Slater determinant with N one-electron functions and if we then reevaluate the total energy for a Slater determinant of $N - 1$ one-electron functions (assuming that the individual one-electron functions are the same in both cases), the difference in these two total energies will simply be the parameter ϵ_i for the state that has been omitted. Thus under the assumption that the electron wavefunctions do not change as an electron is removed, we conclude that the ionization energy of the crystal with respect to any given electron state is simply the Hartree-Fock parameter.

It is further argued that since the removal of one electron makes a change in the potential of only one part in N we may neglect this change. Koopmans' theorem, then, states that *the Hartree-Fock parameter in a solid is the negative of the ionization energy for the corresponding state in the crystal, computed in the Hartree-Fock approximation.*

We may immediately see that the argument is incomplete. It is true that the removal of an electron changes the potential by only one part in N; however, there is a resulting change in the ϵ_i for all N electrons and we might expect these corrections to add up to a change in the total energy which was independent of N. Completion of the argument requires use of the variational aspect of the Hartree-Fock calculation. The equations were obtained by requiring that the total energy of the system be stationary with respect to small variations of the Hartree-Fock functions. Thus the error in each

[1] W. Kohn and L. J. Sham, *Phys. Rev.*, **140**: A1133 (1965).

[2] T. Koopmans, *Physica*, **1**: 104 (1933).

Hartree-Fock function of order $1/N$ changes the total energy only to order $1/N^2$. When the errors are added for each Hartree-Fock function the result remains of order $1/N$ and negligible for a large system.[1]

Koopmans' theorem is essential to the interpretation of the properties of solids in terms of the energy-band structures. It follows directly from Koopmans' theorem that the change in energy of the system when an electron is transferred from one state to another is simply the difference in the two Hartree-Fock parameters, since both the initial and final state energies may be related directly to the same ionized state. Thus, Koopman's theorem allows us to regard calculated energy bands as one-electron energy eigenvalues.

Koopmans' theorem specifically relates energies calculated within the Hartree-Fock approximation. It points out a new aspect of Hartree-Fock theory as applied to large systems rather than single atoms, the equality of differences in Hartree-Fock parameters and the corresponding transition energies. It seems likely now, though this opinion is not widely accepted, that to the extent this aspect is new, the Hartree-Fock theory fails in large systems. We may state instead a somewhat vague intuitive guide which purports to apply to real systems rather than Hartree-Fock systems: *In many respects the effects of the electron-electron interaction do not change much in going from the free atom to the solid.* We know that the differences in total energies for different configurations of the free atom, calculated in the Hartree-Fock approximation, agree well with the experimental energies of transition. We would therefore conclude that if the Hartree-Fock term values, the ϵ_i, in the free atom describe to a good approximation the free atom transition energies (calculated or observed), then this will also be true in the solid made up of these atoms; if they are not good approximations in the atom, they will not be in the solid. If and only if calculated term values in the free atom can be treated as one-electron energies, the corresponding calculated term values in a solid made up of these atoms can be treated as one-electron energies. Another way of stating this is to say that Koopmans' theorem will be valid for a crystal only if it is valid for the free atoms that constitute the crystal.

For example, we would obtain the energy of a $3s$ to $3p$ transition in atomic sodium by performing a self-consistent calculation for the two configurations and subtracting the total energy. However, the core states in sodium do not change very much in the transition, so the difference in values of ϵ_i calculated with the same potential will give a good estimate of the transition energy. A one-electron picture is applicable and Koopmans' theorem applies even to the atom. In the solid the transfer of an electron

[1]The author is indebted to Dr. Conyers Herring for pointing out this variational argument.

between states in a band represents an even smaller change in one-electron wavefunction. We expect no problem with Koopmans' theorem in the metal and find none. Similarly, Koopmans' theorem seems to work well in all but solids containing transition-metal atoms. The one-electron picture clearly breaks down in, e.g., the chromium atom. The configuration of atomic chromium is generally taken to be $3d^5 4s^1$. Clearly if we computed one-electron energies for chromium the d state would either lie above or below the s state. If it lay below there would be no s electrons; if it lay above there would be two s electrons. The energies of the different levels depend in an important way upon the occupation of the other states. We may expect this difficulty to carry over to the solid and Koopmans' theorem is suspect in any transition metal. We will see how meaningful one-electron states in transition metals may be obtained when we discuss the crystal potential in Sec. 3.5. We will then also see the relation between that method and a modified description of free-atom states.

It is important to realize that energy bands can be defined for the real system in any case. Again on the basis of translational symmetry we may construct many-electron states of well-defined wavenumber. The ground state, for example, will correspond to $\mathbf{k} = 0$. We can define a band energy as the change in energy when an electron is brought from infinity and added to a N-electron system initially in the ground state. Such an energy change can be given as a function of the wavenumber associated with the $N + 1$ electron state, leading to energy bands which are directly related to experiment. Such bands are called *quasiparticle bands*. They will be discussed in the following chapter in the framework of the Fermi liquid theory. The self-consistent-field calculations are simply efforts to obtain approximate quasiparticle bands.

Even this does not totally avoid the problem raised by Koopmans' theorem. If we seek the change in energy as an electron moves from one state to another, we may remove an electron from the ground state of the system, but in putting it back in a different state we are really adding it to a system that is already excited by the initial removal, and the energy will differ from the energy of adding it to the ground state. The corrections are the *quasiparticle interactions* which will be discussed in Fermi liquid theory.

3.5 The crystal potential The self-consistent potential and exchange terms are well defined in a crystal just as in a free atom, and if we really sought solutions of the Hartree-Fock equations then any approximations could be evaluated by asking how close they correspond to Hartree-Fock. However, it is well known that a true Hartree-Fock calculation for a metal would not give correct results; the difficulty is already apparent in a free-electron gas. As we have indicated, that problem is soluble[1] in the Hartree-Fock approximation and leads to a Slater determinant with all plane-wave states occupied for wavenumbers less than the Fermi wavenumber, k_F, which will be defined more precisely later. The Hartree-Fock parameters depend upon k and contain a term proportional to $(k - k_F) \ln(k - k_F)$. Such a term leads for example to an infinite velocity, $(1/\hbar)\, d\epsilon_k/dk$, at $k = k_F$, contrary to experiment. For reasons such as this the Hartree approximation —with no exchange—has frequently been found to give better results than the Hartree-Fock approximation. Yet the Hartree approximation is found not to be adequate for the calculation of energy bands. The resolution of this dilemma has taken many years and will be discussed here in terms of the potential seen by the electrons.

In the free atom the levels may be divided into the *core levels*, which correspond to states below and including the last filled rare-gas shell, and the higher levels. The core levels will remain essentially unchanged in going from the free atom to the solid. We note in fact that frequently when atomic wavefunctions are calculated for a free ion and a free atom, no distinction is made in the tabulated core wavefunctions. Presumably in the solid the wavefunctions of these low-lying states should be intermediate between the two extremes and again no distinction is appropriate.

We therefore begin our calculation for the solid with a knowledge of the core levels. In sodium, for example, these would be the 1*s*, 2*s*, and 2*p* levels. The task is to compute the states and energies which are modified appreciably in the solid. In sodium this is the free-atom 3*s* state. Similarly, in silicon the core states are taken to be the 1*s*, 2*s*, and 2*p* levels and we compute the levels in the solid which correspond to the 3*s* and 3*p* levels in the free atom. In the noble and transition metals the last filled *d*-shell, or the partially filled *d*-shell, states are appreciably distorted from what they would be in the free atom. Thus in copper, for example, we would regard the 1*s*, 2*s*, 2*p*, 3*s*, and 3*p* levels as core levels; the 3*d* and 4*s* levels need to be calculated. The levels which interest us, and which lie above the core levels, are generally called *valence* or *conduction-electron* states. In the transition metals it is customary to distinguish the *d* states from the conduction states though, as we will see, the distinction is not sharp.

As we begin our energy-band calculation, then, we know the core states and we know the potential arising from the nucleus and the core electrons. In addition the exchange terms coupling valence states and core states may be written in terms of the known core states. These terms are appreciable and must be included. However, they seem to be well approximated by free-electron exchange.

[1] See, e.g., C. Kittel, "Quantum Theory of Solids," p. 89, Wiley, New York, 1963.

In addition the direct interaction between valence electrons is well defined. It may be determined self-consistently with extensive sampling of occupied states or it may be obtained approximately by perturbation theory as we will describe in Sec. 4 of Chap. III. The problem arises with the exchange interaction between valence electrons. If we were to include it in the Hartree-Fock approximation we would obtain the anomalies discussed above. If we neglect it altogether we are constructing a potential significantly different from that used in free-atom calculations. As we have indicated, the electron density in the crystal is approximately a superposition of free-atom densities so the direct coulomb potential in the neighborhood of one atom is close to the potential due to *all* electrons on the atom, while in free-atom calculations we omit the potential due to the electron under consideration. Thus if we include only the direct coupling we assert that the valence electron sees neutral atoms in the crystal, but sees charged ions in the free atom.

This violates our intuitive guide and suggests that exchange must be included. We noted that in the Hartree-Fock approximation we need not omit interactions between an electron and its own contribution to the potential since the direct and exchange contributions to this self-interaction cancel. A simple way out is the use of free-electron exchange for the interaction among valence electrons. This will approximate the omission of self-interactions but does not lead to anomalies at the Fermi wavenumber as does the Hartree-Fock approximation. Thus the approximation to true exchange appears to be an improvement. We may appeal to experiment (or to calculations going beyond the Hartree-Fock approximation, though to date these are less convincing) for a justification of this procedure.

In simple metals the computed valence electron density turns out to be quite uniform so free-electron exchange contributes a potential which is very nearly constant and therefore does not appreciably modify the energy bands. Thus Hartree bands are very nearly the same as those including free-electron exchange; in fact most calculations have omitted exchange among valence electrons, and the results have generally been in good accord with experiment. In semiconductors the electron density is very nonuniform and calculations that include exchange among valence electrons in the free-electron approximation have been exceedingly successful.[1] In the transition metals, copper in particular, attempts to carry over the Hartree approximation used in simple metals led to energy bands in which the *d*-like states were incorrectly located with respect to the *s*-like states.

[1] F. Herman, R. L. Kortum, C. D. Kuglin, and J. P. Van Dyke, in B. Alder, S. Fernbach, and M. Rotenberg (eds.), "Methods in Computational Physics," vol. 8, Academic, New York, 1968.

However, when a potential due to Chodorow[1] was used—a potential which attempted to duplicate the exchange interaction in the free atom—the positions of the different bands were in agreement with experiment.[2] It seems likely that free-electron exchange would also have achieved this. Thus in all cases comparison with experiment seems to support the use of free-electron exchange for the interaction among valence electrons.

This apparent success of self-consistent-field calculations in transition metals would appear to violate our intuitive guide concerning the similarities between potentials in solids and in free atoms, and this is an interesting point. In metallic chromium, which has a free-atom configuration with $3d^5 4s^1$, we would find bands associated with d states and with s states and these bands would be only partially filled. Such a description of the metal seems to account for observed optical transitions in the metal, though we noted that to account for transitions in the free atom it is necessary to recalculate the states and energies in the final configuration in order to obtain the correct energy difference. We might therefore use the converse of our intuitive guide to suggest that a simpler description of the free atom might be possible if we allowed fractional occupation of the free-atom levels, though this would seem an artificial construction. Slater[3] has recently had significant success with such an approach; in this method the energy of a transition is obtained from the incremental change in energy of the free atom as the occupation of one level is reduced by an infinitesimal amount while that of another is increased by an infinitesimal amount. It seems likely that such a description of the free atom will be particularly useful as a starting point for solid state calculations involving transition metal atoms.

4. Energy-Band Calculations

The calculation of the energy bands for any given crystal is a moderately straightforward though extremely complicated calculation once a suitable approximation for the exchange interaction has been decided upon. We must first construct the net potential and then solve the eigenvalue equation to obtain the states and eigenvalues. Some effort may be made to iterate the problem for self-consistency although ordinarily the potential is postulated at the beginning. A number of highly refined techniques have been developed for the calculations themselves but we will only discuss those aspects of the methods which provide particular insight into the nature of solids or which

[1] M. Chodorow, *Phys. Rev.*, **55**:675 (1939).

[2] B. Segall, *Phys. Rev.*, **125**:109 (1962); G. A. Burdick, *Phys. Rev.*, **129**:138 (1963).

[3] J. C. Slater, *Phys. Rev.*, **165**:655, 658 (1968).

provide a basis for our later treatments of the properties of solids. A more complete account of the various methods is given in Alder, Fernbach, and Rotenberg.[1]

4.1 The cellular method Wigner and Seitz[2] considered alkali metals and focused their attention on the lowest-lying state in the band; i.e., the $\mathbf{k} = 0$ state. For this state the wavefunction is simply the Bloch function $u_0(\mathbf{r})$ which has the full symmetry of the lattice. They divided the crystal into atomic cells, the cell associated with each atom containing all points closer to that atom than to any other atom. It may readily be shown by symmetry that the normal component of the gradient of $u_0(\mathbf{r})$ vanishes at all atomic-cell boundaries in simple structures. Then for a given potential the eigenvalue equation reduces to a calculation within a single cell with well-defined boundary conditions on the wavefunction at the cell surfaces. For a potential they took a free-ion potential, the same potential that would be used in the calculation of the atomic state and therefore a reasonable choice from the point of view taken in the preceding section. The free-ion potential which they used in each cell is spherically symmetric. In order to solve the problem more readily they replaced the atomic cell by a sphere of equal volume. Then the calculation of the state becomes a spherically symmetric problem in close analogy with the free-atom calculation. The only difference is the vanishing-boundary condition applied to $du_0(r)/dr$ at the Wigner-Seitz sphere rather than the atomic-boundary condition that the wavefunction vanish at infinity. This enabled them to calculate the difference in energy between the bottom of the first band and the free-atom state. With other appropriate approximations they were then able to make estimates of the cohesive energy of the simple metals.

The approach has been generalized to treat other states in the band but it has long been believed that the difficulty of matching the wavefunction at the boundaries makes the results inaccurate. A group under S. L. Altmann[3] at Oxford appears to have surmounted this problem and made the cellular method a useful approach for energy-band calculations.

4.2 The plane-wave method Given the potential seen by conduction electrons, the plane-wave method is probably conceptually the simplest of all band calculation methods. It is therefore worth describing though we will see that it is not computationally convenient.

[1] Alder, Fernbach, and Rotenberg, *op. cit.*

[2] E. Wigner and F. Seitz, *Phys. Rev.*, **43**:804 (1933); **46**:509 (1934).

[3] E.g., S. L. Altmann and C. J. Bradley, *Proc. Phys. Soc. (London)*, **86**:519 (1965); S. L. Altmann, B. L. Davies, and A. R. Hartford, *J. Phys. C* (Proc. Phys. Soc.), **1**(2):1633 (1968).

We take as the potential a superposition of free-atom potentials, $v(\mathbf{r})$, each of which should include a free-electron exchange potential based upon that free-atom density. The total potential then may be written

$$V(\mathbf{r}) = \sum_j v(\mathbf{r} - \mathbf{r}_j)$$

where the sum is over the positions $\mathbf{r}_j$ of the N atoms present. We expand our wavefunction in terms of normalized plane waves,

$$|\mathbf{k}\rangle = \Omega^{-1/2} e^{ik\cdot r}$$

where Ω is the volume of the crystal, the normalization volume. Any eigenstate is written

$$\psi(\mathbf{r}) = \sum_k a_k |\mathbf{k}\rangle$$

We substitute this form in the Schroedinger equation,

$$-\frac{\hbar^2}{2m}\nabla^2\psi + V(\mathbf{r})\psi = E\psi$$

and multiply on the left by

$$\langle\mathbf{k}'| = \Omega^{-1/2} e^{-ik'\cdot r}$$

and integrate over volume to obtain a set of simultaneous linear equations in the expansion coefficients a_k.

$$\frac{\hbar^2}{2m} k'^2 a_{k'} + \sum_k \langle\mathbf{k}'|V|\mathbf{k}\rangle a_k = E a_{k'} \tag{2.15}$$

To solve these we now take a particular $a_{k'}$ not equal to zero for $\mathbf{k}'$ lying in the Brillouin zone. We may see that Eq. (2.15) couples us only to the a_k for $\mathbf{k}$ differing from $\mathbf{k}'$ by a lattice wavenumber. This may be seen by writing out the matrix element of V.

$$\langle\mathbf{k}'|V|\mathbf{k}\rangle = \frac{1}{\Omega}\int e^{-ik'\cdot r} \sum_j v(\mathbf{r} - \mathbf{r}_j)\, e^{ik\cdot r}\, d\tau$$

It is convenient to interchange the sum and the integral and to multiply and divide each term by $e^{i(k'-k)\cdot r_j}$.

$$\langle\mathbf{k}'|V|\mathbf{k}\rangle = \frac{1}{N}\sum_j e^{-i(k'-k)\cdot r_j} \frac{1}{\Omega_0}\int e^{-i(k'-k)\cdot(r-r_j)} v(\mathbf{r} - \mathbf{r}_j)\, d\tau$$

Here we have also factored the normalization volume into the number of atoms present N and the atomic volume Ω_0. The first factor appears in diffraction theory and is called the *structure factor*. This structure factor will be of central importance in much of our discussion but at this stage we only utilize one of its properties. Note that it is a sum over all atom positions. Thus the $\mathbf{r}_j$ may be written $\mathbf{T}_j + \boldsymbol{\delta}_j$ where $\mathbf{T}_j$ is a lattice translation originating at one atom position; it is an integral linear combination of primitive lattice translations. If there is more than one atom per primitive cell then for each lattice translation we will have two or more vectors $\boldsymbol{\delta}_j$ giving the position of these atoms in the primitive cell with respect to the first. The structure factor may be factored into a sum over lattice translations and a sum over the $\boldsymbol{\delta}_j$ in each primitive cell. The latter factor will be the same for all primitive cells and may be taken outside the sum over lattice translations. The sum over lattice translations may be evaluated by substituting Eq. (2.1) for the lattice translation. The sum then takes the form

$$\left(\sum_{j \text{ in cell}} e^{-i(k'-k)\cdot\delta_j}\right)\left(\sum_{n_1 n_2 n_3} e^{-i(k'-k)\cdot(n_1\tau_1+n_2\tau_2+n_3\tau_3)}\right)$$

Substituting for $\mathbf{k}'$ and $\mathbf{k}$ from Eqs. (2.3) and (2.4) we find that if $\mathbf{k}' - \mathbf{k}$ is a lattice wavenumber then every term in the second sum is unity and we obtain a result equal to the number of cells present. If on the other hand $\mathbf{k}' - \mathbf{k}$ is not equal to the lattice wavenumber, this becomes the product of three geometric series, at least one of which vanishes. We conclude that $\langle\mathbf{k}'|V|\mathbf{k}\rangle$ is equal to 0 except for $\mathbf{k}' - \mathbf{k}$ equal to a lattice wavenumber. Thus for a given $\mathbf{k}'$ in Eq. (2.15) we obtain only terms with a_k such that $\mathbf{k}$ differs from $\mathbf{k}'$ by a lattice wavenumber.

This is of course an immense simplification of Eq. (2.15). It reduces the number of terms in the equation by a factor equal to the number of cells in the crystal. It leaves us nevertheless with an infinite number of a_k.

Fortunately when the difference between $\mathbf{k}'$ and $\mathbf{k}$ becomes large enough, the matrix element becomes small; the sum over $\mathbf{k}$ may be truncated after some several hundred terms. This leaves us with several hundred simultaneous equations and several hundred unknowns that might be solved on a computer. A different set of equations is obtained for each $\mathbf{k}'$ in the Brillouin zone. The solution of these equations would lead to several hundred eigenvalues corresponding to the energies of the first several hundred energy bands. These could, in principle, be solved as a function of $\mathbf{k}'$ to give the energy-band structure. This conceptually simple method has led to an inordinately long numerical calculation. The other methods which we will describe are conceptually similar but seek more rapid convergence by the choice of a more suitable set of functions in which to expand.

4.3 The orthogonalized-plane-wave method An ingenious improvement on the plane-wave method was proposed many years ago by Herring.[1] He

[1] C. Herring, *Phys. Rev.*, **57**:1169 (1940).

noted that the slow convergence arises because the conduction-electron states contain many oscillations in the region of the core. Here they resemble rather closely the atomic wavefunctions corresponding to the valence electrons. In the plane-wave method many plane waves are required to reproduce these oscillations. Convergence can be improved only if we somehow take these oscillations into account in the basis set in which we are expanding.

Herring did just this by noting that the states which we seek must be orthogonal to the core states, which themselves are known. Thus a complete expansion with respect to the conduction-band states may be obtained by using not plane waves, but plane waves which have been made orthogonal to the core states. In orthogonalizing to the core states we will introduce oscillations at the core which should reproduce reasonably well those of the state we seek. The *orthogonalized-plane-wave method*, or *OPW method*, then is very much like the plane-wave method but with the plane waves replaced by orthogonalized plane waves. This approach has been very useful as a band-calculation technique and will be fundamental to the pseudopotential method which will be discussed presently.

It will be convenient again to begin with normalized plane waves, $|\mathbf{k}\rangle$. We may also define normalized core eigenstates centered at the individual ion positions,

$$|t,j\rangle = \psi_t(\mathbf{r} - \mathbf{r}_j)$$

We have used the index t to denumerate the core state; that is, $1s$, $2s$, $2p$, . . . , and the index j again to specify the atom position. Orthogonalized plane waves then may be written

$$OPW_k = |\mathbf{k}\rangle - \sum_{t,j} |t,j\rangle\langle t,j|\mathbf{k}\rangle \tag{2.16}$$

where

$$\langle t,j|\mathbf{k}\rangle = \frac{1}{\sqrt{\Omega}} \int \psi_t^*(\mathbf{r} - \mathbf{r}_j)\, e^{i\mathbf{k}\cdot\mathbf{r}}\, d\tau$$

We may see that these OPWs are in fact orthogonal to the core states by multiplying on the left by a core state and integrating. We assume that the core states on different ions do not overlap each other, which is a good assumption in practice, and we note that the different core states on the individual ion are orthogonal to each other. It follows immediately that this integral vanishes. We might note also that different OPWs are not orthogonal to each other nor are they normalized. However, they are complete for expansion of the conduction-band states and that is what matters.

We now make an expansion of the wavefunction in OPWs and substitute it into the Schroedinger equation. We again find that OPWs are coupled only for wavenumbers differing by a lattice wavenumber. The appropriate matrix elements may be evaluated from the known potentials and core wavefunctions. We again are left with a set of simultaneous equations but in this case the matrix elements coupling $a_{k'}$ with a_k drop much more rapidly with the difference in wavenumber, and a much smaller set of simultaneous equations is required. For many applications we will see that only two or three OPWs are required; for full band calculations some 25 or 30 may be needed and for very accurate calculations sometimes 50 or 60 are used. The use of OPWs has greatly reduced the computational effort required for the solution of the energy bands. We will carry the OPW treatment somewhat further when we describe pseudopotentials.

4.4 The augmented-plane-wave method[1] Slater[2] slightly earlier had proposed another type of function for expansion of the wavefunction, i.e., *augmented plane waves* or *APWs*. In constructing these it is appropriate first to approximate the potential that will be used in the calculation. Near each nucleus we expect the potential to be rather spherical and at positions between the nuclei we expect it to be relatively flat. This suggests constructing a sphere around each nucleus, making the radii of the spheres sufficiently small that they do not overlap each other, and assuming that the potential within the sphere is precisely spherically symmetric. We then assume that the potential between the spheres is precisely constant. (See Fig. 2.7). The most usual procedure is to construct the true potential we expect in the crystal and then to approximate it by this "muffin-tin" form. Prescribing this form for the potential prevents our doing an accurate self-consistent calculation. In addition it becomes a serious difficulty if we attempt to generalize our treatment to distorted crystals or crystals with defects. However, for energy-band calculations this appears to be a very good approximation in itself and it has been possible, with additional effort, to go beyond the muffin-tin approximation.

The APWs then are constructed as plane waves between the spheres. For energies of interest we also construct solutions in the spherically symmetric potential. The coefficients of these are then adjusted to eliminate discontinuities in the wavefunction at the sphere surface, though it is not possible this way to eliminate the discontinuity in the slope of these functions. The electron eigenstates for a given energy may be exactly expanded in these APWs. They replace the plane waves or the OPWs in the calculations. As in the case of OPWs only a limited number of terms are required and

[1] T. L. Loucks, "The Augmented Plane Wave Method," Benjamin, New York, 1967.
[2] J. C. Slater, *Phys. Rev.*, **51**:846 (1937).

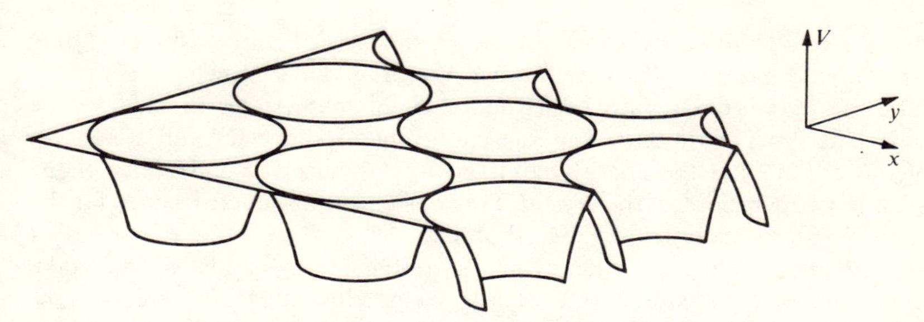

Fig. 2.7 A schematic representation of a "muffin-tin" potential. Each well drops to infinity at the center of the atom, but is truncated in the figure.

this is an efficient method for computation of the bands. Again, however, it requires high-speed computers for accurate calculations.

A third method for energy-band calculation is called the Green's function approach or the Kohn, Korringa, Rostoker (KKR) method. The formulation seems very much different but it has been established that the methods yield the same results using the same potential. We will not describe the details of this method.[1]

4.5 The symmetry of the energy bands Before presenting the results of energy-band calculations it will be desirable to obtain some of the consequences of the crystal symmetry with respect to energy bands. Symmetry information is used extensively in the energy-band calculations themselves and in the presentation of the results. We have used the translational symmetry of the lattice in order to formulate the energy-band picture; we now seek the additional information obtainable through the rotations and reflections making up the point group or the space group in the crystal.

Consider again a crystal with a group of rotations and reflections which take the crystal, and therefore the Hamiltonian, into itself. Let one of these symmetry operations be called R. Let us further say that we have obtained an electron eigenstate ψ_k corresponding to some wavenumber $\mathbf{k}$ in the Brillouin zone. Now let R be an operation that rotates the wavefunction. But rotation of the wavefunction will rotate the wavenumber in just the same way; that is,

$$R\psi_k = \psi_{Rk}$$

The symmetry operation has given us a new wavefunction to be associated with the new wavenumber $R\mathbf{k}$. By applying all of the symmetry operations of the group to this wavefunction ψ_k or to its wavenumber $\mathbf{k}$, we will obtain

[1] For details see F. S. Ham and B. Segall, *Phys. Rev.*, **124**: 1786 (1961).

the *star* of **k**. This set of wavenumbers, in the case of cubic symmetry, may contain as many as 48 wavenumber vectors. The symmetry operations take the Hamiltonian into itself; hence all of these states must have the same energy. This implies that the energy bands have the full symmetry of the crystal; i.e., the energy bands go into themselves under any of the symmetry operations of the crystal. This was true of the energy bands for the square lattice illustrated in Figs. 2.3 and 2.4.

We may note one further symmetry of the energy bands. We may take the complex conjugate of the energy eigenvalue equation $\mathscr{H}\psi_k = E\psi_k$. This clearly takes the Hamiltonian, which is real, into itself but changes **k** to $-$**k**. This is true for any symmetry of the crystal. Thus the energy bands have inversion symmetry even if the group of the crystal does not contain the inversion element. Since taking the complex conjugate of the Schroedinger equation is equivalent to reversing the sign of time, we have made use of the *time-reversal symmetry* of the Schroedinger equation.

For some special choices of **k**, for example **k** lying in a [100] direction in a cubic crystal, some of the operations will take **k** into itself rather than into a distinct wavenumber. These particular operations are called the *group of* **k**. The group of **k** is a subgroup of the full symmetry group of the crystal.

We may use a state corresponding to such a special wavenumber to generate a representation for the group of **k**. If this state of wavenumber **k** is degenerate with other states of the same wavenumber, we may always take linear combinations such that this set $\{\psi_k\}$ transforms according to an irreducible representation of the group of **k**. Thus, knowing the group of **k**, we will know what degeneracies are expected on the basis of symmetry, and we may designate states by the irreducible representation according to which they transform.

Consider, for example, in a cubic crystal a state with a wavenumber in the [111] direction within the Brillouin zone. The group of the wavenumber is the group of the equilateral triangle. (We are ignoring, for simplicity, time-reversal symmetry.) We may expect nondegenerate bands and two-fold degenerate bands along this direction. We would label these bands by Λ_1, Λ_2, or Λ_3 depending upon their symmetry properties.

For an arbitrary wavenumber, of course, the group of the wavenumber is simply E. The only irreducible representation is the unit representation and we expect no degeneracy (except, as it turns out, due to time reversal). Lines in the Brillouin zone for which the group of the wavenumber contains elements other than the unit element are called symmetry lines. Similarly, planes for which the group of the wavenumber contains other than the unit operation are called symmetry planes. Finally, at special points in the Brillouin zone the group of the wavenumber may be larger than that on symmetry lines which thread it; these are called symmetry points. A degeneracy at the symmetry point W was illustrated in Fig. 2.4. Other degeneracies will be apparent in the band structures given in Sec. 4.6.

The evaluation of the determinants in the band calculation can be simplified for states along symmetry lines by use of the symmetry of the crystal to reduce the size of the matrix that needs to be solved. Thus energy-band calculations are most commonly carried out along symmetry lines and at symmetry points in the Brillouin zone.

We may obtain further information about the energy bands in the crystal using methods in direct analogy with the crystal-field splitting of atomic states. We may consider states at a wavenumber corresponding to a symmetry point. We classify these according to the irreducible representation of the group of the wavenumber at the symmetry point. As the wavenumber then moves away from the point, the group of the wavenumber becomes smaller and some of the degeneracies will be split. As in the case of the crystal-field splitting we may determine the irreducible representations into which the original representation will split. Conditions relating the irreducible representations of adjoining points, lines, and planes are called *compatibility relations* and were discussed initially by Bouckaert, Smoluchowski, and Wigner.[1]

In addition it is frequently possible, by following the irreducible representation from line to line through the Brillouin zone, to conclude that bands of certain symmetries must have crossed along some symmetry line. Such degeneracies are called accidental since their precise position is not determined by symmetry. A slight modification of the potential will shift the position of the degeneracy. Frequently a line of such accidental degeneracies will occur on a symmetry plane and this is called a *line of contact* between the two bands which are degenerate along that line.

Again an appreciable amount of information about the energy bands is obtainable from symmetry alone, but to obtain the bands themselves requires a detailed energy-band calculation.

4.6 Calculated energy bands The electronic energy bands in a solid represent the basic electronic structure of the solid just as the atomic term values in the free atom represent the basic electronic structure of the atom. The calculation of these energy bands is very much the task of the specialist, but an understanding of the results is essential to any understanding of the solid. We will therefore present a few characteristic energy bands and discuss them. When we go on later to discuss the properties of solids we will in

[1] L. P. Bouckaert, R. Smoluchowski, and E. P. Wigner, *Phys. Rev.*, **50**:58 (1936). A rather complete discussion may now be found in M. Tinkham, "Group Theory and Quantum Mechanics," p. 284, McGraw-Hill, New York, 1964.

many cases begin again with a crude representation of the band structure which is more tractable for calculation. However, these cruder descriptions of the band structure should always be regarded as approximations to the true structure obtained here.

As we have indicated earlier, the results of energy-band calculations are usually presented for wavenumbers along symmetry lines. In Fig. 2.8 we show the Brillouin zone for the face-centered cubic structure with several symmetry lines and symmetry points indicated. This zone is appropriate for all of the three band structures that we will discuss here. As the energy bands have the full cubic symmetry, a knowledge of the energy along these few lines gives information about very much of the Brillouin zone. There are, for example, eight [111] directions and the energy bands along one of these describes the energy along all of them. At the same time energy bands along symmetry lines can be misleading. We find frequent degeneracies along symmetry lines, whereas for general points in the zone we expect no degeneracies. Note that the group of the wave vector along [111] directions is essentially the group of the triangle. For this reason we chose the symbol Λ when we described the representations of the group of the triangle. Similarly, the symmetry line Δ has essentially the symmetry of the square that was treated in the problem. These groups of the wavenumber are not identical to our examples, however, due to the presence of the time-reversal symmetry. For wavenumbers in a [100] direction, for example, the product of a reflection in a plane perpendicular to that direction and time reversal takes the wavenumber into itself and therefore is an element in the group. Representations are designated with a prime or not depending upon their behavior under such an operation.

The first band structure we consider is that of a simple metal. Aluminum is a prototype simple metal; its energy-band structure is shown in Fig. 2.9. The most striking feature of these energy bands is the smallness of the

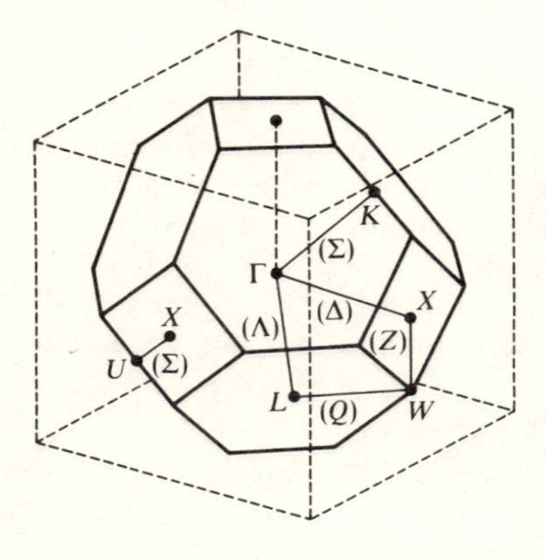

Fig. 2.8 *Symmetry lines and points in the Brillouin zone for the face-centered cubic structure. The* Γ *lies at the center of the zone. Note that the points K and U differ by a lattice wavenumber (in the [111] direction) and are therefore equivalent. The line UX is equivalent to a continuation of the line* ΓK *beyond the Brillouin zone. Symbols for the symmetry lines are shown in parentheses.*

gaps; e.g., the gap between the first and second bands at X. The first band, beginning at Γ, is very close to a free-electron parabola. We will see in fact that all the energy bands in aluminum are quite close to free-electron bands. The complexity of structure shown here arises primarily from the reduction of these free-electron bands to the Brillouin zone. We saw in detail how such a reduction complicated the free-electron bands in one dimension. This is a very important feature of the band structure of the simple metals and one that we will make much use of in the course of our treatments.

Note that the first and second bands are degenerate at W with symmetry W_3. Note also the accidental degeneracy between the second and third bands along the symmetry line Z.

We showed that the number of wavenumber states within the Brillouin zone is equal to the number of primitive cells in the crystal. Each of these levels is doubly degenerate due to spin so each energy band can contain two electrons per primitive cell. Aluminum, in the face-centered cubic structure, has one atom per primitive cell and each atom contributes three electrons (above the core levels). Thus in aluminum there are just enough electrons to fill one and a half bands. The ground state of aluminum will have all states occupied up to an energy, called the *Fermi energy*, or Fermi level, which in aluminum occurs slightly above the third band at W. A horizontal line is drawn at that point to indicate the extent of filling of the bands. The first band is found to be completely filled, the second and third bands partly filled, and the fourth and higher bands are empty. The presence of partly filled bands in the ground state is the characteristic feature of a metal.

The second band structure we consider is that for germanium, a prototype semiconductor. Germanium has the diamond structure which is a face-centered cubic structure with two identical atoms per primitive cell. Thus the same Brillouin zone remains appropriate and the same symmetry lines obtain. The band structure for germanium is shown in Fig. 2.10. In contrast to the case of aluminum, the band gaps are quite large. Again the first band rises from Γ very much as a free-electron parabola but the distortions are very much larger. The bands of diamond and silicon are very similar.

In the diamond structure, with two atoms per primitive cell, each atom (carbon, silicon, or germanium) contributes four electrons. Thus there are just enough electrons (eight per primitive cell) to fill four bands. We see that in the ground state of germanium the first four bands will be completely filled (note Λ_3 and Δ_5 are doubly degenerate bands) while the fifth and higher bands will be completely empty. A finite energy (in the case of germanium about 0.6 volt) is required to take an electron from the ground state of the system and put it into an excited state. These energy thresholds, or band gaps, are larger in silicon and diamond. The presence of these band

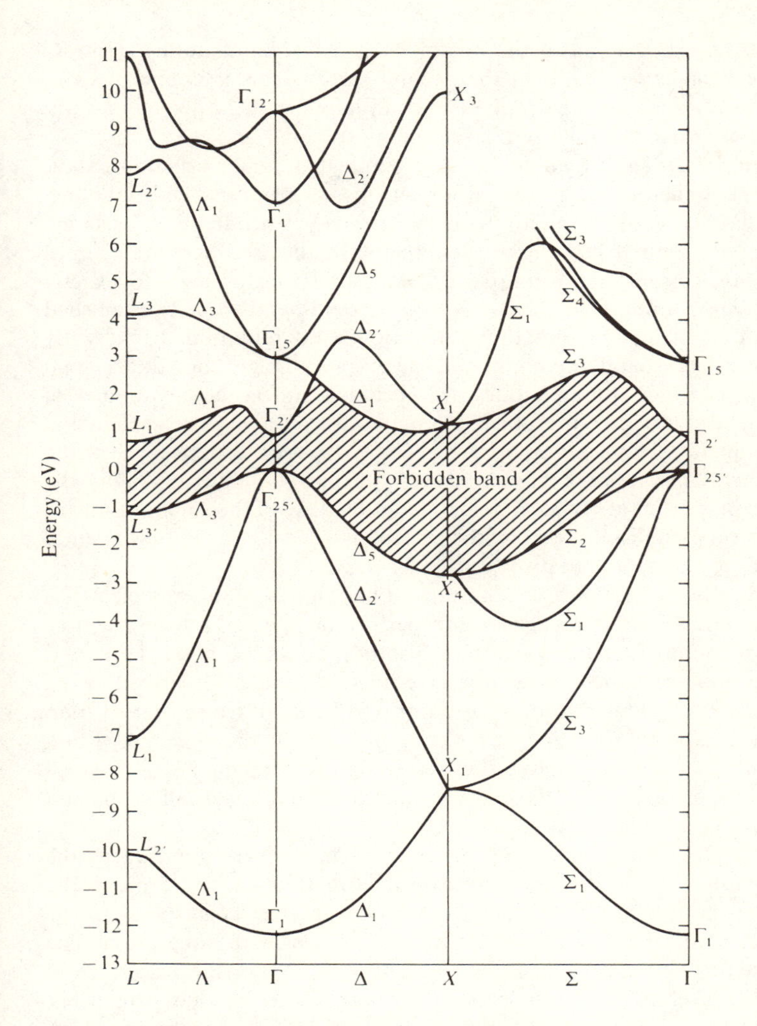

Fig. 2.10 *The energy bands of germanium, with spin-orbit coupling neglected, as calculated by F. Herman, R. L. Kortum, C. D. Kuglin, and J. L. Shay [in D. G. Thomas (ed.), "II-VI Semiconducting Compounds, 1967 International Conference," Benjamin, New York, 1967].*

gaps and the exact filling of the lowest, or *valence*, bands and the empty higher, or *conduction*, bands in the ground state is characteristic of semiconductors. Insulators are simply semiconductors with larger band gaps, and frequently bands with a narrower energy spread.

Semimetal bands tend to be very similar to semiconductor bands. As in semiconductors there are many electrons per primitive cell; graphite, a semimetal, has eight. Bismuth, with two atoms per primitive cell, has ten electrons per primitive cell and is a semimetal. In all cases there are an even number of electrons per primitive cell and therefore enough to exactly fill an integral number of bands. However, semimetals are distinguished from semiconductors in that the conduction-band minimum lies slightly below the valence-band maximum and therefore in the ground state a small number of conduction-band states are occupied and an equal number of valence-band states are unoccupied. Germanium would become a semimetal if the conduction-band minimum, Δ_1 in Fig. 2.10, were dropped below the valence-band maximum Γ'_{25}. Because of the partial filling of bands the electronic properties are those of a metal, but because of the smallness of the overlap a very small number of carriers, about 10^{-5} per atom in bismuth, contribute to the conductivity.

Typically, but not always, a semimetal has an odd number of electrons per atom, but an even number of atoms per primitive cell. Thus in principle an integral number of bands, five in bismuth, could be filled. However, the odd-numbered bands frequently have lines of contact (described in the preceding section, 4.5) with the even-numbered band above so that overlap in energy is unavoidable. Spin-orbit coupling can lift the degeneracy but, as in bismuth, may not remove the overlap. As a consequence almost all elemental semiconductors have even valence and almost all elemental semimetals have odd valence.

In the transition and noble metals the states corresponding to the last-filled rare-gas configuration behave as core states and are much the same as in the free atom. The energy bands of interest correspond to the states that would be *d* states or the highest *s* state in the free atom. The band structure for copper is shown in Fig. 2.11. Again the lowest band rises as a free-electron parabola from Γ; however at energies somewhat above that of Γ_1 it becomes intertwined with a complex set of bands which are frequently referred to as *d* bands. This is only a qualitative description since the *s* and *d* bands are very much intermixed. Above these *d* bands the energies continue very much like free-electron bands but the band gaps are very much larger than those of the simple metals.

Copper, like aluminum, has one atom per primitive cell but 11 electrons per atom beyond the filled argon shells. These electrons are enough to fill five and a half bands. For the copper-band structure this places

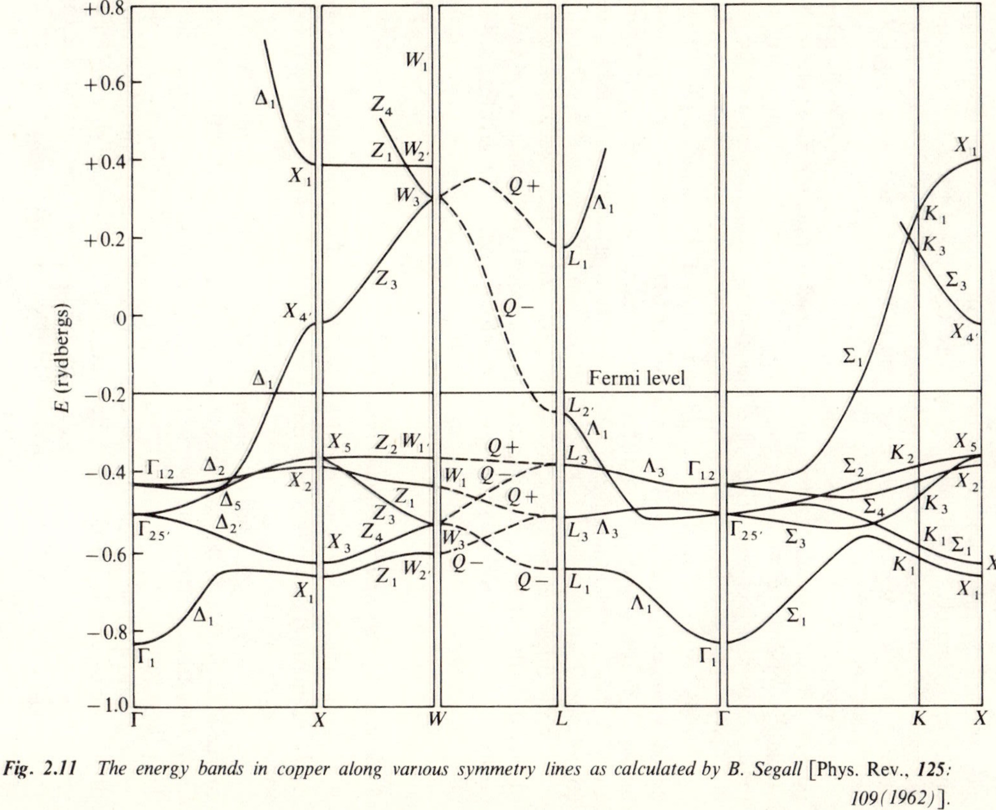

Fig. 2.11 *The energy bands in copper along various symmetry lines as calculated by B. Segall* [Phys. Rev., ***125****: 109(1962)*].

the Fermi energy above the d bands. The bands near the Fermi energy then are very much like free-electron bands corresponding to a single electron per primitive cell but are considerably distorted. If we were to move from copper to zinc, the d bands would narrow appreciably and drop to energies near the Γ_1 minimum. The d bands in zinc can reasonably, although only approximately, be thought of as core levels and the band structure discussed for only the top two electrons per atom. At still one higher atomic number, gallium, the d bands have indeed become narrow and fall far below the free-electron-like bands.

If we drop back in atomic number from copper to nickel, we find that the band structure remains very much the same but the number of electrons available to fill those bands decreases. Thus the Fermi energy cuts through the d bands. This is characteristic of a transition metal and the band structure at the Fermi energy is quite complex.

It should be mentioned that an additional difficulty occurs in nickel: it is ferromagnetic. Electrons with spin parallel to the direction of magnetization have different energy from those with spin antiparallel; the spin degeneracy is split. For ferromagnetic nickel the band structure is indeed quite complicated.

With this brief survey of band structures in solids, we will now go on to describe the band structure of individual materials with much greater care and develop approximate methods that yield greater insight into the nature of these electronic structures.

5. Simple Metals and Pseudopotential Theory

We will be able here only to give the most essential features of the pseudopotential method as applied to simple metals. Details and applications to a wide variety of metallic properties are given by Harrison.[1]

We saw in Fig. 2.9 that the energy bands of aluminum were quite close to free-electron bands. This might at first lead us to believe that the electrons are nearly free and that the potential arising from the ions is quite small. This supposition, however, is untenable. The electron wavefunctions do not at all resemble plane waves near the cores so the effect of the potential upon the *wavefunctions* is very large. At the same time we have seen that the effect of the potential on the *energy bands* turns out to be small. This paradox is very nicely resolved by the idea of a pseudopotential that represents the effect of the potential upon the bands themselves.

We will see that the concept of a pseudopotential and its application

[1] W. A. Harrison, "Pseudopotentials in the Theory of Metals," Benjamin, New York, 1966.

to the simple metals is extremely important in the theory of solids. The smallness of these pseudopotentials allows us for many purposes to treat it as a perturbation and to calculate many more properties of a simple metal than is possible for any other state of condensed matter. We will see in Sec. 9 of this chapter how it may be extended to transition metals. The pseudopotential will also be a useful concept in discussing bands and properties in other solids.

5.1 The pseudopotential In seeking to characterize the effect of the potential on the electron state we are faced with the same convergence problem as in the energy-band calculation. Again we will replace the plane-wave expansion by an expansion in a more suitable set. Just as there are many ways to do band calculations, there are many ways in which pseudopotentials may be formulated. Our formulation here will be based on the orthogonalized-plane-wave method.

We require only two assumptions in our formulation of the pseudopotential, both of which are used in the energy-band calculations. First we replace the many-electron problem by a self-consistent-field problem in which the effects of the interactions between electrons are included as a self-consistently determined potential field which includes a self-consistent field for exchange. Second we divide the electronic states into core and conduction-band states and will assume that the core states are the same as in the free atom. At a later stage we will make additional approximations based upon the fact that the pseudopotential which we define and calculate is small. It is at this point that the method becomes inapplicable to transition metals. The highest d states are not the same as in the free atom and cannot therefore be taken as core states. At the same time the bands arising from the d states do not even resemble free-electron bands (as is clearly seen in Fig. 2.11). Thus the pseudopotential as formulated here cannot be regarded as small. We will see how this difficulty is overcome in Sec. 9. In the simple metals the weakness of the pseudopotential will be used in the self-consistent determination of the potential itself. For the moment we will assume only that a self-consistent potential exists and we will write it as $V(\mathbf{r})$.

We may immediately write down the energy eigenvalue equation satisfied by a conduction-electron eigenstate.

$$-\frac{\hbar^2}{2m}\nabla^2\psi + V(\mathbf{r})\psi = E\psi \tag{2.17}$$

The core states also satisfy this same eigenvalue equation with the same $V(\mathbf{r})$. We will as before index the core states with a t representing the quantum numbers and a j indexing the ion position. Then the eigenvalue equation for the core states may be written

$$-\frac{\hbar^2}{2m}\nabla^2\psi_{t,j} + V(\mathbf{r})\psi_{t,j} = E_{t,j}\psi_{t,j} \tag{2.18}$$

It should be noted here that though the core wavefunctions are assumed to be the same as in the free atom, the core energies will in general not be the same. Potentials arising from neighboring atoms will overlap the core in question. Since the cores are small, however, these contributions will be nearly constant over the core. Thus they shift the energy but not the wavefunction.

We now proceed as in the OPW method. We expand the conduction-band wavefunction in terms of orthogonalized plane waves. Each orthogonalized plane wave may be written as in Eq. (2.16).

$$OPW_k = |\mathbf{k}\rangle - \sum_{t,j}|t,j\rangle\langle t,j|\mathbf{k}\rangle$$

It will be convenient in this section to write the OPW in terms of a projection operator P which projects any function onto the core states.

$$P = \sum_{t,j}|t,j\rangle\langle t,j| \tag{2.19}$$

Then the OPW may be written in the form

$$OPW_k = (1 - P)|\mathbf{k}\rangle$$

and the expansion of the wavefunction in OPWs takes the form

$$\psi_k = (1 - P)\sum_k a_k|\mathbf{k}\rangle \tag{2.20}$$

Note that we have interchanged the sum over $\mathbf{k}$ and the sum in the projection operator.

Substitution of this expansion back in Eq. (2.17) would give us just the OPW method. In the pseudopotential method we depart from this procedure. We note that the OPW expansion is rather rapidly converging; in other words the sum over $\mathbf{k}$ will have important contributions only for relatively small values of $\mathbf{k}$, and the function $\varphi = \Sigma_k a_k|\mathbf{k}\rangle$ will be a smooth function. We call φ the *pseudowavefunction.* Note that the pseudowavefunction is equal (except possibly for normalization) to the true wavefunction outside the cores since P is zero there. Hopefully it will remain smooth in the core region. Then the situation may be represented schematically as shown in Fig. 2.12. In the pseudopotential method we calculate the free-electron-like function $\varphi(\mathbf{r})$ which may reasonably be obtained by perturbation theory from a plane-wave zero-order state. The true wavefunction may be

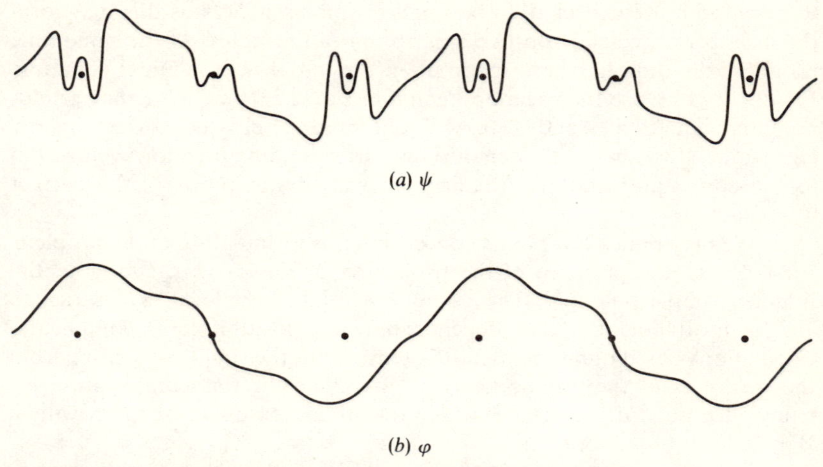

Fig. 2.12 A schematic sketch of a conduction-band wavefunction ψ and the pseudowavefunction φ that corresponds to it. Note that this is the same wavefunction shown in Fig. 2.1b.

calculated simply by orthogonalizing the pseudowavefunction to the core states with the operator $(1 - P)$ and renormalizing.

After substituting φ in Eq. (2.20) we insert this form in Eq. (2.17) to obtain a differential equation satisfied by φ. After rearranging terms we obtain

$$-\frac{\hbar^2}{2m}\nabla^2\varphi + V(\mathbf{r})\varphi - \left[-\frac{\hbar^2}{2m}\nabla^2 + V(\mathbf{r})\right]P\varphi + EP\varphi = E\varphi \tag{2.21}$$

The second, third, and fourth terms on the left are combined and called the *pseudopotential*, W. Thus Eq. (2.21) may be rewritten

$$-\frac{\hbar^2}{2m}\nabla^2\varphi + W\varphi = E\varphi \tag{2.22}$$

From Eqs. (2.18) and (2.19) we see that

$$\left[-\frac{\hbar^2}{2m}\nabla^2 + V(\mathbf{r})\right]P = \sum_{t,j} E_{t,j}|t,j\rangle\langle t,j|$$

so that the pseudopotential may be written in the convenient form,

$$W = V(\mathbf{r}) + \sum_{t,j}(E - E_{t,j})|t,j\rangle\langle t,j| \tag{2.23}$$

Eq. (2.22) is called the *pseudopotential equation.* Since φ is expected to be smooth, we expect W to be small in some sense. Thus this may form the basis for a free-electron-like description of the states. At the same time no approximation has been made upon the initial Schroedinger equation, Eq. (2.17). If we use the pseudopotential of Eq. (2.23) and solve the pseudopotential equation exactly, we will obtain precisely the correct energy eigenvalue. If we then orthogonalize the corresponding pseudowavefunction to the core states and renormalize, we will obtain precisely the correct wavefunction.

We may note a few points concerning this formulation of the problem. First due to the presence of the projection operator the pseudopotential is not a simple potential. The pseudopotential is nonlocal in contrast to $V(\mathbf{r})$, a local potential that depends only on position. This complicates calculations based upon the pseudopotential, but the simplicity arising from the weakness of the pseudopotential far outweighs this complication for many purposes. Furthermore, it will frequently be reasonable to approximate W by a local pseudopotential.

Second we may see from the form of W that it will be small in comparison to the true potential. The potential $V(\mathbf{r})$ is attractive. The second term in the pseudopotential contains the difference $E - E_{t,j}$ which will always be positive. The projection operator is essentially positive so this positive term will cancel to some extent the attractive potential $V(\mathbf{r})$. This feature is frequently given the name *the cancellation theorem*. We came to the same conclusion on the basis of the smoothness of the pseudowavefunction and the cancellation can be seen in other ways. It is not clear that it deserves to be dignified by this title.

We may also note that there is a nonuniqueness in the pseudopotential approach. If we have obtained a solution of the pseudopotential equation then the addition of any linear combination of core states to the pseudowavefunction leads to a pseudowavefunction which, by Eq. (2.20), leads to the same true wavefunction $\psi = (1 - P)\varphi$. Correspondingly there are many forms in which the pseudopotential may be written. We may note in particular that the expression $E - E_{t,j}$ in the definition of the pseudopotential of Eq. (2.23) may be replaced by any function of energy and of t and j and the resulting pseudopotential will lead to the same eigenvalues. This is most easily seen by rewriting the pseudopotential equation with this more general pseudopotential and allowing a modified eigenvalue E'.

$$\left[-\frac{\hbar^2}{2m}\nabla^2 + V(\mathbf{r})\right]\varphi(\mathbf{r}) + \sum_{t,j} f(E,t,j)\,\psi_{tj}(\mathbf{r})\int d\tau'\,\psi_{tj}^*(\mathbf{r}')\,\varphi(\mathbf{r}') = E'\,\varphi(\mathbf{r})$$

We may multiply on the left by the complex conjugate of the true wavefunction ψ^* and integrate over all volume. The true Hamiltonian in the first term is Hermitian and may be applied to the left to give for the first term $E \int \psi^* \varphi \, d\tau$. The second term is identically zero since the conduction-band wavefunctions are orthogonal to all core states $\psi_{t,j}$. Thus we obtain

$$E \int \psi^* \varphi \, d\tau = E' \int \psi^* \varphi \, d\tau$$

We conclude that either the true conduction-band wavefunction is orthogonal to the pseudowavefunction or their energies are identically equal. When they correspond to the same state the two functions will not be orthogonal. Thus we obtain exactly the correct eigenvalue for any choice of $f(E,t,j)$. This is an extremely important point. There is no correct pseudopotential but there are many valid forms. Each of these forms, if used with the pseudopotential equation and solved exactly, leads to precisely the correct energies and wavefunctions. However, when we do calculations based on second-order perturbation theory with the pseudopotential as the perturbation, the results will depend upon the choice we have made. The errors that we make in computed properties should be regarded as errors in the perturbation theory, not as errors in the pseudopotential itself. There is limited meaning to an attempt to "improve" a pseudopotential by adjusting it to fit a result calculated with second-order perturbation theory with experiment. The error did not come from the pseudopotential in the first place. Furthermore, efforts to determine a "more accurate pseudopotential" by carrying perturbation theory to higher order and then comparing with experiment are similarly misguided. Carrying the perturbation theory to higher order inevitably makes the results less sensitive to the pseudopotential, since we know that carrying the calculation to all orders makes the calculated result quite independent of which valid pseudopotential has been selected.

At the same time it *is* possible to speak of a range of pseudopotentials that give rather good answers. One such "optimized" pseudopotential was obtained by use of a variational argument. A form was sought that would yield the smoothest pseudowavefunction (the pseudopotential leading to the minimum value of $\int |\nabla\varphi|^2 \, d\tau / \int |\varphi|^2 \, d\tau$). This was hoped to yield the best convergence of the perturbation expansion. We will use this optimized form in our discussion and will see that it is the same as that of Eq. (2.23) with E replaced by the value that would be obtained in a first-order perturbation calculation. Another approach has been to adjust the pseudopotential to fit a specific property as indicated above. Each property leads to a slightly different pseudopotential but it is found that all of these are quite similar to each other and to the optimized pseudopotential. It has been found by experience that the pseudopotential method is sensible but to be useful it must be approximate.

One trivial choice of pseudopotential is simply $V(\mathbf{r})$. Then the pseudopotential equation becomes simply the Schroedinger equation. Of course the correct eigenvalues are obtained and the pseudowavefunction becomes equal to the true wavefunction. Such a choice, of course, does not gain us anything.

It is interesting to ask for the physical origin of the repulsive term in the pseudopotential which can make perturbation theory appropriate. When we discuss phase shifts we will see that it has the effect of removing integral multiples of π from the phase shifts that describe the scattering. In this way it is seen to leave the eigenstates outside of the ion in question unchanged and to eliminate the oscillatory structure within the ion. Because the repulsive term described above derives from the orthogonality of the conduction band and core states, the effect is frequently attributed to the Pauli principle. However, it is clear that the Pauli principle is quite irrelevant within the one-electron framework we are using. The repulsive effect is in fact classical. The positive-energy electron is accelerated as it moves near an ion; thus it moves more rapidly in this region and spends less time there. Of course if we considered a thermal distribution of classical particles in the neighborhood of an attractive potential, some would be bound and the total density would be higher in the region of an attractive potential. However, by focusing attention on the particles of higher energy (in our case the conduction-band states) the effect is reversed and an attractive potential tends to exclude the particles of interest.

If we use the pseudopotential to perform an energy-band calculation we are quickly led to a secular determinant that is entirely equivalent to that obtained in the OPW method. The pseudopotential will enter through its plane-wave matrix elements $\langle \mathbf{k} + \mathbf{q}|W|\mathbf{k}\rangle$. Here we will instead proceed by perturbation theory but the zero-order states are plane waves so the pseudopotential will again enter only through such plane-wave matrix elements. We will use the optimized form of the pseudopotential which is given by

$$\langle \mathbf{k} + \mathbf{q}|W|\mathbf{k}\rangle = \langle \mathbf{k} + \mathbf{q}|V|\mathbf{k}\rangle + \sum_{t,j} \left(\frac{\hbar^2 k^2}{2m} + \langle \mathbf{k}|W|\mathbf{k}\rangle - E_{t,j}\right) \times \langle \mathbf{k} + \mathbf{q}|t,j\rangle\langle t,j|\mathbf{k}\rangle \quad (2.24)$$

This is not of precisely the form of the general pseudopotential that we discussed earlier; we have exchanged a dependence upon the wavenumber of the right-hand plane wave for the dependence upon energy. We may however verify that this is a valid form by rewriting the $\mathbf{k}$-dependent terms. For example,

$$\frac{\hbar^2 k^2}{2m} |t,j\rangle\langle t,j|\mathbf{k}\rangle = |t,j\rangle\langle t,j| \frac{-\hbar^2 \nabla^2}{2m} |\mathbf{k}\rangle$$

The central point is that the spatial dependence of the corresponding terms in the pseudopotential equation is again simply a sum of core wavefunctions, and therefore the proof that the eigenvalues of the pseudopotential equation are equal to the true eigenvalues again goes through.

This dependence upon the wavenumber of the right-hand state reflects a nonhermiticity of the pseudopotential; that is, $\langle\mathbf{k}|W|\mathbf{k}'\rangle \neq \langle\mathbf{k}'|W|\mathbf{k}\rangle^*$ unless $|\mathbf{k}| = |\mathbf{k}'|$. This feature, like the energy dependence of the form in Eq. (2.23), arises from the nonorthogonality of OPWs and must be taken into account in careful pseudopotential calculations.

It is possible to write the second-order perturbation expansion for the energy such that all matrix elements contain the wavenumber of the zero-order state on the right. Then we see that the optimized form is just what would be obtained by replacing E in Eq. (2.23) by its first-order value.

In the optimized form, the energy $E_{t,j}$ is to be evaluated in the crystal and could differ from atom to atom. It is much more convenient however to neglect this variation and use an average value E_t. This is not an approximation since either value, or even the free-atom value, provides a valid form of the pseudopotential. Such a choice simply takes us slightly away from the optimum form.

We note that in evaluating matrix elements we must first determine $\langle\mathbf{k}|W|\mathbf{k}\rangle$, which appears on the right in Eq. (2.24). This is done by setting $\mathbf{q} = 0$ in Eq. (2.24) and solving for $\langle\mathbf{k}|W|\mathbf{k}\rangle$.

Each term in the sum over j in Eq. (2.24) arises from an individual ion which is in itself spherically symmetric. Similarly the potential V may be written as a sum of spherically symmetric potentials which we write $v(\mathbf{r} - \mathbf{r}_j)$. Thus we may factor the matrix element of Eq. (2.24) into a structure factor depending upon the $\mathbf{r}_j$ and a form factor which is the matrix element arising from an individual ion, just as we did in Sec. 4.2.

$$\langle\mathbf{k} + \mathbf{q}|W|\mathbf{k}\rangle = S(\mathbf{q})\langle\mathbf{k} + \mathbf{q}|w|\mathbf{k}\rangle \tag{2.25}$$

where

$$S(\mathbf{q}) = \frac{1}{N}\sum_j e^{-iq\cdot r_j}$$

and

$$\langle\mathbf{k} + \mathbf{q}|w|\mathbf{k}\rangle = \langle\mathbf{k} + \mathbf{q}|v|\mathbf{k}\rangle + \sum_t \left(\frac{\hbar^2 k^2}{2m} + \langle\mathbf{k}|w|\mathbf{k}\rangle - E_t\right)\langle\mathbf{k} + \mathbf{q}|t\rangle\langle t|\mathbf{k}\rangle \tag{2.26}$$

We have noted that $\langle \mathbf{k}|W|\mathbf{k}\rangle = \langle \mathbf{k}|w|\mathbf{k}\rangle$. This analysis follows closely that given in Sec. 4.2. The matrix elements are written

$$\langle \mathbf{k} + \mathbf{q}|v|\mathbf{k}\rangle = \frac{1}{\Omega_0}\int e^{-i\mathbf{q}\cdot\mathbf{r}}\, v(\mathbf{r})\, d\tau$$

$$\langle \mathbf{k} + \mathbf{q}|t\rangle = \frac{1}{\sqrt{\Omega_0}}\int e^{-i(\mathbf{k}+\mathbf{q})\cdot\mathbf{r}}\, \psi_t(\mathbf{r})\, d\tau$$

$$\langle t|\mathbf{k}\rangle = \frac{1}{\sqrt{\Omega_0}}\int \psi_t^*(\mathbf{r})\, e^{i\mathbf{k}\cdot\mathbf{r}}\, d\tau$$

Again Ω_0 is Ω/N, the atomic volume.

The pseudopotential form factor of Eq. (2.26) depends upon the magnitude of $\mathbf{k}$, the magnitude of $\mathbf{k} + \mathbf{q}$, and the angle between them. (For a spherically symmetric local potential, matrix elements depend only upon magnitude of $\mathbf{q}$.) However, for many purposes we are interested primarily in matrix elements between states of the same energy. The energy bands are spherically symmetric through first order in the pseudopotential. Thus in the ground state of the system all states will be occupied up to an energy, the Fermi energy, which will be evaluated in Sec. 5.3. In discussing electronic properties we will focus our attention on those states with energy near the Fermi energy. The corresponding zero-order pseudowavefunctions are plane waves with wavenumbers that lie on this sphere. We write the magnitude of such a wavenumber as k_F. For matrix elements between two such plane waves the magnitudes of $\mathbf{k}$ and $\mathbf{k} + \mathbf{q}$ are fixed at the sphere radius k_F. Then the form factors depend only on the angle between $\mathbf{k}$ and $\mathbf{k} + \mathbf{q}$, or, equivalently, upon q/k_F. For these matrix elements the form factor is given

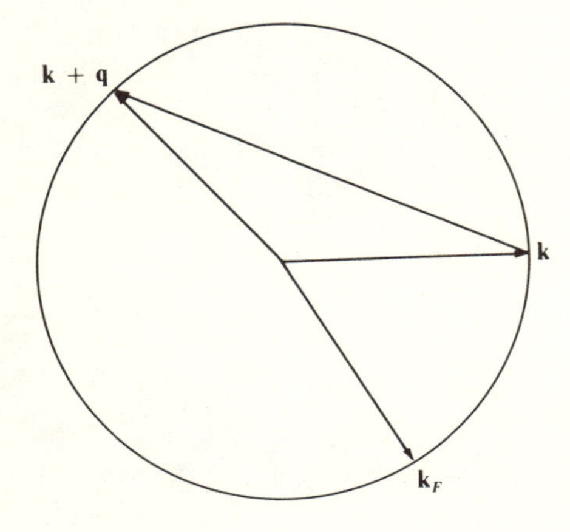

Fig. 2.13 Matrix elements between two plane waves, $|\mathbf{k}\rangle$ and $|\mathbf{k} + \mathbf{q}\rangle$, both of which lie on the Fermi sphere, depend only upon the magnitude of $\mathbf{q}$.

by a single curve as it would be for a local potential. This curve, however, is defined only for q/k_F running from 0 to 2. This curve depends only upon the structure of the ion in question and the atomic volume but not upon the detailed arrangement of the ions. It is called the *OPW form factor* or sometimes just the *pseudopotential.*

A set of these curves for a number of simple metals is shown in Fig. 2.14. Note that in all cases the OPW form factor approaches $-\frac{2}{3}$ of the Fermi energy at long wavelengths. When we treat screening we will see why this is so. At shorter wavelengths the repulsive term in the pseudopotential becomes important and the form factor goes positive. This single curve for a particular metal gives us all of the information about the details of the ion which will be needed for computing a wide range of properties. At this point we are interested only in the band structure itself and the only relevant values will be those for which $\mathbf{q}$ is a lattice wavenumber.

5.2 The model-potential method[1] Heine and Abarenkov[2] have introduced an alternative way of formulating essentially the same problem. They have called their approach the *model-potential method.* In this method the ion potential is replaced by an energy- and wavenumber-dependent square well as illustrated in Fig. 2.15.

It is noted that the calculation of the eigenstate or eigenvalue for an atom, whether it be a free atom or an ion in the lattice, may be divided into two regions: the core region and the region outside the core. A sphere is constructed surrounding the core separating the two regions. Then a calculation of the wavefunction might be carried out by calculating the wavefunction within this model sphere and outside the model sphere and matching them smoothly. Therefore the only information used to construct the wavefunction outside the sphere is the value and slope of the wavefunction at the cell surface as determined by an integration within the sphere. The wavefunctions are expanded in both regions in spherical harmonics and only the logarithmic derivative of each component of the wavefunction at the sphere influences the wavefunction outside of the sphere. Thus the same wavefunction, and therefore the same energy, would be obtained (except for normalization) outside the sphere if the true ion potential were replaced inside the sphere by a flat potential leading to the same logarithmic derivative. This flat potential would necessarily depend upon the angular-momentum quantum number l and the energy. This potential is chosen such that there are no nodes in the wavefunction within the sphere. The procedure then is to replace the strong core potentials by these model potentials within the spheres at every ion. The *model wavefunction* or pseudo-

[1] For a recent discussion and references, see R. W. Shaw, Jr., *Phys. Rev.*, **174**:769 (1968).

[2] V. Heine and I. V. Abarenkov, *Phil. Mag.*, **9**:451 (1964).

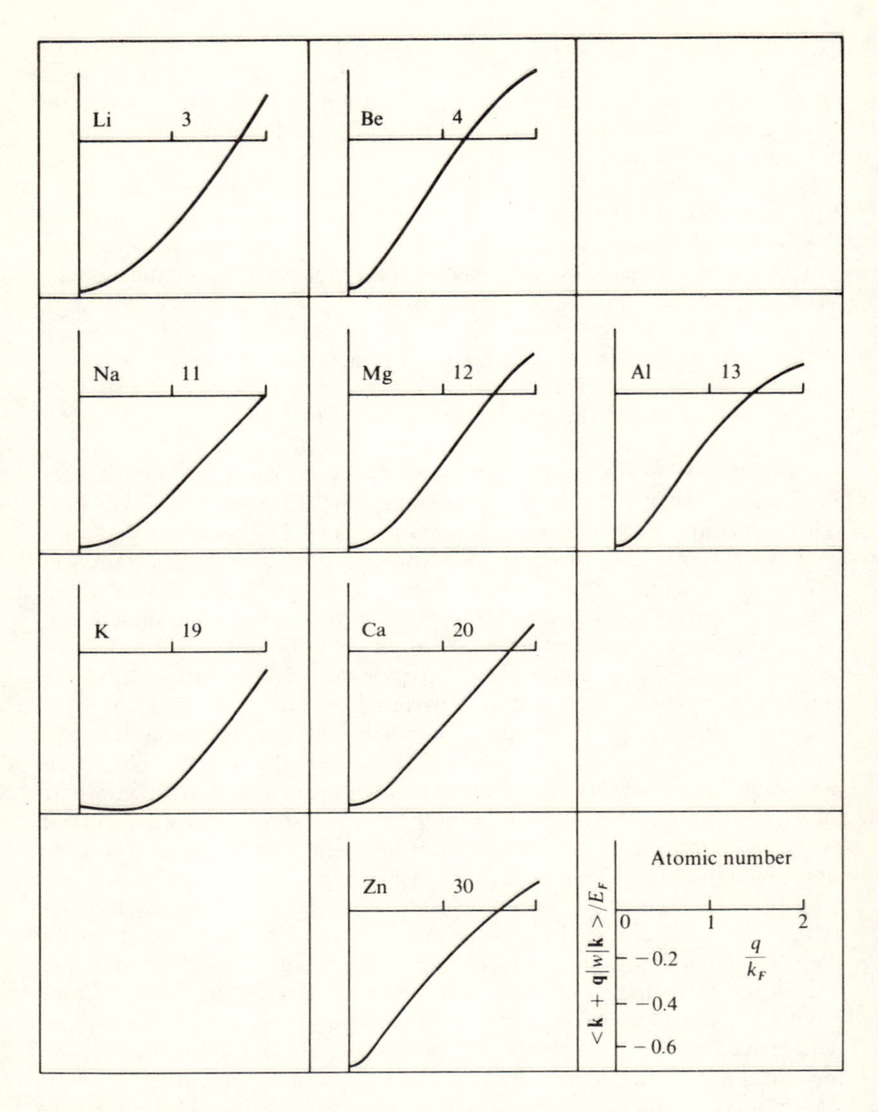

Fig. 2.14 *OPW form factors for the lightest simple metals, divided by the corresponding Fermi energy, as calculated by W. A. Harrison* [Phys. Rev., ***131****:2433 (1963)*]. *The key appears at the lower right. These curves are illustrative, but should be regarded as rough approximations to the optimized form.*

wavefunction obtained, like the pseudowavefunction in the pseudopotential method as we have described it, is smooth and free-electron-like throughout the crystal.

The model-potential method then has very much the same features as the pseudopotential method. We will see now, however, that model potentials may be obtained directly from experiment. The method may first be applied to the free atom. We may determine the value of the constant portion of the model potential by fitting the experimentally observed term values. For each angular-momentum quantum number we thus obtain the

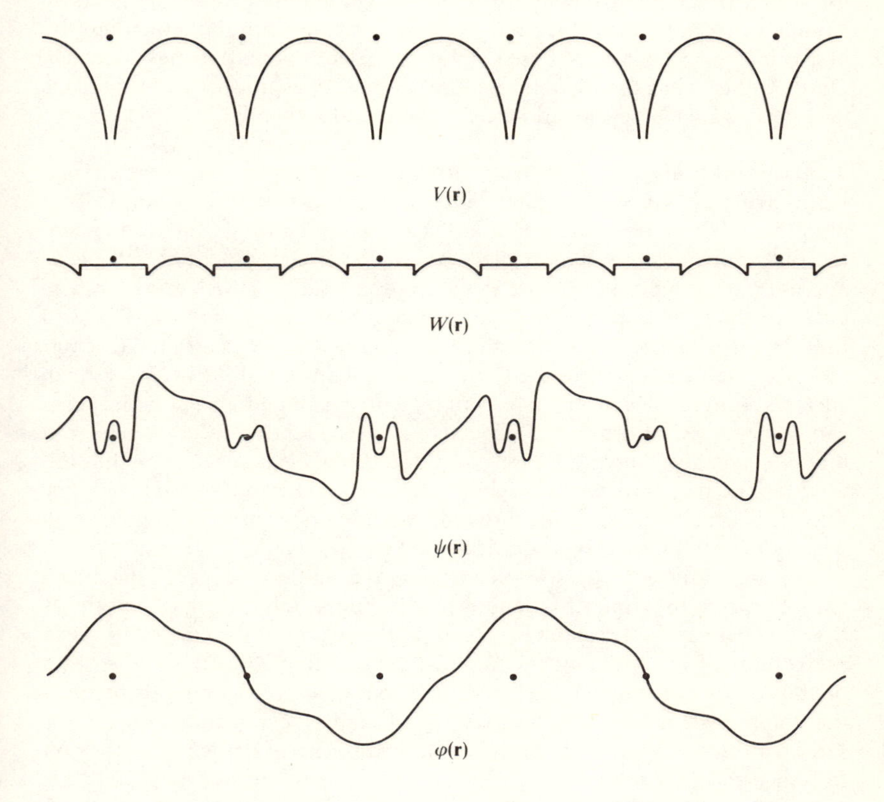

Fig. 2.15 In the model-potential method the true potential $V(\mathbf{r})$ within a sphere surrounding each ion is replaced by a flat potential $W(\mathbf{r})$, adjusted such that the model wavefunction φ has the same slope and derivative at the sphere surface as the true wavefunction ψ but with nodes within the sphere eliminated.

value of the constant for an energy equal to that of the corresponding term values. This value is then interpolated to obtain values at energies appropriate for use in the calculation in the metal. Using this procedure we avoid the complexity of the use of the computed core wavefunctions and core potentials that enter the pseudopotential method. We do not, on the other hand, avoid difficulties associated with, e.g., nonhermiticity of the pseudopotential, though this feature was ignored in the original formulation of the model potential. The use of experimental term values sufficiently simplified the problem so that it became reasonable to determine OPW form factors for all of the simple metals by this method. These tables were computed by Animalu.[1] In most cases they have turned out to be more reliable than the first-principle calculations based upon the pseudopotential method itself. On the other hand the model-potential method does not have the rigorous basis that we have developed here for the pseudopotential.

5.3 Free-electron bands Let us return again to the question of the energy-band structure. As we indicated in Sec. 4.2, the structure factor will be nonzero only when $\mathbf{q}$ is equal to a lattice wavenumber. In the perfect crystal the only nonvanishing matrix elements of the pseudopotential will occur at these wavenumbers. In simple structures the smallest lattice wavenumbers occur in the neighborhood of $2k_F$. We may immediately see from Fig. 2.14 that the form factors in this wavenumber range are quite small; in particular they are small compared to the Fermi energy. Thus the shifts in the electron energies from free-electron values will be quite small and for many purposes can be neglected altogether. This leads us back directly to the free-electron model of metals. This model is very old and has been very successful, but now for the first time we see clearly why it should work so well. In some sense it is accidental since the lattice wavenumbers occur just in the region where the pseudopotential is quite small.

If we really go to the free-electron approximation we ignore the difference between the smooth pseudowavefunctions and the true wavefunctions, which have appreciable structure in the region of the cores. However, the volume of the core tends to be small in the simple metal—of the order of 10 percent of the atomic cell volume—and much of the physics involves the volume where the pseudowavefunction and true wavefunction are the same. In some properties, notably the optical properties, it will be necessary to return to the true wavefunction. We will be dealing extensively with the free-electron approximation when we consider screening and transport properties, and use of the smooth pseudowavefunction will be appropriate there. For the moment we are concentrating on the *eigenvalues* and it is

[1] A. O. E. Animalu, tables appear in W. A. Harrison, "Pseudopotentials in the Theory of Metals," pp. 309ff., Benjamin, New York, 1966. (The cadmium table is in error.)

valid to proceed with the pseudowavefunctions. We are interested in the deviations of the eigenvalues from the free-electron values.

It will be convenient to begin by taking the pseudopotential equal to zero and see step by step the effect of introducing a nonzero pseudopotential. To zero order in the pseudopotential the bands are just free-electron bands, but it will be desirable to retain the pseudopotential terminology.

Beginning with plane-wave-electron pseudowavefunctions, we require that these satisfy periodic-boundary conditions on the crystal surface. The density of states in wavenumber space obtained by satisfying boundary conditions in a volume Ω is simply $\Omega/(2\pi)^3$. This is most readily seen for a rectangular volume in which case the change in wavenumber from one state to the next in the x direction is simply $2\pi/L_1$ where L_1 is the dimension in the x direction. Correspondingly the volume of wavenumber space associated with a single state is $(2\pi)^3/(L_1 L_2 L_3)$; the density is simply the reciprocal of this expression. This result remains correct for more complicated volumes. One electron of each spin is allowed in each of these wavenumber states so that the density of electronic states is just twice the number of wavenumber states.

To zero order (and in fact to first order) in the pseudopotential the energy of the states increases monotonically with wavenumber. Therefore in the ground state of the system all states will be occupied within a sphere in wavenumber space and all states outside will be unoccupied. This is the *Fermi sphere*; its radius k_F is called the *Fermi wavenumber* and its energy E_F measured from the band minimum is called the *Fermi energy*. Using the density of states in wavenumber space found above we see that the Fermi sphere contains states enough for

$$\frac{2\Omega}{(2\pi)^3}\,\frac{4\pi}{3}\,k_F{}^3$$

electrons. This must equal NZ for a metal of N ions of valence Z. We obtain the Fermi wavenumber by solving for k_F.

$$k_F = \left(\frac{3\pi^2 Z}{\Omega_0}\right)^{1/3}$$

We note that the Fermi wavenumber depends on the atomic volume Ω_0 and not, of course, on the size of the system. The wavenumber k_F is of the order of 2π divided by the lattice distance. Thus, as we indicated before, the Fermi wavenumber is of the order of the lattice wavenumbers.

For most electronic properties of metals the electron states very near this Fermi *surface* in wavenumber space determine the behavior, particularly with respect to transport properties. It is only these electrons which can find an unoccupied state at a nearby energy. Correspondingly, when we apply small fields it is only states near E_F for which the occupation changes. The energy bands in the neighborhood of the Fermi surface are of primary importance. We will therefore focus our attention on the states at the Fermi surface.

5.4 The diffraction approximation When the interaction between electrons and the potential is weak, as the small pseudopotential indicates here, it becomes reasonable to describe this interaction as a diffraction phenomenon. We describe each state, then, by a plane-wave pseudowavefunction or a single OPW wavefunction. We would then say that a given electron of wavenumber $\mathbf{k}$ is coupled to all other plane-wave states of wavenumber $\mathbf{k} + \mathbf{q}$ where $\mathbf{q}$ is any lattice wavenumber, but that it will diffract only if that state has the same energy; that is, if $k^2 = |\mathbf{k} + \mathbf{q}|^2$. Solving this equation for $\mathbf{k}$ leads us to the familiar Bragg reflection condition:

$$2\mathbf{k} \cdot \mathbf{q} = -q^2$$

This result means that an electron will diffract only if its wavenumber lies on a plane bisecting a lattice wavenumber. These are the so-called Bragg reflection planes or, in the language of band structure, the Brillouin zone faces. This description of the system is called the *one-OPW approximation* or the *nearly-free-electron approximation.*

We will see that in polyvalent metals these planes do in fact intersect the Fermi surface, but of all the electrons on the Fermi surface it is only a set of measure zero that lies on the intersections with these planes. Thus even taking the pseudopotential into account in these terms we find that almost all of the electrons remain unaffected and behave as free. This gives further justification for our use of the free-electron approximation for treating metals. Nonetheless the electron states that lie on or near the Bragg reflection planes can be important. This can be seen particularly clearly in connection with studies of the Fermi surfaces in metals, and we will consider these next. Our discussion to this point has indicated why the energy bands in simple metals might well be free-electron-like but it will be desirable to discuss Fermi surfaces before carrying this point a little further. This will eventually lead us back to Brillouin zones and the general results of Sec. 2.

5.5 One-OPW Fermi surfaces We have indicated that there is some difficulty in ever seeing the effects of the Bragg reflection in a simple metal. This difficulty however can be circumvented by the application of a magnetic field. Classically, as we have seen in Sec. 2 for the more general case, a free electron's orbit is bent in the presence of the magnetic field and the electron moves in a helical orbit with axis parallel to the magnetic field. An electron at the Fermi surface, correspondingly, will change its wavenumber on a circular path around the Fermi surface. This path represents the intersection of a plane perpendicular to the magnetic field with the Fermi sphere. Thus any given electron will traverse this intersection on the Fermi surface and in many cases may at some point intersect a Bragg reflection plane. If this is the case the electron will be diffracted at that point, changing its direction of motion and correspondingly jumping to a different portion of the Fermi sphere. It will then proceed on its way along another circular path on the Fermi sphere. Thus though in the one-OPW picture the Fermi surface remains spherical, the orbits of the electron within the metal become very complicated. In Fig. 2.16 we see such a possible orbit. Note that the electron orbit may be quite large on the scale of the interatomic distances. If we could see into the metal we would see many complicated orbits as the electrons move in the presence of a magnetic field. The motion of the wavenumber across our spherical Fermi surface is also very complicated and in that case involves discontinuous jumps from one portion of the surface to the other. We now seek a more perspicuous description of the electron states.

The main complexity of the electron orbits in wavenumber space arises from these Bragg reflections or discontinuous jumps in the wave-

Fig. 2.16 *A possible orbit for a nearly-free electron (in real space) in the presence of a magnetic field. Bragg reflections occur at the points a, b, c, and d. The parallel lines represent atomic planes giving rise to the diffraction at d.*

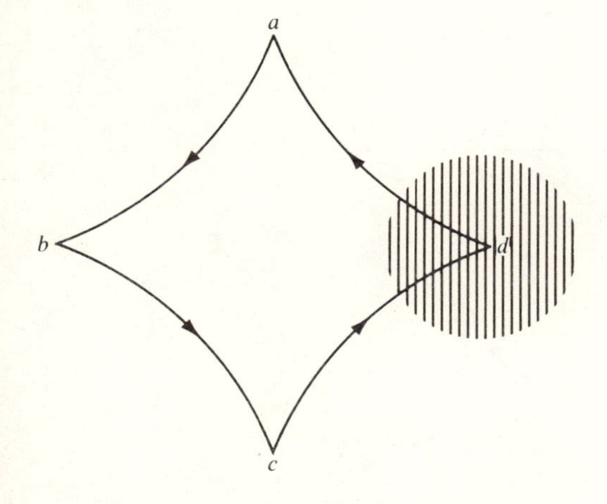

number. In contrast we noted in our discussion of energy bands and Brillouin zones that, if the electron states are represented with wavenumbers entirely within the Brillouin zone, the energy bands are continuous within the zone and therefore the motion of the wavenumber is continuous within the zone. Here we are again considering a system with translational periodicity so the specification of states only within the Brillouin zone is again possible.

We noted that if the translational periodicity of a state may be described by a wavenumber **k**, it may equally as well be described by any wavenumber differing from **k** by a lattice wavenumber. Here we have assumed that the energies are given by $\hbar^2k^2/2m$ but we have allowed the wavenumbers to run beyond the Brillouin zone. Any plane-wave state of wavenumber **k** with **k** lying outside the Brillouin zone can be alternatively described by a wavenumber, differing from **k** by a lattice wavenumber, but lying within the Brillouin zone. Thus to reduce our wavenumbers to the Brillouin zone we need simply translate each of the states by an appropriate lattice wavenumber bringing it into the Brillouin zone. Correspondingly each segment of this spherical Fermi surface can be translated by a lattice wavenumber back into the Brillouin zone to obtain the Fermi surface described in the band picture. This is illustrated for two dimensions in Fig. 2.17.

Such a construction in three dimensions is quite difficult, but there

Fig. 2.17 Translation of segments of the free-electron Fermi sphere by lattice wavenumbers to bring them into the Brillouin zone.

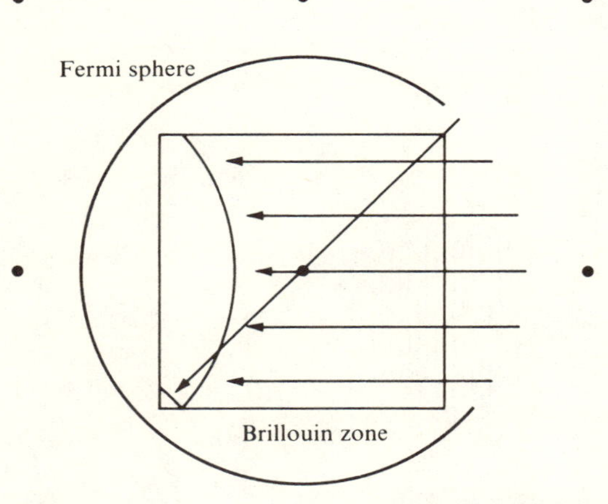

is a short cut. We notice that any spherical surface centered at $\mathbf{k} = 0$, when translated by a lattice wavenumber, becomes a spherical surface centered at that lattice wavenumber. Thus our translation of the segments of the Fermi surface will result in spherical surfaces within the Brillouin zone, each centered at some lattice wavenumber. It is much simpler to construct the wavenumber lattice and Fermi spheres about each lattice wavenumber than to seek out the various translations themselves.

Having done this we obtain a number of intersecting spherical surfaces within the Brillouin zone. We would like next to assign these to their appropriate bands. This is also easily done. We note that each spherical segment divides some band into regions of unoccupied and occupied states; i.e., states with energies above and below the Fermi energy. Thus a point in the Brillouin zone that lies within three spheres in our construction must have the first three bands occupied at that wavenumber. As we follow that wavenumber across a spherical surface to a region occupied in only two bands we have found the segment of surface dividing occupied and unoccupied states in the third band; i.e., we have found a piece of the Fermi surface in the third band. Simply by counting spheres we may assign each of the segments of Fermi surface to its band. Such a construction is illustrated in Fig. 2.18.

We notice that in our construction the Fermi surface corresponding to a given band is repeated over and over in wavenumber space with the periodicity of the wavenumber lattice. This is of course directly related to the ambiguity in the choice of wavenumber describing the translational periodicity of a given state. We can however completely specify the Fermi surface, or the energy bands, by consideration of only a single Brillouin zone centered at $\mathbf{k} = 0$ (see box *a* in Fig. 2.18). This is the description we have used before.

We could of course alternatively construct our Brillouin zone centered on some other wavenumber and this would also completely specify the bands. This will be more convenient if the Fermi surface in the usual Brillouin zone intersects the Brillouin zone extensively, since states on opposite faces of the Brillouin zone are entirely equivalent. If we follow a wavenumber along the Fermi surface until it intersects the Brillouin zone, we must then jump to the opposite zone face and continue to follow the wavenumber along the appropriate Fermi surface. Thus we have only partly accomplished our objective of eliminating the discontinuous jumps in wavenumber. Sometimes, by constructing a Brillouin zone centered at a different point, these additional discontinuous jumps may be eliminated. This was done in Fig. 2.18 for the third and fourth bands by using a zone, box *b*.

Let us now see the relation between our constructed Fermi surface and the diffraction picture. We have indicated that a Bragg reflection will

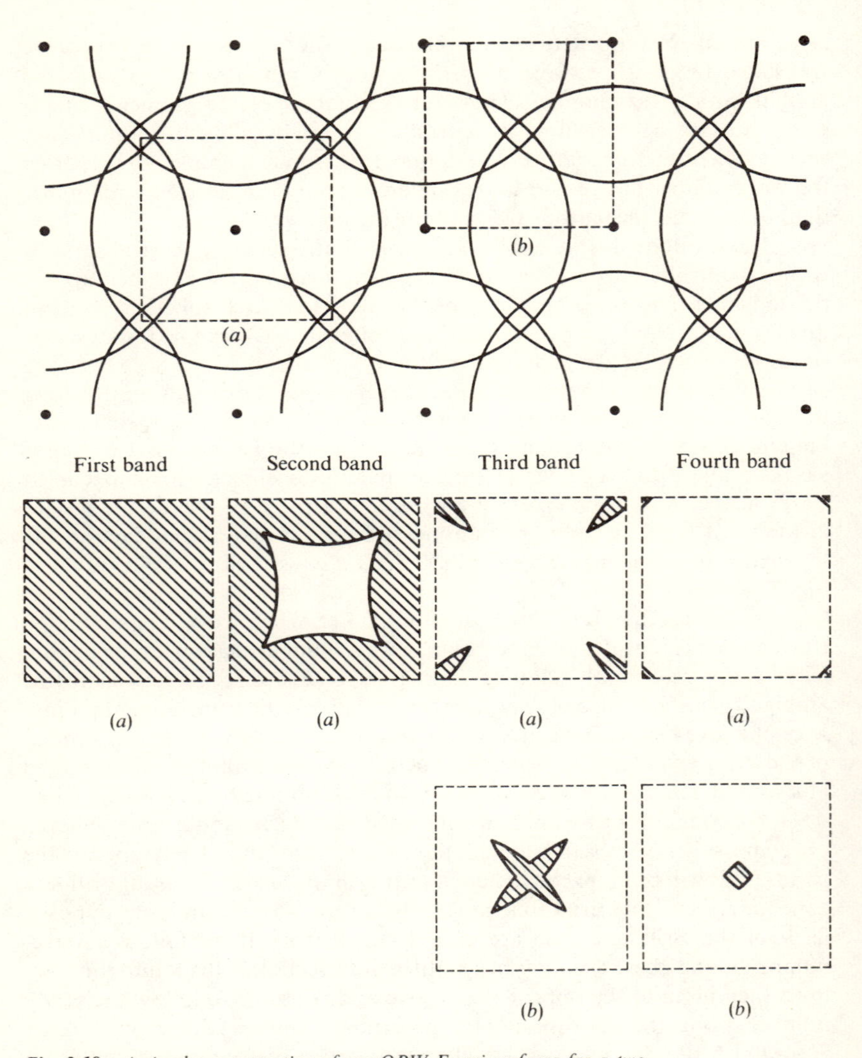

Fig. 2.18 A simpler construction of one-OPW Fermi surfaces for a two-dimensional square lattice. Above, a Fermi sphere is constructed around each lattice wavenumber. Fermi surfaces in the first four bands are identified by counting spheres and redrawn below in the Brillouin zone (a). They may equivalently be redrawn in a Brillouin zone (b) constructed around the point W as shown at the bottom of the figure. For the third and fourth bands this simplifies the visualization of the corresponding orbit shapes. Shaded areas represent occupied states in the bands in question.

occur whenever two states of the same energy differ by a lattice wavenumber. In our construction now we have plotted all states differing by a lattice wavenumber on top of each other in the Brillouin zone. Furthermore, our spheres correspond to states all having the Fermi energy. Thus the intersection of each sphere corresponds to a Bragg reflection. The electron wavenumbers will change spheres at each intersection and therefore will follow the individual-band Fermi surfaces that we have constructed. The diffraction and the zone pictures are intrinsically the same; they are simply different representations.

Because we have obtained the energy bands, or Fermi surface, by reducing the wavenumbers of the states, the Brillouin zone centered at the origin is frequently called the *reduced zone* or the *first Brillouin zone*. Having made this reduction and assigned each segment of Fermi surface to a particular band we could, if we chose, invert the process and translate these segments back to form the original sphere. We would now be able to associate each segment with a particular band. Similarly each region of wavenumber space lying outside the reduced zone may be assigned to some band. That region of wavenumber space giving rise to the second band in our construction is bounded by a polyhedron called the *second Brillouin zone*. Similarly the third, fourth, and fifth Brillouin zones could be constructed. This procedure is called the *extended-zone scheme*. The one-OPW model is an extended-zone representation. In three-dimensional systems the extended Brillouin zones tend to be so complicated that they are not useful; this can be seen from Fig. 2.19, showing the first four Brillouin zones for the face-centered cubic structure.

A third representation of the bands is sometimes made. This involves reconstructing the bands of the first Brillouin zone over and over again with the full periodicity of the wavenumber lattice. Then the third band Fermi surface, for example, is repeated with the full periodicity of the wavenumber lattice as in Fig. 2.18. Such a scheme is helpful in envisaging real-space orbits when the orbit in wavenumber space intersects the zone faces. This scheme is called the *periodic-zone scheme*.

Construction of Fermi surfaces in three dimensions is a direct generalization of that which we have shown in Fig. 2.18 for two dimensions. For a given valence we know precisely the number of electrons per atom and therefore precisely the Fermi radius. The volume of the Fermi spheres is given by one-half the valence times the volume of the first Brillouin zone. The construction then is purely a geometrical exercise and leads to Fermi surfaces such as those shown in Fig. 2.20 for the face-centered-cubic[1] structure. The construction of cross sections of these surfaces is just like the

[1] The corresponding constructions for body-centered cubic and hexagonal close-packed metals may be found in Harrison, *op. cit.*, pp. 83, 84.

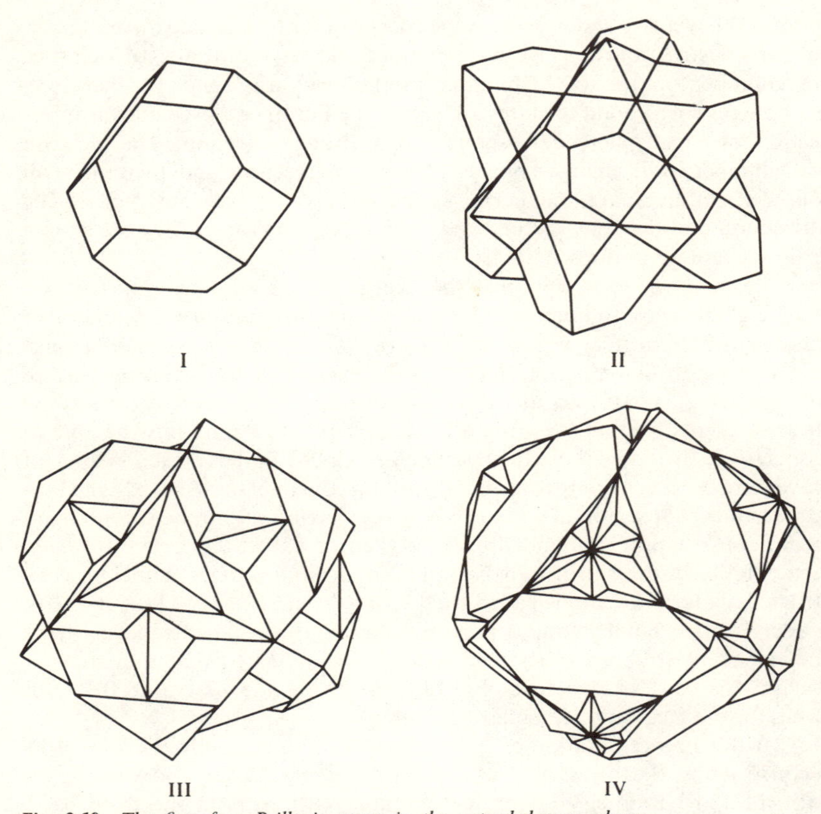

Fig. 2.19 The first four Brillouin zones in the extended-zone scheme for the face-centered cubic structure, first constructed by J. F. Nicholas [Proc. Phys. Soc., *64:953(1951)*].

two-dimensional construction in Fig. 2.18, except of course the circles are smaller for spheres centered at lattice wavenumbers that do not lie in the plane of the cross section. Note that the Fermi surfaces for face-centered cubic metals of valence 3 and 4 are closely analogous to those in the two-dimensional example.

5.6 Experimental studies of Fermi surfaces[1] In order to see the utility of Fermi surface constructions in interpreting experiments we reexamine the

[1] For more extensive discussion, see W. A. Harrison and M. B. Webb (eds.), "The Fermi Surface," Wiley, New York, 1960, or the more recent book, W. Mercouroff, "La Surface de Fermi des Métaux," Masson et Cie, Paris, 1967.

relation between electron orbits in real space and the shape of the Fermi surface. We have already obtained most of the relevant results for general band structures in Sec. 2. However, it may clarify these results to see them illustrated for this simple model.

The velocity now becomes simply the free-electron velocity, $\mathbf{v} = \hbar\mathbf{k}/m$. The equation of motion for $\mathbf{k}$ in the presence of a uniform magnetic field is then simply

$$\frac{d\mathbf{k}}{dt} = -\frac{e}{mc}\mathbf{k} \times \mathbf{H} \tag{2.27}$$

Thus between diffractions the electron moves along an intersection of a plane of constant k_z ($\mathbf{H}$ is taken along the z axis) and the Fermi sphere with constant angular velocity $\omega_c = eH/mc$. If we construct such a plane through our reassembled Fermi surfaces, we may follow the trajectory along such an arc. When it reaches the intersection with another sphere, the electron is diffracted and proceeds along the other sphere; i.e., it follows the Fermi surface for the band in question. It again follows a plane of constant k_z and in fact the same plane in our Brillouin zone since by construction the two wavenumbers connected by diffraction are plotted as the same point. As we found more generally before, the orbit in the Brillouin zone is an intersection of the constant-energy surface (in this case the Fermi surface) and a plane of constant k_z.

We may also follow the electron in real space. As the wavenumber swings through a given arc, the projection of the electron position on a plane of constant z swings through the same arc. It is rotated by 90° and distances are scaled by the factor $\hbar c/eH$. Similarly, when it diffracts it changes direction and follows an arc rotated again by 90° from the wavenumber arc and scaled again by the same factor. Thus we see in detail how the projection of the electron orbit in real space is of precisely the same shape as the corresponding section of the Fermi surface.

The scale factor is known for any experiment. Thus if we have made this simple geometric construction of the Fermi surface, we have a well-defined prediction of the shape and size of all possible electron orbits when a magnetic field is applied to the crystal. In making this prediction we have assumed that the pseudopotential is weak and its effect can be represented as simple diffraction. We have seen that for real band structures, the pseudopotential is not infinitesimal and there will be corresponding distortions of the Fermi surfaces and orbits. Our predictions cannot agree precisely with experiment but they are absolutely unambiguous and therefore subject to precise comparison.

Of course in order to compare with experiment we will need to insert

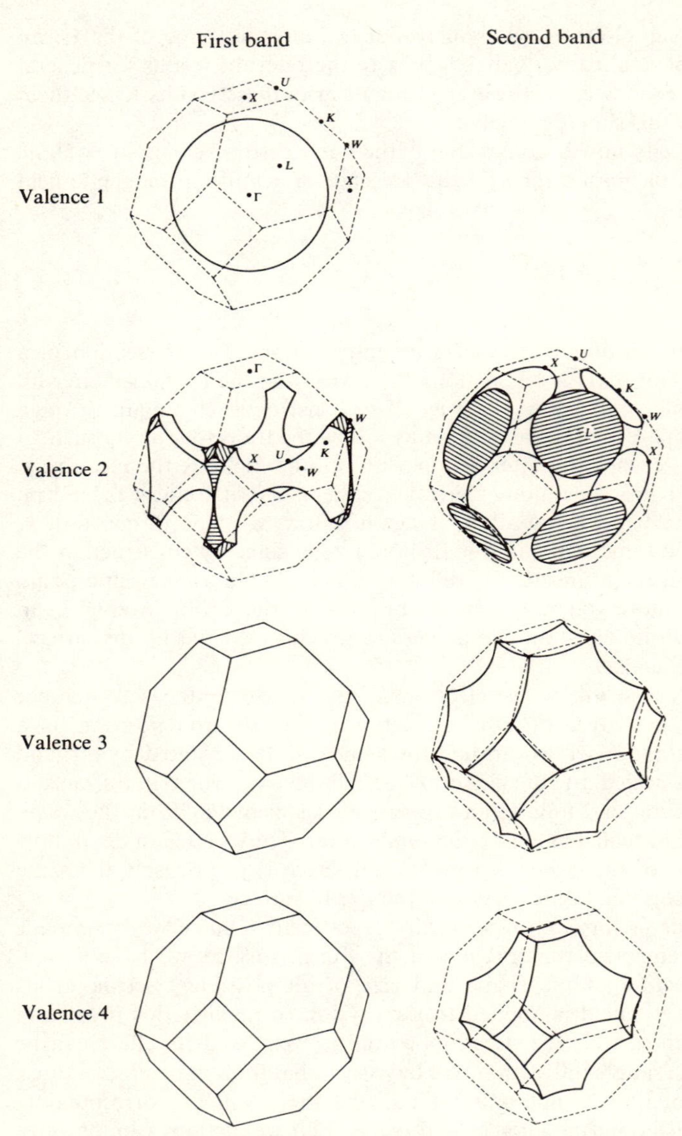

Fig. 2.20 One-OPW Fermi surfaces for face-centered cubic metals of valence 1 through 4. Regions with convex surfaces are occupied; those with concave surfaces, unoccupied. Note that the first-band zones are centered at Γ for valence 1 and X for valence 2. Second-band zones are

Third band **Fourth band**

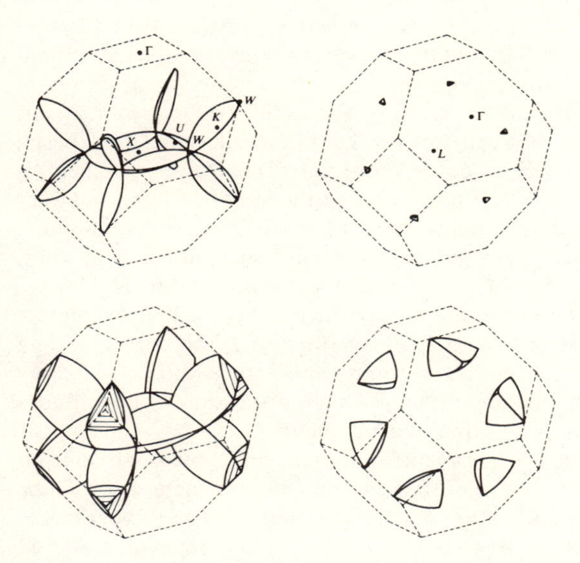

centered at Γ, third-band zones at X, and fourth-band zones at L. Examples of the four valences in this structure are copper, calcium, aluminum, and lead, respectively.

numbers in the formulas and this can be troublesome, especially when magnetic fields are involved. It may therefore be helpful to say a word about units before proceeding. This will also enable us to estimate the magnitudes of quantities such as orbit sizes which are of physical interest. We will use centimeter-gram-second units. The magnetic-field intensity H is ordinarily given in gauss. If we take the electronic charge in electrostatic units and the field in gauss, then the product eH will be in dynes. (We may see from Eq. (2.27), for example, that eH has the units of force.) Thus we measure H in gauss, e in electrostatic units, and all other quantities in cgs units. The fundamental constants become

$$\hbar = 1.054 \times 10^{-27} \text{ erg-sec}$$

$$m = 9.1 \times 10^{-28} \text{ gram}$$

$$c = 3 \times 10^{10} \text{ cm/sec}$$

$$e = 4.8 \times 10^{-10} \text{ esu}$$

The electron wavelength at the Fermi energy will be of the order of the interatomic spacing which is a few angstroms. Thus k_F will be of the order of 10^8 cm^{-1} and equal to 1.75 10^8 cm^{-1} for aluminum. The conversion factor to obtain orbit sizes in real space is $\hbar c/eH$, which is 0.66×10^{-10} cm^2 at 1000 gauss. Thus at this field the free-electron orbit radii corresponding to the aluminum Fermi energy are 1.15×10^{-2} cm, many orders of magnitude larger than the lattice spacing. The cyclotron frequency at this field is $eH/mc = 17.6 \times 10^9$ rad/sec, or 2.8 kilomegacycles. The Fermi velocity in aluminum, which of course does not depend upon the field, is 2.03×10^8 cm/sec. This is small compared to the speed of light, but very large compared to the speed of sound, which is 5.1×10^5 cm/sec in aluminum.

Energies are usually more conveniently measured in electron volts, 1 ev $= 1.602 \times 10^{-12}$ erg. For aluminum, the Fermi energy is 11.6 ev, much larger than KT at room temperature which is $1/40$ ev. In semiconductors, where typical electron energies are of order KT, the corresponding electron velocity is 94×10^5 cm/sec at room temperature, still considerably larger than the speed of sound. With this much discussion of magnitudes we will proceed to the consideration of experiments.

There are a number of experiments that make direct measurements on the size and shapes of the orbits of the electrons in a magnetic field. These give us a direct check on our picture of the Fermi surface. These experiments work only if an electron is able to complete its orbit before being scattered by collision with a defect or an impurity in the crystal. Thus experiments are possible only at low temperatures where the vibrations of the crystal are minimum and they are possible only in very pure materials.

One technique that is particularly easy to understand is the measurement of orbits with magnetoacoustic attenuation. An ultrasonic wave is propagated through the crystal. For simplicity we will think of this as a transverse wave and note that it will give rise (through Maxwell's equations) to a sinusoidal transverse electric field with the periodicity of the motion of the ions. The ions move slowly in comparison to the electrons so we may think of these fields as constant in time. This is illustrated in Fig. 2.21.

A magnetic field is applied perpendicular to the direction of propagation so that the electrons execute orbits such as A and B shown in the figure.

Fig. 2.21 *The magnetoacoustic effect. A transverse sound wave propagating to the right causes local electric fields represented by arrows. With a magnetic field applied perpendicular to the direction of sound propagation, electrons move in orbits such as A and B. Since the accelerations on the two vertical legs of the orbits add, the electron response is strong. When the field is decreased by a factor of 2, the accelerations on the two legs subtract and the response is weak.*

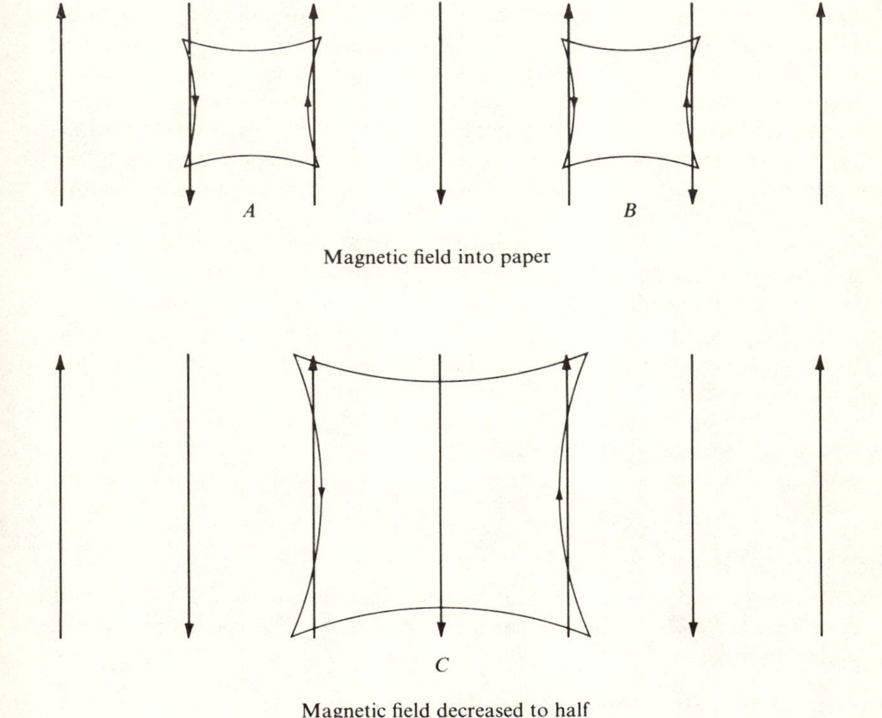

The precise size of these orbits, of course, changes with the magnetic field; the size of the Fermi surface does not depend on the magnetic field and our scale factor is inversely proportional to the magnetic field. Thus as the field is increased, the orbits shrink. Because of this the strength of the interaction of a given orbit shape with the field associated with the sound wave will vary with the magnetic field. For the orbit size shown in the figure we see that an electron may move parallel to the electric field on both legs of its orbit where its velocity is predominantly along field lines (A). In another orbit it may move against the electric field at both sides of its orbit (B). In either case the electrons have a very strong net interaction with the fields present. At half this field these electron orbits will move with the field on one leg of the orbit and against it on the other. The effects tend to cancel and the net interaction is very much weakened. Thus as we vary the magnetic field the sound wave will alternately feel a highly conducting and a weakly conducting medium. Attenuation depends directly on the effective conductivity of the medium; it turns out that the attenuation is highest when the medium responds most weakly. Fluctuations in the attenuation as a function of magnetic field thus lead directly to a knowledge of a dimension of important orbits in the metal or, correspondingly, to important cross sections of the Fermi surface. It can be shown with a stationary-phase argument that these tend to be extreme dimensions. Thus as we perform the measurements as a function of magnetic field at successive different orientations of the magnetic-field direction, we obtain successive calipers of the Fermi surface. Such a plotting out of the Fermi surface in aluminum is shown in Fig. 2.22. The slightly different scale in the comparison with the free-electron Fermi surface appears to be experimental. Such an experiment gives a very striking verification of the picture we have developed.

Perhaps the most precise measurements of the Fermi surface have been made with the de Haas-van Alphen effect, which is a periodic fluctuation in the magnetic susceptibility as the magnetic field is varied. This is a quantum-mechanical effect arising from the quantization of the electron orbits in the magnetic field. We can obtain the effect of this quantization intuitively from our semiclassical results of Sec. 2. We found there that the classical electron-orbit frequency was

$$\omega_c = \frac{2\pi eH}{\hbar^2 c} \left(\frac{\partial A}{\partial E}\right)^{-1}$$

where A was the area of the orbit in wavenumber space and $\partial A/\partial E$ was at constant k_z. We know that at a particular k_z and near a particular energy the quantum-mechanical states must have energies

$$\hbar\omega_c(n + \gamma)$$

where γ is a constant and n an integer. This follows from the correspondence

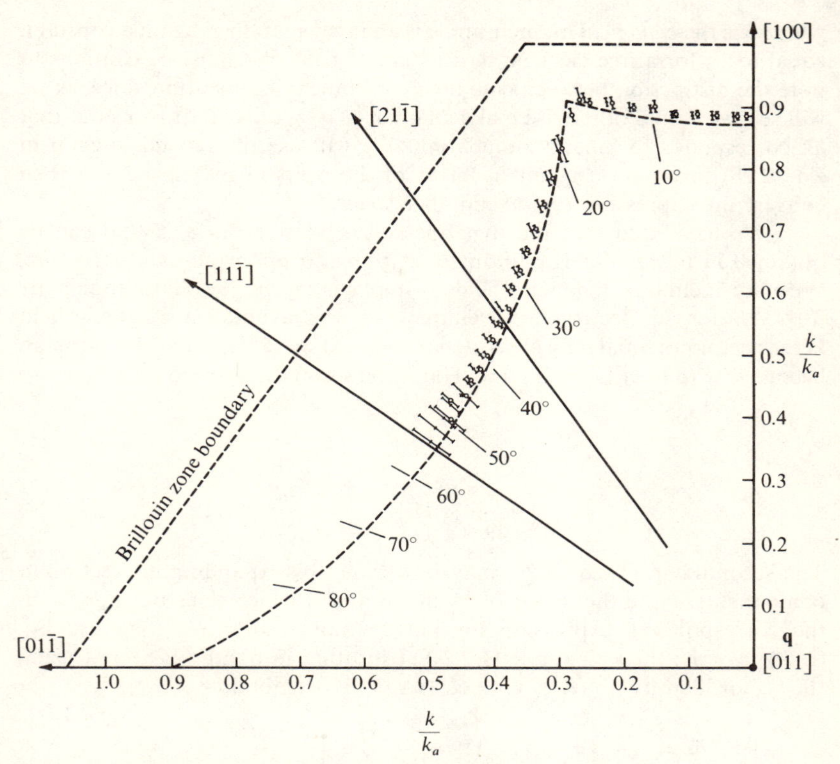

Fig. 2.22 Magnetoacoustic measurements of the Fermi surface of aluminum by G. N. Kamm and H. V. Bohm [Phys. Rev., *131:111(1963)*]. *The dashed line represents the central cross section, a (110) plane, of the second-band one-OPW Fermi surface.*

principle, the necessity of our being able to construct packets from these states which execute classical motion. Thus the difference in area between neighboring orbits in wavenumber space is

$$\Delta A = \frac{\partial A}{\partial E}\hbar\omega_c = \frac{2\pi eH}{\hbar c}$$

The allowed orbits occur in equally spaced increments of area.

It is difficult to derive this result with full rigor, but we can obtain it using the Bohr-Sommerfield quantization rule,

$$\int \tilde{\mathbf{p}} \cdot d\mathbf{r} = \int \mathbf{r} \cdot d\tilde{\mathbf{p}} = (n + \gamma)h \tag{2.28}$$

where $\tilde{\mathbf{p}}$ is the *canonical* momentum, n is an integer, and γ is again a constant, equal to $\frac{1}{2}$ for a free electron in a magnetic field. We must be cautious to note the distinction between canonical and kinetic momentum since, as we will see, we would otherwise make an error of a factor of 2. It turns out that $\hbar\mathbf{k}$ corresponds to kinetic momentum. We will use the second integral in Eq. (2.28) to obtain a result in terms of the orbit in real space and then convert the expression to wavenumber space.

We have seen that the dynamics of an electron in a crystal can be obtained in terms of a Hamiltonian $\mathscr{H}(\mathbf{p})$ based upon the band structure. We may include a magnetic field by replacing $\mathbf{p}$ in the Hamiltonian by $\tilde{\mathbf{p}} + e\mathbf{A}/c$ for an electron, with charge $-e$. For a uniform magnetic field $\mathbf{H}$ the vector potential may be written $\mathbf{A} = -\frac{1}{2}\mathbf{r} \times \mathbf{H}$. Thus the Hamiltonian becomes $\mathscr{H}(\tilde{\mathbf{p}} - e\mathbf{r} \times \mathbf{H}/2c)$ and Hamilton's equations become

$$\frac{d\mathbf{r}}{dt} = \frac{\partial \mathscr{H}}{\partial \tilde{\mathbf{p}}}$$

$$\frac{d\tilde{\mathbf{p}}}{dt} = -\frac{\partial \mathscr{H}}{\partial \mathbf{r}} = -\frac{e}{2c}\left(\frac{\partial \mathscr{H}}{\partial \tilde{\mathbf{p}}} \times \mathbf{H}\right) = -\frac{e}{2c}\left(\frac{d\mathbf{r}}{dt} \times \mathbf{H}\right) \tag{2.29}$$

The second step in Eq. (2.29) may be verified by expanding all vectors in components. Note the factor of $\frac{1}{2}$ in Eq. (2.29) which does not appear in the corresponding expression for kinetic momentum. We may use Eq. (2.29) to write $d\tilde{\mathbf{p}} = -e\,d\mathbf{r} \times \mathbf{H}/2c$. Substituting into Eq. (2.28) and using the vector identity $\mathbf{A} \cdot (\mathbf{B} \times \mathbf{C}) = \mathbf{C} \cdot (\mathbf{A} \times \mathbf{B})$, we obtain

$$\int \mathbf{r} \cdot d\tilde{\mathbf{p}} = \frac{-e}{2c}\int \mathbf{r} \cdot (d\mathbf{r} \times \mathbf{H}) = -\frac{e\mathbf{H}}{c} \cdot \int \frac{\mathbf{r} \times d\mathbf{r}}{2} = (n + \gamma)h$$

Because of the dot product only the components of $\mathbf{r}$ perpendicular to $\mathbf{H}$ enter the integration, which is simply equal to the area of the orbit as projected on a plane perpendicular to $\mathbf{H}$. This is conveniently rewritten in terms of the cross-sectional area of the Fermi surface using our scaling factor, $eH/\hbar c$. Writing the corresponding area of the Fermi surface as A we have

$$A = \frac{2\pi e H}{\hbar c}(n + \gamma) \tag{2.30}$$

which is consistent with the result we obtained before.

This is the principle result of the calculation. We see that the quantum condition only allows a certain discrete set of orbit areas in wavenumber space. The size of these steps is directly proportional to the magnetic field. In the absence of the quantum condition any intersection of a plane perpendicular to $\mathbf{H}$ with a constant-energy surface corresponded to an allowed orbit in wavenumber space. With the quantum condition we find that for a given plane perpendicular to **H**, states are allowed only on the intersection of some of the constant-energy surfaces. The allowed states then lie on tubes in wavenumber space each of which has constant cross section in the planes perpendicular to the magnetic field. Such sets of tubes are shown in Fig. 2.23. We have cut off each of the tubes at a constant-energy surface corresponding to the Fermi energy. The states on these tubes are highly degenerate (ultimately this degeneracy arises from orbits of the same size and shape located at different points in the crystal) such that the number of states within, for example, the Fermi sphere is only very slightly modified. If we substitute for the constants we find that in an ordinary metal at fields of the order of kilogauss there are many thousands of tubes within the Fermi surface. Nonetheless the situation is qualitatively as shown in the figure.

As we increase the magnetic field the cross section of each of the tubes increases continuously, and ultimately each moves out through the Fermi sphere which remains essentially unchanged. At the same time of course the degeneracy associated with each of the tubes within the sphere increases

Fig. 2.23 Orbit quantization in a magnetic field. Allowed states lie on cylinders in wavenumber space and are occupied for those portions of the cylinder within the Fermi surface. This is illustrated above for a free-electron gas at three different magnetic-field intensities. The corresponding quantization for an ellipsoidal Fermi surface is shown below, again with the field vertical.

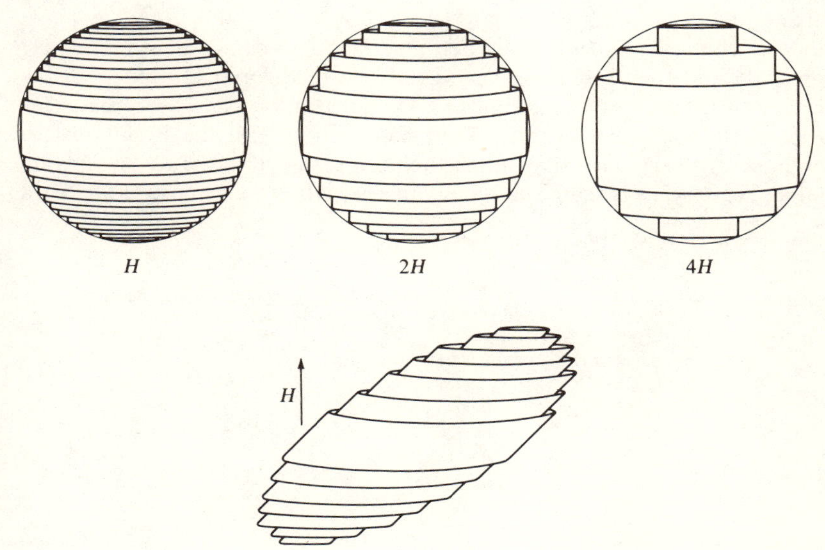

so that the total number of states within the sphere is conserved. Let us now see how this affects the properties of the metal and gives rise to the de Haas-van Alphen effect.

A number of properties, including the magnetic susceptibility, depend upon the density of states at the Fermi energy, which is proportional to the number of states lying between two close constant-energy surfaces at the Fermi energy. This in turn depends directly on the area of the tubes which lie between the two constant-energy surfaces. We note that this area will increase very rapidly as one of the tubes becomes nearly tangent to the equator of the Fermi surface but the area then drops suddenly when the tube pops out of the Fermi surface. Thus the density of states will show a sharp fluctuation as each successive tube pops out of the Fermi surface. The magnetic susceptibility in turn will show sharp fluctuation as a function of magnetic field.

Because of the close spacing of the tubes, the experiment requires quite homogeneous, and frequently very high, fields over the sample. The experiment may conveniently be done by pulsing a magnetic field through a tiny sliver of material (held at very low temperature) and observing the response in a pickup coil surrounding the sample. The fluctuating component of the magnetic field arising from the fluctuating component of the susceptibility is picked up directly and displayed.

We may readily see from Eq. (2.30) that the fluctuations are periodic in $1/H$ and that the period gives a direct measure of the cross-sectional area of the Fermi surface.

Such measurements of the Fermi surface area can be made extremely precisely and they provide a very good test of theoretical models of the Fermi surface. One very complete test, for example, has been made for zinc by Joseph and Gordon.[1] Their comparison with the Fermi surface obtained with the one-OPW approximation is shown in the accompanying table. There is always some ambiguity in associating a given oscillation with

Cross-section areas of the Fermi surface of zinc (in $Å^{-2}$)

Section	One-OPW	Observed
(a)	0.00030	0.00015
(b)	0.0046	0.0025
(c)	0.045	0.0043
(d)	0.06	0.0426
(e)	0.079	0.061
(f)	0.25	0.22

[1] A. S. Joseph and W. L. Gordon, *Phys. Rev.*, **126**:489 (1962).

a particular section of Fermi surface. However, the orientation of the field with respect to the crystal and the variation of the area as the field is rotated give assistance in this association, and those which they have made for zinc seem quite well established.

We note that the agreement is quite good for the larger sections of Fermi surface but the errors are appreciable for the smaller sections. These discrepancies arise primarily from treating the diffraction as infinitely weak. We will soon see how these may be accounted for.

There are a number of other experiments that give direct information about the geometry of the Fermi surface. Notable among these is the Gantmakher resonance which depends upon the use of thin specimens and the adjustment of the magnetic field such that the orbits just fit between two opposing faces of the specimen. Magnetoresistance and anomalous skin-effect experiments also give information about the geometry. More will be said about these effects in our discussion of transport and optical properties.

Another category of experiments gives information about the velocity of the electrons at the Fermi surface as well. The *Azbel-Kaner effect*, or *cyclotron resonance* in metals, is such an experiment. In such an experiment a microwave field is applied parallel to the surface of the metal. At such high frequencies the electromagnetic field penetrates only a very short distance into the metal. This is known as the skin effect and is the same effect that tells us that light will be reflected from a metallic surface. The form of the penetration will be discussed in Sec. 5.1 of Chap. III, but will not be needed here. Penetration to the skin depth is illustrated in Fig. 2.24. A uniform magnetic field is then applied parallel to the surface but perpendicular to the electric-field vector of the microwave field. At low fields the electron orbits will be large in comparison to the skin depth. Most electrons orbiting within the interior of the metal will never feel the microwave field. Some orbiting electrons, however, will dip briefly into the region of the field before diving back into the field-free interior of the metal. They will then, on a second orbit, reenter the field briefly and return to the interior. If the orbiting of the electron is synchronized with the microwave field we will have a resonant interaction between the microwave field and the orbiting electron. If many electrons have the same orbit frequency as the microwave field for some magnetic-field strength, the metal will appear to be highly conducting. This shows up in the experiments as a sharp reduction in the microwave absorption at the metal surface. Thus by varying the field and observing the microwave absorption, we will measure important orbit frequencies in the metal. Let us now return to our one-OPW Fermi surfaces to interpret these orbit frequencies.

In Sec. 2, we related the cyclotron frequencies to the Fermi surface geometry through the derivative of the orbit area (in k space) with respect to energy Eq. (2.10). For the one-OPW model, it is simpler to note that the wavenumber of free electrons rotates with the free-electron cyclotron frequency

$$\omega_c^\circ = \frac{eH}{mc}$$

For a free-electron gas we obtain resonance when the magnetic field takes on the value such that the microwave frequency is equal to this cyclotron frequency. We may also note readily that resonance will also be obtained if the microwave field is any integral multiple of the cyclotron frequency.

If we now consider the periodic potential and allow the electron

Fig. 2.24 *Cyclotron resonance in metals. An applied microwave field $\mathscr{E}$ penetrates only a short distance into the metal. When a magnetic field is applied parallel to the surface, in this case into the figure, most electrons (a) never feel the microwave field. However, some electrons (b) are accelerated in one portion of their orbit. If the orbit frequency is synchronous with the microwave field, they will resonate.*

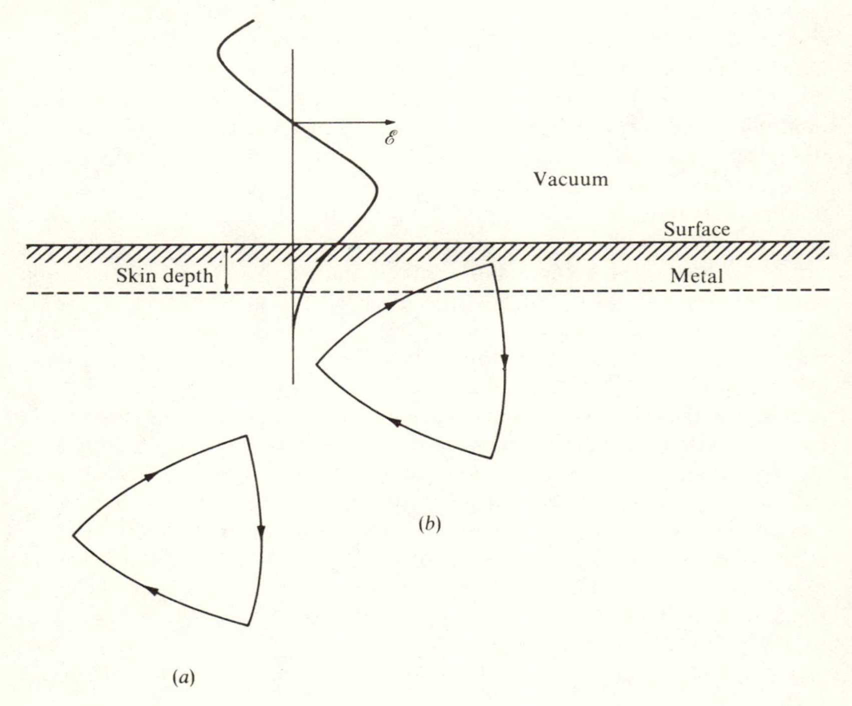

to be diffracted, the orbit will be modified but on each leg of the orbit the angular motion of **k** will be the same. Thus we may readily add up the total angle through which an electron moves during its orbit and obtain directly the orbit frequency; it is given simply by 2π times the free-electron cyclotron frequency divided by the total angle traversed (see Fig. 2.25).

In this way then we may make direct comparison between our model of the electron states and the observed cyclotron frequencies. This comparison is usually made in terms of m_c, the cyclotron mass $m_c/m = \omega_c^\circ/\omega_c$. In this case the agreement is not so good. In many metals the observed cyclotron mass is consistently larger than that calculated by a factor of as much as 2. We might at first try to attribute this discrepancy to finite diffraction strength; i.e., to band-structure effects. It is easily seen, however, that these effects have only a small effect on the cyclotron frequency in comparison to that which they have on the Fermi surface geometry. The discrepancy here is somewhat more fundamental.

In obtaining the cyclotron frequency we have used an assumption which we did not require in discussing Fermi surface geometries. That assumption is that the velocity is given by $(1/\hbar)\nabla_k E$. For free electrons this is given by $\hbar\mathbf{k}/m$, but anything which distorts the energy band modifies the result. In addition to effects that might occur from the periodic potential, there may also be a shift arising from the electron-electron interaction

Fig. 2.25 The calculation of cyclotron frequencies for orbits on one-OPW Fermi surfaces.

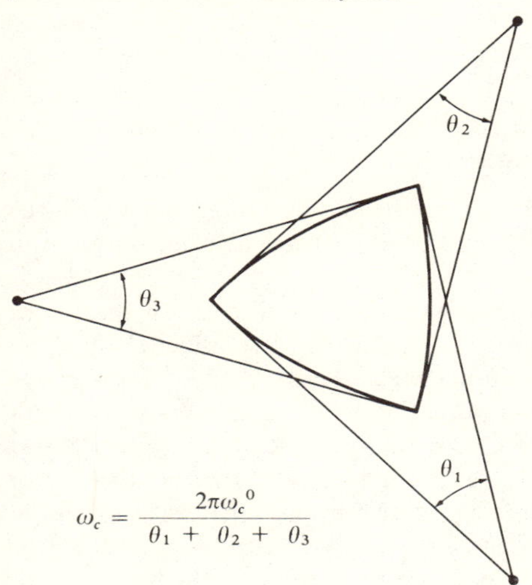

which we have treated only approximately. This is generally believed to be a small effect in metals. In addition there may be a shift in velocity arising from the dynamic interaction between the electron and the ions present, i.e., the electron-phonon interaction. We will discuss this in more detail later; at this point we may note that the electron moving through the crystal will jostle the ions and carry a wake with it. This has the effect of modifying the electron velocity. Such effects are now generally believed to be responsible for the main discrepancies between observed and calculated cyclotron frequencies.

This same shift in $\partial E/\partial \mathbf{k}$ affects also the density of states at the Fermi surface and thereby the electronic specific heat (to be described in Sec. 1.1 of Chap. III). The shift is presumably not the same at all points on the Fermi surface and the density of states gives an average over the entire Fermi surface. The ratio of the observed electronic specific heat to the free-electron value gives then a measure of this average shift. This may also be specified by giving an effective mass that would yield the observed specific heat using the free-electron formula. In the accompanying table we list specific heat masses for a number of metals obtained from experiment.

Specific heat masses for metals

Li, 2.4†	Be, 0.46		
Na, 1.3†	Mg, 1.33	Al, 1.6	
K, 1.1†	Ca, 0.8		
Cu, 1.5	Zn, 0.9	Ga, 0.4	
Ag, 1.0	Cd, 0.75	In, 1.3	Sn, 1.2
Au, 1.0†	Hg, 2	Tl, 1.15	Pb, 2.1

†Taken from specific heats given by K. Mendelssohn, "Cryophysics," p. 178, Interscience, New York, 1960. All other values were taken from J. G. Daunt, "Progress in Low Temperature Physics," p. 210, North-Holland, Amsterdam, 1955.

Some of these corrections arise from the periodic potential, but to a large extent differences from unity are thought to arise from the electron-phonon interaction.

Experimental effects that depend only on the geometry of the Fermi surface and the electron wavenumbers are not directly affected by these electron-electron and electron-phonon effects. They enter only indirectly inasmuch as they modify the pseudopotential, which is not known with all that much precision in any case. Effects, on the other hand, that depend directly upon the electron velocity or the density of states at the Fermi surface are significantly shifted. This has shown up in our discussion of the cyclotron frequencies and of the electronic specific heat. In some cases it is not easy to see whether shifts should occur. For example, in the free-electron theory of the electrical conductivity we will find a conductivity proportional to the product of the electron-scattering time and the electron velocity. If we assume the scattering time is independent of these interactions we would still expect the velocity to be modified and therefore the conductivity to be shifted. On the other hand, we may note that the product of the scattering time and the velocity is simply the mean free path. If we assume the mean free path is independent of interactions, then we conclude that the conductivity should not be affected. The latter turns out to be correct and the conductivity is not believed to be affected by electron-electron[1] and electron-phonon[2] interactions.

Both of these shifts in electron velocity are calculable in principle but at the present stage it appears that only the effects of electron-phonon interaction can be reliably included. For the most part we will ignore these complications.

5.7 Multiple-OPW Fermi surfaces The one-OPW treatment is of course an approximate treatment of the band structure. We were able to go from the Schroedinger equation to the pseudopotential equation without approximation, but we then made the approximation of treating the pseudopotential as extremely small. This enabled us to obtain a reasonable account of the band structure of the simple metal. If we retain the pseudopotential and treat its effect exactly, we obtain the true band structure such as described in Sec. 4. It is sometimes convenient, particularly in studying Fermi surfaces, to do an intermediate treatment between these two; i.e., we include the effect of the pseudopotential to higher order than in the one-OPW approximation, but again we make use of the fact that it is small.

Physically we can describe the effect of a finite pseudopotential by saying that an electron begins to diffract before it quite reaches the Bragg reflection plane. Thus its orbit in wavenumber space, and in real space, is slightly distorted and the sharp corners that we obtained in the one-OPW Fermi surfaces are rounded off. This is illustrated for a pair of diffraction planes in Fig. 2.26. We see that the Fermi surface actually approaches the Bragg reflection planes (or Brillouin zone planes) normally. The distortions are largest right at the Bragg reflection plane and become smaller as we move away from the plane. The second picture can be obtained by describing the states near a Bragg plane as linear combinations of two OPWs (or two plane pseudowavefunctions). Hence it may be called a *two-OPW approximation.* Since the distortions are only large in the neighborhood of the plane, the essential behavior can be obtained by considering only this small number of OPWs in the expansion of the wavefunction. This result depends upon the

[1] J. S. Langer, *Phys. Rev.*, **120**: 714 (1960); **124**: 1003 (1961).

[2] R. E. Prange and L. P. Kadanoff, *Phys. Rev.*, **134**: 566 (1964).

fact that $\langle \mathbf{k} + \mathbf{q}|w|\mathbf{k}\rangle$ becomes smaller at large q. Unfortunately if we make an approximate treatment with a small number of plane waves, using different plane waves at different parts of the Fermi surface, we find that the Fermi surface does not match up properly between. Thus a full-band calculation involving many plane waves is essential to obtain very accurately the Fermi surface.

Let us consider, however, the energy bands very close to the Bragg reflection plane, which is the plane bisecting the lattice wavenumber $\mathbf{q}$.

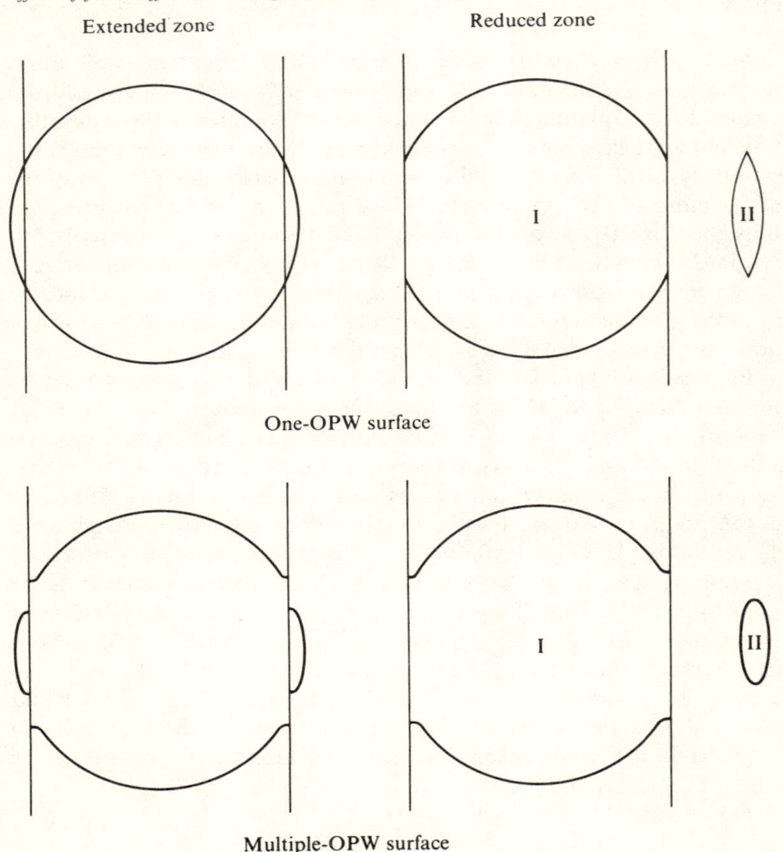

Fig. 2.26 The Fermi surface in the presence of two diffraction planes. Above is the one-OPW surface, in the extended-zone scheme on the left and in the reduced-zone scheme on the right (bands I and II). Below the effect of finite diffraction strength is shown.

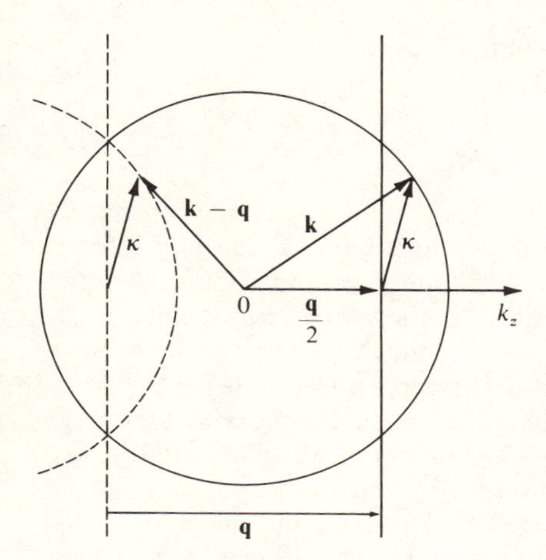

Fig. 2.27 Coordinate system used in the two-OPW approximation for states with k_z near $q/2$. The $\boldsymbol{\kappa}$ is the wavenumber measured from the center of the plane, $\mathbf{q}/2$. The wavenumber of the second OPW lies on the dashed circle.

Let us select our coordinate system such that $\mathbf{q}$ lies in the z direction. We are interested in states with k_z approximately equal to $q/2$. The two OPWs needed to describe the state are $\mathbf{k}$ and $\mathbf{k} - \mathbf{q}$, representing the states between which diffraction may occur. We let the pseudopotential matrix element that couples these states be given by $w = \langle \mathbf{k} - \mathbf{q}|w|\mathbf{k}\rangle$. We select our zero of energy to absorb the effect of the average value of the pseudopotential, $\langle \mathbf{k}|w|\mathbf{k}\rangle$. Then the submatrix of the Hamiltonian based upon these two states is given by

$$\mathscr{H} = \begin{bmatrix} \tfrac{1}{2}k^2 & w \\ w^+ & \tfrac{1}{2}(\mathbf{k} - \mathbf{q})^2 \end{bmatrix}$$

Here we have expressed all quantities in atomic units; i.e., we take the Bohr radius as the unit of length, the electron mass as the unit of mass, and seconds as the unit of time. In these units $\hbar$, m, and e are 1 and energies are in atomic units, 27.2 ev. It is convenient to measure our wavenumbers from the point $\mathbf{q}/2$ as in Fig. 2.27. The Hamiltonian matrix becomes

$$\mathscr{H} = \begin{bmatrix} \mu + \dfrac{q\kappa_z}{2} & w \\ w^+ & \mu - \dfrac{q\kappa_z}{2} \end{bmatrix}$$

with $\mu = \tfrac{1}{2}[(q/2)^2 + \kappa_z{}^2 + \kappa_x{}^2 + \kappa_y{}^2]$. We can obtain the eigenvalues of this matrix by writing the secular equation

$$\left(\mu - E + \frac{q\kappa_z}{2}\right)\left(\mu - E - \frac{q\kappa_z}{2}\right) - ww^+ = 0$$

which we solve immediately for E to obtain

$$E = \mu \pm \sqrt{\left(\frac{q\kappa_z}{2}\right)^2 + w^2}$$

where we have written the number ww^+ as w^2, and neglected the dependence upon $\mathbf{k}$. We could directly calculate E as a function of $\boldsymbol{\kappa}$ and plot it for κ_z small and negative (so that $\mathbf{k}$ is near the zone face and within the zone). There are two solutions corresponding to the two energy bands in the Brillouin zone as seen in Fig. 2.28. It will be much more informative, however, to fix the energy and solve for the transverse wavenumber $\sqrt{\kappa_x^2 + \kappa_y^2}$ in terms of κ_z. This will give us directly a cross section of the constant-energy surface. If we again retain only negative values of κ_z we will obtain the two sections of Fermi surface in the Brillouin zone. It may be preferable, however, to plot one of the surfaces with κ_z greater than zero and one with κ_z less than zero such that the surfaces approach the free-electron sphere as w goes to zero. This is the extended-zone scheme. Solving for the transverse wavenumber in terms of the energy and κ_z we obtain

$$\sqrt{\kappa_x^2 + \kappa_y^2} = \sqrt{2E - \left(\frac{q}{2}\right)^2 - \kappa_z^2 \mp \sqrt{(q\kappa_z)^2 + 4w^2}} \tag{2.31}$$

The resulting surfaces are shown in Fig. 2.29 for three choices of energy. If we have selected the energy equal to the Fermi energy this figure shows the distortion of the Fermi surface at the Bragg reflection plane.

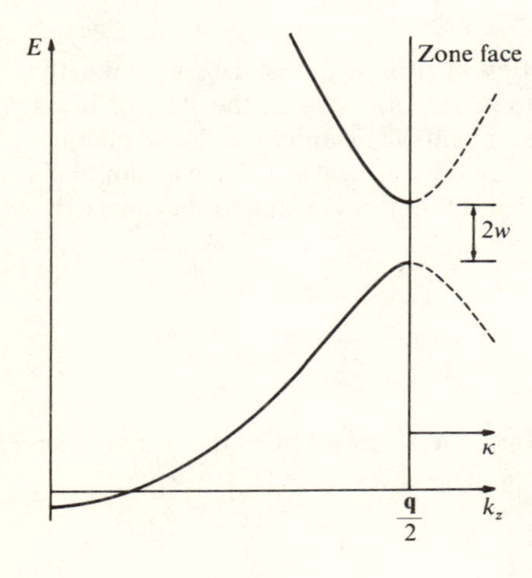

Fig. 2.28 The energy bands near a Brillouin zone face, computed in the two-OPW approximation, taking the matrix element $\langle \mathbf{k} - \mathbf{q}|w|\mathbf{k}\rangle$ *as independent of* $\mathbf{k}$.

In a real crystal, of course, there will be a number of Bragg reflection planes. In some cases we can treat each independently as we have treated this plane. Where two planes cross the Fermi surface close to each other it may be necessary to use a 3-by-3 Hamiltonian matrix. The cross section of the Fermi surface in aluminum constructed in this way is shown in Fig. 2.30.

Such figures are of course very close to what we would obtain had we done the full energy-band calculation. These distortions will modify the Fermi surface dimensions and areas and are the next step required if we wish to compare with experimentally observed Fermi surfaces. Such a comparison gives us a check on the calculated strength of the pseudopotential. Of course if we wish to study Fermi surfaces in detail, it is appropriate to return again to the reduced-zone scheme. Calculated constant-energy surfaces in aluminum are compared with those obtained in the one-OPW approximation in Fig. 2.31.

We have shown sections of the Fermi surface in the second and third bands; these may be readily identified in the valence-3 Fermi surfaces for face-centered cubic structures shown in Fig. 2.20. That figure also shows tiny segments in the fourth band. When the true pseudopotential is introduced, however, the fourth band is lifted entirely above the Fermi energy and these segments therefore disappear; if the pseudopotential were slowly turned on, these segments would shrink around the point W before disappearing. This raising of the fourth band above the Fermi energy may also be seen in the aluminum band structure shown in Fig. 2.9. The disappearance of smaller one-OPW segments of Fermi surface occurs frequently in metals. Even the much larger fourth-band segments for valence-4 metals in the face-centered cubic structure shown in Fig. 2.20 are found to disappear

Fig. 2.29 *Calculated constant-energy surfaces in the two-OPW approximation. Energies of 0.8, 1.0, and 1.2 times $\hbar^2(q/2)^2/2m$ were used for (a), (b), and (c), respectively. The w was taken equal to $0.1\hbar^2(q/2)^2/2m$ for all three cases.*

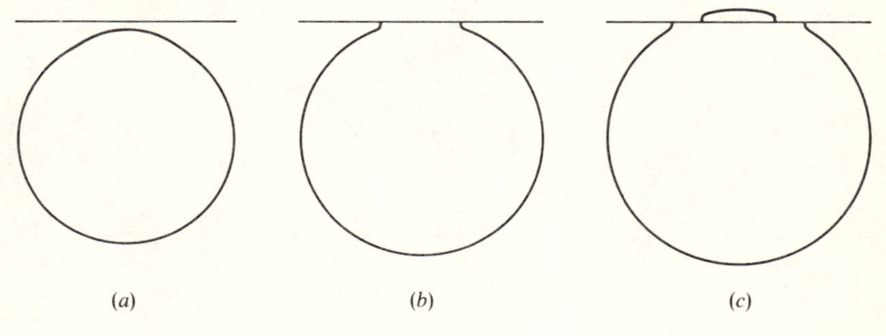

in lead. Similarly, the *contact* with a zone face shown in the two-OPW surface of Fig. 2.26 reduces the total Fermi surface area. In the extreme case, when *all* of the Fermi surface disappears into the zone faces, we have a semiconductor. Metallic properties derive from the existence of Fermi surface, and semiconductors may be viewed as simple metals in which all of the Fermi surface is gone.

Multiple-OPW calculations have improved the agreement between the theoretical and experimental Fermi surfaces. In addition it has been possible in some cases to work backward from the observed experimental surfaces and to obtain experimental estimates of the OPW form factors.

We may note that the shape of the surface which we have obtained in the two-OPW approximation is independent of the sign of w. However, when three or more plane waves are involved, the detailed shape of the Fermi surface will depend on the sign of the matrix elements. Thus it may be possible to obtain the signs from such analyses of experiment. In any case the general form of the pseudopotential, as shown in Fig. 2.14, is quite well

Fig. 2.30 *A (110) section of the Fermi surface in aluminum in the extended-zone scheme. Dashed lines are Bragg reflection planes. Calculations were made in a three-OPW approximation.*

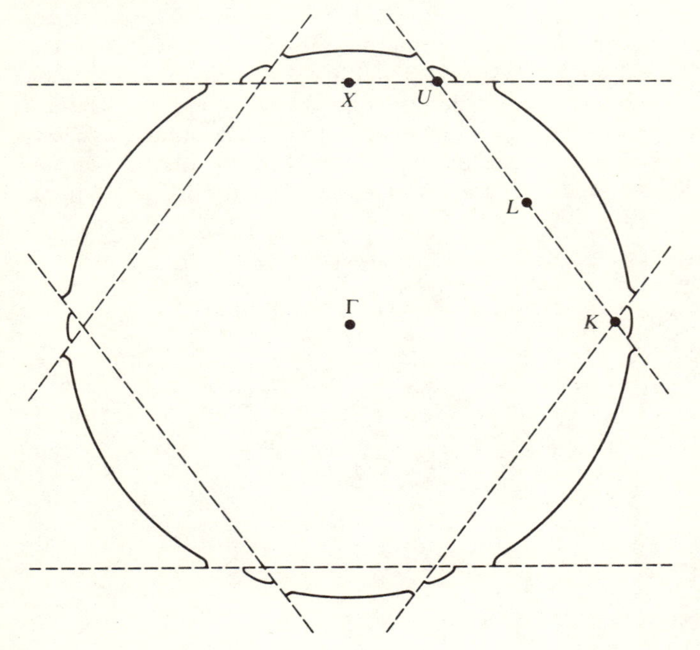

Fig. 2.31 *(a) Intersection of constant-energy surfaces in the second band in aluminum with a (110) plane through* Γ. *(b) Intersection of constant-energy surfaces in the third band in aluminum with a (110) plane through K. In both cases energies are in rydbergs and may be compared with the one-OPW Fermi energy of 0.86 rydberg. Sections on the left are three-OPW surfaces; those on the right are one-OPW surfaces. The scale in (b) is greatly expanded in comparison to that in (a).*

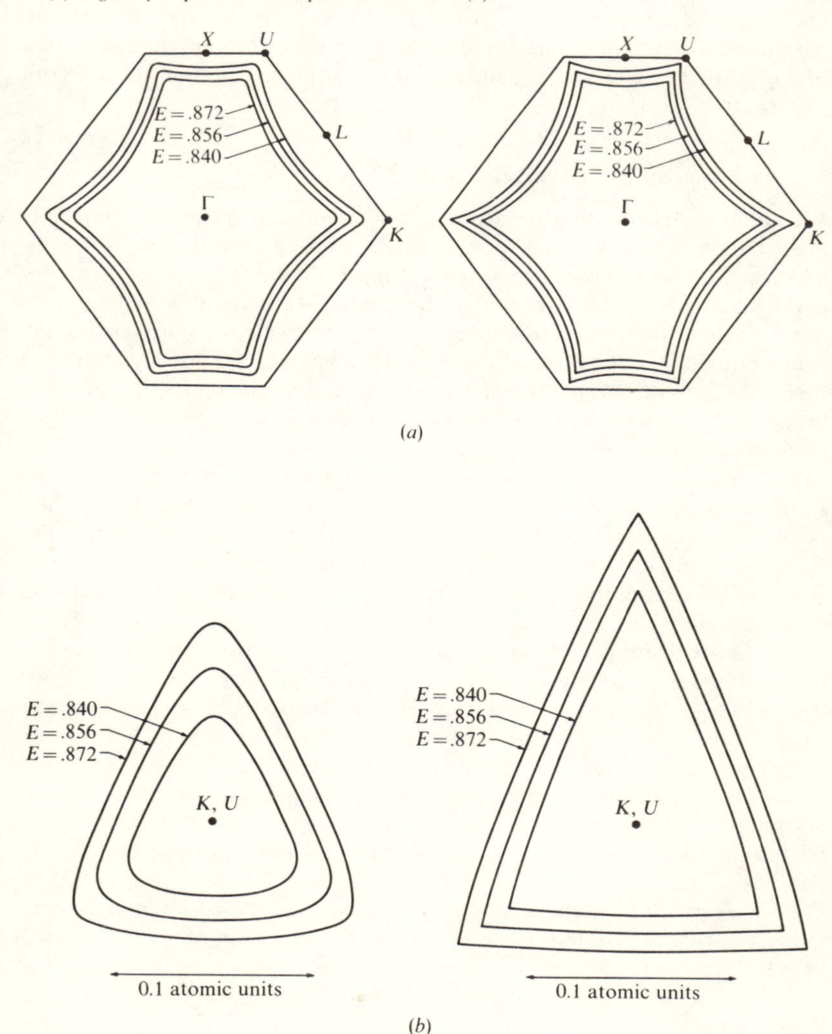

known, and knowing the magnitude of two or three matrix elements will usually enable us to guess the sign.

When we go on to calculate other properties of the simple metals we will find that it is ordinarily most convenient to begin again with the Hamiltonian matrix and include band-structure effects at the same time we consider the property itself. A knowledge of the true band structure frequently is not very useful in the treatment of a property. It is difficult to begin with a mass of tabulated information and obtain a meaningful calculation of the property except in special cases such as the calculation of the Fermi surface itself.

6. *Semiconductor and Semimetal Bands*[1]

We noted in Sec. 4.6 that semiconductors are characterized by completely filled bands, which we call *valence* bands, and completely empty bands, which we call *conduction* bands. This complete filling obtains only at zero temperature where the system is in the ground state. At finite temperature some states in the conduction bands will be occupied and correspondingly some states in the valence band will be unoccupied. Thus at finite temperature there will be some electrons which can contribute to a conduction process; hence the term *semiconductor*. We will see how impurities can also allow electrical conductivity.

For the purposes of our discussion here we may regard semimetals simply as semiconductors in which the conduction-band edge lies slightly below the valence-band edge. Thus even at absolute zero of temperature some electrons will occupy the conduction band and some valence-band states will be unoccupied. We will return in Sec. 6.3 to a specific discussion of semimetals.

The distinction between semiconductors and insulators is simply a quantitative one, involving the magnitude of the energy gap between occupied and unoccupied bands. In semiconductors the gap is sufficiently small (of the order of a volt) that they become reasonably good conductors under ordinary circumstances. This is true of germanium and silicon and various three-five compounds such as indium antimonide and gallium arsenide. In insulators the gap is sufficiently large (several volts) that they no longer conduct. This is the case in many ionic compounds such as sodium chloride.

We can see how quantitative difference leads to qualitative difference in behavior by noting that the probability of occupation of a state is given by a Boltzmann factor $Ae^{-E/KT}$ where E is of the order of the gap and KT is

[1] For a complementary discussion at a similar level, see C. Kittel, "Quantum Theory of Solids," chaps. 14, 15, Wiley, New York, 1963.

of order 0.025 volts near room temperature. In semiconductors the Boltzmann factor may be of order 10^{-10}, but it may drop to 10^{-30} or 10^{-40} in an insulator. Since there are only 10^{22} states in the conduction band we expect no electrons to occupy states in that band in the insulator. When we consider insulators we will see that there are complications associated with these large gaps in addition to the elimination of carriers. These reduce the mobility of whatever carriers may be generated in other ways.

Unfortunately there is no simple approximation that leads to a good description of semiconducting energy bands. To obtain any reasonably complete description of the bands, we must resort to a full energy-band calculation. If we were to attempt to derive the energy bands by making corrections to the free-electron bands, as we did in the simple metals, we see that the matrix elements of the pseudopotential do not simply distort the Fermi surface but they eliminate it entirely. We cannot obtain the states using one or two OPWs but must always consider several, which brings us back in essence to a band calculation.

This does not mean that pseudopotentials are inappropriate for the treatment of semiconductors; in fact the pseudopotential method was developed by Phillips and Kleinman[1] in order to study the energy bands in semiconductors. It turns out that the pseudopotential form factors which enter the treatment of semiconductors are not significantly larger than those which enter the simple metals. Complexity has arisen because of the large number of bands involved. In the diamond structure, for example, there are two atoms per primitive cell and four electrons for each atom; thus there are eight electrons per primitive cell, a sufficient number to occupy four bands. A construction of the free-electron Fermi surface would lead to a large number of very tiny fragments and with the inclusion of the pseudopotential these all disappear into the Bragg reflection planes just as the tiniest segments in aluminum disappeared when a finite pseudopotential was included. At the high wavenumbers involved in the eight-electron sphere, no state can adequately be treated unless several plane waves are included in the expansion of the pseudowavefunction. Thus we must return to the solution of secular equations of rather high order. Nevertheless it is possible to isolate the important matrix elements in this larger Hamiltonian matrix in order to understand the main features in the bands. This is essential to an understanding of the optical properties and of the nature of the tetrahedral, or covalent, bonding in semiconductors. We will return to such a discussion in Sec. 6.3 of Chap. IV, when we discuss atomic properties.

Fortunately in much of the theory of semiconductors, it is not necessary to consider all of these complex bands. As in metals the electronic properties will be determined by those states lying nearest to the Fermi energy. In

[1] J. C. Phillips and L. Kleinman, *Phys. Rev.*, **116**:287 (1959).

metals these are the states at the Fermi surface. In semiconductors these are the highest states in the valence bands and the lowest states in the conduction band.

We will in fact be interested only in the states which lie within an energy of order KT of the band edges. Again with KT only $1/40$ volt in comparison to the band gaps of the order of a volt, it becomes appropriate to consider the extremal states in each band and to compute the small corrections in energy as we move away from these. This is readily done with the $\mathbf{k} \cdot \mathbf{p}$ method.

6.1 $\mathbf{k} \cdot \mathbf{p}$ ***method and effective-mass theory*** We are again interested in solving the energy eigenvalue equation which we now write in the form

$$\frac{p^2}{2m}\psi_k + V(\mathbf{r})\psi_k = E_k\psi_k$$

where we have written $\hbar\nabla/i = \mathbf{p}$, the momentum. We write the solutions in the Bloch form $\psi_k = u_k e^{i\mathbf{k}\cdot\mathbf{r}}$ and obtain

$$\frac{1}{2m}(\mathbf{p} + \hbar\mathbf{k})^2 u_k + V(r)u_k = E_k u_k \tag{2.32}$$

We will consider states with wavenumber near zero. These will ordinarily (though as we will see not always) be extremal points in E_k. The method is directly generalizable to consideration of states in the neighborhood of other points in the Brillouin zone, but the states near $\mathbf{k} = 0$ are the simplest.

To zero order in $\mathbf{k}$ the eigenvalue equation becomes simply

$$\frac{1}{2m}p^2 u_k + V(r)u_k = E_k u_k$$

The solutions are the $\mathbf{k} = 0$ Bloch states for the various bands. There will be an infinite number of such bands and we write these solutions $u_0^{(1)}$, $u_0^{(2)}, \ldots$, and the corresponding energies $E_0^{(1)}, E_0^{(2)}, \ldots$. The $\psi_k^{(n)}$ for all $\mathbf{k}$ are a complete set for the expansion of any function satisfying periodic boundary conditions in the crystal. The $\psi_{k=0}^{(n)} = u_0^{(n)}$ are a complete set for the expansion of any such function having the full translational periodicity of the lattice and therefore for expansion of the $u_k^{(n)}$ with $k \neq 0$. We may therefore use the $u_0^{(n)}$ as the basis for a perturbation expansion for the $u_k^{(n)}$.

We now consider states for $\mathbf{k}$ not equal to zero. We may expand the kinetic energy in Eq. (2.32) and consider the perturbation which is first order in $\mathbf{k}$,

$$H_1 = \frac{2\hbar}{2m}\mathbf{k} \cdot \mathbf{p}$$

and a second-order perturbation which is simply $H_2 = \hbar^2 k^2/2m$. We now proceed with ordinary perturbation theory. The first-order energy shift is simply $(\hbar/m)\mathbf{k} \cdot \langle u_0^{(n)}|\mathbf{p}|u_0^{(n)}\rangle$. If the crystal has a center of symmetry then the $u_0^{(n)}$ will have a well-defined parity with respect to that center. It follows that the matrix element, and therefore the first-order energy shift, for such a crystal will vanish. On the other hand, in indium antimonide, for example, there is no center of symmetry and we obtain a first-order term. In such cases it follows that the extremal energy, i.e., the band maximum or minimum, may be shifted away from the point $\mathbf{k} = 0$. Such a shifted maximum in the valence band for indium antimonide is shown in Fig. 2.32.

In any band structure we may seek the band extremum and make our expansion about the corresponding wavenumber. Then the first-order shifts will vanish by construction. In our present treatment we will restrict our consideration to a case where the band extremum occurs at $\mathbf{k} = 0$ and therefore there is no first-order shift.

We may proceed now to second order to obtain an energy given by

$$E_k^{(n)} = E_0^{(n)} + \frac{\hbar^2 k^2}{2m} + \frac{\hbar^2}{m^2} \sum_m \frac{|\mathbf{k} \cdot \langle u_0^{(m)}|\mathbf{p}|u_0^{(n)}\rangle|^2}{E_0^{(n)} - E_0^{(m)}} \tag{2.33}$$

We could of course carry this to any order we choose but this will be enough to allow us to obtain the curvature at the band extremum.

We see that to obtain the energy we need to sum over all bands at $\mathbf{k} = 0$. However, the increasing energy denominators reduce the importance of the contributions of the higher bands and frequently only the nearest few are important. In addition we frequently can reduce significantly the number of contributing terms by taking into account the symmetry of the states in the individual bands. By noting the irreducible representations

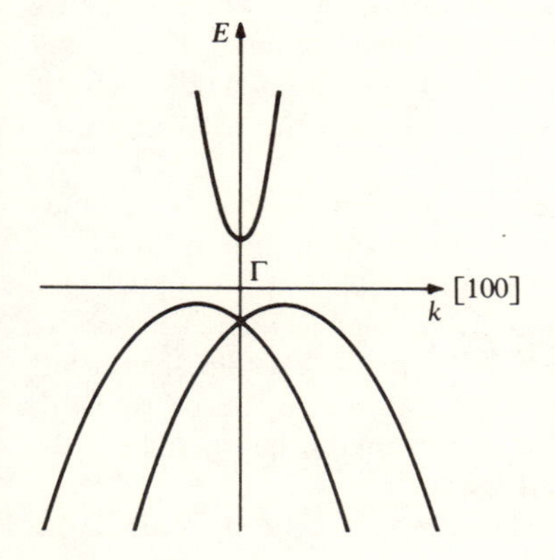

Fig. 2.32 A schematic representation of the energy bands in indium antimonide showing a split valence-band maximum and the low-mass conduction-band minimum. [For a full description, see E. O. Kane, J. Phys. Chem. Solids, *1:249(1956)*].

according to which each of the states transforms, as well as the irreducible representation according to which the momentum operator transforms, we may obtain *product representations* from which we can determine which matrix elements will vanish by symmetry. This is an application of group theory which we did not discuss in Chap. I but which may be found in any of the standard group-theory texts. Note as a specific example that two even states are not coupled because the momentum operator is odd.

We may note that when two bands are coupled which do not differ greatly in energy a very large contribution to the energy is obtained. We may introduce an effective mass by writing the energy of Eq. (2.33) in the form

$$E_k = E_0 + \frac{\hbar^2}{2m}\mathbf{k}\cdot\overleftrightarrow{\left(\frac{\mathrm{m}}{m^*}\right)}\cdot\mathbf{k} \tag{2.34}$$

and the elements of the effective-mass tensor may be obtained from the formula

$$\left.\frac{m}{m^*}\right|_{ij}^{(n)} = \delta_{ij} + \frac{2}{m}\sum_l \frac{\langle u_0^{(n)}|p_i|u_0^{(l)}\rangle\langle u_0^{(l)}|p_j|u_0^{(n)}\rangle}{E_0^{(n)} - E_0^{(l)}} \tag{2.35}$$

Equation (2.35) is known as the *f-sum rule*. It is of some importance in the discussion of optical properties since, as we will see, the strength of interband transitions is closely related to these same matrix elements of **p**.

We may note that when two bands are coupled the matrix element $\langle u_0^{(n)}|p_i|u_0^{(l)}\rangle$ will be of the order of $\hbar/a$, where a is the interatomic spacing,

$$\frac{m}{m^*} \sim 1 + \frac{2\hbar^2}{ma^2\,\Delta E}$$

and $\hbar^2/ma^2$ is of the order of 10 volts. If ΔE is of the order of 0.2 volts, we obtain an effective mass of the order of 0.01 of the electron mass. Thus it will not be surprising in semiconductors to find that in many cases the electrons behave as free particles with masses of much less than the true electronic mass.

This result is of course applicable also to the simple metals. We may consider, for example, the conduction-band minimum in lithium at $\mathbf{k} = 0$. The only band lying below is the 1*s* core band, which is not coupled by a matrix element of **p**; therefore all bands that are coupled lie at energies above the conduction-band minimum. It follows that $1/m^*$ is less than $1/m$ or m^* greater than m. As we go to the heavier alkalis, with more and more bands below, the core bands ultimately dominate and in the heavy alkalis the effective masses are less the true electron mass.

A calculation of the effective masses can be carried out in any band structure if we know, or are willing to guess, the wavefunctions in other bands. We can even estimate the effective masses in many cases by simply knowing the order of magnitude of the matrix elements and using the measured value of the band gaps.

We have obtained for states near a band edge something very much like the pseudopotential theory for simple metals. This effective-mass theory is much older than the pseudopotential method. Again, restricting consideration to band extrema at $\mathbf{k} = 0$, we have given the energy as a function of $\mathbf{k}$ in Eq. (2.34). We may use this expression for E_k to define an effective Hamiltonian $H(\hbar\mathbf{k})$ which yields a Schroedinger-like equation,

$$-\frac{\hbar^2}{2m}\sum_{ij}\frac{\partial}{\partial x_i}\left(\frac{m}{m^*}\right)_{ij}\frac{\partial}{\partial x_j}\varphi + E_0\varphi = i\hbar\frac{\partial\varphi}{\partial t}$$

Like the pseudopotential equation, this will lead to correct eigenvalues (at small k) if solved exactly. Also, like the pseudopotential equation, it will yield a rather smooth pseudowavefunction φ. For a plane-wave solution, we obtain the true wavefunction by multiplying φ by the Bloch function, $u_k^{(n)}$ (or its first-order expansion in $\mathbf{k}$), rather than by orthogonalizing to the core. If φ is normalized in the crystal, $u_k^{(n)}$ should be normalized in a cell of volume Ω_0 by

$$\int u_k^{(n)*}\, u_k^{(n)}\, d\tau = \Omega_0$$

We have an added generality in that we may also obtain solutions corresponding to linear combinations of degenerate band states. Then the true wavefunction is obtained from φ by replacing $\mathbf{k}$ by $-i\nabla$ in the expression for $u_k^{(n)}$.

Unfortunately this approach does not give us directly the form which the pseudo-Hamiltonian takes when a perturbation is added since the derivation was based upon translational periodicity. It is only easily generalizable when the perturbations upon the perfect crystal are slowly varying.

We noted in Sec. 2 that the electrons in a perfect crystal behave dynamically under applied forces as particles with a Hamiltonian $H(\hbar\mathbf{k})$. This was demonstrated by constructing wavepackets and is therefore valid if the applied force varies slowly in space. In particular any applied potential must vary slowly over an atomic distance. For such a slowly varying potential $V(\mathbf{r})$ applied to the crystal, the electron dynamics will therefore be described properly by

$$-\frac{\hbar^2}{2m}\sum_{ij}\frac{\partial}{\partial x_i}\left(\frac{m}{m^*}\right)_{ij}\frac{\partial}{\partial x_j}\varphi + (E_0 + V(\mathbf{r}))\varphi = i\hbar\frac{\partial\varphi}{\partial t} \tag{2.36}$$

This is intuitively quite plausible. A slowly varying potential simply raises or lowers the bands in different parts of the crystal. We will use this form (with $i\hbar(\partial/\partial t)$ replaced by E) in the treatment of impurity states in semiconductors in Sec. 8.2. Replacing the true Schroedinger equation by Eq. (2.36) is generally called *effective-mass theory*. We see that it is quite analogous to pseudopotential theory.

It may also be of interest to consider the effect of slowly varying strains in a semiconductor—e.g., long wavelength lattice vibrations. The extension to this case is not nearly so straightforward. We can see three effects that must be included.

1. The band edge may shift with strain. If the extremum occurs at $\mathbf{k} = 0$ only dilatations can give a shift linear in strain. (By symmetry the coefficient for a linear shift with shear strain must vanish.) If the extremum occurs at $\mathbf{k} \neq 0$, shear strains may also give linear shifts. Such shifts clearly are just like applied potentials and can be included by letting E_0 be a function of position in Eq. (2.36).
2. The effective-mass tensor may be modified by the application of a strain. This may be included by letting m/m^* be a function of position in Eq. (2.36). Note that the factors in the first term of Eq. (2.36) are arranged such that the term remains Hermitian when m/m^* becomes a function of position. (Hermiticity is demonstrated by multiplying on the left by φ'^* and integrating. Two partial integrations are then performed using periodic boundary conditions.
3. We may imagine matching the wavefunctions calculated in two regions of slightly different strain; the true wavefunction, not the pseudowavefunction, is to be matched. Now the Bloch function $u_k(\mathbf{r})$ between atoms (or specifically at cell boundaries) may depend upon strain and therefore may be discontinuous between the regions. Thus a discontinuity in φ must be introduced in the matching of φ to give a continuous wavefunction. We may define β to be some average value of $u_k(\mathbf{r})$ at the cell boundaries; it will be a function of position if the strain varies. This may be inserted in Eq. (2.36) in such a way as to lead to a continuous $\beta\varphi$ and to lead to a Hermitian pseudo-Hamiltonian,

$$-\beta^* \frac{\hbar^2}{2m} \sum_{ij} \frac{\partial}{\partial x_i} \left(\frac{m}{\beta^* m^*} \right)_{ij} \frac{\partial}{\partial x_j} \beta\varphi + \beta(E_0(\mathbf{r}) + V(\mathbf{r}))\varphi = i\hbar \frac{\partial}{\partial t} \beta\varphi \tag{2.37}$$

Hermiticity has been required to guarantee charge conservation through the continuity equation and to assure real eigenvalues. This rather complicated form may be derived as an approximation by use of a cellular method[1] and therefore may reasonably be extended to strains that are rapidly varying with position. It may even be reasonable for treating states in alloys.

If we neglect variations of m/m^* and β with position and take m^* isotropic, we may commute $\partial/\partial x_i$ with β and m/m^* to obtain simply

$$-\frac{\hbar^2}{2m^*}\nabla^2\varphi + [E_0(\mathbf{r}) + V(\mathbf{r})]\varphi = i\hbar\frac{\partial\varphi}{\partial t}$$

This simple form will be used in describing the interactions between electrons and longitudinal lattice vibrations in semiconductors.

We have seen in Eq. (2.37) that a description of semiconductors in terms of pseudowavefunctions is much more complicated than the corresponding description of simple metals. The complexity comes first from the necessity of treating bands in which $1/m^*$ differs greatly from the free-electron value; use of only the first term in Eq. (2.35) would be meaningless in most semiconductors. The second difficulty, which is related, is the structure of the Bloch function, $u_k(\mathbf{r})$, which cannot be taken as constant outside the cores (as the $1 - P$ factor in pseudopotential theory is). In the diamond structure it will drop nearly to zero in the empty spaces between atoms. This was the origin of the β which complicates Eq. (2.37) and also makes it difficult to go from the pseudowavefunction to the true wavefunction. Finally it prevents a clean formulation of the problem for arbitrary atom arrangements. Certainly it is a crude approximation to describe the effect of the $u_k(\mathbf{r})$ by our parameter β.

Nonetheless the effective-mass theory encompassed by Eq. (2.37) is the best available method for handling many of the properties of semiconductors and will form the basis for our study of semiconductors.

In the following section we discuss electron dynamics in semiconductors. For this problem we consider perfect crystals and wavepackets could be constructed. Thus we return to the semiclassical theory and do not require the generality of Eq. (2.37).

6.2 Dynamics of electrons and holes in semiconductors In a pure semiconductor, as we have indicated, we will have just enough electrons to fill the valence bands and no electrons left over to occupy the conduction bands. However, at finite temperature (or, as we will see later, when impurities are present) some states will be occupied in the conduction band and some states unoccupied in the valence band. Excitations of the first type are called *electrons;* those of the second type are called *holes.* The transfer of one electron across the band gap gives rise to one electron and one hole. We are interested now in the dynamics of these excitations.

[1] W. A. Harrison, *Phys. Rev.*, **110**:14 (1958); **123**:85 (1961).

We consider first an electron in the conduction band. We look in particular at the lowest-lying excitations, which will occur near the minimum in that band. Near the minimum we may, as in the $\mathbf{k} \cdot \mathbf{p}$ method, expand the energy around the band minimum. We write the wavenumber, measured from the minimum, as $\boldsymbol{\kappa}$. For a general case the result may be a general quadratic form in wavenumber such as given in Eqs. (2.34) and (2.35). However, by proper choice of the orientation of three orthogonal axes this form may be diagonalized to

$$E_k = E_0 + \frac{\hbar^2}{2m} \sum_i \left(\frac{m}{m^*} \right)_{ii} \kappa_i^2$$

In most materials at least two of these coefficients are equal and the energy is customarily written in the form

$$E_k = E_0 + \frac{\hbar^2}{2m_t} \kappa_t^2 + \frac{\hbar^2}{2m_l} \kappa_l^2$$

where m_t and m_l are called the *transverse* and *longitudinal effective masses;* κ_l is the component of $\boldsymbol{\kappa}$ along the symmetry axis and κ_t is the corresponding transverse component.

We have found that for general band structures

$$\hbar \dot{\mathbf{k}} = \hbar \dot{\boldsymbol{\kappa}} = \mathbf{F} = -e\left(\mathscr{E} + \frac{\mathbf{v} \times \mathbf{H}}{c} \right)$$

$$\mathbf{v} = \frac{1}{\hbar} \nabla_k E = \frac{1}{\hbar} \nabla_\kappa E$$

Thus we may define an electron momentum $\mathbf{p} = \hbar \boldsymbol{\kappa}$ and we obtain Newton's equation $\dot{\mathbf{p}} = \mathbf{F}$ and the velocity is related to the momentum by

$$\mathbf{v} = \overleftrightarrow{\left(\frac{1}{\mathrm{m}^*} \right)} \mathbf{p}$$

where we have defined an effective-mass tensor related to that in Eq. (2.35) by a factor $1/m$. The dynamics of the electrons is simply the dynamics of free particles with an anisotropic mass. Note that in general the velocity will not be parallel to the momentum.

In silicon there are six minima in the conduction band, all occurring at the same energy but lying in the six different [100] directions as shown in Fig. 2.33. The effective masses are given by $m_t = 0.19$ m and $m_l = 0.98$ m. For many purposes it is adequate to replace these six sets of carriers with anisotropic effective masses by a single set with a suitably chosen average isotropic mass. In germanium the minima in the conduction band occur at

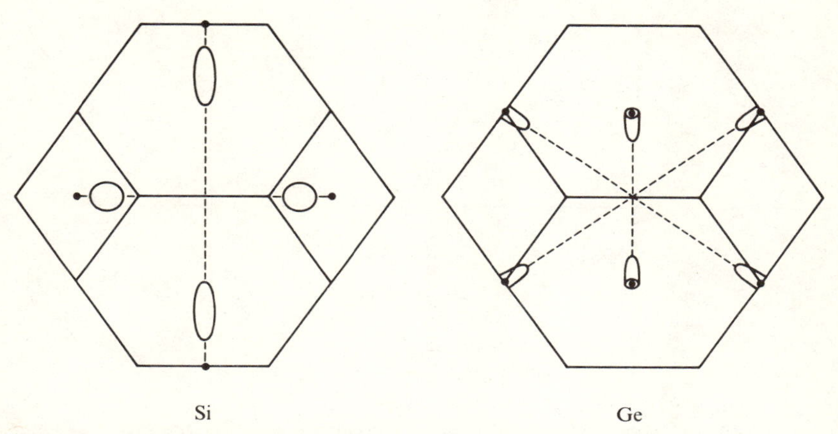

Fig. 2.33 Disposition of the conduction-band minima in the Brillouin zone for silicon and germanium. The ellipsoids represent constant-energy surfaces in the conduction band. The zones are viewed in a [110] direction.

the intersection of the [111] directions with the zone faces, i.e., at the point L. This is also shown in Fig. 2.33. The bands at any two opposite faces of the Brillouin zone may be taken together to form a single valley; thus we have four valleys with a mass tensor given by $m_t = 0.082$ m and $m_l = 1.64$ m. Both in germanium and silicon the electrons contribute more effectively to the conductivity than do holes and therefore dominate when equal numbers are present.

The dynamics of holes in semiconductors is not nearly so simple to understand. It is well known, and intuitively reasonable, that unoccupied states in the valence band behave as positively charged particles. It is not nearly so clear, however, that they should behave as positive particles with a positive mass, and we will consider such systems carefully.

In both germanium and silicon the valence-band maximum occurs at $\mathbf{k} = 0$ and is degenerate. In our discussion we will neglect that degeneracy and consider only the behavior within a single band. We remove a single electron from the state of wavenumber $\mathbf{k}$ near the valence band maximum as shown in Fig. 2.34*a*. We then consider the behavior of the system as a whole, again writing e as the magnitude of the electronic charge, a positive number.

The current in the system was zero before the removal of the electron; after the removal the current becomes

$$\mathbf{j} = 0 - (-e)\frac{1}{\hbar}\nabla_k E_k$$

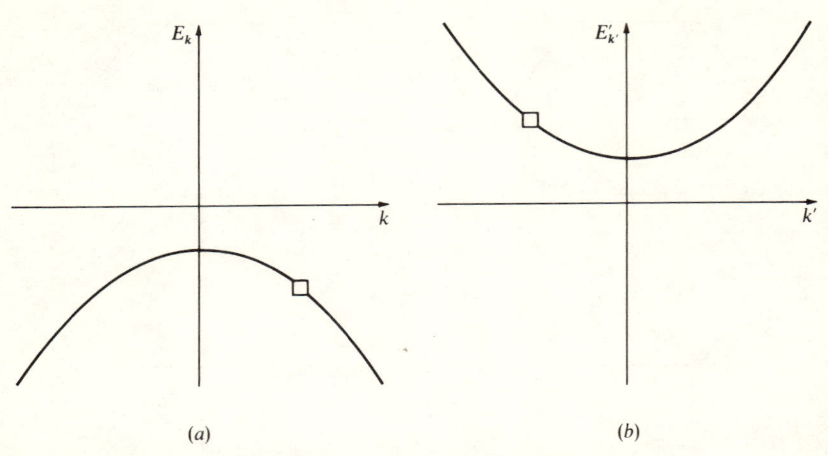

Fig. 2.34 (a) A valence-band maximum is shown with a single state, denoted by the square, unoccupied. (b) The same band and state represented by the modified wavenumber and energy which are convenient for describing the dynamics of holes.

since the current carried by the state in question would be $(-e)(1/\hbar)\nabla_k E_k$ if it were occupied.

If we apply fields to the system, all electron states move in wavenumber space according to the formula we wrote before.

$$\hbar\dot{\mathbf{k}} = (-e)\left(\mathscr{E} + \frac{\mathbf{v}}{c} \times \mathbf{H}\right)$$

The unoccupied state of course moves also according to the same formula. Similarly, if we construct packets for each of the states, the velocity of the packet will be given by

$$\mathbf{v} = \frac{1}{\hbar}\nabla_k E_k$$

and a packet associated with the unoccupied state will move with this same velocity.

Finally we note that we may define an energy for the removal of this electron as the energy to take the electron from the state $\mathbf{k}$ to some zero-energy reservoir. (We would similarly then define the energy to create an electron state as the energy required to take an electron from this same zero-energy reservoir and put it into an appropriate state in the conduction band.) The total change in energy of the system required to create a hole is given by

$$\Delta E_{\text{tot}} = -E_k$$

The formulas that we have derived governing the behavior of the hole are the same as those for the behavior of an electron except for changes in sign for the current and for the energy required to create the excitation.

We would correctly calculate the behavior of holes in applied fields by using these formulas. However, there is a real advantage in rewriting these formulas in a way that makes the behavior more intuitive. We do this by defining a wavenumber $\mathbf{k}' = -\mathbf{k}$ and we may define an excitation energy $E'_{k'} = -E_k$. In terms of these parameters, which give a perfectly well-defined specification of the system, we may rewrite a momentum $\mathbf{p}'$, the current, and the rate of change of $\mathbf{p}'$.

$$\mathbf{p}' = \hbar\mathbf{k}'$$

Then

$$\mathbf{v} = \frac{1}{\hbar}\nabla_k E = \frac{1}{\hbar}\nabla_{k'} E'$$

$$\mathbf{j} = e\mathbf{v}$$

$$\dot{\mathbf{p}}' = -\hbar\dot{\mathbf{k}} = e\left(\mathscr{E} + \frac{\mathbf{v}}{c} \times \mathbf{H}\right)$$

Thus $E'_{k'}$ is the energy of an excitation that propagates, carries current, and has the dynamics of a positively charged particle with wavenumber $\mathbf{k}'$ and momentum $\mathbf{p}'$. Making this correspondence we have simply modified our coordinate system and described the states as shown in Fig. 2.34*b*.

We note that a hole wavepacket will bend to the right in a magnetic field as would a positively charged particle. This would not be the case if we made a wavepacket for an unoccupied state in a band with positive curvature such as the conduction band. We may note also that if we construct a hole packet it will cause a field in the system like a positively charged particle, though that field must ultimately arise from the ions that compensate the charge of the electron gas. Finally we may note that the energy $E'_{k'}$ may be expanded in wavenumber near the band edge, $E'_{k'} = \hbar^2 k'^2/2m^*$, for an isotropic band, and the effective mass is a positive number.

We have found that the excitations in a semiconductor may be divided into the two distinct categories, electrons and holes, and that each behaves in a simple and intuitive manner much as electrons and positrons. We may note that when an electron in the conduction band drops back to the valence band, this corresponds to the annihilation of an electron-hole pair with the release of an energy equal to the sum of the two excitation energies as we have defined them above. When we consider the transport properties of semiconductors we will make full use of this simple intuitive description of the system.

In spite of the complexity of the semiconducting bands, for many properties we may restrict our attention to the band edges where the simple intuitive picture given above obtains. For properties that require knowledge of the bands away from the band extrema we must go to the results of a band calculation or to the approximate description to be given in Sec. 6.3 of Chap. IV.

6.3 Semimetals As we suggested at the beginning, much of what is said about semiconductor bands applies also to semimetals. A nearly-free-electron approach cannot be used directly since almost all of the free-electron Fermi surface disappears into zone faces and several OPWs would be required to describe a single state.

Most electronic properties, however, are determined entirely by states at the Fermi energy, and in semimetals the Fermi energy lies very close to the conduction- and valence-band edges, within 0.02 ev in bismuth. Thus the bands may be expanded around the extremal points and an effective-mass theory is applicable though deviations from parabolicity tend to be somewhat larger than in most semiconductors.

At low temperatures all states in the conduction band which lie below the Fermi energy are occupied and well-defined conduction-band Fermi surfaces exist. Within the effective-mass approximation these would be ellipsoidal surfaces, and this is found to give a good approximation to the observed shapes. Similarly there are ellipsoidal hole surfaces and the total volume of hole surface must equal that of electron surface. The semimetal behaves as a metal with tiny slivers of Fermi surface; the various experimental techniques for their study, described in Sec. 5.6, are applicable. Many such studies have been made, particularly on bismuth.[1]

7. *Insulator Bands*

As we have indicated in the preceding section, insulators are distinguished from semiconductors by their larger band gaps. We expect a totally negligible number of electron or hole excitations to be generated thermally, though they may be generated by shining light on the insulator.

The prototype insulator is an ionic crystal such as sodium chloride. We expect these to be insulating since they contain full bands and, though the electrons are accelerated by applied fields as we have described before,

[1] For a series of such studies, see W. A. Harrison and M. B. Webb (eds.), "The Fermi Surface," Wiley, New York, 1960.

the same states remain occupied; therefore no current flows. We may say, using the framework of the theory of simple metals, that electrons are being accelerated by the fields and by diffraction such that the total acceleration vanishes.

It now seems likely, however, that this description of insulators may be misleading. A band calculation for an ionic crystal, for a molecular crystal, or for a rare gas crystal (which is generally considered a molecular crystal) leads to very narrow valence bands which are full in the ground state. When an electron is removed from such a narrow band very little energy is required to "reorganize" the remaining states in order to lower the electrostatic energy of the hole. The entire band description may break down for the valence electrons and a description in terms of localized single-ion states may be more appropriate.

A survey[1] of properties for which intuitive theories lead to different behavior depending upon whether one envisages localized states or band states suggests a rather sharp division between the two types of materials. Compounds can be ordered on the basis of electronegativity difference of the components, or other measures of "ionicity." Those which lie below a certain value of ionicity are much more naturally interpreted in terms of valence bands; those above, in terms of localized valence states.[2] Labeling compounds by the columns in the periodic table of their components, the survey found that almost all 3-5 compounds (e.g. InSb) should be envisaged with valence bands; all 1-7 compounds (e.g., NaCl) and all compounds containing oxygen were considered to have localized states. The 2-6 compounds were divided between the two depending upon their ionicity. In all of these materials the distinction concerns the valence bands; any conduction electrons are thought to occupy propagating Bloch states.

We will nevertheless proceed with a band description of both electrons and holes in insulators and then in Secs. 7.3, 7.4, and 7.5 see how this picture should be modified for valence bands which are sufficiently narrow. We begin by describing the tight-binding method for constructing energy bands. This method was mentioned briefly for the one-dimensional case in Sec. 5.3 of Chap. I. We noted then that if the states may be described approximately by atomic states localized on individual atoms, the corresponding energy bands will be very narrow. Correspondingly the effective masses will be very large. This behavior is characteristic of calculated valence bands in ionic crystals, though the conduction band tends to be rather broad.

7.1 The tight-binding approximation Perhaps the *tight-binding method* is the simplest approach conceptually for describing energy bands. For this reason, it has probably been used much more than its reliability would

[1] S. Kurtin, T. C. McGill, and C. A. Mead, *Phys. Rev. Letters*, **22**: 1433 (1969).

[2] More recent studies (e.g., M. Schlüter, *Phys. Rev.*, **B17**: 5044 (1978)) suggest that there is no such sharp distinction.

indicate. Nonetheless it is a great conceptual aid and the terminology that derives from it is widely used particularly in discussing insulators.

The approach is very old and has been used in chemistry for some time under the name of *linear combination of atomic orbitals*, or LCAO method. The idea is that we know the wavefunctions and energies for electron states in the free atom. If we bring a collection of atoms together, whether this be in a molecule or in a solid, the description of the states when the atoms just begin to overlap should be obtainable by correcting the free-atom states. In particular, we write a wavefunction for the system as a linear combination of atomic orbitals centered on the individual atoms. In order to obtain an exact expansion of the wavefunction, it would of course be necessary to include in the expansion those states which correspond to dissociated electrons, i.e., positive-energy states. However, in the LCAO method we use only bound states in the expansion and frequently only those of a single term value. For simplicity we will begin by thinking about an elemental crystal—only one atomic type present. The formation of the state is illustrated in Fig. 2.35.

We write the normalized atomic wavefunction for energy and angular-momentum quantum numbers t centered on the jth atom as $\psi_t(\mathbf{r} - \mathbf{r}_j)$ and focus our attention on the occupied levels of highest energy. We will seek a state that can be approximated as a linear combination of a single atomic state; for example, $3s$ for sodium. If the level is degenerate in the atomic case it would be appropriate to take linear combinations of all the degenerate states; for example, $3p$ for aluminum. (This expansion is in fact

Fig. 2.35 *The formation of tight-binding states. In* (*a*) *the states are sufficiently separated that local atomic states, each with arbitrary phase, may be taken as eigenstates. In* (*b*) *the atomic states overlap but if they are superimposed with coherent phases, approximate eigenstates are obtained; shown is the real part of such a state. The sixfold degeneracy of* (*a*) *is lifted as shown schematically to the right.*

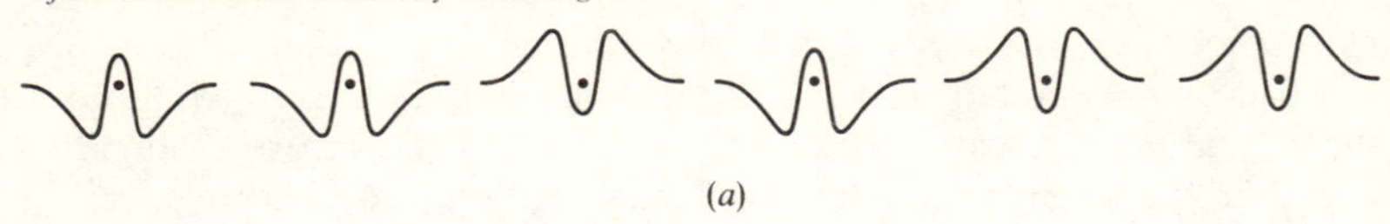

(*a*)

E_t E_k

(*b*)

not a very good one for metals but becomes more appropriate when these metals are part of an ionic compound.) We seek an expansion of the form

$$\psi = \sum_j c_j \psi_t(\mathbf{r} - \mathbf{r}_j)$$

Now we have shown that the eigenstates for a periodic system can be constructed to have translational periodicity given by

$$T\psi(\mathbf{r}) = e^{-ik \cdot T} \psi(\mathbf{r})$$

from which it follows that c_j will be proportional to $e^{ik \cdot r_j}$ for a state transforming according to the wavenumber **k**. Thus we will write the wavefunction

$$\psi_{k,t} = \frac{1}{\sqrt{N}} \sum_j \psi_t(\mathbf{r} - \mathbf{r}_j)\, e^{ik \cdot r_j} \tag{2.38}$$

where N is the number of atoms present. Note that if the atomic wavefunctions did not overlap each other, this wavefunction would be normalized. Because they do overlap, the wavefunction as written is not accurately normalized.

Note that the use of translational periodicity has brought us immediately to the band picture. The wavenumbers are restricted to the Brillouin zone since the specification of any wavenumber outside the Brillouin zone gives precisely the same coefficients, and therefore the same wavefunction, as a state which differs by a lattice wavenumber but which lies within the Brillouin zone. Furthermore, we have been led to one energy band arising from each atomic state. Thus, using this approach we can associate each energy band with the atomic state from which it derives (in addition to numbering the bands according to energy as we have done before). If a more accurate description of the states is sought by including terms in more than one atomic wavefunction, the distinction becomes less clear, but frequently it is still meaningful to associate a particular atomic state with each band.

Having assumed this wavefunction we may now directly compute the expectation value of the energy

$$\langle E \rangle = \frac{\int \psi^*_{k,t}(\mathbf{r})[-(\hbar^2/2m)\nabla^2 + V(\mathbf{r})]\psi_{k,t}(\mathbf{r})\, d\tau}{\int \psi^*_{k,t}(\mathbf{r})\psi_{k,t}(\mathbf{r})\, d\tau}$$

We again write the potential as a superposition of potentials centered on the individual atoms, $V(\mathbf{r}) = \sum_j v(\mathbf{r} - \mathbf{r}_j)$, which in this case is quite consistent with our assumed wavefunctions. Furthermore we may write

$$\left[-\frac{\hbar^2}{2m}\nabla^2 + v(\mathbf{r} - \mathbf{r}_j)\right]\psi_t(\mathbf{r} - \mathbf{r}_j) = \epsilon_t \psi_t(\mathbf{r} - \mathbf{r}_j)$$

where ϵ_t is the energy eigenvalue for the corresponding state in the free atom. Then

$$\left(-\frac{\hbar^2}{2m}\nabla^2 + V\right)\psi_{k,t} = \frac{1}{\sqrt{N}} \sum_j \left[\epsilon_t + \sum_{i\neq j} v(\mathbf{r} - \mathbf{r}_i)\right]\psi_t(\mathbf{r} - \mathbf{r}_j)e^{ik\cdot r_j}$$

and the expectation value of the energy becomes

$$\langle E\rangle = \epsilon_t + \frac{1/N \sum_{j,l} e^{ik\cdot(r_j - r_l)} \int \psi_t^*(\mathbf{r} - \mathbf{r}_l) \sum_{i\neq j} v(\mathbf{r} - \mathbf{r}_i)\psi_t(\mathbf{r} - \mathbf{r}_j)\,d\tau}{\int \psi_{k,t}^* \psi_{k,t}\,d\tau} \tag{2.39}$$

Notice that the corrections to the free-atom values will be small if the neighboring wavefunctions and potentials do not overlap greatly. In the tight-binding method we treat the overlap of the wavefunctions as a small correction. To lowest order in this overlap, then, we may set the denominator in Eq. (2.39) equal to 1. Furthermore it is reasonable to expect that three-center integrals will be small compared to two-center integrals and to drop them from the summation. Then we will have terms only for $l = i$ or $l = j$. This may not always be a very good approximation but it is one customarily made in the tight-binding method.

Let us consider first the terms for which $l = j$.

$$\frac{1}{N}\sum_j \int \psi_t^*(\mathbf{r} - \mathbf{r}_j) \sum_{i\neq j} v(\mathbf{r} - \mathbf{r}_i)\psi_t(\mathbf{r} - \mathbf{r}_j)\,d\tau$$

This is simply the expectation value (based upon our atomic wavefunctions) at each ion of the potential arising from all of the neighbors. For a perfect crystal this value will be independent of j so that the sum over j divided by N is equal to the value taken by one term

$$\int \psi_t^*(\mathbf{r} - \mathbf{r}_j)\psi_t(\mathbf{r} - \mathbf{r}_j) \sum_{i\neq j} v(\mathbf{r} - \mathbf{r}_i)\,d\tau$$

The potential due to each atom is attractive and therefore this is a negative term which is independent of the wavenumber of the state being considered. This term contributes to the binding of the crystal and is important in that regard but has little bearing on the energy-band structure itself.

We consider next the terms for which $l = i$.

$$\frac{1}{N}\sum_{\substack{i\neq j\\ j}} e^{ik\cdot(r_j - r_i)} \int \psi_t^*(\mathbf{r} - \mathbf{r}_i)v(\mathbf{r} - \mathbf{r}_i)\psi_t(\mathbf{r} - \mathbf{r}_j)\,d\tau$$

Again these terms are independent of j so we can replace the sum over j by a factor of N and take our origin at a given r_j.

$$\sum_{r_i \neq 0} e^{-ik\cdot r_i} \int \psi_i^*(\mathbf{r} - \mathbf{r}_i) v(\mathbf{r} - \mathbf{r}_i) \psi_t(\mathbf{r}) \, d\tau$$

These are the k-dependent terms that give rise to interesting band-structure effects. In the tight-binding approximation it is frequently assumed that the nearest-neighbor contributions dominate and only these are included.

Let us consider now the simplest case, that of an energy band arising from atomic s states. Furthermore we will simplify the problem by letting the atoms lie in a simple cubic structure. We look in particular at the wavenumber-dependent terms arising from nearest-neighbor overlaps. The integral in each term is given by

$$\lambda = \int \psi_i^*(\mathbf{r} - \mathbf{r}_i) v(\mathbf{r} - \mathbf{r}_i) \psi_t(\mathbf{r}) \, d\tau$$

and for s states takes on the same value for all neighbors. The value of λ may be obtained from the atomic wavefunctions, the atomic potential, and the near-neighbor distance. Then the wavenumber-dependent correction takes the form

$$\lambda \sum_{\substack{\text{nearest} \\ \text{neighbors}}} e^{-ik\cdot r_i} = 2\lambda(\cos k_x a + \cos k_y a + \cos k_z a)$$

Note that if we included contributions for more distant neighbors this would give us additional corrections to the energy bands which would also be sums of cosines. The important part of the integral will ordinarily occur beyond the last node in both of the atomic wavefunctions. Thus, since the potential is attractive, λ will ordinarily be negative. In Fig. 2.36 we show this result plotted along three symmetry lines in the simple-cubic Brillouin zone. Note that the origin of energy is taken to be the atomic-energy eigenvalue shifted by the wavenumber-independent contribution to our overlap corrections.

It may be noted that the energy rises quadratically with wavenumber from the center of the zone as we found in the $\mathbf{k} \cdot \mathbf{p}$ method. In this case the effective-mass tensor is isotropic. We may write

$$E = 2\lambda\left(3 - \frac{k^2a^2}{2} + \cdots\right)$$

$$= 6\lambda + \frac{\hbar^2k^2}{2m^*} + \cdots$$

where

$$\frac{m^*}{m} = -\frac{1}{\lambda} \frac{\hbar^2(1/a)^2}{2m}$$

If the overlap of the atomic functions is very small then λ will be small and the effective mass will be large. This result is very plausible physically. If the overlap of the wavefunctions is very small, the electrons are quite immobile. If the overlap is very large then the mass becomes small and the electrons become quite mobile. At the same time, as we approach this mobile condition, the whole tight-binding approach becomes inaccurate. Thus in simple metals, where we know that the electrons behave as quite free, the tight-binding approximation will be a very poor expansion of the wavefunctions. On the other hand, in the ionic crystal the overlap of atomic valence states is very small and atomic orbitals provide a good starting point.

Within the framework of the tight-binding approximation we are able to obtain the effective-mass tensor which is given exactly by the $\mathbf{k} \cdot \mathbf{p}$ method. In addition we obtain the terms of higher order in k. However the two approaches give very different though complementary pictures of the origin of effective masses. The $\mathbf{k} \cdot \mathbf{p}$ method relates effective masses to band gaps and makes clear how tiny masses may arise. The tight-binding method relates effective masses to atomic overlap and makes clear how very large masses arise.

Fig. 2.36 *Tight-binding s bands in a simple-cubic structure computed including only nearest-neighbor overlaps. The horizontal line is at the center of gravity of the band; if there were one electron per atom, this would be the Fermi energy.*

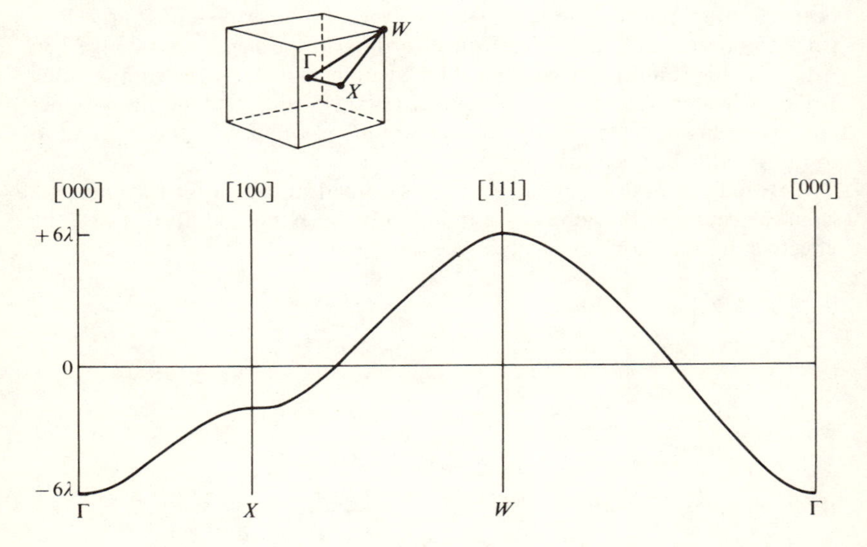

7.2 Bands and binding in ionic crystals We consider for example the electron states in sodium chloride. Imagine first an array of sodium and chlorine atoms arranged as in the sodium chloride structure but with a very large separation between atoms as illustrated in Fig. 2.37. In the ground state both the sodium and chlorine atoms will be neutral; an appreciable energy is required to remove an electron from the sodium and carry it against the electric field of the sodium nucleus to a chlorine atom to produce a chlorine ion. On the other hand, as we decrease the separation between atoms, the energy required becomes smaller simply because the distance over which the electron must be carried is smaller. In addition, by ionizing all sodium atoms at the same time we will gain some work in moving the electron to the chlorine site which is then surrounded entirely by positive sodium ions. Thus, even before the wavefunctions begin to overlap and the states begin to form bands, the more stable electron configuration becomes that of sodium ions and chlorine ions. This is illustrated schematically in the shift of the electron-energy levels as a function of atom separation a shown in Fig. 2.38.

Note that the chlorine ion corresponds to an argon electron configuration. The Cl^- line in the figure represents a $6N$-fold degeneracy

Fig. 2.37 *Sodium and chlorine atoms in a sodium chloride configuration but widely spaced. Net work is required to carry an electron from an Na to a Cl if the spacing is sufficiently large; net work is gained if the spacing is sufficiently small.*

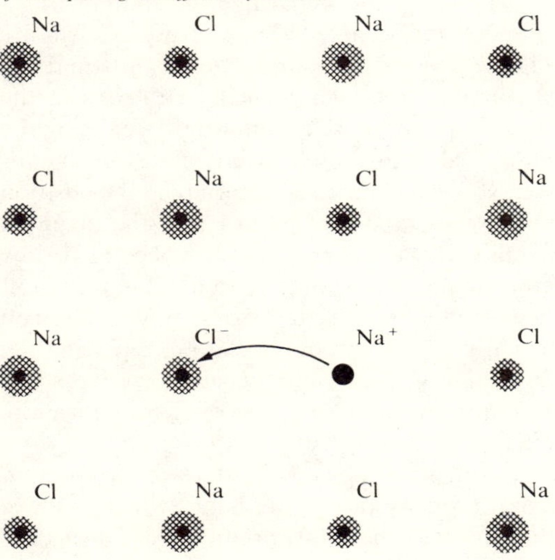

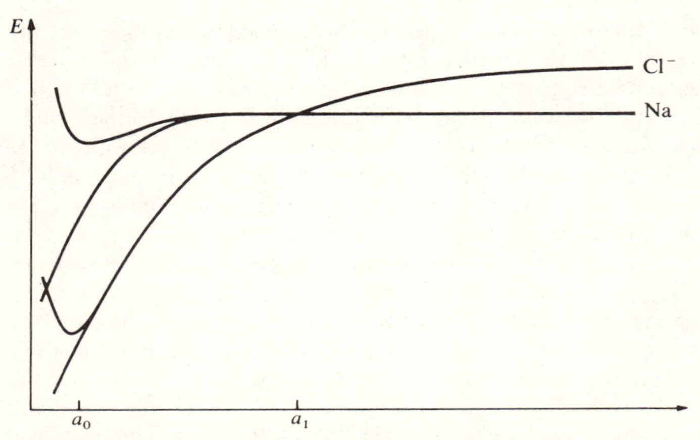

Fig. 2.38 A schematic representation of the formation of bands in NaCl. At spacings less than a_1 the formation of ions is favored. At somewhat smaller spacings the atomic wavefunctions overlap and tight-binding bands are formed. The observed spacing is a_0.

(with N chlorine ions in the crystal) and there are just enough electrons to fill the corresponding states once the sodium valence electron from each atom is transferred to the chlorine.

As a is decreased further, the wavefunctions on the chlorine atoms begin to overlap and the states broaden into bands. These become the valence bands and all states are filled. At the same time the sodium wavefunctions, which now correspond to unoccupied states of the system, are also broadened into bands. These become the conduction bands and are empty in the ground state. Ultimately the energy of the electrons in the chlorine atoms will begin to rise, simply because by sufficiently reducing the volume the kinetic energy associated with the electronic states must ultimately become large. The real crystal will come to equilibrium at the position of minimum total energy. We have marked that point a_0 on the diagram. The binding energy of the crystals is just the energy that has been gained by bringing the free atoms together to form the ionic crystal. This turns out to a good approximation to be just equal to the energy of taking an electron from a free sodium atom, putting it on a free chlorine atom, replacing the sodium and chlorine ions by point positive and negative charges, and bringing them together from infinity to the observed crystal configuration. This is consistent with the sketch given in Fig. 2.38.

The energy band which has arisen from the 3s state in the sodium atom may be approximately described by the s-state band which we have constructed in Fig. 2.36 with, however, the appropriate crystal structure.

In the ionic crystal this (unoccupied) sodium $3s$ band will lie several volts above the filled $3p$ chlorine bands. A measure of this gap can be obtained from optical absorption in sodium chloride. Absorption occurs when electrons are lifted from the valence band to the conduction band. After this excitation we speak of an electron in the conduction band and a hole in the valence band.

The three p states which are degenerate in the free atom split into three bands as found in Prob. 2.20 for the simple cubic crystal. More accurate calculation of either band would mix the s and p states somewhat. Nonetheless a full calculation in an insulator such as sodium chloride would yield narrow valence bands which have predominantly the character of single atomic states on the appropriate set of ions.

7.3 Polarons and self-trapped electrons Even if the energy states are calculated by accurate methods, there are complications in insulators which raise other questions concerning the validity of the results in the real system. In the first place an electron interacts very strongly with the ions in the crystal through an electrostatic interaction. Thus the lattice itself will become distorted by the presence of electrons. We will discuss the system classically here and we will see how a quantum-mechanical treatment can be made when we discuss the electron-phonon interaction in Chapter IV. We will then see how the distortions are described as the virtual emission and reabsorption of lattice vibrations. For our present purposes we may envisage an electron in a sodium chloride conduction band drawing the neighboring sodium ions toward it and pushing the neighboring chlorine ions away. The change in electrostatic energy is linear in such displacements, while the elastic energy is quadratic in the displacement (since the lattice was in equilibrium before the introduction of the extra electron). Thus we can always gain energy by allowing the distortion.

There are two rather distinct ways in which this distortion may occur. If the displacements of the ions are all small and are spread over a large number of ions, as is characteristic for electrons in conduction bands, it is clear that the total energy (including displacement) of an extra electron will be almost independent of position. Thus the electron in the conduction band may move under applied electromagnetic fields and the displacement field will move with it. Such an electron is called a *polaron*. In contrast, if the displacements of the nearest ions are sufficiently large, the lowering of energy may depend significantly upon whether the electron lies between ions (still viewed classically) or at an ion site. This can lead to the binding (or self-trapping) of the electron to a given region of the crystal; the details of such a binding require a quantum-mechanical description.

The most familiar case of self-trapping is that of a hole bound to two adjacent chlorine ions. The ions may move together and, with the hole, form what amounts to a Cl_2^- ion. The motions of the electrons present (or equivalently, the hole) are very fast compared to characteristic frequencies for the ions so that an electron sampling the neighboring environment sees undistorted ions and correspondingly higher energy. This *self-trapped hole* is immobilized by the deformation of the lattice and cannot contribute to electrical conduction.

The self-trapping of an electron or a hole is an interesting case where symmetry arguments and group theory can be misleading. In the band picture we begin with a system with translational periodicity, suggesting a

Fig. 2.39 *Lattice distortions in insulators. In the case above, the displacements vary slowly with position and we may imagine an electron wavepacket spread over many lattice distances moving, with its accompanying distortion, through the crystal; this is a polaron if the distortion is in a polar crystal. Below, the distortion is strongly localized and the packet becomes immobilized at the center of distortion. Such self-trapping is common in insulator valence bands which, because they have small spread in energy, allow localization with only small increase in band energy.*

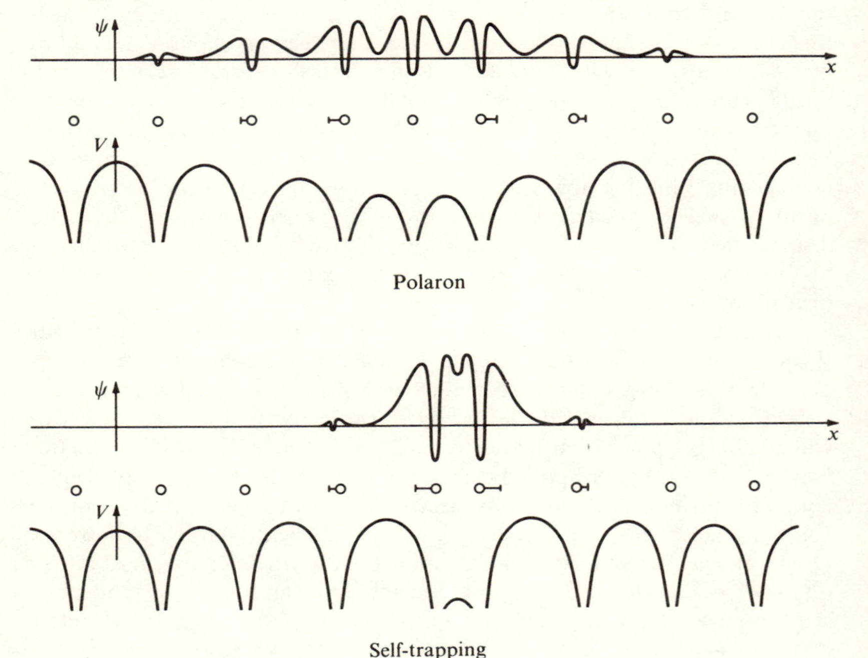

Hamiltonian with that periodicity, which leads to Bloch states that are uniformly spread through the crystal, and a consequent charge density with the initial translational periodicity. This is self-consistent, but it is also possible for the lattice to distort, destroying the periodicity and leading to localized states that maintain the distortion. Similarly a nonperiodicity of purely electronic origin gives rise to antiferromagnetism in some systems. Both are considered cases of *broken symmetry*.

When deformations are sufficiently small and diffuse that the excitation of the system is mobile, we may retain the usual energy-band description. The polaron, however, has lower energy than the pure electronic excitation and therefore a smaller activation energy for formation. Furthermore, the magnitude of this lowering of energy will depend upon the electron velocity and hence its wavenumber. Thus the "band structure" of the excitation is modified by the distortion of the crystal lattice. This is plausible on physical grounds in that the moving electron carries a polarization *dressing* with it and we should expect its dynamics to be modified.

Electrons in all solid state systems are dressed by similar distortions of the lattice. However, only in ionic crystals is the term polaron used. In other systems "dressed electrons" are discussed. The polaron case corresponds to strong coupling with the lattice "field" and requires special methods of analysis. These suggest shifts in effective mass by factors of 2 or 3. In nonpolar semiconductors the coupling is much weaker and the shifts are of little interest. In simple metals the dressing gives rise to the shifts in specific heat mass which we have discussed. For many purposes we may think of the result of dressing of electrons as simply a modification of the band structure for the excitations. We will consider the coupling problem in more detail when we discuss interactions between electrons and phonons.

7.4 The Mott transition*[1] *and molecular solids There is a second difficulty in our treatment of states in insulators. It arises from the use of the self-consistent-field approximation which was fundamental to our formulation of energy bands. This approximation may break down in a fundamental way when the energy bands are sufficiently narrow and the corresponding effective masses sufficiently large. This is characteristic of molecular crystals such as solid CO_2 and rare-gas solids. We will see that these are most appropriately described not in a band picture but as free molecules bound together by van der Waal forces. Similarly the apparent localization of valence states which we discussed in Sec. 7.1 might arise from this breakdown in the band picture.

[1] A recent paper on this subject is N. F. Mott and R. S. Allgaier, *Phys. Status Solidi*, **21**:343 (1967).

We may see the source of the difficulty by considering an extreme tight-binding case. A rare-gas solid would be a suitable example, but the discussion will be simplified by considering a system with one electron per atom. Consider hydrogen atoms placed on a cubic array with such large spacing that the overlap may be neglected. We could write one-electron states that were single-atom atomic functions, $\psi_{1s}(\mathbf{r} - \mathbf{r}_j)$. These are then orthogonal on the different sites because of the large atomic spacing. We could instead take tight-binding states such as in Eq. (2.38). These are also orthogonal, and all have the same energy when overlap is negligible. From a one-electron point of view these are simply a different linear combination of degenerate states and an equally good representation.

However, let us now write the wavefunction for N electrons for the crystal with N sites. The expectation value for the number of electrons on any one atom is the same (one) with either choice of wavefunction; so is the expectation value of any one-electron operator. If on the other hand we ask for the probability of finding at least *two* electrons on a given site, the answer will depend upon which choice of wavefunction we have made. Using atomic wavefunctions, we obtain exactly zero probability. Using tight-binding wavefunctions, we find the probability is significant. Similarly we obtain different expectation values for any two-electron operator.

A particularly important two-electron operator is the electron-electron interaction (which is replaced by a one-electron potential in the self-consistent-field method). The total Hamiltonian for our system of electrons contains the kinetic energy and electron-ion interaction (which are one-electron operators) and the electron-electron interaction which is a two-electron operator. We can ask the question: Which is the best many-electron state for the real system with the proper (as opposed to self-consistent-field) electron-electron interaction? This can be answered using a variational argument and thereby choosing the wavefunction with the lowest expectation value of the Hamiltonian, $\langle\psi|H|\psi\rangle/\langle\psi|\psi\rangle$. Clearly the contributions from the one-electron operators will be the same in this lattice with no overlap and electron-electron interaction will be less with the atomic wavefunction, which prevents two electrons from occupying the same site.

The correct ground state of the many-electron system is obtained with products of atomic one-electron functions, not tight-binding one-electron functions. In this extreme tight-binding limit, the tight-binding approximation breaks down.

If we now, however, decrease the spacing until overlap occurs, we will begin to gain energy from the one-electron terms and the wavefunctions in the Bloch form. Ultimately these will win out and the ground state of the many-electron system will be made of products of Bloch functions, not of atomic functions. Mott has suggested that this transition will be sharp (though this is not established), and it is called the *Mott transition.*

Note that on the low-density side of the Mott transition we cannot define an effective mass. Furthermore, no current can flow except by the ionization of one of the atoms. Finally, the binding of the system comes only from van der Waal's interaction. On the high-density side of the transition the mass may be high but it exists. Current can flow without an activation energy and the conductivity is therefore nonzero. Finally the binding may be thought of as arising from the overlap terms which we considered in the tight-binding discussion.

The electronic states in the rare-gas solids and in the core states in all solids are generally thought to be of the atomic type. In addition, a Mott transition in the valence band may well account for the properties of the strongly ionic insulators discussed at the beginning of Sec. 7.

7.5 Excitons[1] One further complication arises in insulators and semiconductors from the breakdown of the self-consistent-field method. This complication leads to a more familiar phenomenon. When an electron is raised to the conduction band, it leaves behind an unoccupied state in the valence band. This hole in the valence band is positively charged as we found in our discussion of semiconductor bands. Thus the excited electron moves in the field of this positively charged hole. If the crystal is appropriately described by Bloch states, we might imagine the electron and hole orbiting around each other as in positronium or in the hydrogen atom. Such a bound pair is called an *exciton.* It is an intrinsically many-body effect that does not arise in the self-consistent-field approximation where each electron is assumed to see an average potential which has the translational periodicity of the lattice.

It is interesting to carry this picture of an exciton a little further in order to see the magnitude of the binding of an exciton in a typical semiconductor, germanium. Our picture is of a light electron (the geometric mean of the two identical transverse and one longitudinal masses is 0.22) orbiting around a somewhat heavier hole. Since the orbit will be large compared to the interatomic spacing, it is appropriate to reduce the electron-hole interaction by the macroscopic dielectric constant, 16 for germanium. (We see that this is consistent by noting that the Bohr radius of the lowest-lying state is $\hbar^2\kappa/m^*e^2$ if κ is the dielectric constant. Taking $m^* = m/5$, we find a radius of 80 atomic units (42 Å) which is many atomic distances.)

We will see in more detail how the wavefunctions for such bound states are described when we consider impurities in semiconductors. How-

[1] For a comprehensive discussion, see R. S. Knox, "Theory of Excitons," Academic, New York, 1963.

ever, we can already see qualitatively the form of the excitonic states. Because of the large scale of the exciton we may think in terms of wavepackets and define a relative coordinate **r** which is the separation between the electron and the hole. The bound state then will be a hydrogenic wavefunction $\Phi(\mathbf{r})$. In addition we may define a center of mass coordinate **R** and write our eigenstates in the form

$$\Psi_{ex} = \Omega^{-1/2}\,\Phi(\mathbf{r})e^{ik\cdot R}$$

This state has the proper translational symmetry for eigenstates of the perfect crystal. The state $\mathbf{k} = 0$ corresponds to a stationary exciton; states for **k** not equal to 0 correspond to propagating excitons. The stationary $1s$ state is the lowest-energy exciton.

The binding energy is lower than the hydrogen binding energy by $(m^*/m)\kappa^{-2}$, which leads to a binding of only 0.01 ev. In a typical semiconductor excitons exist in principle but will be completely ionized except at very low temperatures. For this reason they tend to be quite unimportant in semiconductors and the band description of electronic states which we gave is appropriate.

In a typical insulator, on the other hand, the masses are large and the dielectric constants are typically of the order of 5. Thus the radius of the lowest-lying state calculated as above is of the order of the interatomic distance, and the use of the dielectric constant becomes inappropriate. We must use a quite different approach in discussing them. This other approach is closely related to our tight-binding picture, and the tightly bound state is called the *Frenkel exciton.*

We consider the simplest case, which is that of the rare-gas solid. The filled state in the rare-gas atom is not seriously distorted when the atoms are combined in a crystal. In this case, as we have indicated, the binding comes not from the overlaps but from a simple van der Waal's interaction. (Incidentally this interaction itself is an intrinsically many-electron effect and is not therefore included in our self-consistent-field approximation.) We may therefore think of the ground state of a rare-gas solid as consisting of individual electrons occupying the atomic orbitals on the individual rare-gas atoms. We might make an excitation of the system by raising an electron to an excited state on a single atom; e.g., raising an electron to a $3s$ orbital on a neon atom. We cannot describe this excitation with a one-electron wavefunction since the absence of a $2p$ orbital on the excited atom is essential. For simplicity we will write the many-electron wavefunction as a product wavefunction although it is only a slight generalization to construct from such product wavefunctions the properly antisymmetrized linear combination. The state with a single electron excited on the jth neon atom may be written

$$\Psi = \psi_j^{(*)} \prod_{i \neq j} \psi_i$$

where $\psi_j^{(*)}$ is the ten-electron wavefunction for the neon atom in an excited configuration at the jth site and the ψ_i are ten-electron neon wavefunctions in the normal configuration at the ith sites. This state does not have the required translational symmetry. However, we may obtain the proper symmetry by taking tight-binding-like combinations of these excited states to obtain an exciton wavefunction given by

$$\Psi_{ex} = N^{-1/2} \sum_j \left(\psi_j^{(*)} \prod_{i \neq j} \psi_i \right) e^{i\mathbf{k}\cdot\mathbf{r}_j}$$

As in the case of excitons in the semiconductor these may be propagating states and the energy will depend upon the wavenumber, though only slightly.

Note that these are linear combinations of atomic wavefunctions which on each atom are the wavefunction for all of the electrons on that atom. In contrast in the tight-binding method we constructed linear combinations of one-electron wavefunctions. Thus if we were to make a wavepacket of tight-binding wavefunctions these would yield a concentration of charge in the region of the packet. On the other hand if we were to take a linear combination of our exciton wavefunctions we would have a localization of energy in the region of the packet but not a localization of charge. The exciton itself therefore cannot carry current but can carry energy through the crystal, whereas the tight-binding one-electron states may carry either or both. Excitons contribute to thermal conductivity but not to electrical conductivity in insulators.

In analogy with the tight-binding result, the energy to form an exciton will be roughly equal to the atomic excitation energy. The energy to form a free electron in the conduction band will be of the order of the ionization energy of the atom. By treating the van der Waal's dipole-dipole interaction between atoms we could obtain dispersion curves for excitons in rare-gas solids in analogy with the tight-binding approach. Excitons in ionic crystals are more akin to those in the rare-gas solid than to those in a semiconductor but their theory is not so straightforward.

Recently there has been interest in the possibility of a new phase in solids called the *excitonic insulator*.[1] We imagine a semiconductor with a small gap. The energy required to make an exciton is equal to the band gap minus the binding energy of the exciton. However, if the gap is sufficiently small the difference may be negative and the system becomes unstable

[1] An extensive treatment of excitonic insulators has been given by D. Jerome, T. M. Rice, and W. Kohn, *Phys. Rev.*, **158**:462 (1967).

against the formation of such excitons. Such an instability is very much like the instability against the formation of Cooper pairs, which will be discussed in connection with the superconducting state; the ground state of the system may be constructed much as the superconducting state is constructed. This new state, the excitonic insulator, is again an insulating state but it may make a phase transition to a semiconducting or semimetallic state as the pressure or temperature is changed. Several systems have been considered which may be excitonic insulators, but none has been established as an excitonic insulator.

7.6 Wannier functions[1] Though the tight-binding method is conceptually very simple, there are serious difficulties when we attempt to use it as a quantitative method. Part of these difficulties arise because the tight-binding states, Eq. (2.38), are not orthogonal to each other. In particular, if we construct two states of the same wavenumber but based upon different atomic states we may easily see that because of overlap terms these states are not orthogonal. This difficulty could be avoided by taking appropriate linear combinations of the atomic orbitals on different atoms such that the overlap integrals would vanish. There is a systematic way to achieve this which is the method of Wannier functions. We define them here as an aside; they will not be used in our subsequent discussion.

These functions are obtained by beginning again with the exact solution for the eigenvalue problem in the crystal. We write such a solution for the nth band as a Bloch function

$$\psi_k{}^{(n)}(\mathbf{r}) = u_k{}^{(n)}(\mathbf{r})e^{ik\cdot r}$$

Note that in this case the $u_k{}^{(n)}$ are normalized in the entire volume rather than in an atomic cell as in Sec. 6. Then we construct a Wannier function associated with the jth atom and the nth band in the form

$$a_n(\mathbf{r} - \mathbf{r}_j) = \frac{1}{\sqrt{N}} \sum_k u_k{}^{(n)}(\mathbf{r})e^{ik\cdot(r - r_j)}$$

or

$$a_n(\mathbf{r}) = \frac{1}{\sqrt{N}} \sum_k u_k{}^{(n)}(\mathbf{r})e^{ik\cdot r} \tag{2.40}$$

where the sum is over all $\mathbf{k}$ in the band. In the second form we have taken our origin of coordinates to lie at an atomic site and noted that $u_k{}^{(n)}(\mathbf{r})$ goes into itself when displaced by a lattice distance. (These Wannier functions

[1] For a more complete discussion, see G. Weinreich, "Solids, Elementary Theory for Advanced Students," pp. 127ff., Wiley, New York, 1965.

are in fact associated with primitive atomic cells but in a simple structure with one atom per cell they may be associated with that one atom.)

We may of course transform back to obtain the eigenstates in terms of Wannier functions.

$$\psi_k^{(n)}(\mathbf{r}) = \frac{1}{\sqrt{N}} \sum_j a_n(\mathbf{r} - \mathbf{r}_j) e^{ik \cdot r_j}$$

These are of the form of tight-binding states with the $a_n(\mathbf{r} - \mathbf{r}_j)$ playing the role of atomic states but, in contrast to the tight-binding method, these solutions are exact and therefore orthogonal. We may also immediately show that Wannier functions centered on different sites, or arising from different bands, are orthogonal to each other, again in contrast to the tight-binding method.

$$\int a_n^*(\mathbf{r} - \mathbf{r}_j) a_m(\mathbf{r} - \mathbf{r}_i)\, d\tau = \frac{1}{N} \sum_{k,k'} e^{-i(k \cdot r_i - k' \cdot r_j)} \int \psi_{k'}^{(n)*}(\mathbf{r}) \psi_k^{(m)}(\mathbf{r})\, d\tau$$
$$= \delta_{ij} \delta_{nm}$$

Wannier functions are localized but not so well localized as atomic wavefunctions. We might attempt to gain some idea of the localization by constructing a Wannier function, first approximating the Bloch functions (which depend on wavenumber) by functions that do not depend on wavenumber. We need then integrate Eq. (2.40) over all **k** in the Brillouin zone. If we replace the Brillouin zone by a sphere of radius k_0 such that $\frac{\Omega}{(2\pi)^3} \frac{4}{3} \pi k_0{}^3 = N$ we may readily compute $a_n(\mathbf{r})$,

$$a_n(\mathbf{r}) = \frac{1}{\sqrt{N}} \frac{\Omega}{(2\pi)^3} u^{(n)}(\mathbf{r}) \int_0^{k_0} d^3k\, e^{ik \cdot r}$$
$$= 3\sqrt{N}\, u^{(n)}(\mathbf{r}) \left(-\frac{\cos k_0 r}{(k_0 r)^2} + \frac{\sin k_0 r}{(k_0 r)^3} \right)$$

where r is measured from the central point of the Wannier function. The resulting Wannier function is proportional to $u^{(n)}(\mathbf{r})$ at small r and at large r decreases as $-\cos k_0 r/(k_0 r)^2$ modulated by $u^{(n)}(\mathbf{r})$.

This represents quite a long-range oscillating wavefunction. The main part of this spreading arises, however, from our approximate treatment of the wavefunction (neglecting the variation of $u_k{}^{(n)}$ with **k**). From our treatment of the translation group we know that the two states at opposite faces of the Brillouin zone which differ by a lattice wavenumber are equivalent. Thus the true eigenstates ψ_k smoothly approach the same function as we approach the two opposing zone faces. By neglecting the variation of $u_k^{(n)}$ with **k** we have introduced a sharp discontinuity in the behavior of the wavefunction at the zone faces. It is this sharpness which has given rise to the long-range oscillations. Accurately determined Wannier functions should be rather well localized to a single atomic cell [Weinreich[1] has shown that they fall off faster than any finite power of $(1/r)$], though some residual spreading into neighboring cells must always occur if the corresponding atomic functions overlap.

For each band we have found one Wannier function associated with a given atomic cell. For a crystal with two atoms per primitive cell, such as silicon, we might wish to take linear combinations of Wannier functions from pairs of bands in order to obtain states associated with individual *atoms*. For silicon we might instead wish to take linear combinations of the four occupied valence bands to obtain the counterpart of "bonding orbitals" of theoretical chemistry, and linear combinations of empty bands to obtain "antibonding orbitals."

It is interesting to note that if we construct a Slater determinant including all Wannier functions from the occupied valence bands, this many-electron wavefunction is mathematically equivalent to a Slater determinant of the corresponding band wavefunctions. Thus if we discuss full bands in perfect crystals it is entirely a matter of choice which basis we take.

If we remove a single electron, there is a distinction. The eigenstates of the system (in the Hartree-Fock scheme) are Slater determinants with a single *band* state omitted. Since the Wannier functions are linear combinations of band states of different energies, they are not themselves energy eigenstates.

If on the other hand we remove a single atom, the state of the system might be better described by omitting the corresponding Wannier functions from the Slater determinant. This would nevertheless be an approximation.

Wannier functions provide us with a means for the development of rigorous theorems which are analogous to the crude theorems we might derive with a tight-binding approximation. However, a good and accurate Wannier function would be extremely difficult to calculate accurately and its structure would be quite complicated.

8. *Impurity States*

It would seem natural at this point to move on to a discussion of transition metals, the fourth category of crystals for which we will discuss energy states. One of the central ideas, however, that will be useful in describing

[1] *Ibid.*, p. 133.

the energy bands in transition metals can be much more directly introduced through a discussion of impurity states. We therefore proceed to that problem.

Our discussion to this point has been directed primarily at crystals with translational symmetry and therefore with electron states that may be written in terms of Bloch functions. Of course any real crystal contains defects and surfaces that destroy this perfect periodicity. It will be interesting to see the change in the electron states in the presence of such defects.

8.1 Tight-binding description The simplest situation is that which occurs in an insulator when one of the ions present is replaced by an impurity ion. Then we might reasonably expect to proceed with a tight-binding approach.

Let us first assume that the atomic state in the impurity ion differs in energy sufficiently from the states of the solvent ions that we expect the impurities state to lie between the tight-binding bands of the pure material. Then in the spirit of the tight-binding approximation we would seek a state associated with the impurity which was simply the atomic wavefunction associated with that impurity as a free ion. The energy of that state, however, will be shifted by the presence of the crystal in which it is embedded. The energy will be shifted by the potentials arising from neighboring ions, since these potentials will overlap the impurity ion. This shift may be obtained in just the same way that it was in the tight-binding method. We note further that if the atomic state were degenerate (e.g., if it were an atomic *d* state) then this degeneracy may be partially lifted by the anisotropic potential arising from the neighbors. This is simply the crystal-field splitting that we dealt with using group theory in Chap. I; the symmetry alone told us what the splitting would be. In order to calculate the magnitude of the splitting it is necessary of course to use expressions for the actual potential. In an ionic crystal it is a reasonable approximation to replace the potential arising from the neighboring ions by simple coulomb potentials arising from the net charge of these ions.

We write the potential arising from the neighbors as

$$V_{\text{cf}}(\mathbf{r}) = \sum_{j \neq 0} (\pm)_i \frac{e^2}{|\mathbf{r}_j - \mathbf{r}|}$$

where the sign of each term depends upon the sign of the ion and $\mathbf{r}_j$ is measured from the impurity site. This is an approximation to the *crystal field* seen by the impurity. If the impurity states in question arise from atomic *d* states of energy E_d, we may write each impurity state ψ_n as a linear combination of the five degenerate atomic states,

$$\psi_n = \sum_m a_{nm} \psi_d^{(m)}(\mathbf{r}) \tag{2.41}$$

Substituting this in the Schroedinger equation

$$\left[-\frac{\hbar^2}{2m}\nabla^2 + V_d(r) + V_{cf}(\mathbf{r})\right]\psi_n = E_n\psi_n$$

where $V_d(r)$ is the free-atom potential of the impurity, we multiply on the left by $\psi_d^{(m')*}(\mathbf{r})$ and integrate over volume to obtain

$$\sum_m H_{m'm}a_{nm} = E_n a_{nm'}$$

with a Hamiltonian matrix

$$H_{m'm} = E_d\,\delta_{m'm} + \langle m'|V_{cf}|m\rangle$$

Diagonalization of this matrix gives the energy levels in the crystal.

This calculation could be improved by allowing more terms in the expansion of ψ_n given in Eq. (2.41). In particular orbitals on the neighboring atoms might be allowed. This would simply increase the size of the Hamiltonian matrix, with off-diagonal elements which are precisely of the form of the overlap integrals that entered the tight-binding method,

$$O_{jn} = \langle j|V_d(r) + V_j(r)|n\rangle$$

where V_j is the potential associated with the jth neighbor and $\psi_j = |j\rangle$ is the corresponding atomic orbital with an energy which we write E_j. If these corrections are indeed small, they might be incorporated in perturbation theory, rather than by diagonalizing a larger matrix. Then the impurity state becomes

$$\psi_n' = \psi_n + \sum_j \frac{O_{jn}}{E_n - E_j}\psi_j$$

representing an impurity state that extends over the neighbors. This correction will give an additional shift in the energy E_n which is second order in the overlap integrals.

The expansion is valid only if $O_{jn} \ll E_n - E_j$, which also gives the criterion for the one-orbital impurity state being approximately correct. When the unperturbed impurity level is close to the tight-binding levels of the insulator (close on the scale of the overlap integrals) this approach breaks down and we must return to the large matrix. This becomes particularly true when the impurity level lies within a band in the insulator solvent. Then the impurity state should be regarded as a resonant state, as will be discussed in Sec. 8.5.

8.2 Donor and acceptor levels in semiconductors We consider first the simplest type of impurity in a semiconductor. Consider, for example, an arsenic atom substituted for a germanium atom in the germanium crystal.

As a crudest approximation we might neglect the difference between the potential due to the arsenic atom and that due to a germanium atom. Then the energy-band structure would be unchanged by the substitution. Arsenic has one more electron than germanium so there would be one additional electron that would necessarily go into the conduction band. However, the potentials are not the same; the arsenic ion has one more nuclear charge so there would be an additional positive charge localized at the arsenic. This recalls the picture we developed for the exciton in germanium; in fact the binding of the electron to the positive charge is essentially the same in the two cases.

In addition the core states in arsenic are slightly different from those in germanium and therefore the interaction between electron and core is slightly different. Precise calculation of the binding of this additional electron to the arsenic would include also this small difference in core potential, but the main difference remains the coulomb potential arising from the extra positive charge. As in the exciton case, we find a bound state for the electron with a binding energy of about 0.01 ev.

Again, at all but the lowest temperatures, these states will be ionized and our initial picture of one electron added to the conduction band for each arsenic substitution is quite close to the truth. For this reason arsenic, or in fact any atom in the fifth column of the periodic table, when substituted for an element in a column-4 semiconductor, is called a *donor*. It donates an electron to the conduction band. We see that by *doping* a semiconductor with donors we introduce carriers which then give rise to electrical conductivity. By varying the doping we may vary the concentration of electrons in the conduction band. This of course gives rise to the flexibility of semiconductors and their usefulness in electronic devices.

The same argument may be carried through for the substitution of a gallium atom for a germanium atom. We then have one less nuclear charge on the gallium site than on the germanium and one hole in the valence band. We may carry through the same argument to find that the hole may be bound to the gallium site by a very small energy, a few hundredths of an electron volt. At ordinary temperatures these bound-hole states will be entirely ionized. Thus the column-3 elements are called *acceptors* in column-4 semiconductors. Each gallium atom accepts an electron from the valence band and therefore provides a hole. By doping a semiconductor with an acceptor we introduce holes which also provide conductivity; we can vary the concentration of carriers by varying the doping.

A semiconductor which has been doped with donors is called an *n-type* semiconductor; the carriers are negatively charged. A semiconductor which has been doped with acceptors is called a *p-type* semiconductor; the carriers have positive charge. Semiconductors with equal numbers of donors and acceptors (or with a negligible number of either) have equal numbers of positive and negative carriers and are called *intrinsic* semiconductors. The properties are determined by the parent material rather than by impurities. We will return to the properties of these different types of semiconductors at various points in our discussion.

This treatment, as we indicated in our discussion of excitons, is meaningful only if the electron is very weakly bound to the center so that its orbit extends many lattice distances. If we were to substitute an element from column 2 or column 6 of the periodic table and begin with the same approach, we note that the orbit's size is reduced by a factor of 2. The use of the dielectric function becomes more suspect and our entire treatment is in some trouble.

Note that the bound-electron levels (donor levels) have energy very slightly below those for the free electron in the conduction band. These are conventionally represented schematically as shown in Fig. 2.40. Electron energy is customarily plotted as the ordinate; the abscissa represents some spatial distance. The conduction bands are represented as a band extending throughout space while the donor levels are represented by short lines lying just below the conduction band. Similarly the acceptor levels are drawn as short lines lying just above the filled valence band. Such diagrams will be useful in discussions of semiconductor systems in Sec. 3 of Chap. III.

We could represent states due to other impurities similarly. The

Fig. 2.40 *A conventional diagram of electron states in a semiconductor. Six localized donor levels are shown, four of which are occupied and two of which are unoccupied. The two remaining electrons are in the conduction band.*

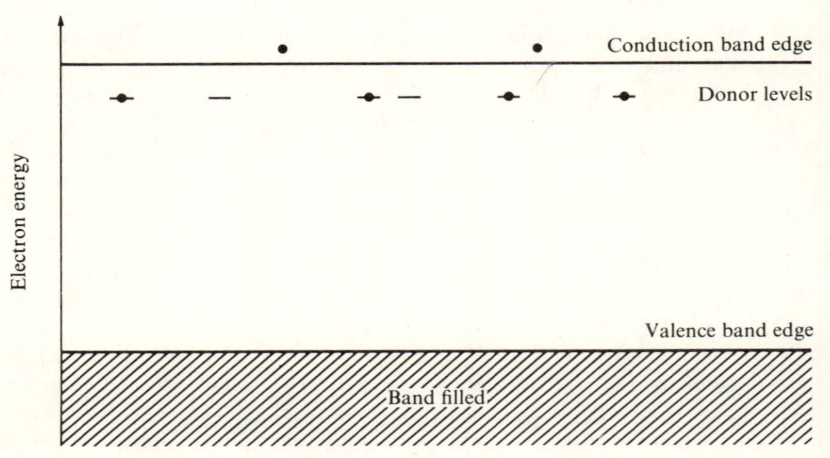

corresponding levels may lie deep within the forbidden region between the valence and conduction bands and may trap electrons that would otherwise be free, thereby removing them from the conduction process. Such levels are called *electron traps*. To treat them properly it is desirable to seek a more direct relationship between these localized states and energy bands. We will do this next.

8.3 Quantum theory of surface states and impurity states We have discussed impurity states in semiconductors as electrons in Bohr orbits around the impurity atom. The corresponding wavefunction is of course one that decays exponentially away from the impurity. It is reasonable to seek such exponentially decaying states which are close to a band minimum using the $\mathbf{k} \cdot \mathbf{p}$ method just as we did for small-wavenumber states. We will do this first for states that decay in one direction only. We seek a wavefunction which can be written as a Bloch function having the full periodicity of the lattice modulated by an exponential function $e^{-\mu \cdot r}$. The treatment proceeds just as in the usual $\mathbf{k} \cdot \mathbf{p}$ treatment. We now find every $\mathbf{k}$ replaced by $i\boldsymbol{\mu}$. This is simply an extension of the energy $E(\mathbf{k})$ into the plane of complex $\mathbf{k}$.

The states that we have obtained decay in only one direction. They are solutions of the Schroedinger equation in the pure material but do not satisfy periodic boundary conditions and therefore were not obtained in the band calculation. Also they are not normalizable in an infinite system since they grow exponentially in one direction. They may be valid solutions, however, in a semi-infinite system or, more particularly, in the neighborhood of a crystal surface. If the potential at the surface of the crystal is such as to allow it (there must be a region of positive kinetic energy to connect the wavefunctions decaying into the crystal and into free space), we may expect such *surface states* to occur; they are frequently called *Tamm* states. They arise in metals and insulators as well as in semiconductors. The energies at which they occur are of course sensitive to the details of the potential at the surface.

We note that the $\mathbf{k} \cdot \mathbf{p}$ calculation might also be carried out (again with no approximation except the expansion in $\mathbf{k}$) for wavenumbers with one component imaginary and two components real. When surface states such as we described above occur there will in addition be surface states decaying exponentially into the crystal but propagating parallel to the surface. Such states may carry surface currents but there will of course be no component of current perpendicular to the crystal surface.

We noted that the generalization of the $\mathbf{k} \cdot \mathbf{p}$ formula into the forbidden band was equivalent to an analytic continuation of the parabolic energy band into complex wavenumber space. This analytic continuation of the bands remains appropriate even when we cannot assume that the energy bands are parabolic; i.e., as we move further into the forbidden energy region and must therefore consider higher terms in a $\mathbf{k} \cdot \mathbf{p}$ expansion. The validity of such analytic continuation follows from the validity of the $\mathbf{k} \cdot \mathbf{p}$ expansion.

We note that with analytic continuation we can consider states which lie deep within the forbidden band as well as those which lie near the band edge. A surface state lying near the band edge will decay with distance from the surface only very slowly. As we move deeper into the forbidden energy region the decay will become more and more steep and the state correspondingly more and more localized. As we move still further down in energy to the neighborhood of the valence band the decay must again become more gradual and will correspond more closely to a decaying valence-band state. For a given analytic form for the energy bands we may span the entire forbidden energy region. This may be illustrated for the simple case of bands given by $E = \pm\sqrt{\Delta^2 + \alpha k^2}$; see Prob. 2.22.

The continuation into the complex wavenumber plane has brought us very close to the effective-mass approach that we discussed in Sec. 6.1. There we wrote for the perfect crystal an effective Hamiltonian given by

$$H = -\frac{\hbar^2}{2}\nabla\overleftrightarrow{\left(\frac{1}{m^*}\right)}\nabla + E_0(\mathbf{r}) + V(\mathbf{r})$$

which for E_0 constant and $V(\mathbf{r})$ equal to zero will have plane-wave eigenstates for a positive kinetic energy and eigenvalues equal to the true band eigenvalues; it will also yield the decaying surface states that we discussed above. Finally it can give more general negative energy solutions which decay, for example, in all directions.

We could imagine obtaining impurity states by matching the corresponding decaying states to a solution in the neighborhood of the impurity, but this does not appear to be a practicable approach in most cases. Instead, the difference between the impurity potential and that of the host is simply added to the Hamiltonian as $V(\mathbf{r})$ in order to obtain approximate impurity states. This $V(\mathbf{r})$ should be something akin to the difference in pseudopotential rather than in true potential if the effective Hamiltonian above is to be used. If, for example, the impurity were of a lower row in the periodic table than that of the solvent, the difference in true potential would be so strong that an extra node or nodes would be introduced in the state at the impurity site and such nodes could not be reasonably treated in an effective-mass approximation. The states obtained are in any case only the envelopes of the true states just as the pseudowavefunction is only an envelope for the true wavefunction.

It is interesting to apply this approach, for example, to donor states in silicon or germanium, taking the extra potential due to the impurity to be simply the coulomb potential. If the effective mass were isotropic the coulomb potential would lead simply to hydrogenlike pseudowavefunctions. The lowest-lying donor state would be of the form $e^{-\mu r}$. (To obtain the approximate true wavefunction, at least outside the impurity cell, we could multiply this envelope function by the true conduction-band wavefunction at the band minimum.) For an anisotropic mass we may obtain an approximate eigenstate by taking a solution of the form

$$\varphi \sim e^{-\mu_1|z|}\, e^{-\mu_2(x^2+y^2)^{1/2}}$$

where the z axis is taken along the longitudinal axis of the band minimum.[1] Note that this form is necessarily approximate. Even for an isotropic band it cannot give the correct isotropic solution, $e^{-\mu r}$. Furthermore it contains an unphysical cusp over the entire $z = 0$ plane. However, it may be meaningful in a variational calculation with the μ_1 and μ_2 taken as variational parameters. In germanium this yields a pancake-shaped pseudowavefunction with ratio of axes of about 3:1. The pseudowavefunction extends far out in the directions for which the mass is small.

In germanium there are four minima in the conduction band and therefore four equivalent impurity states associated with the four valleys that are degenerate. If we look more carefully we find that these degeneracies are split by deviations from the effective-mass theory. By symmetry it can be shown that they are split into a single symmetric nondegenerate state and threefold degenerate p-like states. The ground state turns out to be the symmetric state which we might write

$$\varphi = \tfrac{1}{2}(\varphi_{\bar{1}11} + \varphi_{1\bar{1}1} + \varphi_{11\bar{1}} + \varphi_{111})$$

where the subscripts indicate the direction of the band minimum from which the individual states derive. The resulting pseudowavefunction is a superposition of four pancakes oriented perpendicular to the four [111] directions.

Our discussion of impurity states so far has envisaged single isolated impurities. As long as the density of impurities is sufficiently low that the impurity state pseudowavefunctions do not overlap each other, this description is appropriate. However, once they do overlap we expect the degenerate impurity states to broaden out into bands. At low concentrations these will represent widely spaced atomiclike states and they can very reasonably be treated in a tight-binding approximation. At low concentrations we will expect to lie at the low-density side of the Mott transition and that the states will not be conducting. Any current carried by such impurity states must go by a hopping mechanism. At higher concentrations we expect to be on the high-density side of the Mott transition and to have ordinary conduction

[1] This approach was introduced by H. Fritzsche, *Phys. Rev.*, **125**: 1560 (1962).

within the *impurity band*. At sufficiently high densities impurity bands will overlap the adjacent conduction band and the two will merge into a single band.

It is interesting to consider the intermediate-density range in germanium at sufficiently low temperatures that the impurity states are occupied. We note that the individual impurity states are quite anisotropic and that the overlap will be predominantly in directions perpendicular to the [111] directions. If we now shear the lattice, we lift the initial degeneracy of the states associated with the different valleys and change the linear combinations that form the ground state. Thus, by lowering the energy of the states associated with the valleys in a [111] direction we may form a ground state which is predominantly made from the pancake perpendicular to that direction. Conductivity along that same [111] direction becomes drastically reduced by the elimination of overlap in that direction. Such modifications of the conduction by distortion of the crystal are called *piezoconductivity* effects. They can be extremely large and can lead to very large anisotropies, factors of the order of 10^7 with attainable stresses.[1]

8.4 Phase-shift analysis Our discussion of impurity states to this point has dealt entirely with states that lie within the forbidden band of the pure material. The presence of an impurity, however, will modify all of the states in the system by destroying the translational periodicity. We will see that in some cases an impurity potential will not be sufficiently deep to pull a level out of one of the pure material bands. We may proceed nevertheless, at least in principle, to construct states by solving the Schroedinger equation within the region of impurity potential and then matching to a solution for the pure material outside of the impurity potential. This is most conveniently done using phase shifts.

This analysis will be of most interest in metals and it may be particularly conveniently formulated in terms of the pseudopotential approach. In addition the results will shed new light on the nature of the pseudopotential. Thus we focus attention upon the simple metals and return again to the pseudopotential equation of Sec. 5.

$$-\frac{\hbar^2}{2m}\nabla^2\varphi + W\varphi = E\varphi \tag{2.42}$$

In our previous discussion of the pseudopotential we took advantage of the translational periodicity of the lattice and periodic-boundary conditions to obtain zero-order solutions which were propagating plane waves. When we consider a system with an impurity it becomes more convenient to consider a spherical crystal and to apply a vanishing boundary condition

[1] *Ibid.*

on the wavefunction at the surface. We will eventually introduce a single impurity at the center of this sphere.

For a spherically symmetric system we may of course factor the pseudowavefunction into an angular and a radial part and obtain angular-momentum quantum numbers l and m from the solution of the angular equation. The radial equation becomes

$$-\frac{\hbar^2}{2m}\frac{1}{r^2}\frac{\partial}{\partial r}r^2\frac{\partial}{\partial r}R_l + W_l(r)R_l + \frac{\hbar^2}{2m}\frac{l(l+1)}{r^2}R_l = ER_l \tag{2.43}$$

We have introduced a subscript l on W since, because of the projection operator, the pseudopotential depends upon the angular-momentum quantum number. It is most convenient mathematically to obtain an equation for $P_l(r)$ which is related to $R_l(r)$ by

$$P_l(r) = rR_l(r)$$

By direct substitution into Eq. (2.43) we obtain

$$-\frac{\hbar^2}{2m}\frac{\partial^2}{\partial r^2}P_l + W_l'(r)P_l + \frac{\hbar^2}{2m}\frac{l(l+1)}{r^2}P_l = EP_l \tag{2.44}$$

where $W_l'(r) = rW_l(r)(1/r)$. (This complication again arises from the operator nature of W. If W were a simple potential then $W_l'(r)$ would be identical to $W_l(r)$.)

P_l may also be preferable to R_l for intuitive arguments and for that reason we will use it in our illustrative curves. Eq. (2.44) is a one-dimensional pseudopotential equation with a centrifugal potential and vanishing-boundary condition at $r = 0$ whereas the kinetic-energy term in Eq. (2.43) is complicated. Furthermore the probability of an electron lying in an interval dr is proportional to $R_l^2r^2\,dr = P_l^2\,dr$ so that in some sense P_l^2 tells more directly the electron position. On the other hand, the customary notation for the solutions is given in terms of R_l so we will return to that description once we see the form of the solutions.

The asymptotic form at large r of the solutions of Eq. (2.44) in the presence of a pseudopotential localized near $r = 0$ can be written immediately. When r is sufficiently large the centrifugal term becomes negligible and the general form of the solution is

$$P_l(r) \sim A\sin kr + B\cos kr \tag{2.45}$$

with k related to the energy by $E = \hbar^2k^2/2m$. Correspondingly the asymptotic form of the solutions R_l is given by

$$R_l(r) \sim A\frac{\sin kr}{r} + B\frac{\cos kr}{r} \tag{2.46}$$

For the particular case of $l = 0$ and a vanishing pseudopotential, the asymptotic forms, Eqs. (2.45) and (2.46), become exact solutions of Eq. (2.44). However, in that case the second term in Eq. (2.46) is not regular at $r = 0$ and the correct solution is

$$R_0(r) = A \frac{\sin kr}{r}$$

For general l the solutions of Eq. (2.43) for a vanishing pseudopotential are the spherical Bessel functions $j_l(kr)$.[1] These are all regular at the origin and at large distances approach

$$j_l(kr) \sim \frac{\sin(kr - l\pi/2)}{kr} \tag{2.47}$$

which is consistent with Eq. (2.46).

Thus by taking our origin at the center of a spherical crystal of radius R and by selecting k such that $kR - l\pi/2 = n\pi$ (with n an integer), we may satisfy the vanishing-boundary condition at the surface of the crystal. These spherical Bessel functions become the radial part of the zero-order solutions of Eq. (2.42). This is an equivalent solution of the zero-order problem only with a modified-boundary condition. The propagating waves that we obtained before may be written as an expansion in spherical Bessel functions in the form

$$e^{ikr\cos\theta} = \sum_l (2l + 1)i^l j_l(kr) P_l(\cos\theta)$$

If we were now to introduce a pseudopotential with translational periodicity we would find this set of basis states an extremely awkward framework in which to work. However we may make a clear physical separation of the problem when we add a perfect lattice into which a single impurity has been substituted. We add first the pseudopotential corresponding to a perfect lattice and second the difference in pseudopotential between the lattice with the impurity and that without. The first addition introduces band structure; the corresponding corrections to the zero-order states have been discussed previously and will not be of interest here. We will instead focus our attention on the difference in pseudopotential due to the addition of an impurity. In terms of perturbation theory the two effects are of the same order in the pseudopotential. When we consider scattering by impurities we will see how cross terms between the two perturbations may be included when perturbation theory is carried to higher order. For our present purposes we will be adding a spherically symmetric pseudopotential at the

[1] Discussion and properties of the spherical Bessel functions are given by L. I. Schiff, "Quantum Mechanics," pp. 76ff., McGraw-Hill, New York, 1949, and on p. 85 in 3rd edition, 1968.

center of our spherical crystal and this will correspond to the addition of one impurity to the system. The treatment here will be exact rather than a perturbation expansion.

We will assume that the pseudopotential difference is restricted to a single atomic cell and will construct a sphere around the impurity site which contains that potential. We could obtain the exact pseudowavefunction within the cell for any energy by integrating the pseudopotential equation from the origin to the cell surface, taking a regular solution at the center. This would then be matched to solutions obtained outside the cell. The general solution outside the cell, where $W(\mathrm{r})$ is zero, is a linear combination of the appropriate spherical Bessel function and the spherical Neumann function, $n_l(kr)$. The $n_l(kr)$ are the solutions of Eq. (2.43) with $W_l = 0$ which are singular at $r = 0$. At large distances these approach

$$n_l(kr) \sim -\frac{\cos(kr - l\pi/2)}{kr} \tag{2.48}$$

Thus the general solution outside the cell for fixed l may be written

$$\varphi_l = A_l[\cos \delta_l j_l(kr) - \sin \delta_l n_l(kr)] \tag{2.49}$$

The adjustable parameters needed for matching are the A_l and δ_l. They have been introduced in this way so that the large r behavior takes the convenient form

$$\varphi_l(r) \sim A_l \frac{\sin(kr - l\pi/2 + \delta_l)}{kr} \tag{2.50}$$

The parameter δ_l is called the *phase shift*.

The solution and its derivative obtained by integration within the cell at any energy may be matched to the exterior form, Eq. (2.49), at the cell surface. The energy then must be adjusted to satisfy the vanishing-boundary condition, $kR - l\pi/2 + \delta_l = n\pi$. The phase shift may depend of course upon the energy as well as upon the quantum number l. This represents an exact solution of this spherically symmetric problem subject to the given boundary conditions.

Clearly if no perturbing potential had been introduced Eq. (2.49) would give a valid solution both inside and outside the cell with $\delta_l = 0$. As a perturbing potential is introduced the phase shift grows and the solutions are modified. This is illustrated in Fig. 2.41*a* and *b*.

We are particularly interested here in the energies of the states. We note that from our matching condition the energy, measured from the band minimum in the solvent, is given by

$$E = \frac{\hbar^2 k^2}{2m} = \frac{\hbar^2}{2m}\left(\frac{n\pi + l\pi/2 - \delta_l}{R}\right)$$

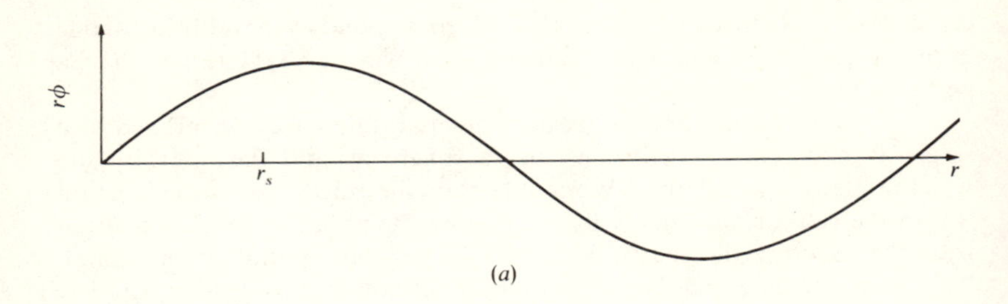

(*a*)

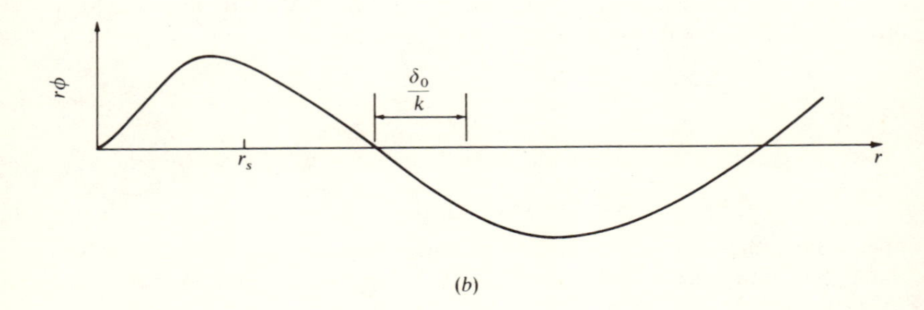

(*b*)

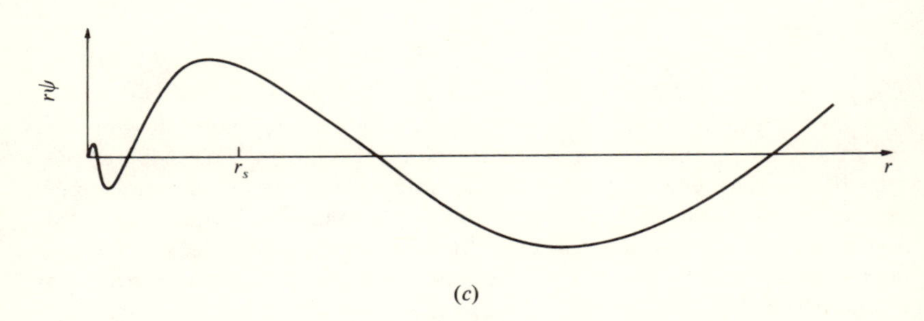

(*c*)

Fig. 2.41 *The modification of the radial wavefunction due to the addition of a sodium atom at* $r = 0$. *(a) The solution* $P_0(r) = rj_0(kr)$ *for* $l = 0$ *and a vanishing pseudopotential. (b) A sodium pseudopotential has been introduced at* $r = 0$ *and the pseudopotential equation integrated to the atomic-cell boundary* r_s, *where it is matched to the general solution outside, introducing a phase shift* δ_0. *(c) The pseudopotential* $W'_0(r)$ *is replaced by the true sodium potential within the cell and the corresponding Schroedinger equation integrated to* r_s *and matched. By following the deformation of the wavefunction as the potential is introduced, we may see that the phase shift in (c) is* $2\pi + \delta_0$.

where the wavenumber was obtained from the boundary condition. Different states are obtained for each positive integer n. The presence of a nonvanishing phase shift has shifted the energy of each state slightly. However we note that as long as the phase shifts are less than π, which was the case in Fig. 2.41b, no level is shifted beyond the energy of the neighboring unperturbed states; i.e., the nth perturbed state will have energy between the $(n - 1)$st and the $(n + 1)$st unperturbed states. Since the unperturbed states are so closely spaced in a large system (forming a quasicontinuum) these are indeed extremely small shifts in energy.

This form of solution to the problem can give an interesting picture of the relation between the true potential and the pseudopotential. In Fig. 2.41 we imagined introducing a sodium pseudopotential in a free-electron gas and seeking the pseudowavefunction corresponding to an eigenstate with energy near the atomic 3s energy. Because of the weakness of the pseudopotential the pseudowavefunction was only slightly deformed and the phase shift was less than π as illustrated in part b of the figure. We might instead return to the Schroedinger equation, introduce the true sodium potential, and again seek the eigenstate. We must obtain the same energy and a wavefunction that is equal to the pseudowavefunction outside the sodium core if the latter was obtained with a valid pseudopotential. However, because the sodium potential is quite strong the wavefunction within the core will be greatly deformed and will resemble the atomic 3s wavefunction. The result of such a calculation is illustrated in Fig. 2.41c. By following the deformation of the wavefunction as the sodium potential is slowly introduced we may see that the phase shift goes through 2π as two nodes move into the core region leading to a phase shift of 2π plus the value obtained in the pseudopotential calculation.

In this context we see that a well-chosen pseudopotential simply eliminates integral multiples of π from the phase shifts, leaving a small shift which may be estimated in perturbation theory and yet yielding the correct solution outside the core. As long as we are discussing simple metals we need only consider relatively small phase shifts.

We may also discuss the existence of bound impurity states in terms of this exact analysis. We will do this for s states for which the spherical Bessel functions and spherical Neumann functions are given exactly for all r by the limiting form, Eqs. (2.47) and (2.48). We focus our attention on the lowest-lying unperturbed state, $n = 1$. If a small positive phase shift is introduced the wavenumber of this state is determined by the matching $kR + \delta_0 = \pi$. We let the potential increase and the phase shift increase toward π and we see that the wavenumber approaches zero. Through this whole range the energy shift remains very small as it did for the states we discussed above. However, if we continue to follow the solution as the potential increases, corresponding to a phase shift greater than π at the bottom of the band, the nature of the state changes abruptly. It becomes an exponentially decaying localized state (corresponding to imaginary wavenumber). This occurs only for an attractive potential (for which the phase shifts are found to be positive). At this finite strength of potential we say that a bound state is formed; its energy will drop rapidly below the band minimum as the potential is increased further. It is important to notice this qualitative change in behavior when the potential becomes sufficiently strong to bind a state. At that point perturbation theory, in the sense we have used it in the pseudopotential method, has broken down. The phase-shift analysis remains valid and can be used either with the true potential or the pseudopotential.

We expect that in the simple metals pseudopotentials can always be constructed which will prevent the appearance of phase shifts as large as π. Correspondingly we do not expect bound impurity states to occur among alloys of the simple metals. We cannot of course rule out the possibility of bound states in semiconductors since the addition of an impurity of a different valence will inevitably give rise to a coulomb potential. Such a long-range potential, which cannot occur in metals because of the mobility of the electrons, *will* in the semiconductor lead to bound states.

There is a very powerful relation, the *Friedel sum rule*,[1] between the phase shifts and the number of electrons accumulated in the region of the impurity. The sum rule itself is quite important and some very informative physical insights arise in its derivation. We will therefore proceed with that. The general approach is to convert a volume integral of the square of the wavefunction to a surface integral which can be evaluated in the region where the asymptotic form is applicable. This will lead to the desired relation between the localized electron density and the phase shifts.

When an impurity potential is introduced and the wavefunction modified, there is a change in the probability density in the region of the center. Such a density ultimately must be computed from the true wavefunction rather than from the pseudowavefunction, so we will proceed with the Schroedinger equation rather than with the pseudopotential equation. We will note in the course of the proof how this analysis is changed if the pseudopotential equation is used.

The radial Schroedinger equation may be written in direct analogy with the radial pseudopotential equation, Eq. (2.43).

$$-\frac{\hbar^2}{2m}\frac{1}{r^2}\frac{\partial}{\partial r}r^2\frac{\partial}{\partial r}R_l + V(r)R_l + \frac{\hbar^2}{2m}\frac{l(l+1)}{r^2}R_l = ER_l \tag{2.51}$$

[1] J. Friedel, *Phil. Mag.*, **43**: 153 (1952), see Appendix.

Without loss of generality we may take R_l real. We also write this equation for a second state of slightly different energy, E'.

$$-\frac{\hbar^2}{2m}\frac{1}{r^2}\frac{\partial}{\partial r}r^2\frac{\partial}{\partial r}R_l' + V(r)R_l' + \frac{\hbar^2}{2m}\frac{l(l+1)}{r^2}R_l' = E'R_l' \tag{2.52}$$

Equation (2.51) is multiplied on the left by R_l' and Eq. (2.52) is multiplied by R_l. Both are then multiplied by $4\pi r^2$ and integrated from the origin to a large radius M (which, however, is smaller than the radius of the system). Equation (2.52) is then subtracted from Eq. (2.51) to obtain

$$-\frac{4\pi\hbar^2}{2m}\left(\int_0^M R_l'\frac{\partial}{\partial r}r^2\frac{\partial}{\partial r}R_l\,dr - \int_0^M R_l\frac{\partial}{\partial r}r^2\frac{\partial}{\partial r}R_l'\,dr\right)$$

$$= (E - E')\int_0^M 4\pi r^2 R_l' R_l\,dr$$

Note that we have taken advantage of the fact that V is a simple potential; thus the two terms involving it cancel. If we used the pseudopotential method the corresponding terms would not have canceled because of the energy dependence of the pseudopotential. The difference in the charge density which would be computed from $\varphi^*\varphi$ as opposed to $\psi^*\psi$ is the orthogonalization hole, and is intimately connected with the energy dependence of the pseudopotential. An analysis similar to that given here can lead to a direct relation between them.[1]

Each of the integrals on the left may be integrated by parts once. The resulting integrals cancel and we are left only with the surface terms evaluated at the origin (where they are zero) and at the large radius M.

$$-\frac{4\pi\hbar^2}{2m}M^2\left(R_l'\frac{\partial}{\partial r}R_l - R_l\frac{\partial}{\partial r}R_l'\right)_M = (E - E')\int_0^M 4\pi r^2 R_l' R_l\,dr$$

At this stage we let E' approach E. Thus we may write $R' = R + (\partial R/\partial k)\delta k$. Finally noting $E' - E = \hbar^2 k\delta k/m$ we obtain

$$\int_0^M 4\pi r^2 R_l^2\,dr = \frac{2\pi M^2}{k}\left(\frac{\partial R_l}{\partial k}\frac{\partial R_l}{\partial r} - R_l\frac{\partial^2 R_l}{\partial r\,\partial k}\right)_M$$

We have now succeeded in obtaining the integral of $\psi^*\psi$ within a sphere in terms of the wavefunctions at the sphere surface.

[1] R. W. Shaw and W. A. Harrison, *Phys. Rev.*, **163**:604 (1967).

We now take the radius M to be sufficiently large that the asymptotic form, Eq. (2.50), is applicable. The differentiations may be performed directly noting in particular that the phase shift is a function of energy or wavenumber. After some algebra we obtain

$$\int_0^M 4\pi r^2 R_l^2\, dr = \frac{2\pi}{k^2} A_l^2 \left[M + \frac{\partial \delta_l}{\partial k} - \frac{\sin 2(kM - l\pi/2 + \delta_l)}{2k} \right] \quad (2.53)$$

Normalization factors are obtainable from this expression. We note that if M becomes the radius L of the system, the left-hand side becomes the normalization integral and must be unity. The term $M = L$ dominates on the right and therefore $2\pi\, A_l^2/k^2$ must be equal to $1/L$.

We are specifically interested in the change in local electron density due to the introduction of the phase shifts, and therefore we subtract from Eq. (2.53) the corresponding equation with vanishing phase shifts. Writing the fraction of an electron which has been localized as δn, we obtain after some rearrangement

$$\delta n = \frac{1}{L} \left[\frac{\partial \delta_l}{\partial k} - \frac{\sin \delta_l}{k} \cos(2kM - l\pi + \delta_l) \right]$$

This is the localization due to a single state. It may be summed over all occupied states allowed by periodic-boundary conditions. For a given l there are $2l + 1$ values of m; $L\, dk/\pi$ values of wavenumber in the range dk are allowed by the boundary conditions and two spin states occur for each of these. Thus the total number of electrons accumulated is given by

$$\Delta N = \frac{2}{\pi} \sum_l (2l + 1) \int_0^{k_F} dk \left[\frac{\partial \delta_l}{\partial k} - \frac{\sin \delta_l}{k} \cos(2kM - l\pi + \delta_l) \right] \quad (2.54)$$

The oscillating term was discarded in early derivations of the Friedel sum rule. It is now recognized to be real and to be quite important. We cannot perform the integration over k explicitly but we may obtain an asymptotic series for large M for the oscillatory electron density, $\rho(M) = (\delta \Delta N/\delta M)/(4\pi M^2)$, by successive integration by parts. The leading term is

$$\rho(r) = -\frac{1}{2\pi^2 r^3} \sum_l (2l + 1)(-1)^l \sin \delta_l \cos(2k_F r + \delta_l) \quad (2.55)$$

where the phase shifts are again evaluated at the Fermi surface. These density fluctuations are the so-called *Friedel oscillations.* Their physical origin is of course the discontinuous change in the occupation of states at the Fermi surface.

Note that this calculation is not self-consistent. We have assumed a well-localized potential but have found charge-density fluctuations and therefore potentials at large distances. When we consider screening in the next chapter we will redo this problem self-consistently but in the Born approximation. The results there differ from the small δ_l limit of Eq. (2.55) by a factor $1/\epsilon(2k_F)$, the reciprocal of the static Hartree dielectric function discussed there.

In addition to these Friedel oscillations we obtain immediately from Eq. (2.54) a localization of ΔN electrons given by

$$\Delta N = \frac{2}{\pi} \sum_l (2l + 1)\, \delta_l \tag{2.56}$$

This is the Friedel sum rule which we sought. Note that it is an exact expression and did not depend upon any perturbation expansion.

In this evaluation we have dropped the lower-limit contribution $-(2/\pi) \sum_l (2l + 1)\delta_l(k = 0)$. For small potentials these phase shifts will be seen to vanish. If the potential is large enough to bind a single state (actually $2l + 1$ degenerate states) of quantum number l, the lowest positive-energy phase shift will approach π. Our evaluation applies only to positive-energy states, but by dropping the lower limit we include the $2(2l + 1)$ bound states in ΔN. (The 2 is for spin.) Thus Eq. (2.56) gives both the localized bound states and the localization of the positive-energy states. Since we have based our analysis on the true potential the phase shifts may well contain integral multiples of π. These represent directly the bound core states. If the phase shifts are evaluated using a pseudopotential (and this is a perfectly valid method of evaluating the phase shifts at the Fermi energy), these multiples of π are eliminated and ΔN gives us only the localization of the conduction-band states.

We will not make specific use of the actual phase shifts that occur in solids. It may be of interest, however, to illustrate this discussion by giving the phase shifts for aluminum. The phase shifts may be obtained by a detailed integration of the Schroedinger equation (or pseudopotential equation) in the region of the atom. They may also be obtained approximately to lowest order in perturbation theory. This is simply the Born approximation.

$$\begin{aligned} \delta_l &= -\frac{2mk}{\hbar^2} \int_0^\infty r^2 j_l(kr) w_l(r) j_l(kr)\, dr \\ &= -\frac{2mk\Omega_0}{4\pi\hbar^2} \frac{1}{\Omega_0} \int_0^\infty 4\pi r^2 j_l(kr) w_l(r) j_l(kr)\, dr \end{aligned} \tag{2.57}$$

This is a direct generalization to pseudopotentials of the familiar formula[1] for potentials. They become identical if the true potential is used to compute the phase shifts. Note that for attractive potentials this leads to positive phase shifts as we indicated above and for weak pseudopotentials to small phase shifts.

These approximate phase shifts are directly obtainable from the OPW form factor which we discussed previously. The OPW form factor,

$$\langle \mathbf{k} + \mathbf{q}|w|\mathbf{k}\rangle = \frac{1}{\Omega_0}\int e^{-i(\mathbf{k}+\mathbf{q})\cdot\mathbf{r}}\, w(r)\, e^{i\mathbf{k}\cdot\mathbf{r}}\, d\tau$$

is just the matrix element of a single ionic pseudopotential between two plane waves, both with wavenumbers on the Fermi surface. We may expand the two plane waves in the integral in terms of spherical harmonics and perform all angular integrations.[2] We obtain

$$\langle \mathbf{k} + \mathbf{q}|w|\mathbf{k}\rangle = \sum_l (2l + 1)P_l(\cos\theta)\frac{1}{\Omega_0}\int 4\pi r^2 j_l(kr) w_l(r) j_l(kr)\, dr$$

where θ is the angle between $\mathbf{k}$ and $\mathbf{k} + \mathbf{q}$ and the P_l are Legendre polynomials. Thus

$$\langle \mathbf{k} + \mathbf{q}|w|\mathbf{k}\rangle = -\frac{2\pi\hbar^2}{mk\Omega_0}\sum_l (2l + 1)\,\delta_l P_l(\cos\theta) \tag{2.58}$$

The δ_l may be easily obtained by integration of tabulated form factors. This leads to values for aluminum at the Fermi energy of

$$\begin{aligned} \delta_0 &= +0.57 \\ \delta_1 &= +0.81 \\ \delta_2 &= +0.24 \\ \delta_3 &= +0.05 \end{aligned} \tag{2.59}$$

The phase shifts will be smaller at lower energies. Using the Born approximation formula Eq. (2.57) and noting that $j_l(kr)$ varies as $(kr)^l$ at small kr, it may be seen that the phase shifts vary as k^{2l+1} at low energy. An approximate description of the phase shifts at all energies might be obtained by scaling such a form to give our computed phase shifts at the Fermi energy. Such a model is sketched in Fig. 2.42.

The use of perturbation theory, or the Born approximation, is justified only if the resulting phase shifts are small compared to π. The values we have

[1] Schiff, *op. cit.*, p. 165 in 1st edition, p. 330 in 3rd edition.

[2] These integrations are of the form of those discussed in W. A. Harrison, "Pseudopotentials in the Theory of Metals," p. 278, Benjamin, New York, 1966.

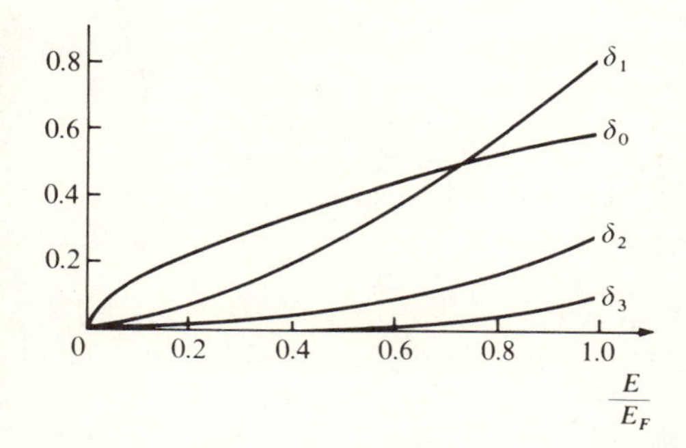

Fig. 2.42 *The phase shifts in aluminum, taking the energy dependence from the Born approximation.*

found are in fact somewhat less than π but not sufficiently so that we could be guaranteed good quantitative accuracy. It would be a mistake however to attempt a more accurate calculation of the phase shifts from existing pseudopotentials which have been computed using the linear screening described in the next chapter. A potential due to the electrons has been constructed which yields *in perturbation theory* the correct total screening. In the case of aluminum a localized charge of three electrons is required to balance the ionic charge of 3. Then there will be no long-range coulomb potential in the electron gas in which the aluminum ion is inserted. It may be readily verified that the phase shifts given in Eq. (2.59) lead, according to the Friedel sum rule, to a localized charge of 2.9. The addition of phase shifts for higher l would bring this number to 3. Thus the pseudopotentials that we have calculated have a built-in correction for the errors in perturbation theory and exact calculation of phase shifts in terms of them would lead to incorrect screening of the potential.

We have discussed here the states in simple metals for which perturbation theory may be a reasonable approximation. We have also discussed the origin of bound states. An additional complication arises when there are scattering resonances. This is the case for the d states in transition and noble metals; we will turn to that next.

8.5 Scattering resonances The concept of resonant states has become very prominent in solid state physics during the past year or two. This has been particularly directed at resonant d states. The theory of scattering resonances is very well developed.[1] We will not attempt to redevelop that analysis nor will we use very heavily the results given there. We will instead develop a physical picture of resonances as they occur in solids and see how the concepts are used in solid state calculations. We will first consider

[1]See, e.g., A. Messiah, "Quantum Mechanics," pp. 396ff., Wiley, New York, 1965.

the more familiar bound states of a free atom and see how these evolve into resonant states by a modification of the environment of that atom. We will then examine the same problem from the phase-shift point of view and see what the structure of the band states becomes in a simple metal when a transition metal atom is added. In Sec. 9 we will return to resonant states using a different point of view and discuss their relation to energy bands in pure transition and noble metals.

We consider, to be specific, an isolated silver atom. We will consider the electronic states with d symmetry, using a self-consistent-field method. Thus we may write down the radial Schroedinger equation, Eq. (2.51), for the $l = 2$ state. We might obtain the eigenstates by integrating the Schroedinger equation from the origin at different energies and seeking solutions and energies such that the resulting wavefunction approaches zero at infinity and is therefore normalizable. In silver two such states are of interest here, the $3d$ and $4d$ states. We have sketched these in Fig. 2.43a along with the sum of the potential and centrifugal terms. We have also sketched schematically the result of integrating the Schroedinger equation radially at an energy intermediate between these two. This wavefunction grows exponentially at large distances. Finally, we have sketched the result of integration at an energy above E_{4d} (and in fact, above the ionization energy for the atom). This positive-energy state oscillates at large distance and has the asymptotic form given in Eq. (2.50). These are scattering states for electrons incident on the silver atom or ion.

We now insert this silver atom at the center of the spherical crystal of a simple metal described in the preceding section. To be specific let this be aluminum. The potential within the silver atom is unchanged (except for an approximately constant shift) but is joined onto the potential in aluminum outside the atom itself. We let this be the constant potential E_0 equal to the aluminum conduction-band minimum. This is shown in Fig. 2.43b. We might now seek the solution by integrating the radial equation within the silver atom and then joining the solutions in the neighborhood of the atomic-cell boundaries to the solution of the Schroedinger equation (or the pseudopotential equation) for aluminum. This external solution is simply the proper linear combination of spherical Bessel and Neumann functions for energies greater than E_0 and the corresponding decaying solutions for energies less than E_0.

We may again compute the $3d$ state which will again be bound but may have slightly different energy, E'_{3d}. If, on the other hand, we again perform the integration at the energy E_{4d} we obtain a wavefunction that looks very much like the atomic wavefunction within the silver atom but at large distances this function will not decay exponentially since it corresponds to an energy greater than E_0. Since the $4d$ state is quite well localized in the free silver atom, however, the wavefunction will have decayed to a small value by the time we leave the silver cell, and therefore the oscillating tail will be quite small. We would call this a *resonant d state*.

Note that this is precisely the situation that occurs in alpha decay from nuclei. The alpha particle is very nearly bound to the nucleus by the combined effect of nuclear forces and the coulomb repulsion, but is nevertheless at an energy above the zero of energy far from the nucleus. Therefore the wavefunction at large distances oscillates and corresponds to a nonzero probability of the alpha particle leaving the nucleus. Similarly, the *d* electron is very nearly bound to the silver atom but may decay with a probability proportional to the square of the amplitude of the oscillating tail of the wavefunction. At the same time, of course, incident electrons are scattering into the *d* state such that no changes occur with time. In this case it is not so fruitful to describe the problem in terms of probable decays but it is more appropriate to consider the eigenstates themselves.

Let us look at the tail of this resonant state. It will again have the form $\sin(kr + \delta)/r$ at large distances. The amplitude is small but nonetheless it is seen from Eq. (2.53) that the wavefunction is not normalizable in an infinite system. It is normalizable in our metal crystal but then the probability of the electron lying within the silver atomic cell becomes of the order of the ratio of the cell radius to the crystal radius, r_s/L. Here the probability is much larger than it would be for a free-electron state but it is nevertheless infinitesimal in a large system.

Let us next consider a state with energy in the *neighborhood* of E_{4d}. Integration of the radial Schroedinger equation within the silver atom will not have led to quite as small a wavefunction at the cell boundary, and therefore the tail will be slightly larger. Thus a normalized wavefunction will have slightly lower probability of lying within the silver atom than does the resonant state at energy E_{4d} but still a significantly enhanced probability. Thus we find a range of states in the band which have enhanced probability of lying within the silver atom. If we move still further from the resonant energy, to the energy E_{int} of the free-atom diagram, we see that the integrated wavefunction has had exponential growth by the time we are to match it to the oscillating tail. Thus the tail becomes very large and the normalized wavefunction leads to a very small probability of the electron lying at the silver atom. These again correspond to electrons scattering from the silver atom, which behaves as a repulsive potential.

This leads to a very important insight into the nature of electron states in solids. All of this discussion is consistent with the energy-band point of view in which each electronic state is associated with the entire system. Nonetheless we have found that the probability of finding an electron with 4*d* angular momentum at the silver atom with an energy close to the

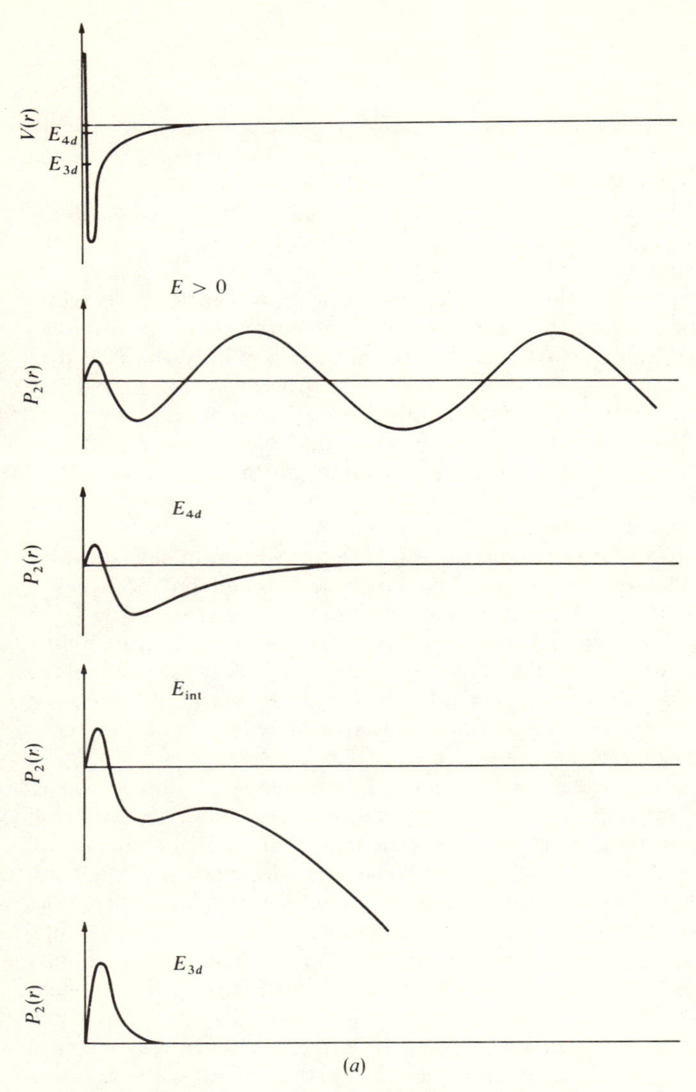

Fig. 2.43 *(a) The sum $V(r)$ of the potential and centrifugal terms, $\frac{\hbar^2}{2m}\frac{l(l+1)}{r^2}$, for a free atom of silver. In the four curves below are shown the $P_2(r) = rR_2(r)$ obtained from integrating the radial Schroedinger equation at a series of energies, E slightly greater than the zero of energy, E_{4d} equal to the free-atom energy of the 4d state, E_{int} slightly less than E_{4d}, and E_{3d} equal to the free-atom energy of the 3d state.*

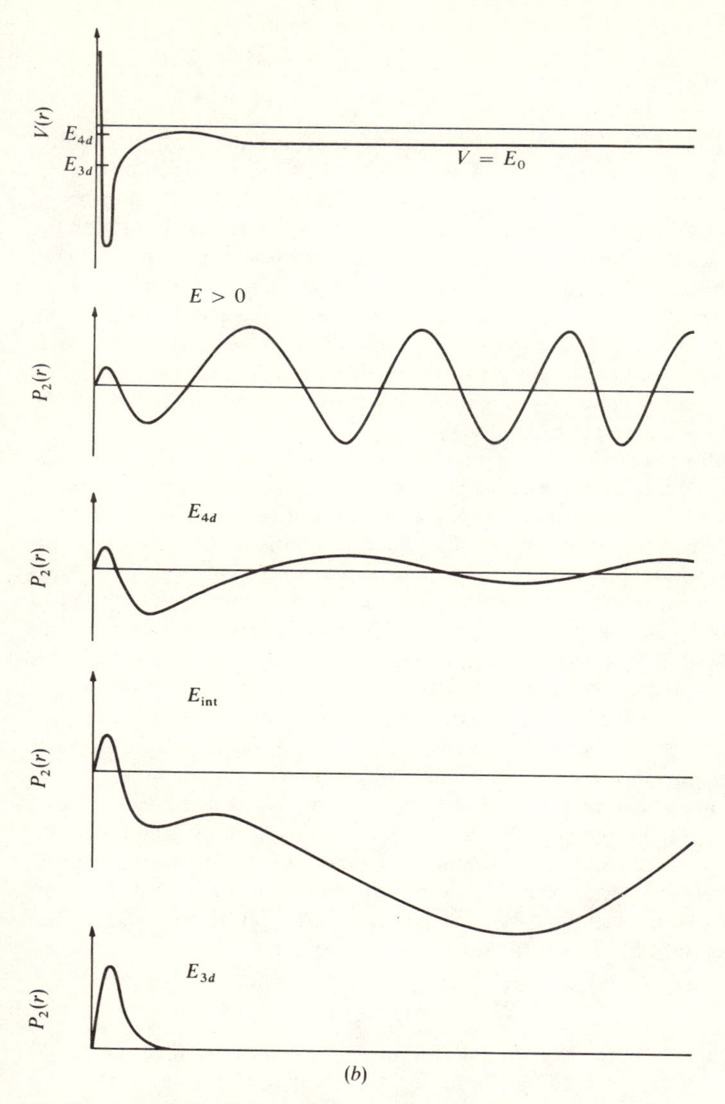

(*b*) *The corresponding* $V(r)$ *modified by the insertion of the silver atom in aluminum. This modification is made by lowering* $V(r)$ *at large* r *to a value* E_0 *equal to the conduction-band minimum in aluminum. This value is taken to lie between* E_{3d} *and* E_{int}. *Below are shown the results of integrating the radial Schroedinger equation using the modified potential but at the same energies as in* (*a*). *The atomic 4d state becomes a resonant state in the metal.*

free-atom energy is very high. We have even found that the probability of finding an electron with some energy different from the free-atom silver energies at the silver atom is very small. We have reduced the number of aluminum conduction electrons in the system by the three that went with the aluminum atom which was removed, and this reduction has occurred mainly within the silver cell. To a good approximation the distribution of the electrons in energy and space is very much the same as we would guess if we simply superimposed free-atom distributions, though we have also found some broadening of the levels. Thus we see that physical arguments from an atomistic point of view in solids are not completely inconsistent with the band picture that we have been developing.

Our discussion has been directed at d states where the resonances tend to be narrow and the concept to be particularly fruitful. Of course, it is possible to have resonances arising from states of any angular momentum and the discussion remains applicable. Such resonances can occur in semiconductors where the conduction band, for example, may have a local minimum. We might then expect a localized state in the neighborhood of an impurity as we discussed in Sec. 8.2. However, if the conduction band drops below this local minimum at some other point in the Brillouin zone, then the state in question remains essentially a positive-energy state though it may well be a resonant state in the sense that we have described here.

It has been suggested[1] that a resonant p state appears in lithium when one of the $1s$ core electrons is removed. Such a resonant state would cause an important modification in the x-rays which are subsequently emitted as electrons drop from the conduction band to fill this hole. In general we do not expect such resonances to occur in simple metals for reasons closely related to those which we discussed in connection with bound states in alloys of the simple metals. We may compute the phase shifts using the pseudopotential obtained from Eq. (2.23) in Sec. 5.1.

$$w(r) = v(r) - \sum_t (E - E_t)\,|t\rangle\langle t|$$

The removal of a core electron will add a deep potential well to the initial $v(r)$ and one might expect this to produce the resonance. However, the core energies E_t will also be lowered by an amount approximately equal to the depth of that well. The core states $|t\rangle$ remain relatively unchanged and appear in the pseudopotential even though one of those states is unoccupied. Thus the change in the repulsive term of the pseudopotential will tend to cancel the effect of the deep potential well making a resonance unlikely. The argument does not apply, however, to p resonances for elements in the lithium row of the periodic table since there are no core p states, the $l = 2$

[1] F. K. Allotey, *Phys. Rev.*, **157**:467 (1967).

component of the pseudopotential is identical to the true potential, and there is no cancellation. Thus the proposed resonant states could explain a well-known anomaly in the x-ray emission spectrum of lithium, though an alternative explanation based upon many-body effects will be discussed in Sec. 5.8 of Chap. III.

To sharpen the resonance picture somewhat it is appropriate to reconsider the same problem using the phase-shift formalism. In contrast to the localized states which we found in preceding sections to be pulled out of bands, we here have additional states inserted in the band. We should emphasize again, however, that each state in the resonant region yields only a small probability of an electron occupying the local site and that many states through the resonant region contribute together to give the counterpart of the truly localized state.

It is possible to see qualitatively how the $l = 2$ phase shift behaves in the neighborhood of the resonance by reexamining Fig. 2.43. At an energy slightly below resonance, the energy E_{int}, the wavefunction becomes very large and negative at the central cell surface. As we move up in energy through the resonance the wavefunction at the cell surface goes through zero and then becomes very large and positive. The wavefunction at the matching surface has changed sign as a node has moved through the surface. Thus in the very narrow energy range at resonance the phase shift has increased through approximately π. It can be shown[1] that the phase shift in the neighborhood of the resonance is given by

$$\tan \delta_l = \frac{\Gamma}{2(E_r - E)}$$

where E_r is the resonant energy and Γ is called the width of the resonance. Note that this does give the expected change in phase shift by π in the resonant region. We may note by applying the Friedel sum rule at a succession of energies through resonance that 10 d electrons have been localized at the silver site in an energy range of order Γ.

The Γ may be related approximately to the decay time of the state through the uncertainty principle. It is also therefore related directly to the amplitude of the oscillatory tail of the resonant state.

All of our discussion of resonant states has been based upon a self-consistent-field approximation. We might raise again, however, the question of the Mott transition with respect to this resonant state. We might construct a wavepacket from the resonant states to yield a truly localized wavefunction and similarly make appropriate adjustments in the band states in order to make them orthogonal to this. We could then construct many-electron wavefunctions based upon the two kinds of wavefunctions

[1] Messiah, *op. cit.*, p. 399.

and question, as we did before, which of these many-electron functions would yield the lowest expectation value of the electron-electron interaction. For a very sharp resonant state it may well turn out that the localized state yields the lowest energy. Thus we might have a truly localized state and energy near the corresponding free-atom energy, though this lies in the middle of a conduction band. Such a situation is thought to describe *f* states in the rare earths but not the *d* states in iron-series metals. Even when local magnetic moments are formed (to be discussed in Sec. 7 of Chap. V) the states are viewed as resonant.

In the light of these considerations of resonant *d* states we will go on in Sec. 9 to discuss electronic states in pure transition and noble metals.

8.6 Electron scattering by impurities We have examined the shifts in the energies of states when a defect or impurity is added to an otherwise perfect crystal. It is of interest also to interpret the change in the structure of the states in terms of scattering theory.

We may describe a scattering event due to a single impurity in terms of pseudowavefunctions consisting of an incoming wave, $e^{i\mathbf{k}\cdot\mathbf{r}}$, and an outgoing scattering wave which at large distances must approach $f(\theta,\varphi)\, e^{ikr}/kr$, where $f(\theta,\varphi)$ is the scattering amplitude.

$$\psi \sim e^{i\mathbf{k}\cdot\mathbf{r}} + f(\theta,\varphi)\frac{e^{ikr}}{kr}$$

If we select our coordinate axis along $\mathbf{k}$, there will be no dependence upon φ and we may expand both terms in spherical harmonics. This can be equated

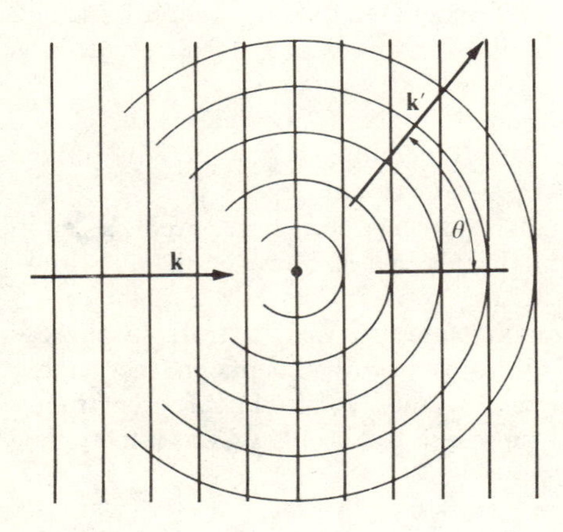

Fig. 2.44 *Scattering of an incident electron of wavenumber* $\mathbf{k}$ *by an impurity. The scattered wave is spherical and at large distances is proportional to* $f(\theta)\, e^{ikr}/kr$. *The scattering may alternatively be viewed as a set of transitions to states of wavenumber* $\mathbf{k}'$ *with a probability per unit time proportional to* $|f(\theta)|^2$ *if* θ *is the angle between* $\mathbf{k}$ *and* $\mathbf{k}'$.

to the solutions of the eigenvalue problem which we found previously and $f(\theta)$ obtained in terms of phase shifts. The result[1] is

$$f(\theta) = \frac{1}{2i}\sum_{l=0}^{\infty}(2l + 1)(e^{2i\delta_l} - 1)\, P_l(\cos\theta)$$

Thus the results we obtained before may be related directly to the scattering amplitude.

This result is again exact. If the phase shifts are small compared to π we may expand the exponent and make use of Eq. (2.58) to obtain

$$\begin{aligned} f(\theta) &= \sum_l (2l + 1)\,\delta_l P_l(\cos\theta) \\ &= -\frac{mk\Omega_0}{2\pi\hbar^2}\langle \mathbf{k} + \mathbf{q}|w|\mathbf{k}\rangle \end{aligned}$$

where w is the pseudopotential giving rise to the scattering. The differential scattering cross section is given by $\sigma(\theta) = |f(\theta)|^2/k^2$. This was derived by taking the ratio of incoming and outgoing current densities. These could be evaluated between cores where the pseudowavefunction is equal to the true wavefunction and therefore the use of the pseudowavefunction is justified. The result is

$$\sigma(\theta) = \frac{m^2\Omega_0{}^2}{4\pi^2\hbar^2}|\langle \mathbf{k} + \mathbf{q}|w|\mathbf{k}\rangle|^2 \tag{2.60}$$

Equivalent formulas may be obtained with the direct use of time-dependent perturbation theory and the pseudopotential equation. This is the more usual procedure and the one that will be followed here. Within this framework we will also be able to see more clearly the separation between the effects of the impurity and those of the periodic lattice. It will be convenient to return again to the crystal with periodic-boundary conditions.

The time-dependent perturbation theory will be applicable in semiconductors when we consider the interaction of positive-energy states with donor atoms even though a negative-energy-bound state may also have been formed. In such a case the pseudopotential must be selected to give all phase shifts less than π. It will be applicable in general whenever a non-transition element is added either to a semiconductor or to a simple metal. It will even be appropriate, as we will see in the following section, when the impurity is a transition metal as long as the states of interest happen not to be close to resonance.

Let us consider first the scattering of an electron by a simple metal impurity in another simple metal; e.g., magnesium as an impurity in alumi-

[1] Schiff, *op. cit.*, p. 103 in 1st edition, p. 119 in 3rd edition.

num. We may again write a pseudopotential for the metal containing the impurity. It can be written as a superposition of pseudopotentials centered on the individual ions, but in this case the pseudopotential for the impurity ion will be different from that for the host. If, for example, we substitute an impurity of type b for a host atom of type a at the atomic site $\mathbf{r}_0$, the pseudopotential becomes

$$W(\mathbf{r}) = \left[\sum_{r_j \neq r_0} w^a(\mathbf{r} - \mathbf{r}_j)\right] + w^b(\mathbf{r} - \mathbf{r}_0) \tag{2.61}$$

The individual pseudopotentials may be constructed as before. The potential that enters the individual pseudopotentials may again be imagined as a free-atom potential, or the appropriately screened form to be discussed in Sec. 4.4 of Chap. III may be used.

We begin with zero-order pseudowavefunctions which are simply plane waves. These correspond to zero-order *wavefunctions* which are plane waves orthogonalized to the core states (in the region of the impurity they are of course orthogonalized to impurity core states). We seek now matrix elements of the pseudopotential based upon these zero-order plane-wave states. For this purpose it is convenient to rewrite the pseudopotential of Eq. (2.61) in the form

$$W(\mathbf{r}) = \left[\sum_{r_i} w^a(\mathbf{r} - \mathbf{r}_i)\right] + w^b(\mathbf{r} - \mathbf{r}_0) - w^a(\mathbf{r} - \mathbf{r}_0) \tag{2.62}$$

The first term is of course simply the pseudopotential of the perfect crystal. It has matrix elements connecting only states that differ by a lattice wave-number and these matrix elements we have found before. The remaining term is just the perturbation introduced by the addition of the impurity. Its matrix elements are given by

$$\langle \mathbf{k} + \mathbf{q}|W|\mathbf{k}\rangle = \frac{e^{-i\mathbf{q}\cdot\mathbf{r}_0}}{N}[\langle \mathbf{k} + \mathbf{q}|w^b|\mathbf{k}\rangle - \langle \mathbf{k} + \mathbf{q}|w^a|\mathbf{k}\rangle] \tag{2.63}$$

The matrix element on the left, with a capital W, is between plane-wave states normalized in the volume of the crystal. The matrix elements on the right are simply the OPW form factors which we defined earlier and which correspond to a normalization within an atomic cell. That change in normalization has given rise to the factor $1/N$ on the right. In any case, there are matrix elements of this perturbing potential between all states.

This perturbing pseudopotential will be only a weak perturbation for the same reasons that the form factors themselves are quite small. The difference in *potential* between the impurity and the host ion on the other hand may be quite large. If, for example, we are to dissolve a gallium ion in aluminum, the difference in potential will be sufficiently large that an extra node will be introduced in the wavefunction at the impurity site. This corresponds to the fact that the gallium valence electrons are $4s$ and $4p$ states while the aluminum valence electrons are $3s$ and $3p$ states. It would be incorrect to treat the difference in *true* potentials between the perfect and defective lattice as a perturbation since the phase shifts would be greater than π and the expansion in the phase shifts would not be valid. On the other hand, in the pseudopotential method this extra node is built in by the orthogonalization procedure and the pseudopotential difference is in fact small.

We would like now to regard the pseudopotential as a source of scattering of the electrons. Thus we may write the scattering probability given by the Golden Rule

$$P_{k,k'} = \frac{2\pi}{\hbar} |\langle \mathbf{k}' | W | \mathbf{k} \rangle|^2 \, \delta(E_{k'} - E_k) \tag{2.64}$$

The pseudopotential W is that given in Eq. (2.62). The first term in Eq. (2.62) has matrix elements which are very much larger (by a factor of N) than the second. Each state $\mathbf{k}$ is coupled to every state $\mathbf{k}'$ differing from $\mathbf{k}$ by a lattice wavenumber. However, these will contribute to the scattering in Eq. (2.64) only if that final state has the same zero-order energy. This is of course just the requirement that $\mathbf{k}$ lie on a Bragg reflection plane, or a Brillouin zone face. For any such electron the computed scattering rate would indeed be huge. However, if we are interested in electrical conductivity, this rapid scattering would involve only an infinitesimal fraction of the electrons present. Even eliminating them from the conduction process altogether would not appreciably modify the conductivity. These of course are the interactions that are more reasonably described in terms of energy-band structure. For an arbitrary electron that does not lie on a Bragg reflection plane these matrix elements do not contribute to the scattering. We turn then to the matrix elements given in Eq. (2.63). The scattering probability is given by

$$P_{kk'} = \frac{2\pi}{\hbar N^2} |\langle \mathbf{k} + \mathbf{q} | w^b | \mathbf{k} \rangle - \langle \mathbf{k} + \mathbf{q} | w^a | \mathbf{k} \rangle|^2 \, \delta(E_{k'} - E_k) \tag{2.65}$$

This probability contains a factor of the square of the number of atoms present in the denominator, but we will see that this is nevertheless the term of interest.

We might seek the total scattering probability per unit time for a particular electron; i.e., we might sum Eq. (2.65) for all final states $\mathbf{k}'$. Since all states are coupled we should convert this to an integral. Thus we multiply by the density of final states (of the same spin since only those are coupled by the matrix element), $\Omega/(2\pi)^3$. To be specific we let $\mathbf{k}$ lie on the Fermi surface and we may write the integral over wavenumber in the form

$$\int 2\pi k'^2 \, dk' \sin\theta \, d\theta$$

where θ is the angle between $\mathbf{k}$ and $\mathbf{k}'$. We have already performed the integration over φ since the matrix elements do not vary as φ varies. We may perform the integration over the magnitude of $\mathbf{k}'$ first, using the delta function of Eq. (2.65).

$$\sum_{k'} P_{kk'} = \frac{2\pi}{\hbar N^2} \frac{\Omega}{(2\pi)^3} \int 2\pi k'^2 \sin\theta |\langle \mathbf{k}'|w^b|\mathbf{k}\rangle - \langle \mathbf{k}'|w^a|\mathbf{k}\rangle|^2 \frac{m}{\hbar^2 k'} \delta(k' - k) \, dk' \, d\theta \quad (2.66)$$

$$= \frac{\Omega_0 k m}{2\pi\hbar^3 N} \int_0^{\pi} |\langle \mathbf{k}'|w^b|\mathbf{k}\rangle - \langle \mathbf{k}'|w^a|\mathbf{k}\rangle|^2 \sin\theta \, d\theta$$

where Ω_0 is the atomic volume. We may now note that because of the delta function both initial and final states in each form factor lie on the Fermi surface, and therefore this is precisely the OPW form factor that we described earlier. Using the cosine rule we may rewrite the angular integral as an integral over wavenumber q to obtain

$$\sum_{k'} P_{kk'} = \frac{\Omega_0 m}{2\pi\hbar^3 N k} \int_0^{2k} |\langle \mathbf{k} + \mathbf{q}|w^b|\mathbf{k}\rangle - \langle \mathbf{k} + \mathbf{q}|w^a|\mathbf{k}\rangle|^2 q \, dq \quad (2.67)$$

This is equivalent to Eq. (2.60) obtained with phase shifts. We may now readily compute the integral from form factors such as we have given earlier. All constants are known and we obtain a total scattering rate.

It may be readily verified that the expression on the right in Eq. (2.67) has the appropriate units of reciprocal time. We may note, however, that we still have a factor of the reciprocal of the number of atoms present. If we were to add a number of impurities we may expect that if these are far enough apart the electrons will scatter from each independently and we may simply add the total scattering rates from each of the atoms. Combining this factor with the reciprocal number of atoms present gives simply the fraction of atoms that are impurity atoms. We find the scattering, as expected, proportional to the atom fraction of impurities.

For calculation of electrical resistivity in metals, we are ordinarily interested in a weighted average of the scattering time. We ask in particular for the rate at which the momentum $\hbar\mathbf{k}$ is randomized. Thus we ask for the rate of change of the average value of $\mathbf{k}$ for an electron initially in state $|\mathbf{k}\rangle$,

$$\frac{d\langle\mathbf{k}\rangle}{dt} = \sum_{k'} (\mathbf{k}' - \mathbf{k})P_{kk'}$$

but $d\langle\mathbf{k}\rangle/dt$ will be parallel to $\langle\mathbf{k}\rangle$ so we may write

$$\frac{d\langle\mathbf{k}\rangle}{dt} = -\left[\sum_{k'} (1 - \cos\theta)P_{kk'}\right]\langle\mathbf{k}\rangle \equiv -\frac{\langle\mathbf{k}\rangle}{\tau} \tag{2.68}$$

This factor, $1 - \cos\theta$, can be included in the integral of Eq. (2.66). This leads to an additional factor of $q^2/2k^2$ in Eq. (2.67). We have written the resulting scattering rate as $1/\tau$. The parameter τ is called the relaxation time and is used directly in transport theory. Note that the solution of Eq. (2.68) is simply $\langle\mathbf{k}(t)\rangle = \langle\mathbf{k}(0)\rangle e^{-t/\tau}$.

We may also, if we choose, carry the calculations to high order. In the next order the matrix element in Eq. (2.64) is replaced by

$$\langle\mathbf{k}'|W|\mathbf{k}\rangle \rightarrow \langle\mathbf{k}'|W|\mathbf{k}\rangle + \sum_{k''} \frac{\langle\mathbf{k}'|W|\mathbf{k}''\rangle\langle\mathbf{k}''|W|\mathbf{k}\rangle}{E_k - E_{k''}} \tag{2.69}$$

In this case there are two different types of higher-order terms that need to be considered. First, there are terms in which all pseudopotential matrix elements entering are the matrix elements of Eq. (2.63) arising from the impurity itself. Such corrections generally will be small and will not appreciably change the picture we developed in lowest order. Carrying these terms to all orders presumably would correspond to an exact calculation of the scattering by the impurity pseudopotential, just as does the phase-shift calculation of the scattering.

There are also terms in which one of the matrix elements in the sum of Eq. (2.69) is a matrix element arising from the perfect crystal. Such terms will contribute for arbitrary $\mathbf{k}$ because the energy delta function again connects the states $\mathbf{k}'$ and $\mathbf{k}$ and the intermediate state need not be of the same energy. It can be seen rather easily that these contributions enter to the same order in N as the higher-order terms of the first type and as the first-order terms. Matrix elements from the perfect lattice give corrections to the scattering because of the energy-band structure. The inclusion of such terms is a perfectly tractable calculation and represents a tremendous simplification over what we might have envisaged without the use of pseudopotentials. In principle we might calculate the full energy-band structure and then attempt to determine scattering from these tabulated wavefunctions and energies. That, in contrast, would be an extremely complicated calculation. By using perturbation theory of higher order, as in Eq. (2.69), we systematically keep terms in the pseudopotential to any given order and should easily be able to obtain meaningful results for the simple metals. Only a very limited amount of effort has gone into such calculations to date.

The calculation of scattering in semiconductors cannot be cast in such a neat form. We may again visualize the perturbation as being a difference in pseudopotential between the host and the impurity atoms, but we must now take as unperturbed states the appropriate Bloch states at the band edges and these are not simple plane waves. If we believe the impurity potential to be sufficiently spread out (as we did for the donor states) we might well use the effective-mass approximation and then the calculation proceeds in a direct and simple way. If the perturbation is strongly localized to the region of the impurity we have the same difficulties that arose in describing deep impurity states. We may formulate the scattering in terms of phase shifts or we may proceed with perturbation theory. The entire difficulty is that of obtaining appropriate matrix elements. Once these are obtained or guessed the calculation of relaxation times proceeds just as it did for the simple metal.

9. *Transition-Metal Bands*

The transition metals, metals with partially filled d or f shells, constitute most of the elements in the periodic table. In the iron series the energy bands are all very similar to those which we showed earlier for copper, copper being the element that closes the series. Because of this similarity between the noble metals and the transition series preceding them, we regard the noble metals also as members of the series with respect to energy-band structure.

Very recently[1] it has proven possible to generalize the pseudopotential method for simple metals to the transition metals. We will begin by outlining the formulation of the transition-metal pseudopotential in Sec. 9.1. In Sec. 9.2 we will describe earlier approximate treatments of the band structure but in the framework of this recent theory. In the transition metals as in the simple metals the ultimate aim of the introduction of a pseudopotential is the use of perturbation theory and the direct calculation of properties rather than of band structures. We will therefore return in Sec. 9.3 to a discussion of the transition-metal pseudopotential and its use in perturbation theory.

9.1 Transition-metal pseudopotentials It is convenient to reformulate the simple-metal pseudopotential in a slightly different, but equivalent, way before generalizing it to the transition metals. We may think of the formulation of the pseudopotential as an attempt to make an expansion of the true wavefunction. Noting that the true conduction-band wavefunctions contain

[1] W. A. Harrison, *Phys. Rev.*, **181**, 1036 (1969).

sharp oscillations in the neighborhood of the atoms and therefore that a plane-wave expansion will converge slowly, we seek an expansion in an overcomplete set including not only plane waves but atomic core functions. This point of view includes the method of expansion of the wavefunction in OPWs since each OPW is simply a sum of a plane wave and atomic core wavefunctions. From the success of the OPW method and the pseudopotential method in simple metals we know that such an expansion can be made rapidly convergent. We need not carry through this formulation of the simple-metal pseudopotentials in detail; it will be a special case of the transition-metal pseudopotential which we are to develop.

In the treatment of simple metals we required that the core states be eigenstates of the Hamiltonian in the metal. For a metal like copper we could not include atomic *d* states with the cores since the atomic states were not solutions in the metal, yet the *d*-like states were sufficiently strongly localized that their expansion in plane waves would be slowly convergent. Thus the essential benefit of the pseudopotential method was lost for a construction of *d*-like states. When we now view the formulation in terms of expansion in an overcomplete set it becomes natural to attempt a treatment of transition metals by using an overcomplete set including not only plane waves and core states but also atomic *d* states. Though the atomic *d* states are not eigenstates of the Hamiltonian they would seem to be just the additional terms needed for a rapidly converging expansion of the *d*-like wavefunctions. Such a method has in fact been successfully used by Deegan and Twose[1] to generalize the OPW method for calculation of transition-metal band structures.

It will actually be more convenient, though not essential, to use eigenstates determined from the free ion rather than the free atom. Thus in copper we will use a *d* state for Cu^+. The *d* states we use are eigenstates of the Hamiltonian containing the corresponding free-ion potential $V^i(r)$.

$$\left[\frac{-\hbar^2\nabla^2}{2m} + V^i(r)\right]|d\rangle = E_d{}^i|d\rangle \tag{2.70}$$

We may also of course take the core states to be eigenstates of the same Hamiltonian and therefore orthogonal to the $|d\rangle$. Now in the metal the true potential $V(\mathbf{r})$ in the neighborhood of any atom will differ from the free-ion potential by a quantity we write $-\delta V$; that is, $V(\mathbf{r}) = V^i(r) - \delta V(\mathbf{r})$. We will need to know the effect of the operation of the Hamiltonian in the metal upon the atomic *d* state. From Eq. (2.70) we see immediately

$$\left[\frac{-\hbar^2\nabla^2}{2m} + V(\mathbf{r})\right]|d\rangle = E_d{}^i|d\rangle - \delta V(\mathbf{r})|d\rangle \tag{2.71}$$

[1] R. A. Deegan and W. D. Twose, *Phys. Rev.*, **164**:993 (1967).

It will be more convenient to use as a parameter the expectation value of the Hamiltonian in the metal with respect to the d state, $E_d = \langle d| \frac{-\hbar^2\nabla^2}{2m} + V(\mathbf{r})|d\rangle$, rather than $E_d{}^i$. Note that $E_d = E_d{}^i - \langle d|\delta V|d\rangle$. The final form was obtained by multiplying Eq. (2.71) on the left by $\langle d|$. Then Eq. (2.71) may be rewritten as

$$\left[\frac{-\hbar^2\nabla^2}{2m} + V(\mathbf{r})\right]|d\rangle = E_d|d\rangle - \Delta|d\rangle \tag{2.72}$$

where Δ is defined by

$$\Delta|d\rangle = \delta V|d\rangle - |d\rangle\langle d|\delta V|d\rangle \tag{2.73}$$

Note that δV arises from the charge density due to the valence states in the metal and due to the distortion of the d states in the metal, as well as from the overlap of potentials from neighboring ions.

No approximation has been made as yet; these are simply definitions. The same definitions could be made for the core states. However, there we assume that the states are sufficiently small that the potential δV does not vary appreciably over the state; δV becomes a constant in Eq. (2.73) and Δ is seen to be zero.

We may now proceed to the derivation of a pseudopotential equation in a manner similar to that in the simple metals. The only difference is the existence of a nonvanishing Δ. Since the d states in the ion are very nearly eigenstates in the crystal we may expect the operator Δ to be small. We return to the eigenvalue equation in the metal,

$$\left[\frac{-\hbar^2}{2m}\nabla^2 + V(\mathbf{r})\right]|\psi\rangle = E|\psi\rangle \tag{2.74}$$

Now we expand the wavefunction in an overcomplete set including core states $|\alpha\rangle$ and free-ion d states $|d\rangle$ as well as plane waves.

$$|\psi\rangle = |\varphi\rangle + \sum_\alpha a_\alpha|\alpha\rangle + \sum_d a_d|d\rangle \tag{2.75}$$

The sum over plane waves has been written as a pseudowavefunction φ, just as in the simple-metal pseudopotential method. Eq. (2.75) is substituted in the Schroedinger equation, Eq. (2.74). We use Eq. (2.72) to rewrite the result of operation on the states $|d\rangle$ by the Hamiltonian and the corresponding equation without the Δ for operation upon the core states. Finally we collect all the terms in $|\alpha\rangle$ in $|d\rangle$ on the left to obtain

$$\frac{-\hbar^2}{2m}\nabla^2|\varphi\rangle + V|\varphi\rangle + \sum_\alpha a_\alpha(E_\alpha - E)|\alpha\rangle + \sum_d a_d(E_d - E)|d\rangle - \sum_d a_d\,\Delta|d\rangle = E|\varphi\rangle \tag{2.76}$$

To evaluate a_α we multiply on the left by $\langle\alpha|$. In the first two terms we use the Hermiticity of the Hamiltonian to let it operate to the left. We note further that since the variation of Δ over the core states can be neglected it may be taken out of the integral $\langle\alpha|\Delta|d\rangle$ and this integral is found to vanish due to the orthogonality of the core and the d states. We obtain immediately

$$a_\alpha = -\langle\alpha|\varphi\rangle \tag{2.77}$$

just as in the pseudopotential for simple metals. We next evaluate the a_d by multiplying on the left by $\langle d|$. For simplicity we make two approximations though neither is essential. First we neglect the overlap of the atomic d functions on adjacent atoms. Second we assume δV to be spherically symmetric so that $\langle d'|\Delta|d\rangle = 0$ if d' and d are different states on the same ion. In most cases the first approximation is justified; the overlap will be small. If it were not we could of course define the ionic potential to rise more rapidly at large r and eliminate the overlap; this would simply modify the definition of Δ. The second approximation also may be avoided by computing the crystal-field splitting of the d states due to the nonspherical components of Δ, resulting in linear combinations of ionic d states that will then yield vanishing matrix elements of Δ by construction. With these two approximations we may immediately obtain a_d,

$$a_d = -\langle d|\varphi\rangle + \frac{\langle d|\Delta|\varphi\rangle}{E_d - E} \tag{2.78}$$

An additional term has been obtained due to the potential correction Δ. Eqs. (2.77) and (2.78) are substituted back into Eq. (2.76) to obtain the *transition-metal-pseudopotential equation*,

$$\frac{-\hbar^2}{2m}\nabla^2|\varphi\rangle + W|\varphi\rangle - \sum_d \frac{\Delta|d\rangle\langle d|\Delta|\varphi\rangle}{E_d - E} = E|\varphi\rangle \tag{2.79}$$

where the *transition-metal pseudopotential* is given by

$$W|\varphi\rangle = V|\varphi\rangle + \sum_\alpha (E - E_\alpha)|\alpha\rangle\langle\alpha|\varphi\rangle + \sum_d (E - E_d)|d\rangle\langle d|\varphi\rangle$$
$$+ \sum_d (|d\rangle\langle d|\Delta|\varphi\rangle + \Delta|d\rangle\langle d|\varphi\rangle) \tag{2.80}$$

For $\Delta = 0$, Eqs. (2.79) and (2.80) become identical to the simple-metal equations with the d states regarded as core states. Also, like the pseudopotential equation for simple metals, these results are exact except for the inessential assumptions described earlier. The presence of Δ has added two repulsive terms to the pseudopotential, Eq. (2.80). In addition, it has given rise to a formally new term in the pseudopotential equation, the term quadratic in Δ. This will be called the *hybridization term* for reasons that will become clear when we discuss the energy bands.

As in the simple metals the point of the formulation is to allow perturbation theory. We may expect the pseudopotential, Eq. (2.80), to be small in the same sense that the simple-metal pseudopotential is. In addition, the hybridizing term will be small when we consider energies that are well removed from the energy E_d which we call the *resonant energy*. The term, however, diverges when the energy approaches resonance, and this is the new feature of the results which distinguishes it from simple-metal theory. We will return to the perturbation calculations but will first use these results for a discussion of the energy bands themselves.

9.2 The energy bands We may see qualitatively the form of the solutions of Eq. (2.79) by multiplying on the left by $\langle\varphi|$. Writing the expectation value $\langle\varphi| -\hbar^2\nabla^2/2m + W|\varphi\rangle$ as E_k we obtain

$$(E_k - E)(E_d - E) - \sum_d \langle\varphi|\Delta|d\rangle\langle d|\Delta|\varphi\rangle = 0 \qquad (2.81)$$

In the spirit of pseudopotential theory we may expect that pseudowavefunction $|\varphi\rangle$ can be approximated by a plane wave. Then $E_k \approx \hbar^2k^2/2m + E_0$. Now let the system be a crystal lattice made up of N transition-metal ions with nonoverlapping d states. For simplicity we include only a single atomic d state from each ion. Then every term in the sum over d in Eq. (2.81) becomes identical,

$$\sum_d \langle k|\Delta|d\rangle\langle d|\Delta|k\rangle = N\langle k|\Delta|d\rangle\langle d|\Delta|k\rangle \equiv \Delta_k^* \Delta_k$$

The factor N cancels against the normalization factor in the plane waves leading to a hybridization form factor Δ_k calculated with plane waves normalized in an atomic volume. Then Eq. (2.81) becomes precisely of the form of the secular equation based upon a 2-by-2 Hamiltonian

$$H = \begin{bmatrix} \dfrac{\hbar^2k^2}{2m} + E_0 & \Delta_k \\ \Delta_k^* & E_d \end{bmatrix}$$

Such a coupling of states is called hybridization in theoretical chemistry. Here we see precisely what the form of the coupling is. We measure energies from E_0 and the solutions of Eq. (2.81) become

$$E = \frac{E_d + \hbar^2k^2/2m \pm \sqrt{(E_d - \hbar^2k^2/2m)^2 + 4\Delta_k^* \Delta_k}}{2}$$

A plot of such solutions is given in Fig. 2.45 with Δ_k taken as a constant, $E_d/5$.

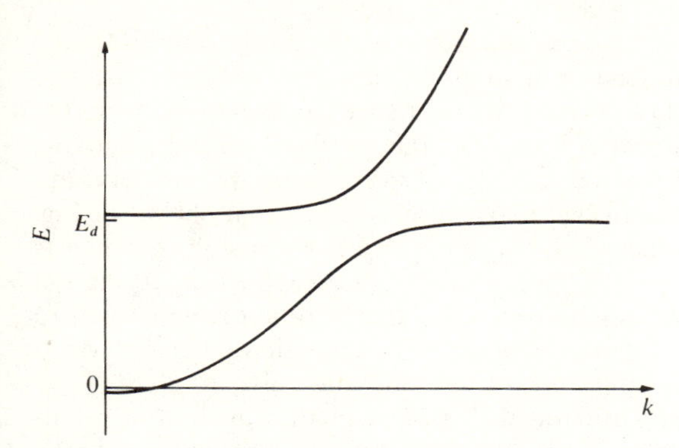

Fig. 2.45 *The energy bands arising from hybridization between a free-electron band and local atomic states.*

Fig. 2.45 is quite reminiscent of the energy bands of copper shown in Fig. 2.11, though in copper there is hybridization with all of the five d states of different angular-momentum components. In order to duplicate the true copper-band structure, we first note that in the treatment given above, the atomic d states could be replaced by tight-binding linear combinations. The plane wave $|\mathbf{k}\rangle$ is then coupled only to the tight-binding state of the same wavenumber and the same solutions are obtained. Second we include tight-binding states based upon each angular-momentum quantum number. Third we incorporate the effect of the overlap between neighboring atomic d states. Thus we begin our treatment of the bands by performing a tight-binding calculation of the d states, precisely as we calculated tight-binding bands for insulators. Fourth, rather than take $|\varphi\rangle$ to be a plane wave, we do a multiple-OPW calculation of $|\varphi\rangle$ based upon the pseudopotential of Eq. (2.80). Finally, when we introduce the hybridization we use matrix elements of Δ based upon the tight-binding d bands and the computed $|\varphi\rangle$.

If such a calculation were carried through in detail it would be a first-principles band calculation, leading to the correct eigenvalues of the initial Schroedinger equation. Of more interest is the fact that this approach provides a convenient framework for approximate interpolation for energy-band calculations carried out at specific points in the Brillouin zone. The tight-binding calculation for the d bands may be carried out approximately in terms of a few overlap parameters. A small number of plane waves, usually four, may be used in the calculation of $|\varphi\rangle$ and the dependence of matrix elements $\langle\mathbf{k} + \mathbf{q}|W|\mathbf{k}\rangle$ upon $\mathbf{k}$ can be neglected. Finally, a simple approximate form for the dependence of Δ_k upon k may be assumed.

This is in fact precisely the scheme that was developed long before the advent of transition-metal pseudopotentials for the fitting of band structures in transition metals.[1] In those treatments the many parameters that enter—e.g., the pseudopotential matrix elements, the overlap integrals, and the hybridization parameters—were regarded as disposable parameters and were adjusted to fit computed band structures such as that shown in Fig. 2.11. With so many parameters it is of course not surprising that it was possible to make a very good fit to the band structure. Such interpolation schemes made possible the summations over the Brillouin zone needed to obtain such properties as the density of states as a function of energy which were not tractable in the framework of the full band calculation.

More recently Heine[2] developed this same picture of the transition-metal band structures from a somewhat different point of view using the Green's function methods of Kohn and Rostoker and of Korringa which we mentioned briefly in Sec. 4 of this chapter. In this approach a muffin-tin potential was used, as in the APW method. All of the information about the atomic potentials enters through the logarithmic derivatives of the wavefunction at the surfaces of the inscribed spheres. These in turn can be written in terms of the phase shifts. The s and p phase shifts are of course described in terms of the simple-metal pseudopotentials; the $l = 2$ phase shift is taken to be of the resonant form, $\tan \delta_2 = \frac{1}{2}\Gamma/(E_d - E)$, and higher phase shifts were neglected. Thus it is possible to specify the band structure approximately in terms of only four parameters, δ_0, δ_1, Γ, and E_d. In this sense it was a major simplification over the earlier approaches. In addition he was able to make a direct estimate of the breadth of the resonance Γ (which in the transition-metal pseudopotential method is directly related to the matrix elements of Δ). In spite of the significant simplification over the earlier treatments, the actual determination of bands using Heine's method remains rather complicated, though some of the complexity is absorbed in structure constants which have been evaluated and published.

We should recognize that transition-metal energy bands themselves are quite complicated and we may expect that their determination can never be made mathematically simple. Their evaluation in terms of transition-metal pseudopotentials leads to a secular equation of the same form as in the earlier treatments and could be considered in that regard an advance only in putting the model on a firmer foundation. For more general problems, however, the transition-metal pseudopotential equation leads to the possibility of the use of perturbation theory, of the calculation of screening and of the

[1] M. Saffren, in W. A. Harrison and M. B. Webb (eds.), "The Fermi Surface," Wiley, New York, 1960; L. Hodges and H. Ehrenreich, *Phys. Rev. Letters*, **16**:203 (1965); F. M. Mueller, *Phys. Rev.*, **153**:659 (1967).

[2] V. Heine, *Phys. Rev.*, **153**:673 (1967).

total energy, and of the treatment of a vast array of properties of transition and noble metals. In such calculations, as in the simple metals, we bypass the detailed calculation of the bands themselves.

Before turning to that, we should note that we have spoken only of d resonances and in terms of the iron series. In this series the d-like states are believed to be describable in the ordinary band picture and the approach we have used is appropriate. The same approach of course is applicable to other transition-metal series but in the rare-earth series additional complications arise. The f levels are extremely well localized and are most usually thought of as being on the low-density side of a Mott transition. In addition, the interaction between the electrons becomes so strong that the energy of each state becomes strongly dependent upon which other states are occupied. Under such circumstances it becomes inappropriate to describe the states in terms of a band picture. It is instead necessary to explicitly treat the electron-electron interactions in the course of studying properties. We will see how some related effects in the iron series can be included in a band picture when we discuss local moments in Chap. V.

9.3 Perturbation theory and properties The formulation of perturbation theory based upon Eq. (2.79) for states well removed from resonance is immediate. When the states of interest lie near resonance, or when we sum over all states, the treatment becomes exceedingly complex. It is not appropriate to reproduce this complex algebra here but it may be helpful to outline the main features.[1]

In zero order the pseudowavefunction is a single plane wave $|\mathbf{k}\rangle$. It seems clear that we should obtain energies to the same order in the hybridization terms (which are proportional to Δ^2) and in the pseudopotential. Furthermore, matrix elements of the form $\langle d|\varphi\rangle$ should be taken as of the same order as $\langle d|\Delta|\varphi\rangle$. This follows from the assertion that all terms in the pseudopotential given in Eq. (2.80) are of the same order since we are taking Δ^2 as of the same order as W.

We may see, in the framework of perturbation theory, how Δ is to be computed. Recall that $-\delta V(\mathbf{r})$ includes the potential arising from the valence-band states and from distortion of the d-like states. To lowest order this is simply the potential due to zero-order valence states and, in fact, due to plane-wave valence states since the orthogonalization terms are of higher order. Lowest order is sufficient for the calculation of Δ, though we will see that higher order terms are needed for the screening calculation. The potential from neighboring ions is of simple Coulomb form and its spherical average will be constant. Thus to obtain $-\delta V(\mathbf{r})$ to lowest order,

[1] Details are given by W. A. Harrison, *Phys. Rev.*, **181**, 1036 (1969).

we simply calculate the average electron density ρ due to valence states, one per atomic volume for copper. Poisson's equation is integrated assuming a spherically symmetric correction potential at any ion to obtain $-\delta V = C - 2\pi\rho e^2 r^2/3$, from which Δ may be computed since the constant C drops out. It is interesting, and easy to show, that if we add the potential due to the valence charge of the ion, the total potential gradient vanishes at the Wigner-Seitz sphere.

It is convenient again to define an energy parameter E_k by

$$E_k = \frac{\hbar^2 k^2}{2m} + \langle \mathbf{k}|W|\mathbf{k}\rangle \tag{2.82}$$

This is a zero-order energy with some of the first-order contributions but not all. In perturbation theory it becomes valid to replace E by E_k in all terms in the pseudopotential of Eq. (2.80). Thus we discard some second-order terms in the pseudopotential which do not affect our results to the order to which we compute. Inclusion of the term $\langle \mathbf{k}|W|\mathbf{k}\rangle$ in Eq. (2.82) makes the results explicitly independent of the particular choice of the origin for energy, whereas with that term omitted the matrix elements of the pseudopotential depend upon the choice of origin. In addition the inclusion of $\langle \mathbf{k}|W|\mathbf{k}\rangle$ here produces in the corresponding contributions a form identical to the optimized pseudopotential for simple metals.

We now proceed systematically to perturbation theory based upon the pseudopotential equation, Eq. (2.79). We obtain the pseudowavefunction to first order in W

$$|\varphi\rangle = |\mathbf{k}\rangle + \sum_q \frac{|\mathbf{k}+\mathbf{q}\rangle\langle \mathbf{k}+\mathbf{q}|W|\mathbf{k}\rangle}{E_k - E_{k+q}} - \sum_q \frac{|\mathbf{k}+\mathbf{q}\rangle\langle \mathbf{k}+\mathbf{q}|\Delta|d\rangle\langle d|\Delta|\mathbf{k}\rangle}{(E_d - E_k)(E_k - E_{k+q})} \tag{2.83}$$

From this we can also obtain the true wavefunction to first order in W.

$$|\psi\rangle = |\varphi\rangle - \sum_\alpha |\alpha\rangle\langle\alpha|\mathbf{k}\rangle - \sum_d |d\rangle\langle d|\mathbf{k}\rangle + \sum_d \frac{|d\rangle\langle d|\Delta|\mathbf{k}\rangle}{E_d - E_k} \tag{2.84}$$

In Eq. (2.84) $|\varphi\rangle$ is given by Eq. (2.83). Similarly we may obtain the energy eigenvalue to second order in W. We obtain

$$E = E_k - \sum_d \frac{\langle \mathbf{k}|\Delta|d\rangle\langle d|\Delta|\mathbf{k}\rangle}{E_d - E} + \sum_q \frac{\langle \mathbf{k}|\left[W - \dfrac{\Delta|d\rangle\langle d|\Delta}{E_d - E_k}\right]|\mathbf{k}+\mathbf{q}\rangle\langle \mathbf{k}+\mathbf{q}|\left[W - \dfrac{\Delta|d\rangle\langle d|\Delta}{E_d - E_k}\right]|\mathbf{k}\rangle}{E_k - E_{k+q}} \tag{2.85}$$

Note that in the second term we have left E in the denominator since its first-order contributions will lead to a contribution to the energy which is second order.

These results are applicable to states well removed from resonance where the hybridization term may be regarded as small. For such states, we compute Fermi surfaces or electron scattering precisely as we did for simple metals. The only difference is that a hybridization perturbation has been included in addition to the pseudopotential.

In copper, for example, the d resonances lie well below the Fermi energy. Thus for states at the Fermi surface we may treat both the pseudopotential and the hybridization terms as a perturbation. We may even construct an OPW form factor in terms of the matrix element of the sum of the two terms between two plane waves with wavenumber at the Fermi surface. As in simple metals this matrix element can be factored into structure factor and a form factor. The form factor, illustrated for copper in Fig. 2.46, enters the calculation of electronic properties precisely as do the OPW form factors for simple metals.

The perturbation approach is no longer justified, however, if the states of interest lie near resonance where the hybridization term diverges. In particular if we seek a self-consistent evaluation of the potential (even in copper) we must sum the charge density of all occupied states, including those near resonance. We might at first hope to evaluate the charge density to first order in W using the wavefunctions of Eq. (2.84). A convergent result is obtained if we replace the sum over wavenumber by an integral and take principal values in the integration through resonance. It is not difficult to see that such a summation over unperturbed states is incomplete. This is easily seen for the simple case of a single transition metal atom dissolved in a simple metal or a free-electron gas. As earlier we use spherical boundaries and consider only a single component of angular momentum so that only one d state is included in the sum over d. Then we examine the wavefunction as we take E_k through resonance. To zero order, the pseudowavefunction (which is equal to the true wavefunction at large distances) is given by the spherical Bessel function $j_2(kr)$. As indicated by the second two terms in Eq. (2.84) this pseudowavefunction is to be orthogonalized to the core and to the d states. In addition, the true wavefunction contains the atomic d state, as represented by the final term in Eq. (2.84). This term becomes large at resonance and changes sign as we pass through resonance. This description can be brought into closer correspondence with our discussion of phase shifts if we select the sign of the wavefunction such that the sign of the atomic d state remains fixed through resonance. Then the amplitude of the oscillating tail, corresponding to the pseudowavefunction $j_2(kr)$ changes sign. This of course represents exactly the introduction of a π phase shift as we cross resonance. Thus though our zero-order state $j_2(kr)$ below resonance represents a true unperturbed state (vanishing phase shift), the zero-order state $j_2(kr)$ above resonance represents a state in which a phase shift of π has already been introduced. In changing zero-order states as we sum through resonance we omit a single perturbed state.

This is perhaps more easily seen in terms of the energy. We see from the second term in Eq. (2.85) that the effect of the hybridization is to slightly lower the eigenvalues below resonance but to slightly raise the eigenvalues above resonance, whereas we know that an attractive interaction has introduced the resonance and an attractive interaction will lower the energy of *every* state. Below resonance every state is slightly lowered and the perturbation theory correctly estimates this lowering. Above resonance every state is lowered very nearly to the unperturbed energy of the next lowest state and our perturbation theory computes the difference between the

Fig. 2.46 The computed pseudopotential form factor for copper, including the effects of hybridization as well as those of the transition-metal pseudopotential for copper. This form factor enters the calculation of electronic properties in the same way as do the simple-metal OPW form factors. From J. A. Moriarty [Phys. Rev., ***B6***, *1239* (*1972*)].

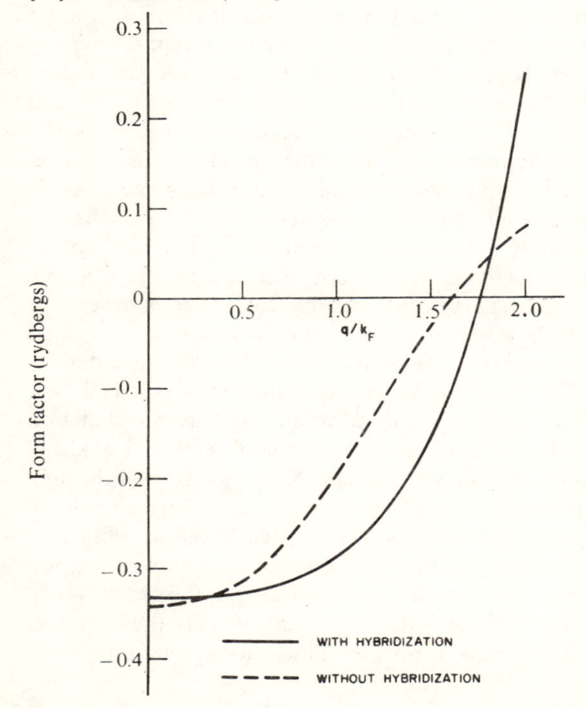

final energy and the energy of the next lowest unperturbed state. This correspondence is illustrated in Fig. 2.47 from which it is clear that in summing over unperturbed states we omit a single state.

We might at first expect that the omission of a single state would not be important. In a transition-metal, however, the number of d resonances introduced is comparable to the number of conduction-band states and clearly these contributions must be added. Even in the case of a single resonant center, it is not difficult to see that the effect of the omitted state either on the screening or on the total energy is comparable to the contribution of the sum of perturbations over the remaining states.

We must therefore supplement our perturbation calculation, whenever we wish to sum over states, by constructing the wavefunction of the omitted d-like state. The natural approach for doing this is to return to the Schroedinger equation, in the form given in Eq. (2.76), and take a set of coefficients, a_d, to be of zero order. We then introduce the plane-wave components as first-order perturbations. When this is done carefully, it can be verified that the added states which are obtained are orthogonal to the appropriate order to those obtained in perturbation theory but based on plane-wave zero-order states. We will return to such a treatment of

Fig. 2.47 *A schematic diagram of the unperturbed and perturbed levels computed in perturbation theory when a resonance is introduced. The quantization of unperturbed levels arises from the crystal boundaries.*

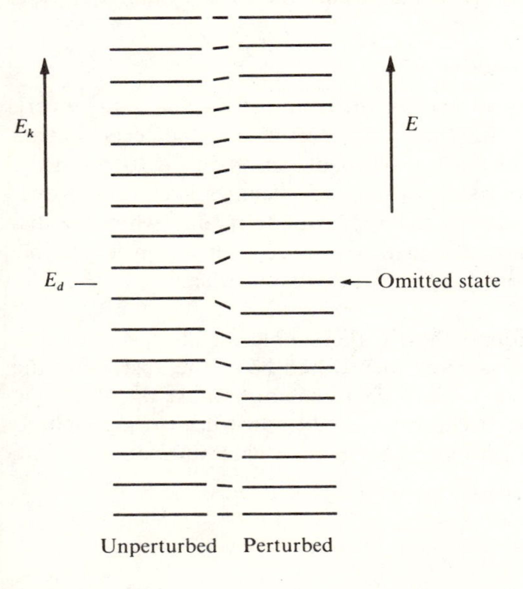

d states when we discuss screening of pseudopotentials in Chap. III, and when we discuss atomic properties in Chap. IV.

Once we have computed both the valence-band states and the *d*-like states by perturbation theory we may directly proceed to the computation of properties. As we have indicated we may sum over states to obtain the charge density self-consistently. This was in fact necessary to obtain the potential that entered the form factor shown in Fig. 2.46. That form factor may then be used in the computation of a wide range of electronic properties.

The perturbation approach to transition metals makes an artificial distinction between *k*-like states and *d*-like states and correspondingly the eigenvalues obtained correspond to a somewhat artificial band structure. In a similar way, second-order perturbation theory for the bands in simple metals led to artificial bands near the zone faces and it was necessary to diagonalize a submatrix in the multiple-OPW method to obtain meaningful results for such states. Nevertheless, most properties of metals depend upon integrals over the states and in many cases meaningful values may be obtained, both in simple and in transition metals, in terms of the results of simple perturbation theory. Furthermore we have seen that in noble metals the Fermi energy is sufficiently far from resonance that the electronic properties can be treated just as simply as in the simple metals once an OPW form factor has been obtained. In both the simple metals and the transition metals the pseudopotential equation is a valid starting point. The divergences arise from the use of perturbation theory and alternative approximations may be made when the perturbation theory fails.

10. Electronic Structure of Liquids

The periodicity of the lattice has given us the band picture and a framework for discussing both perfect crystals and perfect crystals with defects in them. In the liquid, however, all of the translational periodicity is lost and the general framework which we have constructed becomes irrelevant. Nonetheless it is not difficult to see that the approximate models which we discussed for metals and for insulators can be carried over and it is possible to discuss the electronic states even with the loss of order.

10.1 Simple metals[1] We found that in the simple metals the electronic energies were very much the same as they would be in the absence of the periodic potential arising from the ion. This was not because the electronic states were plane waves but because we could construct orthogonalized plane waves, the orthogonalization being to each core state on each ion,

[1] A variety of discussions of liquid metals are given in P. D. Adams, H. A. Davies, S. G. Epstein (eds.), "The Properties of Liquid Metals," Taylor and Francis, London, 1967.

and the expectation value of the energy based upon these was close to the free-electron value.

We may in principle again proceed in this way for the liquid. We might specify the position $\mathbf{r}_j$ for each of the N ions. We can then orthogonalize the plane waves to the core state on every ion, wherever it happens to sit. We may again construct a pseudopotential for the system which may be written as the sum of pseudopotentials arising from the individual ions.

$$W(\mathbf{r}) = \sum_j w(\mathbf{r} - \mathbf{r}_j)$$

where the individual pseudopotentials are the same as those given before. To be specific let us take a form of the pseudopotential from Eq. (2.26) of Sec. 5.

$$w(\mathbf{r}) = v(\mathbf{r}) + \sum_t \left(\frac{\hbar^2 k^2}{2m} + \langle \mathbf{k}|w|\mathbf{k}\rangle - \epsilon_t \right) |t\rangle\langle t|$$

We have replaced the core energy in the metal, E_t, by the value it would have in the free atom, ϵ_t; we could equally as well have taken a value averaged over the liquid system. Nothing has changed upon melting except that it is more difficult to specify the position of the individual ions.

We proceed now to compute the energy of the electronic states in increasing orders of the pseudopotential. We may find first the zero-order energy, which is simply the kinetic energy of free electrons as it was in the perfect crystal. The first-order correction is given by $\langle \mathbf{k}|W|\mathbf{k}\rangle = \langle \mathbf{k}|w|\mathbf{k}\rangle$. If the pseudopotential were a simple local potential this would be a constant determining the energy-band minimum. Actually we may see from the form for the pseudopotential given above that this first-order term does depend on wavenumber and therefore will cause some distortion of the free-electron band. This correction is ordinarily small, corresponding approximately to a shift in the effective mass of a few percent.

Up to this point the configuration of the ions has not entered and our results are the same as for the perfect crystal. The ion positions will enter the calculation as we proceed to second order. We then obtain an additional shift in the energy which we may write

$$E_k^{(2)} = \sum_q S^*(\mathbf{q}) S(\mathbf{q}) \frac{\langle \mathbf{k}|w|\mathbf{k} + \mathbf{q}\rangle\langle \mathbf{k} + \mathbf{q}|w|\mathbf{k}\rangle}{\hbar^2/2m(k^2 - |\mathbf{k} + \mathbf{q}|^2)}$$

as we did for the perfect crystal. Now, however, the structure factor, which is given by

$$S(\mathbf{q}) = N^{-1} \sum_j e^{-i\mathbf{q}\cdot\mathbf{r}_j}$$

is no longer a delta function at the lattice wavenumbers. It will in fact have nonvanishing values at every wavenumber. Fortunately in our calculation it is only the amplitude of the structure factor which enters through S^*S. This is precisely the function of structure which is measured directly in x-ray or neutron-diffraction experiments. A typical measured structure factor is shown in Fig. 2.48. Thus in spite of the fact that we do not know the individual ion configuration in the liquid, we can experimentally obtain precisely the information needed for the calculation of these energies to second order.

It is of course no accident that the structure factor needed for our calculation is precisely that which is measured in diffraction experiments. The energy-band structure itself arises directly from the diffraction of electrons by this same configuration.

In the perfect crystal the evaluation of the energy involved a number of plane waves coupled by a few small lattice wavenumbers. When two coupled states were nearly degenerate it was necessary to diagonalize the appropriate submatrix. In the case of the liquid this sum over wavenumber must be replaced by an integral since we have nonvanishing contributions from every wavenumber. In this integration we will of course cross wavenumbers at which the energy denominator vanishes, but in this case we may simply take principal parts of the integral and the resulting energy, as a function of wavenumber, is not itself singular.

Such a calculation of the energy band for a liquid metal is straight-

Fig. 2.48 The amplitude of the structure factor squared times the number of atoms present for aluminum, taken from x-ray data of C. Gamertsfelder [J. Chem. Phys., ***9:****450*(*1941*)].

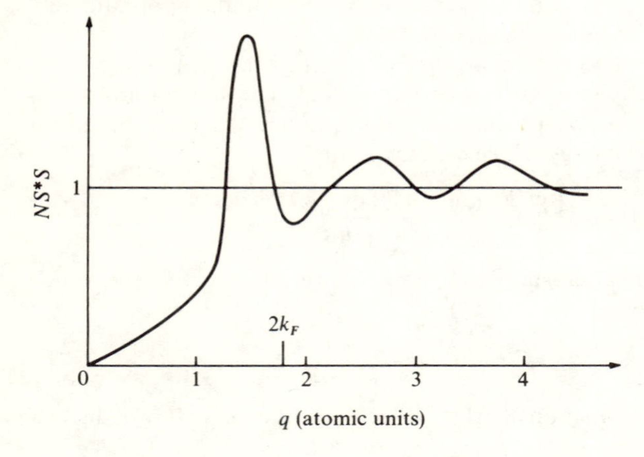

forward. We use experimental structure factors and computed form factors and perform the integrals numerically. Although only a limited number of such calculations have been performed,[1] it has seemed clear that in the simple metals the distortions of the energy bands from free-electron parabolas are only rather minor.

It is legitimate here to question the use of perturbation theory for this problem. We expect that if we could actually solve for the true eigenstates, they would not bear very much resemblance to the one-OPW states which we are using as a starting approximation; certainly wavenumber is not a good quantum number for these states. The rigorous meaning of the dispersion curve which we have computed is under some question.

This same argument, however, could be applied to a perfect crystal with a single vacancy, though it seems intuitively clear for that case that band states are a reasonable basis set. We further note that though we have multiplied the number of states to which a single OPW is coupled by a factor of the order of the number of atoms present, N, we have cut the matrix element connecting them by a factor of $N^{-1/2}$ in comparison to that in the perfect crystal. Thus our computed loss in amplitude from the zero-order state would be comparable to that in the perfect crystal.

We will see in Sec. 10.3 how the theory of liquid metals can be placed on a more rigorous footing, though this formulation will not really invalidate the ideas given here.

10.2 Insulators and semiconductors We have found the tight-binding method useful in the discussion of insulating crystals. It is therefore interesting to see the form this approach takes in a liquid insulator. To be specific we consider the band arising from overlapping $3s$ functions centered on the sodium ions. In contrast to the perfect crystal we cannot treat these overlaps in an orderly fashion. They will differ from ion to ion and will depend upon the configuration of each ion's neighbors. Nonetheless it seems clear that, if the band in the perfect insulator were narrow in comparison to the separation between atomic levels, the net effect of these interactions in the liquid would simply be to broaden the atomic levels into diffuse bands. In this case we could compute statistically the expected spread in the levels from an assumed distribution of neighbor distances.

We restrict our attention to the effect of two-center overlap integrals. The distribution of these will depend only on the pair-distribution function. This again is directly obtainable from the diffraction studies of the liquid. It is well known that the probability of finding two atoms at a separation r (which is simply the pair-distribution function) is simply the Fourier trans-

[1] The most recent such calculations are given by R. W. Shaw and N. Smith, *Phys. Rev.*, **178**:985 (1969).

form of the squared amplitude of the structure factors. We may see this by writing

$$S^*(\mathbf{q})S(\mathbf{q}) = \frac{1}{N^2}\sum_{ij} e^{-iq\cdot(r_i - r_j)} = \frac{1}{N}\langle\sum_r e^{-iq\cdot r}\rangle + \frac{1}{N}$$

where the sum over **r** is a sum over all neighbors of a given atom and this sum is to be averaged over all atoms. The term $1/N$ arises from the terms $i = j$. This expression can be written in terms of the probability per unit volume of finding an atom at a distance r, $P(r)$.

$$\langle\sum_r e^{-iq\cdot r}\rangle = \int P(r)\, e^{-iq\cdot r}\, d\tau = \int P(r)\,\frac{\sin qr}{qr}\, 4\pi r^2\, dr$$

By transforming back we may obtain $P(r)$ in terms of the measured structure factors. The calculation of the electronic energies based upon such a $P(r)$ is moderately straightforward and is outlined in Prob. 2.29.

We might finally ask about the structure of a semiconductor, which in the crystal has bandwidths that are comparable to or larger than the separation between bands. In fact most semiconductors when melted become metallic. For example, liquid silicon and germanium are both metallic and can sensibly be treated as simple metals. However, some semiconductors such as germanium can be obtained in an amorphous, or glassy, state when deposited in films at low temperatures. The density is low, as in the semiconducting phase, but there appears to be no long-range regularity in the structure. Apparently the disorder gives rise to a high density of traps spread throughout the gap. In spite of these many traps the properties of such *amorphous semiconductors* are very much like those of an intrinsic crystalline semiconductor. We will sharpen up our description of the corresponding electronic structure and examine its consequences after discussing electronic properties in general in the following chapter.

10.3 Description in terms of one-electron Green's functions[1] Although we have found that the electronic structure of metals can be described in terms of the techniques we have used for perfect crystals, it is of some interest to see a more rigorous formulation. One-electron Green's functions have become more and more prominent in the theory of solids in recent years, though in many cases the calculations could equally as well have been carried out with the older methods. Thus in order to understand the current literature it is necessary to be familiar with the formalism. Furthermore in the case of simple liquid metals we will see that a more meaningful definition of electronic

[1] This discussion is based upon the treatment given originally by S. F. Edwards, *Proc. Roy. Soc. (London)*, **A267**:518 (1962).

structure can be given. Here we will simply define the one-particle Green's function. The relation to classical Green's functions may be seen in the Appendix to this section.

As in our other discussion of liquids we will not concern ourselves with the fact that ions themselves move in time. The electronic motions are very much more rapid and to a good approximation we should obtain the same results for a "frozen" liquid based upon the positions of the ions in the real liquid at some given instant. Given the ion positions we could in principle write down the Hamiltonian of the system involving superimposed pseudopotentials centered on the ion positions. Furthermore, in principle, we could evaluate all of the electronic eigenstates though this would be impossible in practice. From a one-electron point of view these eigenstates and eigenvalues provide a complete description of the system. From these eigenstates and eigenvalues we can compute any property of the liquid that we choose. We will see that the Green's function provides an alternative description of the system; it contains the same information that the eigenstates and eigenvalues contain and it is possible from the Green's function to compute any property. However, the perturbation expansion of the Green's function can be more easily carried to high order, and the calculation of many properties from it is more direct than that from the eigenstates. This benefit is gained with some loss of conceptual simplicity.

The wavefunction in the Schroedinger representation is determined, of course, from the equation

$$H\psi(\mathbf{r},t) = i\hbar \frac{\partial}{\partial t} \psi(\mathbf{r},t)$$

The Green's function is a function of two spatial variables $\mathbf{r}$ and $\mathbf{r}'$ and two time variables t and t'. This is perhaps the major conceptual complication in comparison to the wavefunction. Two Green's functions are defined, indicated by a plus or a minus, by the equation

$$\left(i\hbar \frac{\partial}{\partial t} - H \mp i\epsilon\right) G_{\pm}(\mathbf{r},\mathbf{r}';t,t') = \delta(\mathbf{r} - \mathbf{r}')\,\delta(t - t') \tag{2.86}$$

The time derivative and the Hamiltonian operate on the coordinates $\mathbf{r}$ and t. The ϵ is a small real parameter which will go to zero in the end. This is simply a formal definition of the Green's function. We will first write down the solutions of this equation which may be verified by substitution. We will then study the Green's function and see how physical properties of the system are determined in terms of it.

We consider only the case for which the Hamiltonian is independent of time. Then there exist eigenstates of the Hamiltonian which we write $\psi_n(\mathbf{r})$

with energies E_n. The Green's function may be written in terms of these in the form

$$G_{\pm}(\mathbf{r},\mathbf{r}';t,t') = -\frac{i}{\hbar}\sum_n \psi_n(\mathbf{r})\psi_n^*(\mathbf{r}')\, e^{-i(E_n \pm i\epsilon)(t-t')/\hbar}\, \Theta_{\pm}(t - t')$$

where the Θ function is defined by

$$\Theta_+(x) = \begin{matrix} 0 & \text{for } x > 0 \\ -1 & \text{for } x < 0 \end{matrix}$$

$$\Theta_-(x) = \begin{matrix} 1 & \text{for } x > 0 \\ 0 & \text{for } x < 0 \end{matrix}$$

This may be substituted directly into Eq. (2.86) noting from the definition of the Θ function that its time derivative is a δ function. $\sum_n \psi_n(\mathbf{r})\psi_n^*(\mathbf{r}')$ is simply $\delta(\mathbf{r} - \mathbf{r}')$ since the ψ_n form a complete set. (This may be seen by multiplying it by an arbitrary function $f(\mathbf{r}')$ and integrating over $\mathbf{r}'$. The volume integral over $\mathbf{r}'$ simply gives the expansion coefficients of $f(\mathbf{r}')$ in terms of the eigenstates and the sum over n reexpands that function in $\mathbf{r}$ space.) The remaining terms cancel and we have demonstrated that both Green's functions are solutions to Eq. (2.86).

We may note immediately that the Green's function takes the wavefunction at one position and time to the wavefunction at some other position and time; that is,

$$\psi(\mathbf{r},t) = \begin{cases} i\hbar \displaystyle\int G_-(\mathbf{r},\mathbf{r}';t,t')\psi(\mathbf{r}',t')\, d^3r' & \text{for } t > t' \\ -i\hbar \displaystyle\int G_+(\mathbf{r},\mathbf{r}';t,t')\psi(\mathbf{r}',t')\, d^3r' & \text{for } t < t' \end{cases}$$

Again the integration over $\mathbf{r}'$ gives us the expansion coefficient of the wavefunction at time t' in terms of the eigenstates. The exponential in time gives the change in phase of each of these coefficients between the time t' and the time t, and finally the sum over n reexpands the wavefunction at the time t. Note that the plus Green's function gives the negative of the wavefunction if the time t is earlier than the time t' and zero otherwise. The minus Green's function gives the wavefunction itself if the time t is later than the time t' and zero otherwise. Thus the Green's function represents an integration of the Schroedinger equation over time. This is a very central aspect of the Green's function but not one that we will use directly here.

Note that the Green's function for this case, in which H is independent of time, depends only upon the difference in time. Thus we may directly make a Fourier transform in the time variable to obtain a Green's function which is a function of ω; that is,

$$G_{\pm}(\mathbf{r},\mathbf{r}';t,t') = \frac{1}{2\pi}\int d\omega\, G_{\pm}(\mathbf{r},\mathbf{r}';\hbar\omega)e^{-i\omega(t-t')}$$

This new Green's function may be obtained directly by performing the Fourier transform (see Prob. 2.30)

$$G_{\pm}(\mathbf{r},\mathbf{r}';\hbar\omega) = \sum_n \frac{\psi_n(\mathbf{r})\psi_n^*(\mathbf{r}')}{\hbar\omega - E_n \mp i\epsilon} \tag{2.87}$$

and is a solution of the equation

$$(\hbar\omega - H \mp i\epsilon)\, G_{\pm}(\mathbf{r},\mathbf{r};\hbar\omega) = \delta(\mathbf{r} - \mathbf{r}') \tag{2.88}$$

Having the form of G given in Eq. (2.87) we may show that the density of states as a function of energy is related to the "trace" of the Green's function; that is,

$$\begin{aligned} n(E) &= \frac{1}{2\pi i}\int [G_+(\mathbf{r},\mathbf{r};E) - G_-(\mathbf{r},\mathbf{r};\mathrm{E})]\, d^3r \\ &= \frac{1}{2\pi i}\sum_n \left(\frac{1}{E - E_n - i\epsilon} - \frac{1}{E - E_n + i\epsilon}\right) = \frac{\epsilon}{\pi}\sum_n \frac{1}{(E - E_n)^2 + \epsilon^2} \end{aligned} \tag{2.89}$$

The final form approaches a sum of δ functions at the energy eigenvalues E_n as ϵ approaches zero; thus the sum becomes the density of states.

This may be readily verified by integrating a single term of Eq. (2.89) over an energy range, say from E_1 to E_2.

$$\int_{E_1}^{E_2} n(E)\, dE = \frac{1}{\pi}\sum_n \int_{E_1}^{E_2} \frac{\epsilon\, dE}{(E - E_n)^2 + \epsilon^2}$$

$$= \sum_n \frac{1}{\pi}\left[\arctan\left(\frac{E_2 - E_n}{\epsilon}\right) - \arctan\left(\frac{E_1 - E_n}{\epsilon}\right)\right]$$

The arc tangents will be plus or minus $\pi/2$ in the limit as ϵ approaches zero depending upon whether the limit in question is greater or less than E_n. Thus if the nth state lies in the energy range E_1 to E_2, the term will contribute a one to the integral. If the energy E_n lies outside the range, no contribution to the integral will be made. Thus the sum in Eq. (2.89) will contribute unity to an integration over energy for each state which lies in that energy range and therefore is precisely the density of states. From the density of states alone we may compute the partition function for the system and therefore all of the thermodynamic properties. The Green's function gives this information in a very direct way.

In order to proceed further it will be desirable to Fourier transform the Green's function in the spatial coordinates as well as time coordinates. If the system were translationally invariant, the Green's function would be a function only of $\mathbf{r} - \mathbf{r}'$ and could be characterized by a single Fourier transform as was true for the time variation. However, neither a liquid nor a solid has full translational symmetry and therefore we must construct a Fourier transform on both of the coordinates. The Green's function may be written

$$G_{\pm}(\mathbf{r},\mathbf{r}';E) = \sum_{k,k'} |\mathbf{k}\rangle\langle\mathbf{k}|G_{\pm}|\mathbf{k}'\rangle\langle\mathbf{k}'|$$

where $|\mathbf{k}\rangle$ is a normalized plane wave in $\mathbf{r}$, and $\langle\mathbf{k}'|$ the complex conjugate of a plane wave in $\mathbf{r}'$. The Fourier transform of the Green's function is frequently written

$$G_{\pm}(\mathbf{k},\mathbf{k}';E) = \langle\mathbf{k}|G_{\pm}|\mathbf{k}'\rangle = \frac{1}{\Omega}\int d^3r\, d^3r'\, e^{-i\mathbf{k}\cdot\mathbf{r}}\, G_{\pm}(\mathbf{r},\mathbf{r}';E)\, e^{i\mathbf{k}'\cdot\mathbf{r}'}$$

Of particular interest is the diagonal term, $\langle\mathbf{k}|G|\mathbf{k}\rangle$. It again contains a δ function in energy and now the squared amplitude of the expansion coefficient of each state in plane waves. Thus if all of the states in the energy range in question are occupied, this Green's function gives the probability of finding an electron with energy E and wavenumber $\mathbf{k}$. This is closely related to what we have called energy-band structure but the Green's function is well defined for any one-electron system (even liquids) while the meaning of the energy-band structure in the liquid is a little tenuous. We will see this a little more clearly when we now attempt to compute the Green's function itself.

We begin by multiplying Eq. (2.88) on the left by $\langle\mathbf{k}|$ and on the right by $|\mathbf{k}'\rangle$ and inserting a $\sum_{k''}|\mathbf{k}''\rangle\langle\mathbf{k}''|$ (which is the identity) in the middle. We obtain

$$\sum_{k''}\langle\mathbf{k}|E - H \mp i\epsilon|\mathbf{k}''\rangle\langle\mathbf{k}''|G_{\pm}|\mathbf{k}'\rangle = \delta_{kk'} \tag{2.90}$$

Now H contains the kinetic-energy operator and the pseudopotential or potential. Here we will write the latter as W since all of our algebra will be consistent with use of a nonlocal pseudopotential. In most literature V appears since a simple potential has usually been envisaged. We note that $\langle\mathbf{k}|\hbar^2\nabla^2/2m|\mathbf{k}'\rangle = \hbar^2k^2/2m\, \delta_{kk'}$. We write $\hbar^2k^2/2m = \epsilon_k$ to simplify notation. Then Eq. (2.90) becomes

$$(E - \epsilon_k \mp i\epsilon)\langle\mathbf{k}|G_{\pm}|\mathbf{k}'\rangle - \sum_{k''}\langle\mathbf{k}|W|\mathbf{k}''\rangle\langle\mathbf{k}''|G_{\pm}|\mathbf{k}'\rangle = \delta_{kk'} \tag{2.91}$$

The Green's function is a particularly convenient formulation for perturbation theory, treating the potential as a perturbation, and we will proceed to that now. (Much of many-body theory is similarly based upon two-electron Green's functions, treating the interaction between particles as a perturbation.) The zero-order Green's function is obtained immediately from Eq. (2.91) by setting W equal to zero and solving for the Green's function.

$$\langle \mathbf{k}|G_{\pm}{}^{0}|\mathbf{k}'\rangle = \frac{\delta_{kk'}}{E - \epsilon_k \mp i\epsilon} \tag{2.92}$$

We note that only the diagonal components, $\mathbf{k} = \mathbf{k}'$, of the zero-order Green's function are nonvanishing. These represent the probability of finding an electron with wavenumber $\mathbf{k}$ and energy E. (We obtained this interpretation by integrating over E.) In zero order we find a nonvanishing contribution only for wavenumbers such that $\hbar^2k^2/2m = E$. We may note, in terms of the structure of the Green's function, that each eigenstate of the system corresponds to a pole in the Green's function. In these terms the purpose of the small parameter ϵ is to shift the pole off the real axis, to tell us on which side of the pole we are to pass in an integration over energy.

For $W = 0$ we have a free-electron gas and we see that in the (k,E) plane the Green's function has its poles along the line $E = \hbar^2k^2/2m$, representing the free-electron band structure.

Let us turn now to the perturbation expansion for the Green's function. For simplicity we will suppress the $\pm$ and the E which are the same for every Green's function in our expressions. Thus we will write $G(\mathbf{k},\mathbf{k}') = G_{\pm}(\mathbf{k},\mathbf{k}';E)$ and again will use a superscript of zero to indicate the zero-order Green's function of Eq. (2.92). In this notation Eq. (2.91) becomes

$$(E - \epsilon_k \mp i\epsilon)\, G(\mathbf{k},\mathbf{k}') - \sum_{k''} \langle \mathbf{k}|W|\mathbf{k}''\rangle\, G(\mathbf{k}'',\mathbf{k}') = \delta_{kk'} \tag{2.93}$$

The perturbation expansion may be written

$$G(\mathbf{k},\mathbf{k}') = G^0(\mathbf{k})\,\delta_{kk'} + G^0(\mathbf{k})\langle \mathbf{k}|W|\mathbf{k}'\rangle\, G^0(\mathbf{k}') + \sum_{k_1} G^0(\mathbf{k})\langle \mathbf{k}|W|\mathbf{k}_1\rangle\, G^0(\mathbf{k}_1)\langle \mathbf{k}_1|W|\mathbf{k}'\rangle\, G^0(\mathbf{k}') + \cdots \tag{2.94}$$

where in the zero-order Green's function we have suppressed the second wavenumber index since the two indices are always the same. This expansion may be verified by substitution back into Eq. (2.93). Zero-order Green's functions appearing in each term are called *propagators* and provide the energy denominators of the perturbation theory. The matrix elements of the potential are called *vertices*. In field theory each of these terms is represented by a Feynmann diagram.

It is striking that the higher-order terms in the perturbation expansion of the Green's function can be written down so systematically. We see in fact that the expansion looks very much like a simple geometric series, which means it is possible to sum the series for $G(\mathbf{k},\mathbf{k})$ to all orders. We see this by defining a Σ which is an arbitrary function of wavenumber and energy noting that

$$\frac{1}{E-\epsilon_k \mp i\epsilon - \Sigma} = \frac{1}{E-\epsilon_k \mp i\epsilon}\left[1 + \frac{\Sigma}{E-\epsilon_k \mp i\epsilon} + \frac{\Sigma^2}{(E-\epsilon_k \mp i\epsilon)^2} + \cdots\right]$$

$$= G^0(\mathbf{k})\{1 + G^0(\mathbf{k})\Sigma + [G^0(\mathbf{k})\Sigma]^2 + \cdots\} \tag{2.95}$$

If we now define

$$\Sigma(\mathbf{k},E) = \langle\mathbf{k}|W|\mathbf{k}\rangle + \sum_{k_1\neq k}\langle\mathbf{k}|W|\mathbf{k}_1\rangle\, G^0(\mathbf{k}_1)\langle\mathbf{k}_1|W|\mathbf{k}\rangle$$
$$+ \sum_{\substack{k_1\neq k\\ k_2\neq k}}\langle\mathbf{k}|W|\mathbf{k}_1\rangle\, G^0(\mathbf{k}_1)\langle\mathbf{k}_1|W|\mathbf{k}_2\rangle\, G^0(\mathbf{k}_2)\langle\mathbf{k}_2|W|\mathbf{k}\rangle + \cdots \tag{2.96}$$

we see that the expression in Eq. (2.95) is precisely the diagonal Green's function of Eq. (2.94). The first term in which a Σ appears in Eq. (2.95) gives the entire Green's function expansion of Eq. (2.94) except for those terms in which one or more of the intermediate wavenumbers is equal to $\mathbf{k}$. The Σ^2 term gives all contributions in which *one* of the intermediate states is $|\mathbf{k}\rangle$, and so forth. Thus we have found that the Green's function may be written in the form

$$G(\mathbf{k},\mathbf{k}) = \frac{1}{E-\epsilon_k - \Sigma(\mathbf{k},E) \mp i\epsilon} \tag{2.97}$$

where Σ (which is called the *proper self-energy*) is given by Eq. (2.96). We have been able to sum all orders in the perturbation theory but unfortunately our final result contains a Σ which requires again a summation in all orders for evaluation.

Nonetheless, if we simply evaluate Σ to second order in W we may substitute the result back into Eq. (2.97), divide it out, and see that in fact this does include the summation of *some* terms in the perturbation expansion for the Green's function to all orders. In the usual language of field theory we would say that we had summed a certain class of diagrams to all orders. In this problem, if we evaluate Σ to second order in the pseudopotential, we expect significant improvement in our results over a straight expansion of the Green's function to second order in the pseudopotential.

As a particular example we might notice that if our system involved only two states, $\mathbf{k}_a$ and $\mathbf{k}_b$, which were coupled by a matrix element

$\langle \mathbf{k}_b | W | \mathbf{k}_a \rangle = W$ and $\langle \mathbf{k}_a | W | \mathbf{k}_b \rangle = W^+$ (we also take $\langle \mathbf{k}_b | W | \mathbf{k}_b \rangle = \langle \mathbf{k}_a | W | \mathbf{k}_a \rangle = 0$), then Σ to second order becomes

$$\Sigma(\mathbf{k}_b, E) = \frac{WW^+}{E - \epsilon_a \mp i\epsilon} \tag{2.98}$$

and Eq. (2.97) becomes

$$G(\mathbf{k}_b, \mathbf{k}_b) = \frac{E - \epsilon_a \mp i\epsilon}{(E - \epsilon_b \mp i\epsilon)(E - \epsilon_a \mp i\epsilon) - WW^+}$$

Though we have carried Σ only to second order the result is exact. This may be seen in two ways. First, all higher-order terms in Eq. (2.96) vanish since there is only one intermediate state. We also see that the result is exact by seeking poles in $G(\mathbf{k}_b, \mathbf{k}_b)$, i.e., parameters such that the denominator vanishes. Setting the denominator equal to zero in fact leads precisely to the secular equation for the 2-by-2 Hamiltonian matrix. We see that the summation to all orders has rectified one defect of ordinary perturbation theory; i.e., the divergence which occurs when two degenerate states are coupled.

In this example we took the diagonal terms equal to zero. If we took them both to be $\overline{W}$, the second-order equation, Eq. (2.98), would be of the form

$$\Sigma(\mathbf{k}_b, E) = \overline{W} + \frac{WW^+}{E - \epsilon_a \mp i\epsilon}$$

and the results would not be exact. However for this simple case we can readily write Σ to all orders in W and $\overline{W}$, sum a geometrical series, and again obtain an exact result (see Prob. 2.31).

We see that the accuracy of the second-order result depends upon our choice of the zero of energy and is best when it is chosen such that $\langle \mathbf{k} | W | \mathbf{k} \rangle$ is zero. It may therefore be desirable in any calculation to make that choice.

In calculating Σ to second order for a liquid metal we require $\langle \mathbf{k} | W | \mathbf{k} + \mathbf{q} \rangle \langle \mathbf{k} + \mathbf{q} | W | \mathbf{k} \rangle = S^*(\mathbf{q}) S(\mathbf{q}) \langle \mathbf{k} | w | \mathbf{k} + \mathbf{q} \rangle \langle \mathbf{k} + \mathbf{q} | w | \mathbf{k} \rangle$. The form factors are known for most metals and, as we noted in Sec. 10.1, the product $S^*(\mathbf{q}) S(\mathbf{q})$ for a liquid is obtained directly from diffraction experiments. The higher-order terms which we have discarded require combinations such as $S(-\mathbf{q}_1 - \mathbf{q}_2)\, S(\mathbf{q}_2)\, S(\mathbf{q}_1)$, or $S(-\mathbf{q}_1 - \mathbf{q}_2 - \mathbf{q}_3)\, S(\mathbf{q}_3)\, S(\mathbf{q}_2)\, S(\mathbf{q}_1), \ldots$. These are not obtained in diffraction experiments as presently performed. The second-order calculation of Σ uses all of the available information about the structure of the liquid and discards only terms which are not known in any case. Furthermore, use of the second-order Σ in the calculation of $G(\mathbf{k}, E)$ incorporates the structural information as completely as possible.

Thus the Green's function method would seem to be the best possible analysis for liquid metals in view of our knowledge of the structure.

The calculation of Σ to second order takes the form

$$\Sigma(\mathbf{k},E) = \sum_q \frac{S^*(q)S(q)\langle\mathbf{k}|w|\mathbf{k}+\mathbf{q}\rangle\langle\mathbf{k}+\mathbf{q}|w|\mathbf{k}\rangle}{E - \epsilon_{k+q} \mp i\epsilon} \tag{2.99}$$

with $\epsilon_{k+q} = \hbar^2(\mathbf{k}+\mathbf{q})^2/2m$. The sum is replaced by an integral and evaluated as a function of E and k. The result leads, through Eq. (2.97), to the diagonal Green's function. As ϵ approaches zero the poles approach the curve

$$E = \epsilon_k + \Sigma(k,E) \tag{2.100}$$

In the case of simple liquid metals the distribution of poles obtained does not differ appreciably from the dispersion curve which we obtained by ordinary perturbation theory, so the gain is not great. Our main purpose has been to introduce the one-particle Green's function and to indicate the manner in which it is used in current solid state theory.

We have given a well-defined procedure for calculating the Green's function which involves a perturbation expansion of the proper self-energy Σ. These calculations are very similar to the ordinary perturbation theory calculations which we described in the section on simple metals, as may be seen from Eqs. (2.99) and (2.100). However, we may use the computed Green's function to calculate properties directly. We have shown above how the density of states is computed from the Green's function. It is also possible to calculate properties such as the conductivity from the Green's function directly using the so-called *Kubo formalism.*[1]

Appendix on Green's functions It is desirable to describe briefly the classical Green's function in order to see its relation to the Green's functions used in quantum mechanics. We follow Mathews and Walker.[2]

We consider a general Hermitian differential operator L which appears in the inhomogeneous differential equation,

$$Lu(x) - \lambda u(x) = f(x) \tag{2.101}$$

where λ is a given constant, $f(x)$ is a given function, and boundary conditions upon $u(x)$ are prescribed. The homogeneous equation might describe the dynamics of a vibrating spring and $f(x)$ might be an applied force, or source term.

We may define a Green's function for the homogeneous equation by

$$LG(x,x') - \lambda G(x,x') = \delta(x - x') \tag{2.102}$$

where L operates on x rather than x' and the same boundary conditions are required as above.

[1] R. Kubo, "Lectures in Theoretical Physics," vol. 1, Boulder, Col., 1958; Interscience, New York, 1959.

[2] J. Mathews and R. L. Walker, "Mathematical Methods of Physics," p. 255, Benjamin, New York, 1964.

It may be verified by substitution that

$$G(x,x') = \sum_n \frac{u_n(x)u_n^*(x')}{\lambda_n - \lambda}$$

where the u_n are eigenfunctions, $Lu_n(x) = \lambda_n u_n(x)$. Note $G(x,x') = G^*(x',x)$.

Having evaluated the Green's function for the vibrating spring, for example, we may now solve Eq. (2.101). We multiply both sides by G and integrate over x.

$$\int [G(x',x)Lu(x) - \lambda G(x',x)u(x)]\, dx = \int G(x',x)f(x)\, dx$$

L is Hermitian and may operate to the left, leading by Eq. (2.102) to a delta function. Thus

$$u(x') = \int G(x',x)f(x)\, dx \tag{2.103}$$

Thus, knowing G, we may compute the solution directly for any source term. Note that the Green's function depends upon the differential operator and the boundary conditions.

In quantum mechanics, the source term is a homogeneous term, perhaps a perturbing potential V. Thus in Eq. (2.101), $f(x)$ becomes $V(x)u(x)$. The Green's function is the same and Eq. (2.103) becomes an integral equation for $u(x)$.

$$u(x') = \int G(x',x)V(x)u(x)\, dx$$

In ordinary scattering calculations boundary conditions are usually taken as requiring incoming or outgoing waves. Different Green's functions are obtained for each.

In our calculation the variable x includes time and we use two sets of boundary conditions, one requiring solutions which decay as $t \to +\infty$, the other requiring solutions which decay as $t \to -\infty$.

10.4 Resistivity in liquid metals Again we have focused our attention on the energy levels in a system but we might alternatively ask for a description of these states in terms of electron scattering. This of course can be done in terms of Green's functions, in which case we find the scattering rate proportional to the imaginary part of the electron self-energy. However, it will be simpler to return to the time-dependent perturbation theory which we used for treating electron scattering by impurities.

We will again compute the scattering according to the Golden Rule, treating the pseudopotential as a perturbation, and basing the calculation on zero-order plane pseudowavefunctions. In this case we cannot, as we did in scattering by impurities, separate the matrix elements into band structure and defect terms. The distinction is no longer sharp and we must write the entire matrix element. This we may do as we did in calculating energies in liquid metals by factoring the matrix element into a structure factor and a form factor. The result is in direct analogy to the second-order perturbation result in Sec. 8.6. We have

$$P_{kk'} = \frac{2\pi}{\hbar} S^*(\mathbf{q})S(\mathbf{q})|\langle \mathbf{k} + \mathbf{q}|w|\mathbf{k}\rangle|^2 \, \delta(E_{k'} - E_k)$$

Again all the quantities that enter the scattering rate are known. The absolute value squared of the structure factor may be obtained directly from experiment and will appear as in Fig. 2.48. The form factors are tabulated for most metals and they may be used directly. The integration may be performed just as it was for the case of impurity scattering to obtain a total scattering time or a momentum-relaxation time τ. The latter is readily found to be

$$\frac{1}{\tau} = \frac{mk_F\Omega_0}{4\pi\hbar^3} \int_0^2 NS^*(q)S(q)|\langle \mathbf{k} + \mathbf{q}|w|\mathbf{k}\rangle|^2 \left(\frac{q}{k_F}\right)^3 d\left(\frac{q}{k_F}\right)$$

It is quite striking that we have been able, so directly and simply, to extend our calculations to liquid metals. That aspect of the problem which gave rise to band structure in the perfect crystal gives rise to scattering in the liquid. It is plausible physically that this structure factor should enter the resistivity calculation. It was obtained, after all, from the scattering of neutrons or x-rays by the same system in which we are interested in the scattering of electrons.

It is perhaps surprising that the resistivity of the metal turns out not to be very much higher than the resistivity of a perfect crystal near the melting temperature. The structure factor, which is illustrated in Fig. 2.48, reaches its first peak in the neighborhood of the wavenumber, $2k_F$, which is the upper limit of integration in the scattering-rate calculation. However, the form factors pass through zero also in this region so the integrand remains rather small.

Unfortunately the calculated resistivity is quite sensitive to the exact shape of the structure factor curve and to the exact position of the zero in the form factors. Thus in many cases it has not been possible to reliably predict the observed resistivity of the liquid metals.[1] On the other hand it is clear that there is no inconsistency in the measured resistivity and the picture which we have described here, and thus no mystery in the fact that the resistivity of the liquid metals is quite low.

[1] Comparison with experiment as well as aids to computation are given by W. A. Harrison, "Pseudopotentials in the Theory of Metals," pp. 153ff., Benjamin, New York, 1966. The calculation for copper using the transition-metal form factor shown in Fig. 2.46 is given by W. A. Harrison, *Phys. Rev.*, **181**: 1036 (1969).

Problems

2.1 Consider a one-dimensional crystal with six identical atomic cells, each of length a, and note its Brillouin zone. (Use periodic-boundary conditions.)

Now assume that the potential due to the crystal is sufficiently small that the energy of each state may be approximated by its kinetic energy alone. Plot the energy as a function of wavenumber in the Brillouin zone for the first 12 states. What is the Bloch function u_k for each of these?

Now let each cell contain one atom which may move only in one dimension. Approximate the normal mode frequencies (angular frequencies) by the wavenumber times the speed of sound (the Debye approximation) and plot the frequencies of all modes as a function of wavenumber in the Brillouin zone.

2.2 Construct the primitive lattice and Brillouin zone for the two-dimensional structure based upon regular hexagons shown below. (If it helps, you can introduce a third primitive lattice translation normal to the plane of the figure.)

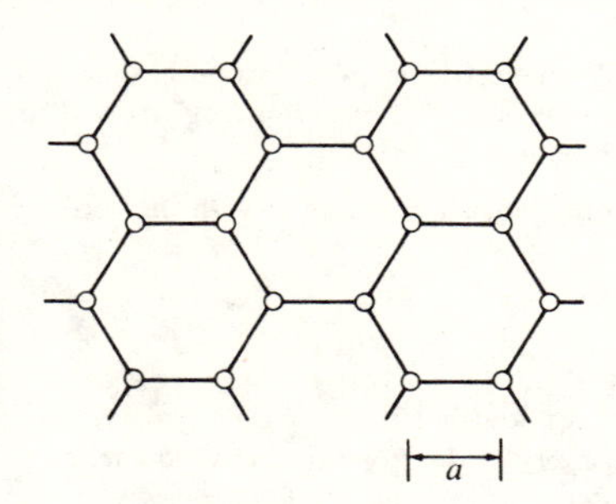

2.3 Can a symmetry degeneracy occur at W for a two-dimensional square lattice, as shown in Fig. 2.3*b*?

2.4 Consider a set of energy bands given by

$$E_k = \pm\sqrt{\hbar^2 k^2 \Delta / m^* + \Delta^2}$$

with all positive-energy states empty, and all negative-energy states full.

a. At time zero add an *electron* with $k_x = k_0$, $k_y = k_z = 0$, and a field $\mathscr{E}_x = \mathscr{E}_y = 0$; $\mathscr{E}_z = \mathscr{E}$. Obtain the current as a function of time. Note particularly the limit as t goes to infinity. (Do exactly rather than by making the band parabolic.)

b. At time zero empty instead a state in the lower band at the same $\mathbf{k}$ as in part *a* and obtain the current as a function of time with the same field applied.

2.5 Consider a one-dimensional crystal with primitive lattice translation a, and let the energy of an electron be given by

$$E(k) = \frac{\hbar^2}{ma^2}\left(\tfrac{7}{8} - \cos ka + \tfrac{1}{8}\cos 2ka\right)$$

a. Sketch the band.

b. Determine the effective mass at the bottom of the band and at the top from quadratic expansion of E in the departure of k from these points.

c. Determine an effective inertial mass for electrons by applying a force F and writing

$$\frac{dv}{dt} = \frac{1}{m_i} F$$

Sketch m_i as a function of k.

d. Determine an effective velocity mass such that

$$v = \frac{1}{m_v} \hbar k$$

and sketch it. (The velocity mass enters expressions for the conductivity.)

e. Note that the number of states per unit length of crystal in the wavenumber interval dk is $dk/2\pi$. Write the density of states per unit *energy* in terms of k and a density-of-states mass, m_d. (The formula should give the correct value for free electrons in one dimension if m_d is taken equal to m.) Is m_d related to either of the above masses?

2.6 Imagine a band for a simple cubic structure (with cube edge a) given by

$$E_k = -E_0(\cos k_x a + \cos k_y a + \cos k_z a)$$

Let an electron at rest ($\mathbf{k} = 0$) at $t = 0$ feel a uniform electric field $\mathscr{E}$, constant in time.

a. Find the trajectory in real space. This can be specified by giving $x(t)$, $y(t)$, and $z(t)$.

b. Sketch the trajectory for $\mathscr{E}$ in a [120] direction.

2.7 The Kronig-Penney model. Consider a one-dimensional crystal with attractive delta-function potentials spaced at a distance a.

$$V(x) = \frac{-\hbar^2}{2m} \Delta \sum_n \delta(x - na)$$

Δ is a measure of the strength of the delta function and has units of L^{-1}. It may readily be shown that these potentials introduce a discontinuity in the slope of the wavefunction at the delta function. Thus the matching through the delta function at $x = 0$ is given by

$$\psi(0^+) = \psi(0^-)$$

$$\frac{\hbar^2}{2m}\left[\left.\frac{\partial\psi}{\partial x}\right|_{0^+} - \left.\frac{\partial\psi}{\partial x}\right|_{0^-}\right] = -\frac{\hbar^2}{2m}\psi(0)\,\Delta$$

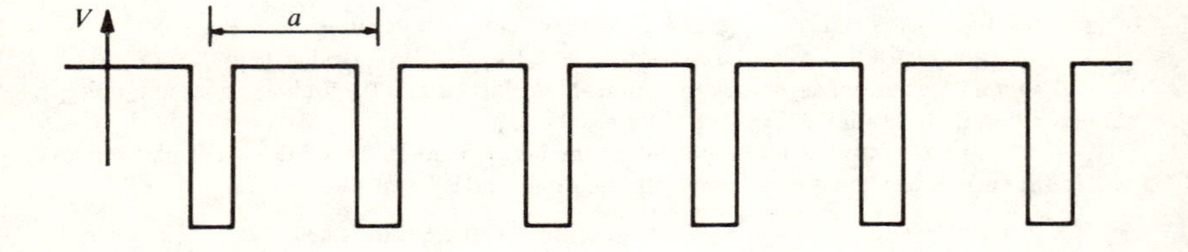

The solutions between any two delta functions are of course

$$\psi = Ae^{\mu x} + Be^{-\mu x}$$

where μ may be real or imaginary ($E = -\hbar^2\mu^2/2m$) and A and B may change from cell to cell.

a. Write two conditions on A, B, and μ in a single cell from the matching conditions for a state of reduced wavenumber k.

b. Solve for B/A from each to obtain an equation for μ. (By plotting both sides as a function of μ (or E) we could graphically find the eigenvalues.)

c. The equation may be written as a simple equation involving cosh μa and sinh μa for states at $k = 0$, or for k at a zone face. Obtain such equations, note that there are solutions for real (for $E < 0$) and imaginary $\mu = iv$ (for $E > 0$), and find all the solutions for the case $\Delta = 0$.

d. See how the energies are shifted as Δ increases slightly from zero (graphically by sketch) and sketch the bands for $\Delta = 0$ and for Δ small.

2.8 Consider the Kronig-Penney model of Prob. 2.7.

a. Obtain the bound states and their energies for isolated atoms (that is, $a \to \infty$).

b. Using the cellular method, obtain equations from which the energy E_0 of the lowest $k = 0$ state may be found for the crystal.

c. Obtain an appropriate value for the energy when $a = 1$, $\Delta = 1$, and compare with the isolated atom value.

d. Obtain the leading term in the energy of part *b* for small a and for large a.

e. Assuming the energy of other states is given by

$$E_0 + \frac{\hbar^2 k^2}{2m}$$

and there is one electron per atom, the average energy per electron is

$$E_0 + \frac{\hbar^2}{2m}\frac{\pi^2}{12a^2}$$

Using the small-a expression for E_0, obtain the equilibrium spacing. (This is the spacing of minimum energy assuming all energy is electronic.)

2.9 If in the Kronig-Penney model we let $\Delta = 20$ and $a = 1$, the free-atom bound states become good solutions. Treating these as core states construct an OPW of wavenumber π/a. Sketch the real and imaginary part. You need not normalize the plane waves (nor OPWs), but the core states must, of course, be normalized.

2.10 Computation of the OPW form factor for a model of lithium. Lithium has a core with two $1s$ electrons. The wavefunctions are approximately given by the form

$$|c\rangle = \left(\frac{\mu^3}{\pi}\right)^{1/2} e^{-\mu r}$$

The potential is taken to be of the form $v(r) = -e^2 e^{-\alpha r}/r$.

a. Obtain an expression for the OPW form factor, which becomes

$$\langle \mathbf{k} + \mathbf{q}|w|\mathbf{k}\rangle = \langle \mathbf{k} + \mathbf{q}|v|\mathbf{k}\rangle + \left(\frac{\hbar^2 k^2}{2m} + \langle \mathbf{k}|v|\mathbf{k}\rangle - E_c\right)\langle \mathbf{k} + \mathbf{q}|c\rangle\langle c|\mathbf{k}\rangle$$

We have noted that electrons $|\mathbf{k}\rangle$ of a given spin are coupled only to the core state of the same spin. The magnitude of both $\mathbf{k} + \mathbf{q}$ and $\mathbf{k}$ is the Fermi wavenumber. Note that

$$\int e^{i\mathbf{k}\cdot\mathbf{r}} f(r)\, d\tau = \int_0^\infty \frac{\sin kr}{kr} f(r) 4\pi r^2\, dr$$

b. For numerical evaluation it is convenient to use atomic units $\hbar = m = e = 1$ atomic unit. Distances are measured in Bohr radii (0.529 Å), masses in units of the electron mass, and energies are in atomic units (1 atomic unit = 2 rydbergs = 27.2 ev). In lithium $\Omega_0 = 140$

atomic units = 140 (Bohr radii)^3. Take $\mu = 2.5$ atomic units^{-1}, $E_c = -2.8$ atomic units, and $\alpha = 0.5$ atomic units^{-1}, and sketch the form factor as a function of q/k_F for $0 \le q \le 2k_F$. For comparison, sketch also $\langle \mathbf{k} + \mathbf{q} | v | \mathbf{k} \rangle$. (Parameters have been chosen to give a reasonable result; actually $E_c \neq \langle \psi_c | -(\hbar^2 \nabla^2/2m) + v | \psi_c \rangle$ for this choice.)

2.11 Redo Prob. 2.6*b* assuming the bands are free-electron-like so the pseudopotential can be included in the diffraction approximation; that is, $E = \hbar^2 k^2/2m$ in the first band and the electron remains in that band.

2.12 *a.* Assuming the structure in Prob. 2.2 is that of a free-electron-like two-dimensional metal with two electrons (one of each spin) per atom, construct the Fermi surface (i.e., the Fermi line) and indicate which band each piece lies in.

b. What is the cyclotron mass for each piece? **H** is, of course, normal to the plane of the figure.

2.13 What is the dimension l of an orbit in *real* space in aluminum in fields of 100 gauss? The cube edge in aluminum is $a = 4.04$ Å.

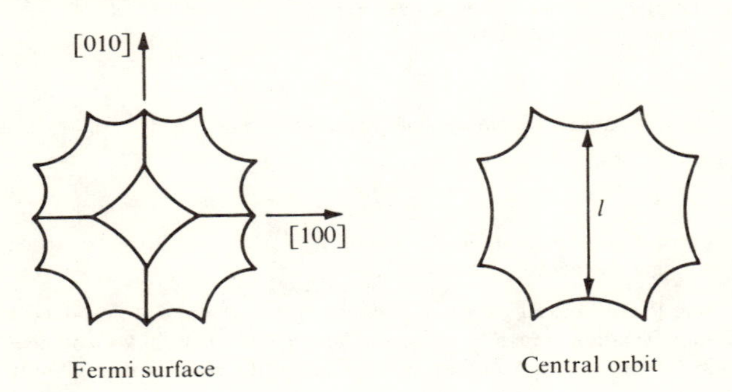

2.14 Assuming a cylindrical Fermi surface of diameter $k_d = 1$ Å^{-1}:

a. Sketch the de Haas-van Alphen period $\delta(1/H)$, which is the difference in $1/H$ between successive peaks in the susceptibility, as a function of the angle between the magnetic field and cylinder axis.

b. Sketch the cyclotron mass at the Fermi energy as a function of this angle, assuming $E = \hbar^2 k_\rho{}^2/2m$, where k_ρ is the component of wavenumber perpendicular to the cylinder axis.

2.15 Beryllium is hexagonal close-packed with lattice constants $c = 6.79$, $a = 4.31$ atomic units (Bohr radii). It has two conduction electrons per atom. Note the one-OPW Fermi surface includes a lens-shaped segment around $\mathbf{k} = 2\pi/c$ in the extended-zone scheme.

a. Using the one-OPW approximation, compute the de Haas-van Alphen periods $\delta(1/H)$ (and the difference in H between successive maxima in the susceptibility when $H = 10^4$ gauss) from these for fields along the c axis and for fields in the basal plane.

b. Compute cyclotron masses for these fields.

c. Taking the corresponding OPW form factor as 0.05 atomic unit and using a two-OPW approximation, estimate the changes in the cross-sectional area for the central section in the plane perpendicular to the c axis. Assume the Fermi *energy* remains at the free-electron value.

2.16 Consider the lowest-lying bands in a hexagonal structure near the center of an edge, K, of the Brillouin zone, as shown. Note that there is a threefold degeneracy of the free-electron

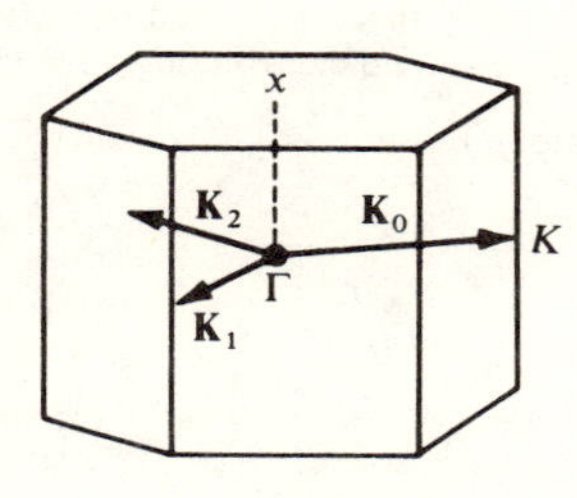

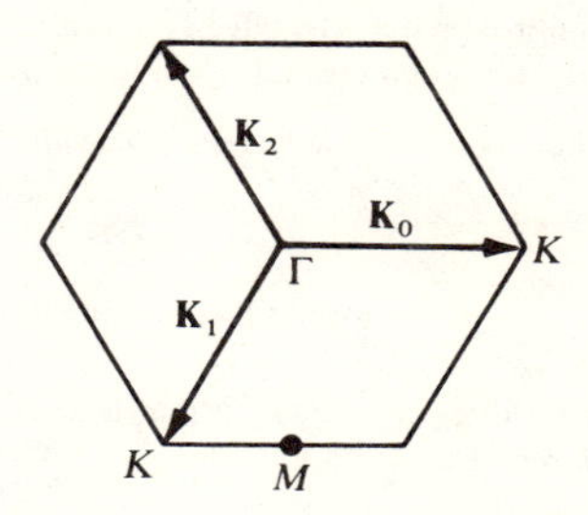

bands at that point. These correspond to plane waves of wavenumber $\mathbf{K}_0$, $\mathbf{K}_1$, $\mathbf{K}_2$. Near $\mathbf{k} = \mathbf{K}_0$ the three plane waves $\mathbf{k}_i = \mathbf{K}_i + \boldsymbol{\kappa}$ are coupled by the small real pseudopotential matrix element $W = \langle \mathbf{k}_0 | W | \mathbf{k}_1 \rangle = \langle \mathbf{k}_1 | W | \mathbf{k}_2 \rangle = \langle \mathbf{k}_2 | W | \mathbf{k}_3 \rangle$; neglect all other matrix elements and approximate the kinetic energy by

$$\frac{\hbar^2}{2m} k_i^2 \approx \frac{\hbar^2}{2m} (K_i^2 + 2\mathbf{K}_i \cdot \boldsymbol{\kappa})$$

Obtain the 3-by-3 Hamiltonian matrix and the bands along ΓK near K. This calculation can be greatly simplified by noting that the group of $\mathbf{k}$ contains the reflection symmetry suggesting propitious linear combinations of $|\mathbf{k}_1\rangle$ and $|\mathbf{k}_2\rangle$. The corresponding transformation of the Hamiltonian matrix reduces the secular equation to second order.

Sketch the bands for small positive and negative κ for $W > 0$. Note $\boldsymbol{\kappa} \cdot \mathbf{K}_0 > 0$ corresponds to the line KM in the reduced Brillouin zone.

2.17 Consider the Kronig-Penney model of Prob. 2.7. The lowest-lying state is given approximately by

$$\psi^{(0)} = \sqrt{\frac{\Delta}{2N}} \sum_n e^{-\Delta|x-an|/2}, \quad E = -\frac{\Delta^2}{8}$$

in atomic units. The $k = 0$ states in higher bands are even or odd around each atom. The odd ones are of the form

$$\psi_m \propto \sin \frac{2\pi x}{a} m \qquad \text{for } m = 1, 2, \ldots$$

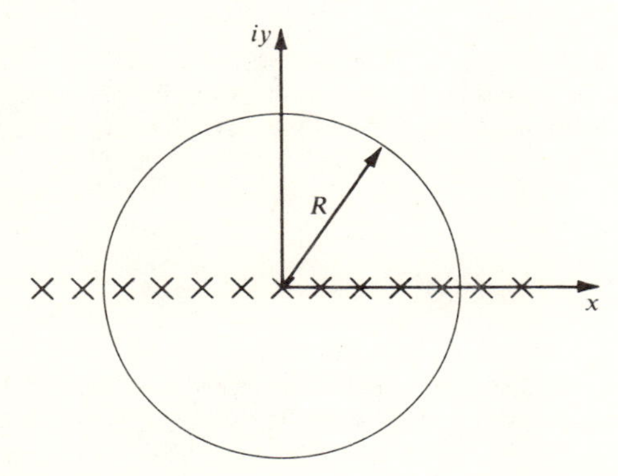

Evaluate approximately the effective mass in the lowest band by the $\mathbf{k} \cdot \mathbf{p}$ method by including only the first term and estimate your error by considering the second contributing term. [Note that the sum you obtained *could* be evaluated by contour integration: To obtain $\sum_{-\infty}^{\infty} f(n)$, consider

$$\int \frac{f(z)\,dz}{1 - e^{2\pi iz}} \qquad \text{around the contour shown.}$$

There are residues at z equal to real integers and at the poles of $f(z)$. As $R \to \infty$, the sum of residues is zero.]

2.18 Consider the energy bands shown below. L is at the center of a zone face $\mathbf{k} = \mathbf{q}/2$. Let

$$\psi_{L1} = \sqrt{\frac{2}{\Omega}} \cos\left(\frac{qz}{2}\right); \qquad \psi_{L2} = \sqrt{\frac{2}{\Omega}} \sin\left(\frac{qz}{2}\right)$$

(We have taken the z axis along $\mathbf{q}$.) The energy of ψ_{L1} in atomic units is $q^2/8 - \Delta$; that of ψ_{L2} is $q^2/8 + \Delta$. Construct all components of the effective-mass tensor, $(m/m^*)_{ij}$ for the lower band at L using the $\mathbf{k} \cdot \mathbf{p}$ method with interaction only between the two bands. You will need to rederive the method for points near $k = q/2$, rather than near $k = 0$.

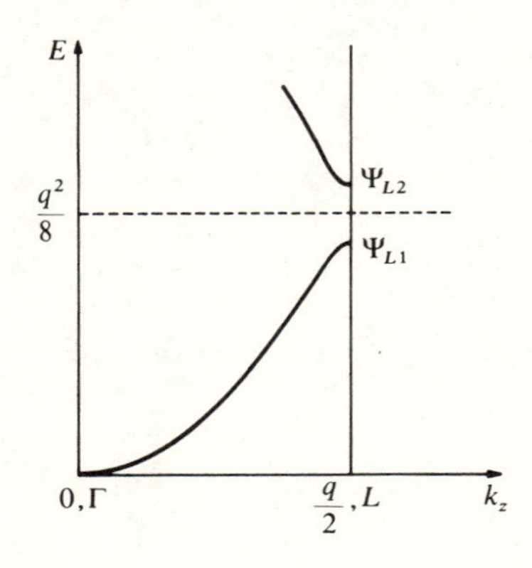

2.19 Obtain tight-binding bands for s states in the structure shown in Prob. 2.2. This will require modification since there are two atoms per primitive cell. Again take

$$\psi_k = \frac{1}{\sqrt{N}} \sum_j \psi_t(\mathbf{r} - \mathbf{r}_j) e^{ik \cdot r_j}$$

and evaluate

$$\frac{\langle \psi_k | H | \psi_k \rangle}{\langle \psi_k | \psi_k \rangle}$$

neglecting three-center integrals and taking only nearest-neighbor overlaps. (Note that the assumed form does not follow from symmetry as it did with one atom per primitive cell, but it is a legitimate guess.)

2.20 Obtain and sketch tight-binding bands along a [100] direction for p states on a simple cubic lattice. Consider only nearest-neighbor overlaps. Introduce as many (and only as many) overlap integrals as you need; replace each by a parameter and obtain its sign by inspection.

2.21 A model-polaron problem. We first construct classically the Hamiltonian of the electron and deformable lattice in one dimension. An electron in the conduction band is described by a Hamiltonian $H = p^2/2m$. The lattice may polarize, and this polarization is described by a single coordinate X. (In a real crystal the polarization is described by many coordinates which we will see are the normal coordinates of the harmonically coupled atoms.) The elastic energy of the lattice is given by $\frac{1}{2}\kappa X^2$; the lattice kinetic energy is given by

$$\frac{M}{2}\dot{X}^2 = \frac{1}{2M}P^2$$

where M has the units of mass and P is the momentum conjugate to X, $P = M\dot{X}$. Finally deformation of the lattice shifts the electron energy by an amount proportional to X and in our model we let the shift be also proportional to the momentum p of the electron, giving a term in the Hamiltonian, ΩpX; this term represents the coupling between the electron and the lattice polarization. (In a real crystal the coupling is again proportional to the coordinate X, but depends upon electron position rather than momentum.)

Quantum mechanically, the system is described by a wavefunction $\psi(x,X)$ satisfying a Schroedinger equation,

$$\left(-\frac{\hbar^2}{2m}\frac{\partial^2}{\partial x^2} - \frac{\hbar^2}{2M}\frac{\partial^2}{\partial X^2} + \frac{\kappa}{2}X^2 + \Omega pX\right)\psi(x,X) = E\psi(x,X)$$

In the absence of coupling ($\Omega = 0$), $\psi(x,X)$ factors to

$$\frac{1}{\sqrt{L}}e^{ikx}\varphi_n(X)$$

where L is the normalization length and the φ_n are harmonic-oscillator wavefunctions. The eigenvalues are $E = \hbar^2k^2/2m + \hbar\omega(n + \frac{1}{2})$ where $\omega = (\kappa/M)^{1/2}$ and n is an integer.

a. Find the eigenstates and eigenvalues with $\Omega \neq 0$. (A similar factorization is possible.)

b. What is the polaron-effective mass?

2.22 Consider conduction and valence bands given by

$$E = \pm\sqrt{\hbar^2\frac{k^2\Delta}{m} + \Delta^2}$$

with $k = |k_x^2 + k_y^2 + k_z^2|^{1/2}$ and $\Delta = 1$ volt. By continuation into the complex plane find the decay rate μ (for states proportional to $e^{-\mu z}$) as a function of energy in the gap.

Estimate the binding energy for a donor level using an effective-mass approximation and a dielectric constant of 16.

Estimate the electric field required to ionize the level at zero temperature (that field where the dipole energy is of the order of the binding energy).

2.23 Consider resonant states for $l = 0$ and for a spherically symmetric egg-shell potential given by $V(r) = \beta\delta(r - r_0)$. The amplitude of the wavefunction for $r > r_0$ is determined by normalization in a large system.

a. For $\beta \gg k$ show that the amplitude inside is small except at discrete resonant energies. The form of spherical Bessel functions may be obtained from Schiff.[1]

[1]Schiff, *op. cit.*, p. 77 in 1st edition, p. 85 in 3rd edition.

b. Obtain and interpret the approximate phase shifts far from resonance.

c. Describe the states of the system as β/k becomes positively infinite for the k of interest.

2.24 Consider an infinitely high spherical potential (hard-core potential) of radius r_0 in an electron gas (uncharged electrons).

a. Obtain the phase shifts for $l = 0$ states.

b. Assuming the electron gas has the density of the sodium conduction electrons and r_0 is the Wigner-Seitz sphere radius (4.0 atomic units for sodium), how many electrons in s states are attracted to or repelled from the region?

c. Assuming the asymptotic form for the Friedel oscillations, what is the shift in s-electron density at $2r_0$, as a fraction of the unperturbed total density?

2.25 Consider a semiconductor with N electrons per unit volume in a conduction band describable by an effective mass m^* (nondegenerate or Boltzmann statistics).

a. Carry the Friedel sum-rule argument as far as you can to obtain the number of electrons localized to a region near a spherically symmetric potential. (Not all integrals can be done.)

b. Evaluate phase shifts by Born approximation for a potential

$$V = \beta\, \delta(\mathbf{r})$$

Note $\int \delta(\mathbf{r}) f(\mathbf{r})\, d^3r = f(0)$ and obtain the number localized in terms of β, T, and N.

2.26 Treat electrons in an effective-mass approximation, $E = \hbar^2k^2/2m^*$. Let a donor potential be $(-e^2/r\epsilon)e^{-\mu r}$ and the electron wavefunctions may be treated as plane waves. Compute the relaxation time

$$\frac{1}{\tau(k)} = \frac{\Omega}{(2\pi)^3} \int d^3k'\,(1 - \cos\theta) P_{kk'}$$

Note τ does in fact depend upon k. Note the behavior of the result as $\mu \to 0$. This difficulty arises from the long-range nature of the coulomb potential.

Obtain the leading term in an expansion in k/μ (i.e., large μ) and evaluate the scattering cross section σ numerically:

$$\frac{\hbar k}{m^*}\tau\sigma = \Omega = \text{the volume of the system}$$

Assume thermal energy ($E = 1/40$ ev): let $m^*/m = 1$, $\epsilon = 16$, $\mu = (0.1/0.529)\text{Å}^{-1}$ ($=0.1$ atomic unit) and $\Omega = 1\ \text{cm}^3$. Not all constants are necessary.

2.27 Consider a simple cubic metal, with one conduction electron per atom. Approximate the pseudopotential by a delta function

$$w(r) = \Omega_0 \beta\, \delta(\mathbf{r})$$

so $\langle \mathbf{k} + \mathbf{q}|w|\mathbf{k}\rangle = \beta$ and β is constant.

a. Introduce a single vacancy and construct the matrix element

$$\langle \mathbf{k} + \mathbf{q}|W|\mathbf{k}\rangle.$$

b. For lowest-order scattering compute the relaxation time τ for an electron at the Fermi surface.

c. An electron on the Fermi surface near a Brillouin zone face may have its τ appreciably modified by higher-order terms. Let the zone face be the plane bisector of the lattice wavenumber $\mathbf{q}_0 = 2\pi/a\,[100]$ and consider a state $|\mathbf{k}\rangle$ on the Fermi surface with $\mathbf{k} \parallel \mathbf{q}_0$. Compute the modification of the scattering rate due to the second-order term for which the intermediate state wavenumber is $\mathbf{k}'' = \mathbf{k} - \mathbf{q}_0$. This is the dominant term. Write the factor by which τ is modified in terms of β/E_F. Assume the Fermi surface remains spherical.

2.28 The transitional-metal pseudopotential equation in atomic units is

$$-\frac{\nabla^2}{2}|\varphi\rangle + W|\varphi\rangle - \sum_d \frac{\Delta|d\rangle\langle d|\Delta|\varphi\rangle}{E_d - E} = E|\varphi\rangle$$

a. Let $\mathbf{q}$ be the smallest lattice wavenumber. Then make a two-OPW approximation for states near $\mathbf{k} = \mathbf{q}/2$, taking $\langle\mathbf{k}|W|\mathbf{k}\rangle \equiv \overline{W}$ and $\langle\mathbf{k} - \mathbf{q}|W|\mathbf{k}\rangle = \langle\mathbf{k}|W|\mathbf{k} - \mathbf{q}\rangle = W$, both independent of $\mathbf{k}$, and take $\Delta = 0$. Evaluate the eigenvalues at $\mathbf{k} = 0$ and at $\mathbf{k} = \mathbf{q}/2$ and sketch the result.

b. In contrast take $W = \overline{W} = 0$ but $\Delta \neq 0$ and make the corresponding sketch of the bands again making approximate determinations of the eigenvalues at $\mathbf{k} = 0$ and $\mathbf{k} = \mathbf{q}/2$. In doing this, let E_d lie between 0 and $\frac{1}{2}(q/2)^2$ and $|\frac{1}{2}(q/2)^2 - E_d| >> |\langle d|\Delta|q/2\rangle|$ and $E_d >> |\langle d|\Delta|q/2\rangle|$. Note that if the angular momentum of the d states is quantized along $\mathbf{k}$, only the $m = 0$ term in the sum over d is nonzero and for this term $\langle d|\Delta|k\rangle = \langle k|\Delta|d\rangle \approx \Delta_0 k^2$, with Δ_0 constant; both of these follow from an expansion of $|\mathbf{k}\rangle$ in spherical harmonics and spherical Bessel functions and the expansion of the latter for small r.

The main interest is in the s- and p-like levels at $\mathbf{q}/2$ relative to the band minimum. Those levels in copper are much more accurately fit by scheme b taking Δ_0 as an adjustable parameter than by scheme a taking W as a adjustable parameter.

2.29 Consider an s band in a *liquid* insulator by the tight-binding method. Again find the energy-expectation value of states

$$\psi(\mathbf{r}) = \frac{1}{\sqrt{N}} \sum_j e^{i\mathbf{k}\cdot\mathbf{r}_j} \psi(|\mathbf{r} - \mathbf{r}_j|)$$

keeping only two-center integrals (for all neighbors).

Assuming that ψ goes as $e^{-\mu r}$ at large r and v is strongly localized, we may write

$$\int \psi^*(\mathbf{r} - \mathbf{r}_i) v(\mathbf{r} - \mathbf{r}_j) \psi(\mathbf{r} - \mathbf{r}_i)\, d\tau = -\lambda_1 e^{-2\mu|r_i - r_j|}$$

for $i \neq j$ and

$$\int \psi^*(\mathbf{r} - \mathbf{r}_i) v(\mathbf{r} - \mathbf{r}_i) \psi(\mathbf{r} - \mathbf{r}_j)\, d\tau = -\lambda_2 e^{-\mu|r_i - r_j|}$$

for $i \neq j$ where λ_1 and λ_2 are positive constants. Sums of overlaps from neighbors may be evaluated assuming the probability per unit volume of a neighbor lying at a distance r is

$$P(r) = \frac{1}{\Omega_0}\left(1 - \frac{\sin ar}{ar}\right)$$

where a^{-1} is a constant of the order of the average near-neighbor distance; that is,

$$\langle \sum_j f(\mathbf{r}_i - \mathbf{r}_j) \rangle = 4\pi \int P(R) f(R) R^2\, dR$$

Noting the high k and low k limits, sketch $E(\mathrm{k})$. (Note the result is well defined at all k but may only be meaningful for $k/a \lesssim 1$.)

2.30 Derive Eq. (2.87) for $G_\pm(\mathbf{r},\mathbf{r}';\hbar\omega)$ and verify that it is a solution of Eq. (2.88).

To go from normalization in a finite system, which we have been using, to the δ-function normalization appropriate here, we expand $f(t)$ in the range $0 < t < T$ with periodic boundary conditions

$$f(t) = \sum_{\omega} g(\omega)\, e^{-i\omega t} \qquad \text{with } \omega = \frac{2\pi n}{T}$$

Then

$$g(\omega) = \frac{1}{T}\int e^{+i\omega t} f(t)\, dt$$

and

$$\int e^{-i(\omega - \omega')t}\, dt = T\, \delta_{\omega\omega'}$$

If we let $T \to \infty$ and replace $\sum_{\omega}$ by $T/2\pi \int d\omega$, and we write $Tg(\omega) = G(\omega)$, then

$$f(t) = \frac{1}{2\pi}\int d\omega G(\omega)\, e^{-i\omega t}$$

$$G(\omega) = \int e^{i\omega t} f(t)\, dt$$

and

$$\int e^{-i(\omega - \omega')t}\, dt = 2\pi\, \delta(\omega - \omega')$$

The last is necessary so that $\sum T\, \delta_{\omega\omega'} = T/2\pi \int d\omega\, [2\pi\, \delta(\omega - \omega')]$. This formula is not useful if the integrand contains a Θ function.

2.31 Rederive the Green's function $G(k_b,k_b)$ for a system with only two states, but taking $\langle k_a|W|k_a\rangle = \langle k_b|W|k_b\rangle \neq 0$ and evaluating $\sum(k,E)$ to all orders in W. The resulting sum in $\sum(k,E)$ may be performed to obtain a simple and plausible form for $G(k_b,k_b)$.

III

ELECTRONIC PROPERTIES

In Chap. II we focused our attention on the electronic states in solids and liquids. We also treated a number of electronic properties on the basis of those states. These treatments were intended to illustrate the nature of the states in question. We now proceed more systematically to consider the electronic properties of solids.

At this stage we will take the one-electron eigenvalues as known and consider the behavior of the many electrons within the known band structure. We will be primarily interested in the occupation of the states rather than in the detailed wavefunctions. The system is then specified by giving the number of electrons (one or zero) occupying each of the states. This is the occupation number representation which is closely associated with the notation of second quantization. That notation will be introduced in detail in Chap. IV.

1. Thermodynamic Properties

At the absolute zero of temperature a system of N electrons would be in the ground state with the lowest-lying N states occupied. At a finite temperature the state of the system will have some of the higher-energy one-electron states occupied and some of the lower-energy one-electron states unoccupied. We will not of course wish to consider a particular excited state of the many-electron system but a statistical distribution of those states. For the moment we will discuss the total energy of the system and therefore will be interested in the probability of occupation of each one-electron state; i.e., an average of the occupation of a one-electron state over a statistical distribution of many-electron states.

We summarize the results of a band calculation by listing all of the Hartree-Fock parameters E_n. Since we will always compare the total energies of similar configurations of the same number of electrons, we may appeal to Koopmans' theorem to write the energy of the system as a sum of one-electron energies E_n given by the corresponding Hartree-Fock parameters, or we may imagine these E_n to be the energies of noninteracting quasiparticles. The probability of occupation of any given one-electron state is given by a *Fermi distribution*; i.e., the probability of a state of energy E_n being occupied is given by

$$f_0(E_n) = \frac{1}{e^{(E_n - \xi)/KT} + 1}$$

The derivation of this form should be familiar[1]; it represents the most probable configuration of a fixed number of indistinguishable particles obeying the Pauli principle with a fixed total energy. The ξ is the *thermodynamic Fermi energy*. It is a constant of the system and is determined by the number of particles present as well as by the distribution of states. We look first at the determination of ξ.

We write the total number of particles

$$N = \sum_n \frac{1}{e^{(E_n - \xi)/KT} + 1}$$

Given the set of energy values E_n and a fixed temperature, the right-hand side depends only upon ξ, which therefore must be selected to yield the appropriate number of particles.

The determination of ξ is most easily seen in the limit of low temperatures. As the temperature approaches zero, $f_0(E)$ approaches a step function, being equal to unity for $E < \xi$ and zero for $E > \xi$. Thus in a metal the Fermi energy is, as we had indicated in our earlier discussions, the energy below which all states are occupied and above which all states are unoccupied. We were able to compute the Fermi energy directly for the free-electron picture simply by selecting a sphere in wavenumber space which

[1] The derivation is given, e.g., by C. Kittel, "Introduction to Solid State Physics," p. 620, 3rd edition, Wiley, New York, 1967.

was large enough to contain the valence electrons. The low-temperature Fermi energy in a metal, measured from the energy of the $k = 0$ state, is customarily written E_F. At finite temperatures the Fermi distribution is smoothed somewhat and the probability of occupation of a state drops from one to zero over an energy range of the order of KT as seen in Fig. 3.1. However, in a metal even at room temperature KT is ordinarily less than $0.005\ E_F$.

In a semiconductor at absolute zero we expect all states in the valence band to be occupied and all states in the conduction band to be unoccupied. We therefore can say that the Fermi energy lies somewhere within the energy gap between the two. At finite temperature the probability of occupation is not exactly one or exactly zero and there will be a small number of electrons excited into the conduction band and a small number of holes in the valence band. The number must be equal for a pure (intrinsic) semiconductor, and this requirement determines the Fermi energy. In particular if the density of states near the conduction-band edge is the same as that near the valence-band edge, the Fermi energy must lie precisely midway between the two band edges. If, on the other hand, the density of states were higher in the valence band, then the Fermi energy must lie nearer to the conduction band. Ordinarily it is necessary to determine the Fermi energy at the temperature of interest and that Fermi energy will depend upon temperature (see Prob. 3.1).

1.1 The electronic specific heat The second problem which we will discuss is the contribution to the specific heat from the electrons. This involves simply a calculation of the change in total energy of the system (with the number of electrons fixed) as the temperature is changed.

This problem is of particular historical interest in connection with the development of quantum mechanics. Classically we expect an energy of

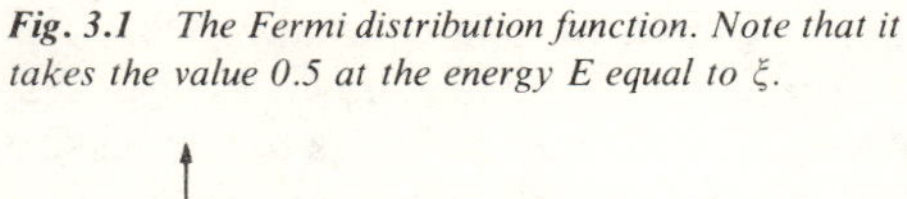

Fig. 3.1 *The Fermi distribution function. Note that it takes the value 0.5 at the energy E equal to ξ.*

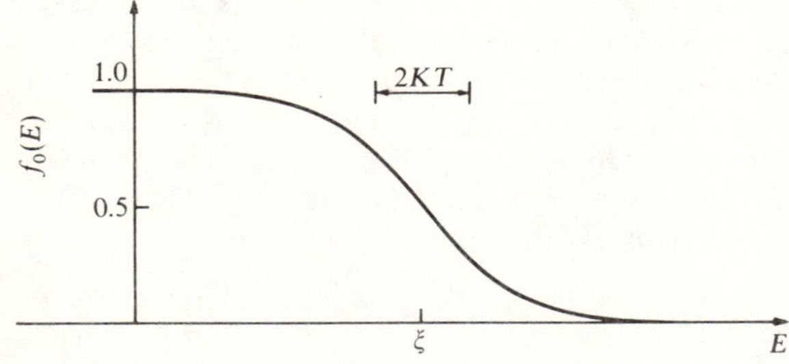

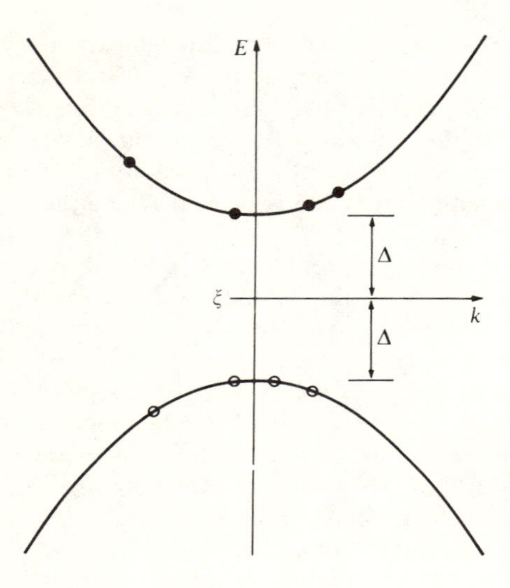

Fig. 3.2 A semiconductor with symmetric electron and hole bands. The solid points represent thermally excited electrons, the empty points represent thermally excited holes. The Fermi energy lies in the middle of the 2Δ energy gap.

KT in every degree of freedom of a system of vibrators. If a crystal is made up of N atoms (each with three degrees of freedom), we might at first expect a thermal energy of $3NKT$, or a specific heat of $3K$ per atom. The $3K$ is the *Dulong-Petit* value of the specific heat and agrees approximately with the specific heat observed in many materials at room temperature. However, it has long been known that the electrons in a metal also behave as free and we might expect an additional energy of $\frac{3}{2}KT$ per electron, as in an ideal gas. For a metal like sodium this yields a specific heat of $\frac{9}{2}K$. However, the specific heat of a metal like sodium is given rather well at room temperature by the contribution of the atoms only. The resolution of this puzzle came only with quantum mechanics and will be seen in our result here.

The simplest case to consider is that of a semiconductor. We will take a particularly simple model of a semiconductor, shown in Fig. 3.2, in which the density of states for electrons is the mirror image of the density of states for holes. Then, as we indicated earlier, the Fermi energy will lie at the midpoint of the band gap. We will select our origin of energy at this midpoint. Then the conduction band lies at an energy Δ and the valence-band edge lies at $-\Delta$. This will ordinarily be of the order of an electron volt and much larger than the thermal energy KT. Thus the Fermi factor determining the occupation of electron states may be approximated in the form

$$f_0(E) = \frac{1}{e^{E/KT} + 1} \approx e^{-E/KT}$$

which is simply the Boltzmann factor. Physically this means simply that the probability of occupation of the states is so small that the probability of two particles occupying a given state is negligible and the Pauli principle becomes of no importance.

In this simple case we have been able to determine the Fermi energy as a constant for all temperatures so we may directly compute the energy of the excited electronic states. This, by the mirror symmetry, will be equal also to the energy of the hole excitations. We may evaluate the total energy of electrons in the conduction band

$$E_{\text{elec}} = \sum_n E_n f_0(E_n) \approx \sum_n E_n e^{-E_n/KT}$$

by converting the sum over states to an integral over wavenumber space and using an effective-mass approximation to describe the energies, $E = \Delta + \hbar^2 k^2/2m^*$.

$$\begin{aligned} E_{\text{elec}} &= \frac{2\Omega}{(2\pi)^3} \int dk\, 4\pi k^2 \left(\Delta + \frac{\hbar^2 k^2}{2m^*}\right) e^{-(\Delta + \hbar^2k^2/2m^*)/KT} \\ &= \frac{\Omega\Delta}{4} \left(\frac{2m^* KT}{\hbar^2 \pi}\right)^{3/2} \left(1 + \frac{3}{2}\frac{KT}{\Delta}\right) e^{-\Delta/KT} \end{aligned}$$

which is of the order of Δ per excited electron. We may add an equal contribution for the excitation energy associated with holes to obtain the total energy. We see that it is proportional to the volume, as it should be. We could take a derivative with respect to temperature to obtain the specific heat.

We note again that ordinarily Δ will be much greater than KT and therefore the exponential will be extremely small. Thus we see, in accord with experiment, that in semiconductors the contribution to the specific heat due to the electrons is extremely small in comparison to the classical value of $\frac{3}{2}K$ per valence electron. It should be totally negligible at all temperatures in comparison to the contributions to the specific heat from the atoms themselves. These other contributions to the specific heat will be considered when we discuss phonons.

We might note in passing that the number of electrons in the conduction band may also be computed in the same way. We obtain the number of electrons per unit volume,

$$N_{\text{elec}} = \frac{1}{4} \left(\frac{2m^* KT}{\hbar^2 \pi}\right)^{3/2} e^{-\Delta/KT}$$

This again will be extremely small in comparison to the number of electrons present. However, if these are the only carriers contributing to the conductivity, they must nevertheless be of importance.

We next consider the electronic contribution to the specific heat of metals. This case is not nearly so simple because we must compute at the same time the shift in the Fermi energy required in order to keep the number of electrons fixed. However, we can see at the outset that in metals the electronic specific heat must be considerably larger than we obtained for semiconductors. Since there is no energy gap, the exponential which appeared above will not appear in the case of metals.

In this calculation we need not restrict ourselves to simple metals. We define a density of states per unit energy given by $n(E)$. In terms of this density of states we may write the total energy and the total number of electrons.

$$E_0 = \int_{-\infty}^{\infty} f_0(E)n(E)E\,dE$$

$$N_0 = \int_{-\infty}^{\infty} f_0(E)n(E)\,dE$$

Due to the form of the Fermi distribution function these integrals are somewhat awkward even if we have a simple analytic expression for the density of states. However, the energy KT will be small in comparison to all other energies of interest in the problem and it is therefore possible to obtain very good approximate evaluations of the integrals by expanding in the temperature. We may consider a general integral of the type that will interest us. Performing a partial integration,

$$I = \int g(E) f_0(E)\,dE = G(E) f_0(E) \Big|_{-\infty}^{+\infty} - \int_{-\infty}^{\infty} G(E)\frac{\partial f_0}{\partial E}\,dE \tag{3.1}$$

where $g(E)$ is any given function of energy but will always contain the density of states as a factor and where $G(E) = \int_{-\infty}^{E} g(E)\,dE$. There are no states at sufficiently low energy and the Fermi factor approaches zero at sufficiently high energy so the surface terms go out.

The $\partial f_0/\partial E$ is sharply peaked at the Fermi energy since, as we have seen, the Fermi function drops rapidly from one to zero near the Fermi energy. Thus $\partial f_0/\partial E$ is approximately equal to a negative delta function. This becomes more accurately true as the temperature is lowered. To lowest order in the temperature we have then $I \approx G(\xi) = \int_{-\infty}^{\xi} g(E)\,dE$. This result is obvious. At absolute zero all states below the Fermi energy are occupied and all those above unoccupied. We will need to do better to obtain the specific heat.

Let us now make an expansion of our function $G(E)$ around the Fermi energy. The $G(E)$ is directly related to our given function $g(E)$ and therefore all derivatives can be evaluated.

$$G(E) = G(\xi) + G'(\xi)(E - \xi) + \tfrac{1}{2}G''(\xi)(E - \xi)^2 + \cdots \tag{3.2}$$

We now evaluate our integral I to various orders in the temperature. We have found the zero-order term $I_0 = G(\xi)$. We can now see from Eqs. (3.1) and (3.2) that the nth-order term is given by

$$I_n = -\frac{1}{n!} G^{(n)}(\xi) \int_{-\infty}^{\infty} (E - \xi)^n \frac{d}{dE}\left(\frac{1}{e^{(E-\xi)/KT} + 1}\right) dE$$

$$= \frac{(KT)^n}{n!} G^{(n)}(\xi) \int_{-\infty}^{\infty} \frac{x^n \, dx}{(e^{x/2} + e^{-x/2})^2}$$

where we have written $(E - \xi)/KT = x$. The integrand will be odd for odd integers so we have only even terms contributing. For even n the integral may be evaluated by contour integration or from tables. Only the second term will be of interest to us here. We obtain

$$I = G(\xi) + \frac{\pi^2}{6}(KT)^2 G''(\xi) + \cdots$$

The integrals that we will need can be obtained from this formula.

At the outset we will determine the dependence of the Fermi energy upon temperature. This will be done by requiring that the total number of electrons present, N_0, remain fixed as the temperature is varied. We write our integral expression for the total number of electrons and expand in temperature. In this case $g(E) = n(E)$, the density of states per unit energy and per unit volume, and $G(E) = \int_{-\infty}^{E} n(E)\, dE$. We have

$$N_0 = \int f_0(E) g(E)\, dE = G(\xi) + \frac{\pi^2}{6}(KT)^2 G''(\xi) \tag{3.3}$$

Now the Fermi energy will differ from its value at absolute zero, ξ_0, by a term which we will see is second order in KT. Therefore we can write $G(\xi) = G(\xi_0) + G'(\xi_0)(\xi - \xi_0) + \cdots$, but the first term is equal to the number of electrons at absolute zero which in turn is equal to N_0. Thus we may solve Eq. (3.3) for ξ to obtain the desired result to second order in KT.

$$\xi = \xi_0 - \frac{\pi^2}{6}\frac{G''(\xi_0)}{G'(\xi_0)}(KT)^2 = \xi_0 - \frac{\pi^2}{6}\frac{n'(\xi_0)}{n(\xi_0)}(KT)^2$$

We may now, in a similar way, obtain the total energy of the system.

$$E_0 = \int f_0(E)n(E)E\,dE = G(\xi) + \frac{\pi^2}{6}G''(\xi)(KT)^2 + \cdots$$

where $G(\xi)$ is given in this case by

$$G(\xi) = \int_{-\infty}^{\xi} n(E)E\,dE$$

We may now obtain the specific heat which is the derivative of the total energy with respect to temperature. This may be evaluated directly.

$$\begin{aligned}\frac{dE_0}{dT} &= K\left[G'(\xi)\frac{d\xi}{d(KT)} + \frac{\pi^2}{3}G''(\xi)\,KT + \cdots\right]\\ &= K\left\{\xi n(\xi)\left[-\frac{\pi^2}{3}\frac{n'(\xi)}{n(\xi)}KT\right] + \frac{\pi^2}{3}[n(\xi) + \xi n'(\xi)]KT\right\} \qquad (3.4)\\ &= \frac{\pi^2}{3}K^2Tn(\xi)\end{aligned}$$

which is our final result. We have kept only terms to lowest order in KT and in the result need not distinguish ξ from ξ_0.

In simple metals the density of states is of the order of the reciprocal of the Fermi energy and each higher derivative of $n(E)$ contains an additional factor of $1/E_F$. Thus our expansion parameter has in fact been KT/E_F which is of the order of $1/200$.

We note first that the electronic contribution to the specific heat is directly proportional to the density of states at the Fermi energy. We note second that it is linear in temperature and therefore approaches zero at low temperatures. Further we note that the electronic specific heat is of the order of KT times the density of states all multiplied by the classical value per electron. Physically this result states that the electronic specific heat involves only those electrons with energies within KT of the Fermi energy. This is a plausible result. Electrons far below the Fermi energy are prevented from being excited by the occupation of the neighboring states.

Again we find that the electronic contribution to the specific heat is very tiny in comparison to the classical value associated with the atomic motion. However, we will find that at low temperatures the atomic contribution to the specific heat drops more rapidly with temperature than does the electronic specific heat. Thus at sufficiently low temperatures the electronic specific heat will dominate and it is observed to be linear in temperature. This is illustrated in Fig. 3.3. All parameters in Eq. (3.4) are known except for the density of states at the Fermi energy; the experiment therefore gives us a direct measure of that density. The density of states obtained in this way was discussed in connection with the simple metals in the previous chapter. These measured densities of states also give important information about transition metals where the unfilled d bands represent very high densities of states; the electronic specific heat in these materials is correspondingly high.

The specific heat is just one thermodynamic property which can be computed directly once we specify the energy levels. The free energy and the entropy are also directly calculable by ordinary methods of statistical mechanics once the energy levels are specified.

Our discussion here is based upon the one-electron approximation in terms of which we can write individual one-electron states with energies E_n and consider statistically their occupation. We will see in Sec. 6 that the Landau theory of Fermi liquid asserts that this remains true if we go beyond the one-electron approximation and allow for the fact that the electrons interact with each other. In that theory the single-particle states that we discuss here will be replaced by the so-called quasiparticle states.

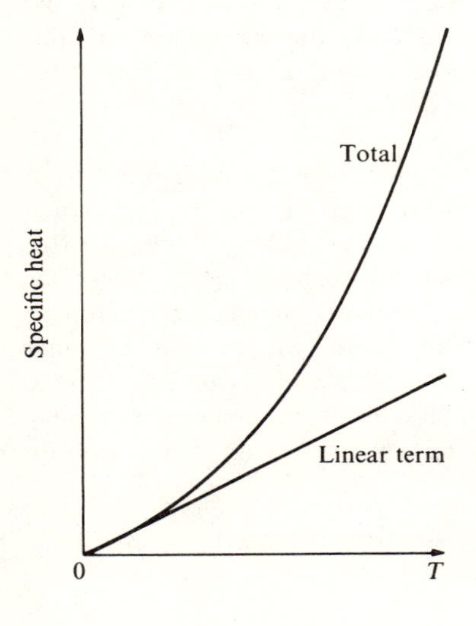

Fig. 3.3 The low-temperature specific heat of a metal. For a metal such as nickel, with a high electronic specific heat, the temperature range shown would be of the order of 20° K.

1.2 The diamagnetic susceptibility of free electrons[1] A particularly interesting thermodynamic property of solids is the contribution of the conduction electrons to the magnetic susceptibility of a material. It is remarkable that the contribution we expect on the basis of *classical* theory is identically zero. This may be shown by noting that the susceptibility is directly proportional to the second derivative of the total energy of the system with respect to magnetic field. Thus to obtain the susceptibility we first evaluate the total energy of the conduction electrons in the presence of a magnetic field.

In classical physics the quantum-mechanical sum over states is replaced by an integral over phase space. Thus the total energy is given by

$$E = \int d^3r\, d^3p\, f_0[\mathscr{H}(\mathbf{p},\mathbf{r})]\, \mathscr{H}(\mathbf{p},\mathbf{r})$$

where $\mathscr{H}$ is the Hamiltonian and f_0 is the equilibrium distribution function, depending only upon the energy of the particle. However, in the presence of a magnetic field the Hamiltonian for an electron (of charge $-e$) is given by

$$\mathscr{H}(\mathbf{p},\mathbf{r}) = \frac{1}{2m}\left[\mathbf{p} + \frac{e\mathbf{A}(\mathbf{r})}{c}\right]^2 + V(\mathbf{r}) \tag{3.5}$$

where $\mathbf{A}(\mathbf{r})$ is a vector potential corresponding to the magnetic field in question, $\mathbf{H}(\mathbf{r}) = \nabla \times \mathbf{A}(\mathbf{r})$. We perform the integration over momentum first and we may change the variable of integration to $\mathbf{p}' = \mathbf{p} + e\mathbf{A}(\mathbf{r})/c$. We obtain a form which is entirely independent of the vector potential and therefore of the magnetic field. Since the energy is independent of the magnetic field, the susceptibility must vanish and the corresponding contribution from the electron gas to the susceptibility must vanish. This result reflects the fact that classically a magnetic field does not change the energy of an electron but simply deflects it.

This result may seem surprising if we think of the susceptibility in terms of the magnetic moment of the electron gas in the presence of a magnetic field. We then imagine each electron executing a helical orbit (each rotating in the same sense for a uniform magnetic field). Therefore each electron gives a contribution of the same sign to the magnetic moment of the system. However, upon closer examination we see that we must consider also electrons at the surface of the system which cannot complete helical orbits because of the boundary. These electrons then move in very much larger orbits around the surface of the system, moving in the opposite

[1] The magnetic properties of simple metals are discussed at length by A. H. Wilson, "The Theory of Metals," pp. 150ff., 2nd edition, Cambridge, London, 1954.

sense to those in the interior as shown in Fig. 3.4. Their contribution to the magnetic moment exactly cancels that of the interior electron. This is seen in more detail by noting that at every point, including points near the surface, there are electrons moving with equal and opposite velocities and the current density is everywhere zero.

This argument breaks down in quantum mechanics since, as we noted in Sec. 5.6 of Chap. II, the orbital motion of an electron in a magnetic field is quantized. Therefore the electronic-energy eigenvalues depend upon the magnetic field, and when the total energy is evaluated it is found to depend upon the magnetic field. The corresponding contribution to the susceptibility is the *Landau diamagnetism*. It is interesting to carry the calculation for free electrons far enough to see the form of the electronic states in the magnetic field.

We rewrite the Hamiltonian of Eq. (3.5) taking the vector potential to be

$$\mathbf{A} = xH\hat{\mathbf{y}}$$

where $\hat{\mathbf{y}}$ is a unit vector in the y direction. This vector potential leads to a uniform magnetic field $\mathbf{H}$ in the z direction. Other forms, which correspond to a different gauge, could be taken for the vector potential (as $-\frac{1}{2}\mathbf{r} \times \mathbf{H}$). Each of these would lead to a different form of the eigenstates, and we will

Fig. 3.4 The motion of electrons under the influence of a magnetic field in a bounded system. Note that the electron skipping along the surface traverses its large orbit in a counterclockwise direction while those in the interior traverse in a clockwise direction. (Specular reflection has been assumed for illustrative purposes.)

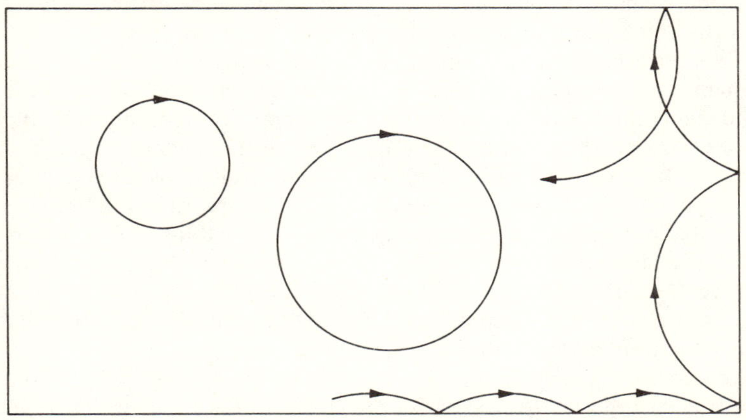

later discuss the meaning of these different forms. The gauge we have chosen is a particularly convenient one. The Hamiltonian then becomes

$$\mathscr{H} = \frac{1}{2m}\left(\frac{\hbar}{i}\nabla + \frac{exH}{c}\hat{\mathbf{y}}\right)^2 \tag{3.6}$$

We have dropped the potential since we are considering a uniform electron gas. By direct substitution we may see that this Hamiltonian has eigenstates of the form

$$\psi(\mathbf{r}) = \varphi(x)e^{ik_y y}\, e^{ik_z z} \tag{3.7}$$

where $\varphi(x)$ satisfies

$$-\frac{\hbar^2}{2m}\frac{\partial^2}{\partial x^2}\varphi(x) + \frac{e^2H^2}{2mc^2}\left(\frac{c\hbar k_y}{eH} + x\right)^2\varphi(x) + \frac{\hbar^2 k_z{}^2}{2m}\varphi(x) = E\varphi(x) \tag{3.8}$$

We have taken periodic-boundary conditions on the crystal surfaces perpendicular to $\hat{\mathbf{y}}$ and $\hat{\mathbf{z}}$. We will need to investigate the boundary conditions on the surfaces perpendicular to $\hat{\mathbf{x}}$.

Now Eq. (3.8) is simply the Schroedinger equation for a harmonic oscillator centered at the position $x = -c\hbar k_y/(eH)$, with spring constant $e^2H^2/(mc)^2$ and with eigenvalue $E - \hbar^2 k_z{}^2/(2m)$. Thus the eigenstates are harmonic-oscillator functions in the x direction and the energies of the electron states in the magnetic field are given by

$$E_{n,k_z,k_y} = \hbar\omega_c(n + 1/2) + \frac{\hbar^2 k_z{}^2}{2m} \tag{3.9}$$

where again we have defined the cyclotron frequency, $\omega_c = eH/(mc)$. This is of the form which we obtained more generally in Sec. 5.6 of Chap. II on the basis of the correspondence principle. We note that the energy is independent of the quantum number k_y, which simply specifies the center on the harmonic oscillator function. We have found a high degeneracy corresponding to the many positions at which the electron may orbit in the magnetic field.

These states may not seem very close to our intuitive picture of states in the magnetic field. However, we may imagine taking tight-binding-like linear combinations of circular orbits (each of which we think of as degenerate), each displaced with respect to the others in the y direction. This leads qualitatively to an eigenstate of the form given in Eq. (3.7). This form arises from our choice of gauge. Each different choice of gauge leads to a different linear combination of degenerate states.

We use the energies of Eq. (3.9) to evaluate the susceptibility of the system. It is customary also to restrict the center of the orbit to lie within

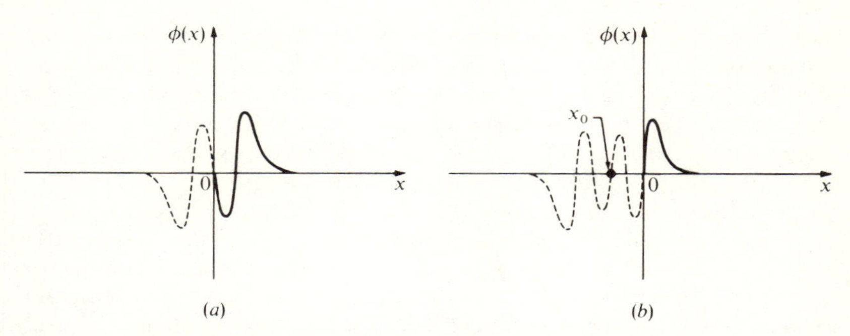

Fig. 3.5 Possible surface states in a magnetic field shown as solid lines for a vanishing boundary condition at $x = 0$. Case (b) shows a skipping orbit; for electrons at the Fermi energy in a metal the dashed curve might have of the order of 10^6 nodes and x_0 would be approximately $-v_F/\omega_c$.

the normalization volume, thus restricting the realm of k_y and the degeneracy of the states. This leads (see, e.g., Seitz[1]) to a susceptibility of

$$\chi = -\frac{e^2}{12\pi mc^2}\left(\frac{3N}{\pi}\right)^{1/3} \tag{3.10}$$

where N is the density of electrons. This is a remarkable result. It arises only in the quantum-mechanical framework and yet the final result is independent of the magnitude of Planck's constant.

We might worry about our cavalier dismissal of the complications of the boundaries because of the importance of the boundaries in a classical discussion. We may see the form of surface states at the surfaces perpendicular to the x direction by applying a vanishing-boundary condition at $x = 0$. The Hamiltonian of Eq. (3.6) remains valid for $x > 0$ and the separation given in Eq. (3.7) remains appropriate. Proper solutions of Eq. (3.8) are now sought for which $\varphi(x)$ vanishes at $x = 0$. Some simple solutions are obtainable immediately For example, with $k_y = 0$ the harmonic-oscillator solutions for odd n become the solutions for $x > 0$ and satisfy the vanishing-boundary condition at $x = 0$. Such a state is shown in Fig. 3.5*a*. For k_y different from zero we can readily see the qualitative form of the solutions for varying numbers of nodes. One such possibility is shown in Fig. 3.5*b*. We can see immediately that these modifications of the surface states cannot affect our calculation of the susceptibility in large systems which was based upon the total energy since the fraction of electrons involved decreases as the dimension of the system in the x direction increases. These surface states must become the origin of the susceptibility if the

[1]F. Seitz, "Modern Theory of Solids," p. 583, McGraw-Hill, New York, 1940.

calculation is based upon a calculation of the magnetic moment of the system instead of the total energy. We see that because of the modification of these surface states we need not expect (as we did for a classical system) the current density near the surface to vanish, though it must in the interior. The origin of the susceptibility can be understood in terms of the corresponding surface currents that arise in a magnetic field.

The magnetic susceptibility of a free-electron gas has been explored by Teller[1] on the basis of such surface states. In addition, these states have been observed as the origin of structure in the absorption of microwaves at the surface of a metal as a function of magnetic field.[2] This structure in the microwave absorption arises from those electrons which in a classical sense are skipping along the surface, moving almost parallel to the surface. In the quantum-mechanical sense these are electrons for which the center of the harmonic-oscillator function is displaced one orbit radius outside of the surface of the metal as shown in Fig. 3.5*b*. The corresponding $\varphi(x)$ within the metal has no nodes and is restricted to lie very close to the surface. Since the microwave field penetrates only a short distance into the specimen, such states (at the Fermi energy) dominate the reflection coefficient of the metal.

1.3 Pauli paramagnetism A second contribution to the magnetic susceptibility arises from the magnetic moment of the electrons. The origin and approximate magnitude of the effect are very easily seen within the simple framework of energy bands we are using here. When no magnetic fields are present, each band state is doubly degenerate, one state having spin up and the other spin down. Thus we may construct two identical band structures for the two spins as shown in Fig. 3.6. Furthermore, both are filled to the same Fermi energy.

When a field H is applied, electrons with spin magnetic moment β parallel to the field are lowered in energy by βH, while those with antiparallel moment are raised in energy by βH. This simply shifts the spin-up and spin-down bands relative to each other by $2\beta H$. However, the Fermi energy is a constant of the system (and is in fact unchanged to first order in H), so electrons are transferred from the spin-down to the spin-up band. The number transferred is clearly $\beta H[n(E)/2]$ per unit volume, where $n(E)/2$ is the density of states of a given spin per unit energy at the Fermi energy. Each electron transferred gives a net moment contribution of 2β so the magnetic moment per unit volume is $\beta^2 n(E)H$, corresponding to a susceptibility of

$$\chi = \beta^2 n(E) \tag{3.11}$$

[1] E. Teller, *Z. für Physik*, **64**:629 (1930).
[2] J. F. Koch and C. C. Kuo, *Phys. Rev.*, **143**:470 (1966).

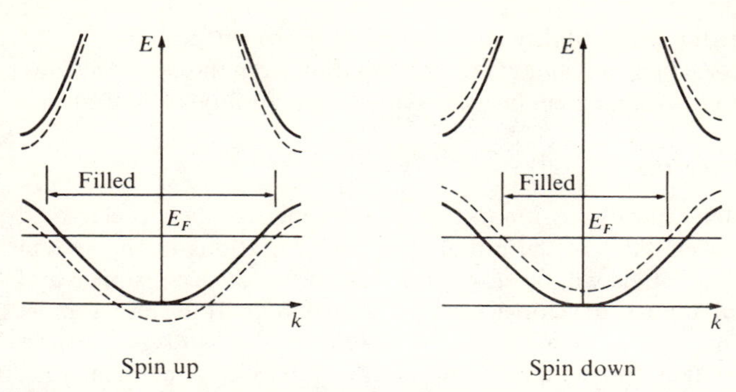

Fig. 3.6 Schematic representation of bands for electrons of spin up and spin down. With no magnetic field (solid lines) there are equal numbers of electrons of each spin. When a magnetic field is applied, the spin-up and spin-down bands are shifted in opposite directions by βH, the interaction energy between a particle of magnetic moment β and magnetic field H. Since the Fermi energy must be the same for both bands, the occupation differs in the two bands and a magnetic moment arises.

This is a paramagnetic contribution, a positive contribution to χ, in contrast to the diamagnetic susceptibility of Eq. (3.10). We may estimate the magnitude of this paramagnetic term by taking β equal to the Bohr magneton

$$\beta = \frac{e\hbar}{2mc} \tag{3.12}$$

appropriate to a free electron, and taking the free-electron density of states, $n(E) = 3N/2E_F = 3Nm/\hbar^2 k_F^2$. We obtain

$$\chi = \frac{e^2}{4\pi mc^2}\left(\frac{3N}{\pi}\right)^{1/3} \tag{3.13}$$

By comparison with Eq. (3.10), we see that the paramagnetic contribution is just three times as large as the diamagnetic contribution. Thus a free-electron gas is found to be paramagnetic. Most simple metals are, in fact, paramagnetic, but Eq. (3.13) is very inaccurate quantitatively. We may think of Eq. (3.11) as valid but there is error both in the use of the free-electron β (which should be corrected for orbital contributions to the magnetic moment) and in the use of a free-electron density of states. In our discussion of the Landau theory of Fermi liquids we will note corrections to $n(E)$ arising from the electron-electron interaction. These are in fact different from the corresponding corrections to the $n(E)$ entering the electronic specific heat.

The magnetic susceptibility of a metal contains, in addition to the two terms discussed here, a diamagnetic contribution from the core electrons. This contribution is calculated just as is the susceptibility of free atoms.

2. *Transport Properties*

We will now extend our discussion to transport properties such as electrical and thermal conductivity. In the initial stages all variations in the system and all applied potentials will be slowly varying over a lattice distance and the semiclassical theory developed in Sec. 2 of Chap. II becomes most appropriate. In most cases we will approximate the energy bands in question by parabolas so that the semiclassical treatment becomes identical to a classical treatment except for the inclusion of the Pauli principle.

As we indicated in that section this is only meaningful when it is meaningful to describe the electron states in terms of packets. The construction of a packet generates an uncertainty in the energy of the particles in question. If we are to discuss the energies of the electrons with an accuracy that is sharp compared to KT, then we must not envisage applied fields that vary appreciably over a distance of the order of δr where $\delta r\,\delta p \approx \hbar$ or $\delta r \approx \hbar v/(KT)$ with v the velocity of the particle. We cannot use this approach, for example, to describe the motion of particles in the fields arising from the individual ions since these fields vary over an atomic distance and constructing a wavepacket on that scale would lead to an energy uncertainty large compared even to the Fermi energy in a metal. We can, however, discuss the motion of electrons in systems that are inhomogeneous if the inhomogeneities are on a macroscopic scale and that is all we will want to do here.

Because of this limitation on the scale of the inhomogeneities, it is possible to define a probability of occupation of the states that is a function not only of momentum and of time, but also of position. We write this distribution function $f(\mathbf{p},\mathbf{r},t)$. The equilibrium function at any temperature will simply be the Fermi function $f_0(E)$. We may obtain the total number of electrons in any given region of momentum and position simply by integrating over momentum space and over real space.

$$dn = \frac{2}{h^3} f(\mathbf{p},\mathbf{r},t)\, d^3p\, d^3r \tag{3.14}$$

Here we have written the density of states in wavenumber space as $2/(2\pi)^3$ per unit volume as before and again have replaced the wavenumber integration by an integration over momentum. By making the restriction $f \leq 1$, we will satisy the Pauli principle.

The definition of this distribution function has required a semiclassical approximation since we are specifying in some sense the momentum and the position of the individual particles. Thus it is required that the variation of f be small except over macroscopic distances in order that the momentum of the particles can be reasonably well defined. In some of our problems there will be no variation of the distribution function with position and this limitation would not arise; in others, however, it will. Note that in a purely classical system there would be no limitation on the variation of the distribution function, nor on its maximum value, and the constant $2/h^3$ would ordinarily be absorbed in f.

2.1 The Boltzmann equation We seek now an equation that will specify the change of the distribution function we have defined with time. This is, of course, the counterpart of the Schroedinger equation which specifies the variation of the wavefunction with time. The resulting transport equation will form the basis of all of our transport calculations. We fix our attention on a particular value of momentum and a particular position in the system and seek the derivative of the distribution function with respect to time. This can perhaps be most systematically done by constructing a cell in phase space and computing the flow of particles in and out.[1] However, we may obtain the result more directly (and perhaps less perspicuously) by a more general argument. We will then interpret each of the terms in our result in a manner consistent with this cell construction.

We consider a particular state that is occupied; that is, $f = 1$. In the presence of applied fields this state will move through phase space according to our semiclassical equations. Of course as we follow the trajectory corresponding to this particular electron, f does not change but remains unity. If we follow an unoccupied state, $f = 0$, the occupation again will not change with time. Similarly the f associated with any state will not change with time; the *total* derivative of f with respect to time must be zero as we follow a trajectory in phase space. Writing this equation, $df/dt = 0$, will give us directly a transport equation.

We wish also, however, to add the possibility that the electrons scatter by a mechanism that is not (and cannot be) included in our applied fields. These are the scattering events that we described earlier. In a scattering process an electron will discontinuously change its momentum and therefore will make a discontinuous jump in phase space. Thus the transport equation that we seek is the equation $df/dt = \partial f/\partial t|_{\text{coll}}$, where the last term is the change in the distribution function due to collisions. We now write this result

$$\frac{\partial f}{\partial t} + \frac{\partial f}{\partial \mathbf{p}} \frac{d\mathbf{p}}{dt} + \frac{\partial f}{\partial \mathbf{r}} \frac{d\mathbf{r}}{dt} = \left.\frac{\partial f}{\partial t}\right|_{\text{coll}}$$

[1]This approach is used by Seitz, *op. cit.*, pp. 168ff.

We should now note the meaning of some of the parameters that enter the equation. The $d\mathbf{p}/dt$ is the rate of change of momentum with time at a given point on the trajectory. This of course is just equal to the applied force at that point and time. The $d\mathbf{r}/dt$ is the rate of change of position on the trajectory for a given momentum. This of course is just the particle velocity **v**. Thus our transport equation may be rewritten

$$\frac{\partial f}{\partial t} = -\frac{\partial f}{\partial \mathbf{r}} \cdot \mathbf{v} - \frac{\partial f}{\partial \mathbf{p}} \cdot \mathbf{F} + \left.\frac{\partial f}{\partial t}\right|_{\text{coll}} \tag{3.15}$$

The interpretation of this equation is now more physical. The rate at which the distribution function at a given position and momentum changes contains three contributions. The first is a drift term; the electrons in question are leaving that region of space with velocity **v** and, if the distribution function varies in space, the number leaving the region will differ from the number entering. The second term similarly represents the change arising because particles at the momentum in question are being accelerated into different momentum states. If the distribution function varies with momentum the number being accelerated out will be different from the number accelerated into the momentum region of interest. Finally there is a change because electrons may be being scattered out at a different rate than they are being scattered in.

The scattering rate written in this way is inconvenient though we could calculate it as we calculated the scattering time due to defects before. Almost universally in transport calculations a *relaxation-time approximation* is made. It is noted that if the distribution function were the equilibrium distribution function there would be no change in f due to scattering. Furthermore if the distribution function differs from the equilibrium function we expect that it will decay exponentially in time to the equilibrium form. This may be written mathematically in the form

$$\left.\frac{\partial f}{\partial t}\right|_{\text{coll}} = -\frac{f - f_0}{\tau}$$

where τ is the relaxation time.

This is consistent with our finding that the momentum of a particular electron is randomized by the characteristic decay time that we computed. The relaxation-time approximation is a very plausible one and agrees with a wide range of experiments; however, it certainly cannot be true in detail. If, for example, the scattering events are primarily elastic they will tend to cause a decay of any current to the equilibrium value of zero but will not so effectively cause the decay of any isotropic deviation from equilibrium of the distribution as a function of energy. Thus it may be necessary to define different relaxation times for different phenomena being studied. In addition we must take care in what we write for the equilibrium distribution function f_0 in $\partial f/\partial t|_{\text{coll}}$. If, for example, the distribution function is inhomogeneous (different total densities of electrons at different points) then in the scattering term we must write the equilibrium distribution function corresponding to the *local* density of particles, $f_0[n(\mathbf{r})]$. Otherwise we will be introducing scattering events that instantaneously transfer electrons from one position to another. However, when the relaxation-time approximation is used carefully it leads to a good description of many properties.

Using this form we may now write the Boltzmann equation in the collision-time approximation

$$\frac{\partial f}{\partial t} + \frac{\partial f}{\partial \mathbf{p}} \cdot \mathbf{F} + \frac{\partial f}{\partial \mathbf{r}} \cdot \mathbf{v} = -\frac{f - f_0(n)}{\tau}$$

Most frequently we will be interested in applying fields to the system and seeking only the linear response; i.e., we write the distribution function $f = f_0(\bar{n}) + f_1$, where $f_0(\bar{n})$ is the equilibrium distribution function based on the *average* electron density (it is not a function of position) and f_1 is the deviation from this equilibrium. This may be substituted in the Boltzmann equation in the collision approximation and only first-order terms in the applied fields retained. The result is the *linearized Boltzmann equation*,

$$\frac{\partial f_1}{\partial t} + \frac{\partial f_0}{\partial \mathbf{p}} \cdot \mathbf{F} + \frac{\partial f_1}{\partial \mathbf{r}} \cdot \mathbf{v} = -\frac{f_1}{\tau} + \frac{\delta_n f_0}{\tau} \tag{3.16}$$

where we have written $\delta_n f_0 = f_0(n) - f_0(\bar{n})$, the difference between an equilibrium distribution function at the local density $f_0(n)$ and the equilibrium distribution function for the average density of the system $f_0(\bar{n})$. This term is very frequently omitted in writing the linearized Boltzmann equation. When the density depends on position it is important that it be included.

2.2 Electrical conductivity Let us now apply the linearized Boltzmann equation to a specific property. In particular we will consider the conductivity measured for a uniform electric field; thus the applied force is given by $\mathbf{F} = -e\mathscr{E}$ for electrons or $+e\mathscr{E}$ for holes. We will do this by computing the distribution function in the presence of the applied field. We will then compute the current in terms of the distribution function and this procedure will give us the proportionality constant between the current and the applied field, which of course is just the conductivity.

We are considering a steady state and the first term in Eq. (3.16) vanishes. In addition the distribution function will be independent of position and the third term also vanishes as does the term in $\delta_n f_0$. We have

$$\frac{\partial f_0}{\partial \mathbf{p}} \cdot \mathbf{F} = -\frac{f_1}{\tau}$$

It is convenient to rewrite the left side noting that

$$\frac{\partial \mathrm{f}_0}{\partial \mathbf{p}} = \frac{\partial \mathrm{f}_0}{\partial \mathrm{E}} \frac{\partial \mathrm{E}}{\partial \mathbf{p}} = \frac{\partial \mathrm{f}_0}{\partial \mathrm{E}} \mathbf{v}$$

This form will be particularly convenient in metals since $-\partial f_0/\partial E$ is then approximately equal to a delta function. We solve now immediately for the first-order term in the distribution function.

$$f_1 = \tau\left(-\frac{\partial f_0}{\partial E}\right)\mathbf{v} \cdot \mathbf{F} = \mp\, e\tau\left(-\frac{\partial f_0}{\partial E}\right)\mathbf{v} \cdot \mathscr{E}$$

where the first sign is for electrons, the second for holes.

We may now directly compute the current flowing by adding up the contribution to the current from each occupied state. No current would be obtained with the equilibrium distribution function so we may simply use the first-order correction in this calculation. We obtain

$$\mathbf{j} = \frac{2}{h^3}\int d^3p f_1 (\mp e)\mathbf{v} = \frac{2e^2\tau}{h^3}\int d^3p\left(\frac{-\partial f_0}{\partial E}\right)(\mathbf{v} \cdot \mathscr{E})\mathbf{v}$$

The sign of the carriers has gone out and the formula is correct for any band structure. We now assume isotropic bands. For convenience we may let $\mathscr{E}$ lie in the z direction. When we integrate over angle the only surviving component of the current will lie in the z direction. We may therefore replace $\mathbf{v}$ by v_z. Then averaging over angle we note that $\langle v_z{}^2 \rangle = v^2/3$. We then obtain the current density in the form

$$j_z = \frac{2e^2\tau\mathscr{E}_z}{3h^3}\int d^3p\, v^2\left(\frac{-\partial f_0}{\partial E}\right)$$

This integral may be readily evaluated by performing a partial integration. We first write $d^3p = 4\pi p^2\, dp$. We now take a parabolic band with $E = p^2/2m^*$ and write the first factor of v as p/m^*, and the second factor of v as $\partial E/\partial p$. Then $(-\partial f_0/\partial E)(\partial E/\partial p) = -\partial f_0/\partial p$. Using the resulting form, a partial integration may be carried out.

$$j_z = \frac{2e^2\tau\mathscr{E}_z}{3h^3m^*}\int_0^\infty dp4\pi p^3\left(\frac{-\partial f_0}{\partial p}\right)$$

$$= \frac{2e^2\tau\mathscr{E}_z}{3h^3m^*}\left(-4\pi p^3 f_0\Big|_0^\infty + 3\int_0^\infty dp4\pi p^2 f_0\right)$$

The first term vanishes at the lower limit where $p = 0$ and at the upper limit where $f_0 = 0$. The final integral, when multiplied by $2/(3h^3)$, is simply the number of electrons (or holes) per unit volume in the band, N. Thus we may immediately identify the conductivity σ which relates the current density to the electric field, $\mathbf{j} = \sigma\mathscr{E}$.

$$\sigma = \frac{Ne^2\tau}{m^*} \tag{3.17}$$

We have assumed isotropic parabolic bands but we have not assumed any limiting form for f_0. The result is equally applicable to simple metals and to electrons or holes in semiconductors.

Equation (3.17) for the conductivity is intuitively very plausible. The $e\mathscr{E}/m^*$ is the acceleration of a particle with a charge e and mass m^* in a field $\mathscr{E}$. It will acquire a velocity $e\tau\mathscr{E}/m^*$ in the course of a scattering time τ. Thus it will carry a current $e^2\tau\mathscr{E}/m^*$. We obtain the total current per unit volume by multiplying by the number of electrons per unit volume. Such a plausibility argument would not account for the numerical coefficient which in this case turns out to be one owing to the way in which the scattering time has been defined.

The electrical conductivity is a sufficiently simple problem that it was possible to solve the Boltzmann equation for the first-order distribution function directly. In any of the more complicated properties it is customary to postulate a first-order distribution function and to verify by substitution that it is correct or at least approximately correct.

2.3 The Hall effect We have noted earlier that if a magnetic field is applied to a system in which a current is flowing, there is a tendency for the charge carriers to be deflected laterally. We may follow the consequences of this step by step. Imagine applying an external electric field $\mathscr{E}_0$ along the axis of a specimen. Electrons will drift in the opposite direction, as seen in Fig. 3.7*a*. If we apply a magnetic field perpendicular to the axis of the specimen the carriers will tend to be deflected to one side. At the surface of the crystal of course they will not drift into space but a surface charge will

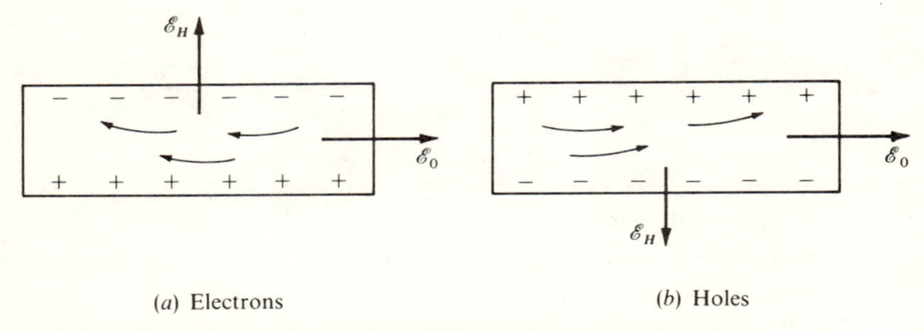

Fig. 3.7 *The physical origin of the Hall effect. (a) Electrons accelerated to the left by the applied electric field $\mathscr{E}_0$, and then deflected upward by a magnetic field into the figure. A resulting accumulation of charge at the surface causes a transverse Hall field $\mathscr{E}_H$ which prevents further accumulation in the steady state. (b) Corresponding construction made for a solid in which the carriers are holes, giving rise to a Hall field of the opposite sign.*

develop. The surface charge then will give rise to a transverse electric field $\mathscr{E}_H$, the *Hall field*, which causes a compensating drift such that the carriers remain within the specimen. At first glance this would seem to be an extremely complicated problem. When formulated in the right way, however, it will become quite simple. In particular we will consider an infinite system with a total electric field $\mathscr{E}$. We will then compute the direction of the current flow. We then imagine a specimen cut with axis along the current flow. The electric field, resolved into appropriate components, will contain both the longitudinal applied field and the transverse Hall field. Thus the details of the surface will never arise.

We begin by writing the Boltzmann equation in the relaxation-time approximation, again for a uniform steady-state system. Thus the derivatives of the distribution function with respect to position and with respect to time vanish.

$$\frac{\partial f}{\partial \mathbf{p}} \cdot \mathbf{F} + \frac{f - f_0}{\tau} = 0$$

The force $\mathbf{F}$ now includes a magnetic as well as an electric force. It turns out that we may include the magnetic field exactly, though we again wish to expand in the applied electric field. If we were to seek an expansion also in the magnetic field, we would find, as will be seen, that the term of particular interest is bilinear in the electric and magnetic field and is in that sense a second-order term. It is simpler to write f_1 as the contribution to the distribution function which is linear in the electric field. Replacing $\mathbf{F}$ by the Lorentz force for a particle of charge $-e$ we obtain

$$\frac{\partial f_0}{\partial E}\mathbf{v}\cdot\left(-e\mathscr{E} - \frac{e}{c}\mathbf{v}\times\mathbf{H}\right) + \frac{\partial f_1}{\partial \mathbf{p}}\cdot\left(-e\mathscr{E} - \frac{e}{c}\mathbf{v}\times\mathbf{H}\right) + \frac{f_1}{\tau} = 0$$

We have divided $\partial f/\partial \mathbf{p}$ into a zero-order and first-order contribution in order to take advantage of the dependence on energy only of the equilibrium distribution function. The first term in the magnetic field vanishes identically since $\mathbf{v}\cdot(\mathbf{v}\times\mathbf{H})$ is identically zero. The second term in the electric field may be dropped since it is second order. We have

$$-e\frac{\partial f_0}{\partial E}\mathbf{v}\cdot\mathscr{E} - \frac{e}{c}\frac{\partial f_1}{\partial \mathbf{p}}\cdot(\mathbf{v}\times\mathbf{H}) + \frac{f_1}{\tau} = 0$$

We cannot simply solve for the first-order distribution function as we did in the case of the conductivity calculation. However, in various ways we may see a plausible guess. One way is to note that the effect of the magnetic field, roughly speaking, is to rotate the distribution. We therefore try a form in analogy with that obtained in the conductivity calculation but with the electric field replaced by a general vector $\mathbf{G}$ which will be determined; i.e., we asume

$$f_1 = e\tau\frac{\partial f_0}{\partial E}\mathbf{v}\cdot\mathbf{G}$$

We may next evaluate the derivative of f_1 with respect to $\mathbf{p}$, assuming now $\mathbf{p} = m\mathbf{v}$.

$$\frac{\partial f_1}{\partial \mathbf{p}} = e\tau\left[\frac{\mathbf{G}}{m}\frac{\partial f_0}{\partial E} + \frac{\partial^2 f_0}{\partial E^2}(\mathbf{v}\cdot\mathbf{G})\,\mathbf{v}\right]$$

The final term will not contribute because it will yield a factor $\mathbf{v}\cdot(\mathbf{v}\times\mathbf{H})$. We substitute our trial distribution function back into the Boltzmann equation. We cancel a factor $(-e)\partial f_0/\partial E$ which appears in each term and obtain

$$\mathbf{v}\cdot\mathscr{E} + \frac{e\tau}{mc}\mathbf{G}\cdot(\mathbf{v}\times\mathbf{H}) - \mathbf{v}\cdot\mathbf{G} = 0$$

or

$$\mathbf{v}\cdot\left(\mathscr{E} + \frac{e\tau}{mc}(\mathbf{H}\times\mathbf{G}) - \mathbf{G}\right) = 0$$

Here we have noted $\mathbf{A}\cdot(\mathbf{B}\times\mathbf{C}) = \mathbf{B}\cdot(\mathbf{C}\times\mathbf{A})$. We note that this will be a solution for all $\mathbf{v}$ if and only if

$$\mathscr{E} = \mathbf{G} - \frac{e\tau}{mc}(\mathbf{H}\times\mathbf{G})$$

We note that the current associated with our first-order trial distribution function may be obtained in precisely the same way it was in the case of electrical conductivity and the result is of the same form, $\mathbf{j} = \sigma\mathbf{G}$. Thus we may write the electric field in terms of the current in the form

$$\mathscr{E} = \frac{\mathbf{j}}{\sigma} - \frac{e\tau}{mc\sigma}(\mathbf{H} \times \mathbf{j}) \tag{3.18}$$

In the absence of the magnetic field the second term vanishes and we obtain precisely the result we had before. The second term is a component of the electric field which is transverse both to the applied magnetic field and to the current. The proportionality constant is called the Hall constant and its magnitude is given by

$$R = \frac{(-e)\tau}{mc\sigma} = \frac{1}{N(-e)c}$$

where we have used the explicit expression for the conductivity.

Note that the sign of the Hall constant is the same as the sign of the carrier which we have taken to be $-e$. If our calculation had been based upon holes, the sign of the charge entering the Lorentz force would have been positive and the corresponding Hall constant would have been positive. This is consistent with the physical argument illustrated in Fig. 3.7*b*. We note further that since the magnitude of the electronic charge and of the speed of light are well known, the Hall constant gives a direct measure of the number of carriers present (if all carriers are of the same sign).

Our derivation of the Hall constant has assumed an effective mass and a relaxation time but otherwise is not restricted specifically to metals nor to semiconductors. In simple metals (at low fields), measurements give values close to the value we would calculate assuming a free-electron approximation based upon the valence electrons. In semiconductors the magnitude gives the appropriate number of electrons or holes in *n*- and *p*-type semiconductors respectively. The measurement of both the electrical conductivity and the Hall constant allows us to determine both the number of carriers and the ratio of the scattering time to the effective mass. This latter ratio determines directly the *mobility*, which is the ratio of average drift velocity to electric field. It turns out that the resulting formula for the Hall constant remains relevant when we consider more complicated and anisotropic band structures. However, the interpretation of N is somewhat more complicated. If we consider, for example, a crystal containing carriers in two bands, N will be a weighted sum of the number of carriers in each which depends upon the effective mass and scattering time of each. It is found also that the transverse electric field is no longer linear in the magnetic field. Distinctly different behavior is obtained in a high-field and a low-field region. Which region obtains depends on whether the product of the cyclotron frequency and scattering time for the various carriers,

$$\omega_c \tau = \frac{eH\tau}{(m^*c)} \tag{3.19}$$

is large or small compared to unity. The criterion can also be written in terms of the *Hall angle* θ, the angle between $\mathbf{j}$ and $\mathscr{E}$. (We may note by comparison of Eqs. (3.18) and (3.19) that $\tan \theta = \omega_c \tau$.)

We noted that at low fields in simple metals N corresponds to the density of valence electrons. At very high fields the Hall constant in simple metals is determined by the difference in the number of electrons and the number of holes arrived at by examining the Fermi surface. If all segments of the Fermi surface are closed this may be computed by subtracting the volume of wavenumber space corresponding to closed surfaces around occupied electronic states from the volume of closed surfaces surrounding unoccupied states. If the Fermi surface is not closed, the behavior of the transverse electric field is more complicated. It depends upon the direction of open orbits on the Fermi surface and upon the relative volume of electron hole and closed surfaces.[1]

We may note that in our calculation the longitudinal electric field was found to be precisely the same with or without a magnetic field. This result is no longer correct when the energy-band structure becomes anisotropic. It is then found that the longitudinal electric field will also depend upon the magnetic field and ordinarily will increase. This extra resistance arising from the addition of a magnetic field is called the *magnetoresistance*.[1] The measurement of magnetic field and Hall effect in metals gives definitive information about the topology of the Fermi surface. We will not go into the details of this method for studying Fermi surfaces.

2.4 Thermal and thermoelectric effects We will consider one further traditional application of the Boltzmann equation, that to effects involving heat transport. In this case we will only formulate the problem and examine the physics of the results. The intermediate algebra is somewhat complicated but may be found in a number of places.[2]

In this case we must specify a system with a temperature gradient. This will be done by ansatz in an intuitively plausible way. For simplicitly we envisage a collection of free electrons (either degenerate as in a simple

[1]A survey of these "galvanomagnetic" effects in metals is given by R. G. Chambers, in W. A. Harrison and M. B. Webb (eds.), "The Fermi Surface," pp. 100ff., Wiley, New York, 1960.

[2]E.g., J. M. Ziman, "Principles in the Theory of Solids," pp. 194ff., Cambridge, London, 1964.

metal or nondegenerate as in a semiconductor). We take our zero of energy to lie at the minimum energy of this band. Then we will assume a distribution function in the form

$$f = f_0\left[\frac{E - \xi(x)}{KT(x)}\right] + f_1(x) \tag{3.20}$$

Here we have allowed both the Fermi energy and the temperature to be a function of position and have let the variation be in the x direction. At any given position the first term represents an equilibrium distribution with a well-defined temperature. In addition we have added a first-order correction to the distribution function. Such a correction is clearly necessary if in addition to the temperature gradient there are fields present. We may immediately see that for most interesting cases there will inevitably be fields present. In a measurement of the thermal conductivity the specimen will ordinarily be electrically insulated from its surroundings. Thus the boundary condition upon our system is a requirement that the current vanish, not that the electric fields vanish. We may expect on physical grounds that if no fields were present the electrons would be boiled downstream in a temperature gradient and currents would flow. Our approach then will be to consider flow both of heat and of current in the presence of a temperature gradient and an electric field. By requiring that the current vanish we may solve for the electric field and then the heat flow and obtain ordinary thermal conductivity. We will also be led naturally to thermoelectric effects.

We are interested again in a steady-state situation and the Boltzmann equation becomes

$$\frac{\partial f}{\partial \mathbf{p}} \cdot \mathbf{F} + \frac{\partial f}{\partial \mathbf{r}} \cdot \mathbf{v} + \frac{f - f_0}{\tau} = 0$$

The f_0 in the relaxation term is the local equilibrium distribution; i.e., the first term of Eq. (3.20). We will allow an electric field in the x direction and this will be the only force entering the Boltzmann equation. We will keep only the lowest-order term in $\partial f/\partial r$, which may be obtained by differentiating the first term in Eq. (3.20) with respect to x. We obtain then for the Boltzmann equation

$$\frac{\partial f_0}{\partial E}\mathbf{v} \cdot (-e\mathscr{E}) - \frac{\partial f_0}{\partial E}\left(\frac{d\xi}{dx} + \frac{E - \xi}{T}\frac{dT}{dx}\right)v_x + \frac{f_1}{\tau} = 0$$

which may be solved immediately for f_1.

$$f_1 = v_x \tau \frac{\partial f_0}{\partial E}\left(e\mathscr{E} + \frac{d\xi}{dx} + \frac{E - \xi}{T}\frac{dT}{dx}\right) \tag{3.21}$$

We may now write the current and the heat flow, noting that the thermal energy is simply the kinetic energy of the electrons.

$$j_x = \frac{2}{h^3} \int d^3p(-e)v_x f_1 \tag{3.22}$$

$$c_x = \frac{2}{h^3} \int d^3p\left(\frac{m}{2}v^2\right)v_x f_1$$

We see that there are contributions to the electric current proportional to the field present and contributions to the heat flow due to the temperature gradient. In addition there is a term in the electric current flow due to the thermal gradient and a term in the heat flow due to the electric field.

By setting the electrical current equal to zero we may eliminate the electric field and $\partial\xi/\partial x$ and obtain the ratio of the flow of thermal energy to the thermal gradient, which is simply the thermal conductivity. For a simple metal we obtain

$$\kappa = \frac{16\pi^3}{9} \frac{mv_F\tau}{h^3} \xi K^2 T = \frac{\pi^2}{3} \frac{N\tau}{m} K^2 T$$

where N is again the density of electrons. The ratio of the thermal conductivity to the electrical conductivity is

$$\frac{\kappa}{\sigma} = \frac{\pi^2}{3} \frac{K^2}{e^2} T$$

The $\pi^2K^2/(3e^2)$ is called the *Wiedemann-Franz* ratio; all of the parameters entering it are fundamental constants. It turns out that to a good approximation this value is observed in many metals.

It is plausible that the thermal conductivity and the electrical conductivity should be related in a manner such as this. Ziman,[1] for example, has argued that in thermal conductivity we may think of an electron as carrying an energy KT and that we may think of an associated force due to the thermal gradient given by $\partial|KT|/\partial x$. By an intuitive argument such as we gave for the electrical conductivity, we obtain the form of the Wiedemann-Franz ratio without, of course, obtaining the numerical coefficient.

Proceeding with these same equations we may also treat thermoelectric effects. We may consider, for example, two wires of different materials as shown in Fig. 3.8. We envisage the temperature being different at the left and at the right with a resulting temperature gradient in both materials. We may first treat the two metals separately, noting that the current vanishes in each; i.e., in each material we set the current given by Eq. (3.22) equal to zero, using Eq. (3.21) for f_1. Performing the integration will give a relation

[1] *Ibid.*

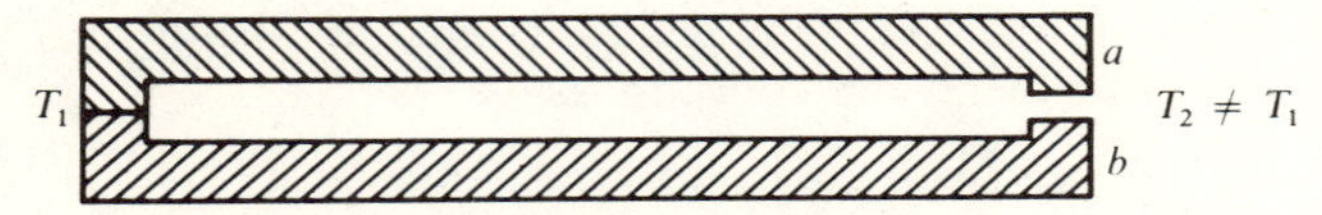

Fig. 3.8 *Two wires made of different materials in a temperature gradient. The junction at the left is a thermocouple. A voltage difference may be measured between the points a and b at the right.*

between $e\mathscr{E} + \frac{d\xi}{dx} = \frac{dV}{dx} + \frac{d\xi}{dx}$ (V is the electrostatic potential energy for the electron as a function of position) in terms of dT/dx. Integrating along each of the wires, we obtain the difference in the Fermi energy, measured from the bottom of the band, at the two ends plus the difference in electrostatic potential energy (which determines the energy of the band minimum) at the two ends. The sum for each metal will be proportional to the difference in temperature at the two ends. The absolute Fermi energy must be the same in the two metals at the left where they are in contact, and the resulting absolute Fermi energy difference at the right is a measurable voltage difference. This is an emf in the circuit which is proportional to the temperature difference at the two ends. We have constructed a *thermocouple* and the emf is the *thermoelectric voltage*.

The study of thermoelectric and thermomagnetic properties has been quite extensive. We have simply outlined the approach for treating one or two of the simplest phenomena.

2.5 Electron tunneling A transport phenomenon which has been of particular interest in recent years is the transport of electrons through thin insulating films. One system that is used to study such effects is obtained by depositing an aluminum strip on a glass slide, letting it oxidize for a few minutes, and then depositing a second strip crossing the first. Such an oxide film on aluminum will ordinarily be continuous and may be of the order of 20 Å thick. The two aluminum strips therefore are not in electrical contact; yet if a voltage is applied between the two a current is found to flow and is in fact proportional to the applied voltage. This is the behavior that would be expected if the film were *not* continuous and the current were conducted through small leaks. However, when the metals were made superconducting it became clear that the current was carried by a tunneling mechanism. We will return to the tunneling in superconductors in Chap. V.

If we consider a conduction-band wavefunction in one of the aluminum strips, it is clear that the wavefunction does not drop immediately to zero at its interface with the aluminum oxide. We expect that it will decay exponentially into the oxide as did the surface states discussed in Sec. 8.3

of Chap. II. The wavefunction will still have a finite value at the opposite edge of the oxide and therefore it is possible for the electron to decay into the second strip. This is the same behavior that would be anticipated if the two aluminum strips were separated by 20 Å of vacuum. The conduction electrons of interest would have energy below the vacuum level, and therefore the wavefunction would have negative kinetic energy within the vacuum and again would decay exponentially, precisely the situation described as quantum-mechanical tunneling. The only difference with the oxide layer is that the exponential decay arises not directly from negative kinetic energy but from the fact that the energies of interest lie within the forbidden band of the oxide. It now seems clear that it would be more appropriate to regard the oxide as an amorphous semiconductor, in the sense discussed in Sec. 7 of this chapter. We will see there that such a material behaves very much the same as a semiconductor or insulator with the Fermi energy fixed near the center of the band gap. This may account for the success of tunneling theory which invariably has assumed an intrinsic crystalline oxide. We also shall make that assumption.

We proceed then to compute the probability of an electron tunneling through the oxide layer. Perhaps the most natural approach would be to construct incident and reflected waves in one strip, to match them to exponentially decaying waves of the same energy in the oxide, and finally to match these to a transmitted wave in the second strip. In the traditional tunneling problem the two sets of matching conditions lead to a unique ratio of transmitted to incident intensity and therefore to a probability of transmission.

The matching conditions here are complicated by the necessity of knowing the value of the Bloch function $u_k(\mathbf{r})$ at the matching surface; this is the parameter β discussed in Sec. 6.1 of Chap. II. If a value for β is assumed, however, the calculation may be carried out.[1] The resulting transmission probability contains the exponential $e^{-2|k|\delta}$ where δ is the thickness of the oxide and k is the complex wavenumber describing the states within the oxide. This is the important factor and leads to transmissions of the order of 10^{-9}. However, additional factors in the result are also of interest.

The physics and the mathematics of this calculation are the same as in the tunneling of photons. We imagine a beam of light totally internally reflected at the surface of a glass prism. It is well known that the electromagnetic fields decay exponentially into the vacuum beyond the prism. If, however, we bring a second prism surface sufficiently close to the first, there will be a small transmission into the second prism. This may be described in terms of classical electromagnetic theory but it may also be

[1] W. A. Harrison, *Phys. Rev.*, **123**:85 (1961).

described as the tunneling of photons. The result of the calculation for electron tunneling is of the same form as that obtained in the electromagnetic case but with refractive indices replaced by parameters proportional to the density of states in each medium. (In the case of the oxide it is the absolute magnitude of the density of states continued into the forbidden region.) These factors favor tunneling when the densities of states in the different metal strips are comparable to the corresponding quantity in the insulator.

From the point of view of studying metals this is an unfortunate consequence but one that is in agreement with experiment. We would like to replace one of the films by a transition metal and to trace out the density of states arising from the d bands. However, the tunneling into the high-mass d bands is very weak, presumably due to the mismatch in the density of states, and it has not been possible to detect them to date. A second effect may also act to suppress tunneling involving d states. We noted in Sec. 9 of Chap. II that atomic d states are strongly localized within the atomic cell and the corresponding band states may not extend appreciably into the oxide. Hybridization with k-like states will enhance the tunneling from d-like states, but even this is cancelled to some extent by the corresponding weakened tunneling by k-like states.

Bardeen[1] has suggested an alternative calculation of the tunneling. He suggested that the appearance of the parameter β in the results is unphysical and it would be preferable to assume a gradual transition from metal to insulator; then it becomes appropriate to compute the wavefunctions using a WKB approximation. The calculation is again equivalent to the electromagnetic case but now with a slowly varying refractive index. In this case the dependence of the transmission upon the density of states cancels out completely as does the dependence on β. Such a form provides even less hope for the study of electronic structure with tunneling.

An approximation is required in both of these treatments and should be mentioned. The metal-oxide interface has been assumed to be planar and the transverse component of wavenumber has been conserved across the boundary. This corresponds both to specular reflection at the boundary and to specular transmission. However, experiments on metal surfaces have indicated that the reflection is quite diffuse. This difficulty also brings into question the validity of the details of these results.

In spite of the uncertainty of the microscopic description of tunneling through insulators, the concept of a transmission coefficient for tunneling may be used to parameterize tunneling experiments. Such experiments involving nonsuperconducting metals have not shown structure that could be associated with variations in the density of states and have not clarified

[1] J. Bardeen, *Phys. Rev. Letters*, **6**:57 (1961).

the details of the tunneling. Correspondingly, it has been possible to describe these experiments without knowing the details.

In most applications of tunneling theory it has been preferable to describe the phenomenon in terms of electronic transitions across the barrier than in terms of transmission of electron waves, though the two descriptions are entirely equivalent. In discussing electronic transitions it becomes necessary to introduce a tunneling term in the Hamiltonian. It is not difficult to see that this is a valid description. We first imagine, as an unperturbed system, two metals separated by a sufficiently thick oxide layer that no tunneling occurs. The total Hamiltonian then contains one term, H_1, which describes the metal that we number 1, and a term, H_2, which describes the metal that we number 2. Each of these has a series of one-electron eigenstates. We denote one of the eigenstates in metal 1 as ψ_1, and one of those in metal 2 as ψ_2. Now let us decrease the oxide thickness to the point where tunneling is possible; i.e., the ψ_1 and ψ_2 overlap each other.

We seek a modified solution, as we did in the tight-binding approximation, by seeking linear combinations of ψ_1 and ψ_2. There are now matrix elements of the Hamiltonian between the states ψ_1 and ψ_2. These correspond directly to the overlap integrals of the tight-binding method. These matrix elements will be nonzero only if the transverse component of the wavenumber in the two states is the same. This corresponds to the assumption of specular transmission discussed above. If two components of transverse wavenumber are the same we write the overlap integral as T. It will clearly be proportional to the exponential $e^{-\delta|k|}$ where again $|k|$ is the magnitude of the imaginary wavenumber in the oxide and δ is the film thickness.

There will also of course be terms corresponding to both one-electron states lying on the same side but overlapping the potential from the other side. However, just as in the tight-binding approximation, these simply shift the bands slightly and are not of interest. We may therefore include the terms of interest by writing a one-electron Hamiltonian

$$H = H_1 + H_2 + H_T$$

where the perturbing term H_T is defined by its matrix elements between the one-electron eigenstates of H_1 and H_2. Labeling these eigenstates by k_1 and k_2, respectively, we write the matrix elements of H_T as $T_{k_1k_2}$.

This Hamiltonian could be used in principle to compute the eigenstates of the system. Each eigenstate would be made up of comparable mixtures of states on the two sides of the junction and this would lead to a slight modification of the energy due to the possibility of tunneling between the two sides. However, it will be much more convenient to treat the tunneling Hamiltonian as a perturbation and calculate the probabilities of transition.

We can immediately relate the tunneling matrix element T to the transmission probability P discussed above. We suppress the transverse components which are the same on both sides and let k_1 and k_2 represent the components of wavenumber normal to the barrier. The probability of transition of an electron in the state k_1 to the metal 2 is given by

$$
\begin{aligned}
P_{12}(k_1) &= \sum_{k_2} \frac{2\pi}{\hbar} |T_{k_1k_2}|^2 \, \delta(E_{k_1} - E_{k_2}) \\
&= \frac{2\pi}{\hbar} \frac{L_2}{2\pi} \int dE_{k_2} \frac{|T_{k_1k_2}|^2 \, \delta(E_{k_1} - E_{k_2})}{\partial E/\partial k_2} \qquad (3.23) \\
&= \frac{L_2}{\hbar} |T_{k_1k_2}|^2 \left(\frac{\partial E}{\partial k_2}\right)^{-1}
\end{aligned}
$$

where in the final expression all parameters are evaluated at the k_2 which conserves energy and L_2 is the length of the metal 2.

We may evaluate the same probability in terms of the transmission coefficient P. In these terms the probability is given by the transmission coefficient times the frequency with which the electron state k_1 strikes the oxide barrier. This frequency, however, is simply the velocity of the electron normal to the barrier divided by the length L_1 of the metal.

$$
P_{1,2}(k_1) = P \frac{1}{\hbar} \frac{\partial E}{\partial k_1} \frac{1}{L_1} \qquad (3.24)
$$

(We could have introduced an erroneous factor of 2 in this expression by using twice the length L_1, which would have seemed natural in terms of our transmission calculation. However, that calculation corresponded to a vanishing-boundary condition in the absence of tunneling rather than the periodic-boundary conditions implied by our numbering of states by wavenumber.) Combining Eqs. (3.23) and (3.24) gives an expression for the magnitude of the tunneling matrix element

$$
|T_{k_1k_2}|^2 = \frac{P}{L_1 L_2} \frac{\partial E}{\partial k_1} \frac{\partial E}{\partial k_2}
$$

We note that this result has the appropriate symmetry between metals 1 and 2 and that it contains the proper exponential dependence upon the thickness of the oxide. The dependence on L_1 and L_2 is also appropriate and will cancel out when we sum over states to obtain the total currents.

We compute the contribution to the current from states with a particular transverse wavenumber. This reduces the problem to one dimension; for a real crystal we must add contributions from all transverse wavenumbers but this does not change the essential features of the results. Transitions from

k_1 to k_2 will contribute to the current only if the state k_1 is occupied and the state k_2 is unoccupied (or the other way around). This follows from the Pauli principle but may also be understood in terms of canceling contributions from k_2 to k_1 if both are occupied. Thus, performing the calculation at $T = 0$, we sum Eq. (3.23) over energies that are less than the Fermi energy on side 1 and above the Fermi energy on side 2. This is just an energy range equal to the applied voltage, as indicated in Fig. 3.9. There we have viewed the metals as free-electron gases separated by a vacuum. In the following section (Sec. 3) we will see how such a diagram arises in terms of energy-band structures. We evaluate the magnitude of the current for an applied voltage of magnitude φ_a; the signs are most easily accounted for by inspection of the figure. The current is equal to the electronic charge times the transition probability per unit time summed over the appropriate states (of both spins). Writing the sum as an integral we have

$$J_{12} = e\int\left(\frac{2L_1}{2\pi}\right)dk_1\,P_{12}(k_1) = \frac{eL_1}{\pi}\int_{E_F-|e\varphi_a|}^{E_F} dE\;P_{12}(k_1)\left(\frac{\partial E}{\partial k_1}\right)^{-1}$$

Fig. 3.9 Potential as a function of position representing a tunnel junction. The two metals are replaced by identical free-electron gases separated by a vacuum (in which the potential is higher than the Fermi energy by an amount called the work function). Case (a) is in equilibrium; no voltage is applied and the Fermi energy ξ is a constant of the system. In case (b) an electrostatic potential difference φ_a is applied, shifting the thermodynamic Fermi energies on the two sides with respect to each other by $e\varphi_a$, giving rise to a field φ_a/δ in the junction and allowing a tunneling current to flow. The cross-hatched region represents occupied states. Note that the density of electrons and the kinetic Fermi energy E_F remain unchanged on the two sides.

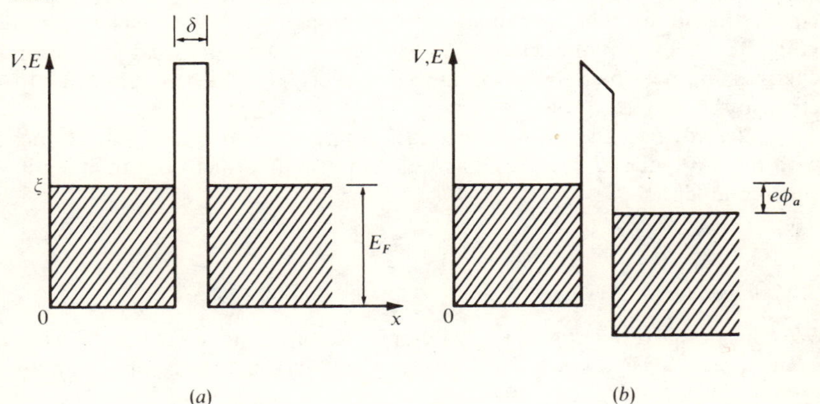

We may use the expression for P_{12} from Eq. (3.23) and write the result in terms of the transmission coefficient to obtain the simple result,

$$J_{12} = \frac{e}{\pi\hbar} \int_{E_F - |e\varphi_a|}^{E_F} P\, dE \tag{3.25}$$

Again, we should sum over transverse wavenumbers (recognizing that P will depend upon transverse wavenumber as well as energy) to obtain a result proportional to the area of the junction.

Note that all density-of-states factors, proportional to $(\partial E/\partial k)^{-1}$, have cancelled out in Eq. (3.25). For small applied voltages the transmission coefficient will not depend appreciably upon applied voltage nor upon energy; it may be evaluated at the Fermi energy and taken out of the integral. We obtain immediately a current proportional to applied voltage and corresponding to a resistance of $\pi\hbar/(e^2 P)$. As we have indicated, this is the behavior observed.

At higher applied voltages we may see from Fig. 3.9*b* that the barrier seen by each state becomes appreciably reduced and the current is expected to rise more rapidly with voltage than linear; this also is observed at voltages approaching a volt.

When we treated simple metals it was natural to assume that P was slowly varying with energy and to evaluate it at the Fermi energy. Equivalently, we could say that $T_{k_1 k_2}$ will be slowly varying with energy. However, the distinction becomes important when we treat superconductors. There we will include the tunneling Hamiltonian and add appropriate terms that give rise to the superconducting state. It is natural to presume that the tunneling matrix elements are not changed as the metal becomes superconducting. If, on the other hand, we were to assume P was unchanged in going to the superconductor, it might seem natural to treat excitation energies in the superconductor as band energies, which would lead to erroneous results.

We will find the tunneling matrix approach convenient for dealing with the *p-n* junction in Sec. 3.2 as well as in the treatment of tunneling in superconductors.

3. Semiconductor Systems

We consider next macroscopically inhomogeneous systems such as occur in semiconductor devices. We will not pursue this very far but simply carry our discussion far enough to see a systematic manner in which we may approach these problems.[1]

3.1 The p-n junction In discussing homogeneous semiconductors we noted that the probability of occupation of any state was given by the Fermi distribution function.

$$f_0(E) = \frac{1}{e^{(E-\xi)/KT} + 1}$$

The value of the Fermi energy ξ was selected in order to obtain the correct number of electrons. In an intrinsic semiconductor it was necessary that the Fermi energy come near the middle of the energy gap. In an *n*-type semiconductor the Fermi energy lay much nearer to the conduction-band edge, which gave rise to an excess number of electrons as opposed to holes. Similarly, in the *p*-type semiconductor the Fermi energy lay very near the valence-band edge.

Let us now consider three semiconductors: one *n*-type, one intrinsic, and one *p*-type, all connected electrically to some external source of electrons. This is illustrated in Fig. 3.10. In this system there is again a unique Fermi energy determining the occupation of the states (since only one Fermi energy is defined for any system in equilibrium). Thus if we draw the energy levels associated with the different semiconductors on the same figure, as in Fig. 3.10, we must construct them with a common Fermi energy.

If these three semiconductors now are put in direct contact, the diagram remains appropriate for the material far from the junctions and it is customarily drawn smoothly between. Thus the Fermi energy is held constant throughout the system and the bands are deformed accordingly. This is illustrated in Fig. 3.11. The ordinate remains the energy of an electron; the abscissa represents the distance through the composite system. The configuration we have constructed is called a *p-n junction*. Such a junction can be made from a single pure semiconductor by diffusing donors in from one surface and acceptors in from the other.

At any position in this *p-n* junction we envisage a distribution of electrons given by the Fermi function. The Fermi energy again is fixed but the band edges vary with position. Let us see now the behavior of an electron in such a system. If we consider, e.g., an electron near the bottom of the conduction band in the *n*-type region drifting toward the junction, it will be reflected when it reaches the junction; it will not have sufficient energy to

[1] For more detail, see J. L. Moll, "Physics of Semiconductors," McGraw-Hill, New York, 1964, or the classic work by W. Shockley, "Electrons and Holes in Semiconductors," Van Nostrand, Princeton, 1950.

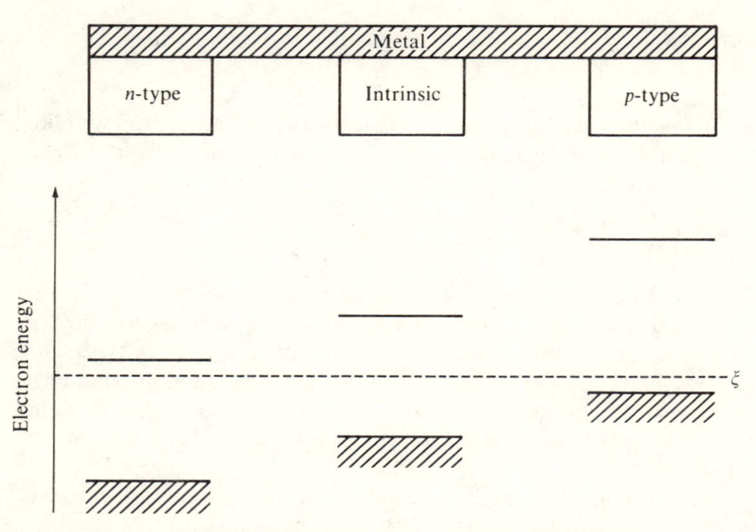

Fig. 3.10 *A diagram representing electron states in three coupled semiconductors showing a unique Fermi energy ξ. Each diagram is analogous to that in Fig. 2.40 with band states represented as horizontal lines in a plot of energy versus position. The cross-hatched line represents the valence-band edge with states below predominantly occupied; the other line represents the conduction-band edge.*

Fig. 3.11 *A diagram representing electron energy as a function of position in a p-n junction. An n-type region on the left is separated from a p-type region on the right by an intrinsic region.*

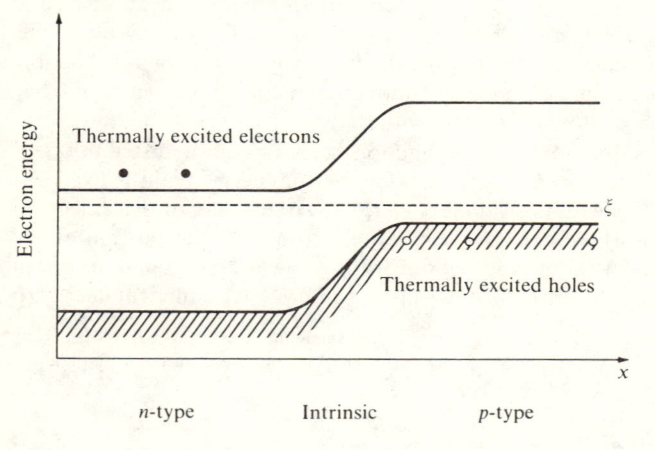

climb the hill and enter the conduction band on the p-type side. Just as in the tunneling system described by Fig. 3.9, we have two conducting regions separated by an insulating region. We may view the tunneling diagram from the same point of view we have given here. Then in Fig. 3.9 the cross-hatched regions represent parts of the conduction band in the metals which are occupied and the line in the insulating (or vacuum) region becomes the conduction-band edge in the insulator. In an ordinary p-n junction the intrinsic region is sufficiently thick that only a negligible current flows by tunneling. In Sec. 3.2 we will discuss junctions in which the intrinsic region becomes sufficiently thin for appreciable tunneling to occur.

We may note that this type of diagram is self-consistent with respect to the fields that we expect to be present. If we consider, e.g., a semiconductor with a fixed density of acceptor levels for $x > 0$ and a fixed density of donor levels for $x < 0$, we construct a diagram as shown in Fig. 3.12. At ordinary temperatures the donor and acceptor levels will all be ionized. Deep within the n-type region there will be as many electrons as there are donor levels. The negative charge on the former just compensates the positive charge on the latter. The region is neutral and there are no electric fields. Near the junction, however, where we have shown the bands bending upward, the band edge becomes increasingly high above the Fermi energy, the density of conduction electron drops, and the donor levels that are present become uncompensated. We have a net positive-charge distribution and the electrostatic-electron-

Fig. 3.12 *The distribution of ion and carrier charges in a p-n junction gives rise to the electrostatic potential which deforms the bands. Here a uniform distribution of donor ions is assumed for $x<0$ and a uniform distribution of acceptors for $x>0$.*

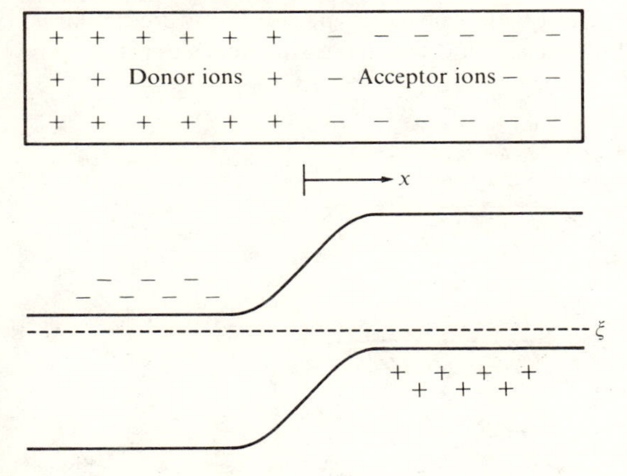

potential energy has a curvature upward as represented by the deformation of the band. Similarly, at the beginning of the *p*-type region, the acceptor levels are uncompensated and the curvature is negative as shown. Finally, deep within the *p*-type region the crystal is neutral and the potential again flattens out. Thus it is simply the electrostatic potential, generated by uncompensated donor and acceptor levels, that deforms the bands in the manner we have shown.

The Fermi energy, which is simply the free energy of the electron, is the same everywhere. This corresponds also of course to the absence of any applied voltage. There is no net diffusion of electrons from one part of the system to another. Only a very tiny number of electrons in the *n*-type region have sufficient energy to flow into the conduction band on the *p*-type side. However, there will be some and this flow will be exactly compensated by the very tiny number of electrons at the bottom of the *p*-type conduction band which flow into the *n*-type region.

Now let us imagine applying a voltage to the system. A voltage that increases the electron energy on the *n*-type side is called a *forward voltage*. It raises the energy of the conduction band on the *n*-type side, reducing the energy difference between the conduction band on the two sides. The semiconductor is a rather good conductor in both the *n*- and *p*-type regions while it is very nearly insulating in the intrinsic barrier region. Therefore the application of the voltage will simply modify the distortion that was introduced by the uncompensated donors and acceptors. With the application of a voltage the system is no longer in equilibrium. The Fermi energy is no longer a constant. Thus we may show on our diagram a "bent" Fermi energy as shown in Fig. 3.13. The difference in the Fermi energies on the two sides is simply the applied voltage times the electronic charge.

Now as the conduction-band minimum on the *n*-type side is raised, (or equivalently the conduction band on the *p*-type side is lowered) a larger number of electrons on the *n*-type side will have sufficient energy to flow

Fig. 3.13 *A p-n junction with a forward voltage V applied, raising the Fermi energy ξ_n in the n-type region above the Fermi energy ξ_p in the p-type region.*

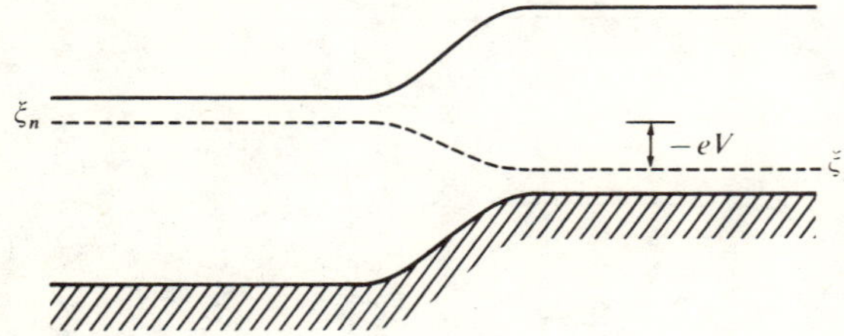

into the p-type side. The density of conduction electrons on the p-type side, however, is not appreciably modified so the backflow is unchanged. We have produced therefore net current by the application of this forward voltage. We may similarly examine the behavior of the holes and we find that the forward voltage will increase the flow of holes into the n-type region making an additional contribution to the forward current. The current flowing across the junction can be quite sizable. For a large applied voltage, the electrons far down in the Fermi distribution are allowed to flow. Ultimately with a voltage applied sufficient to equalize the band edges on the two sides, the current flow would be comparable to that for a uniform n-type semiconductor.

Let us now consider the results of application of a *reverse* voltage. The p-type conduction band is then raised still further above that on the n-type side. The forward flow of current drops while the reverse current remains the same. Thus again we find a current flowing, this time in the reverse direction. However, that current will always be extremely small. It will be limited to the very tiny current corresponding to *electrons* in a p-type semiconductor. Similar examination of the flow of holes indicates that there also the current flow in the reverse direction is extremely tiny.

Thus we see that the p-n junction acts as a rectifier and this is the use to which it is put as a semiconducting device. The current voltage characteristic is shown schematically in Fig. 3.14.

This same approach will enable us to understand the behavior of more complicated semiconductor devices. For example, we may construct a semiconductor that is n type on both sides and p type in the middle as shown in Fig. 3.15. The flow of electrons between the two n-type regions is inhibited by the potential barrier associated with the p-type region. However, by raising and lowering the voltage in the p-type region, we may increase

Fig. 3.14 *Current versus voltage for a p-n junction. Forward voltage is to the right. Only a small current flows with reverse voltages giving rectifying properties.*

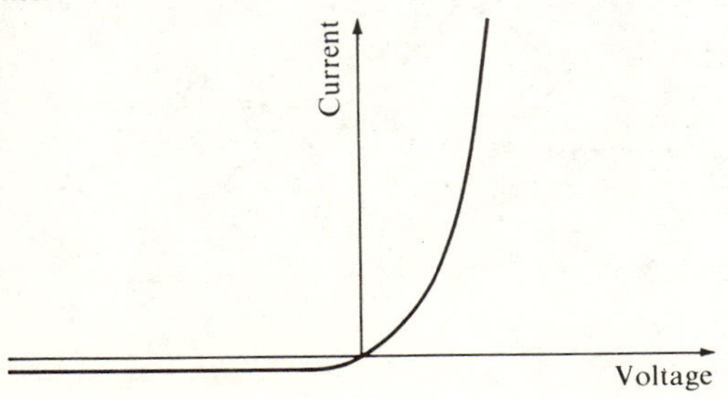

and decrease the flow of electrons between the *n*-type regions. This operates in much the same way that the raising and lowering of the potential on a vacuum-tube grid modulates the current flowing between the plate and the cathode. Such a device is called an *n-p-n junction transistor.*

This is only a very brief introduction to the manner in which devices may be constructed by suitable doping of semiconductors. However, it should serve to illustrate the general ideas behind such devices.

3.2 The tunnel diode It is interesting that, by increasing the doping to a sufficiently high degree in a *p-n* junction, we can obtain the fundamentally different behavior of the *tunnel diode.* By diffusing in a large concentration of donors and acceptors on the two sides of a wafer of semiconductor, we ultimately push the Fermi level above the conduction-band edge on the *n*-type side and below the valence-band edge on the *p*-type side. Thus we have degenerate electron and hole gases very much as in a metal but, of course, with a much smaller density of carriers. In addition to modifying the occupation of levels by increasing the doping, we also greatly reduce the thickness of the intrinsic region. Higher densities of donors and acceptors lead to higher curvatures of the band and thus to the narrow intrinsic region. The diagram for such a tunnel diode is shown in Fig. 3.16. This region in fact can be made so narrow that an appreciable number of electrons will tunnel through the barrier region when a voltage is applied.

To compute the transmission probability or the tunneling matrix element for this case it is most natural to proceed with the WKB approximation. This approach was mentioned in Sec. 2.5. The energy of the conduction-band and valence-band edges are assumed to vary slowly through the junction (small variations over an electron wavelength). The

Fig. 3.15 An n-p-n junction transistor. Current flow between the n-type regions may be modulated by raising and lowering the potential in the p-type region.

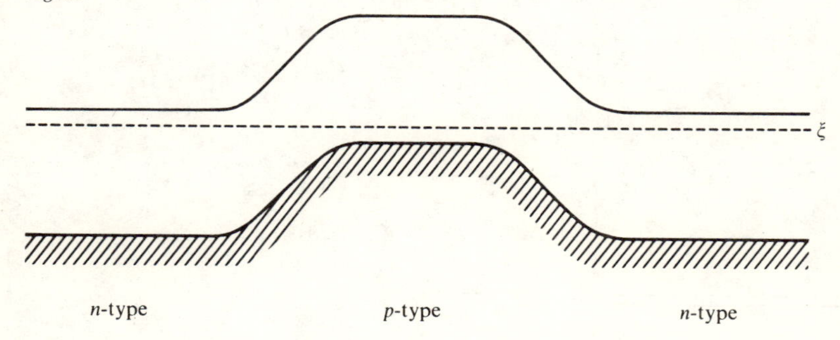

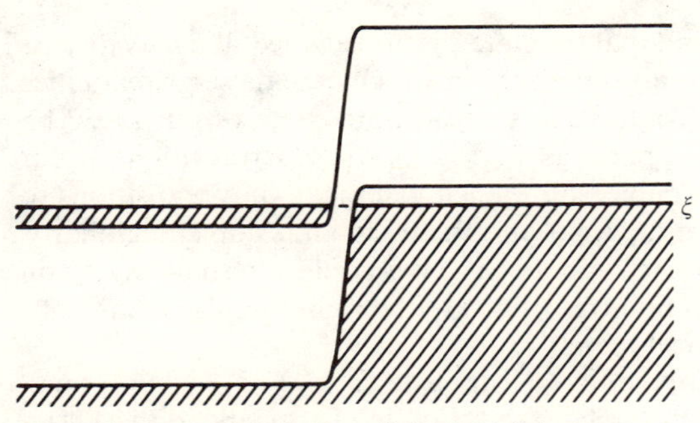

Fig. 3.16 A tunnel diode. By sufficiently high doping the carriers become degenerate in the conducting regions and the intrinsic region becomes very narrow. Then the predominant mechanism for conduction may become tunneling. Shaded regions represent occupied states.

conduction-band edge crosses the energy of the state in question at the left and the valence-band edge crosses it at the right. The calculation is somewhat complicated but ultimately leads to a tunneling matrix element T which depends upon energy and in fact upon applied voltage. In our qualitative discussion of the tunnel diode we may ignore these dependences. Thus we are concerned with transitions between states on side 1 and side 2 as shown in Fig. 3.17.

The tunneling probability for a given state will be proportional to the P_{12} discussed in Sec. 2.5. It will also be proportional to the probability

Fig. 3.17 An expanded view of the tunnel diode diagram of Fig. 3.16 showing the shifting of levels with applied voltage.

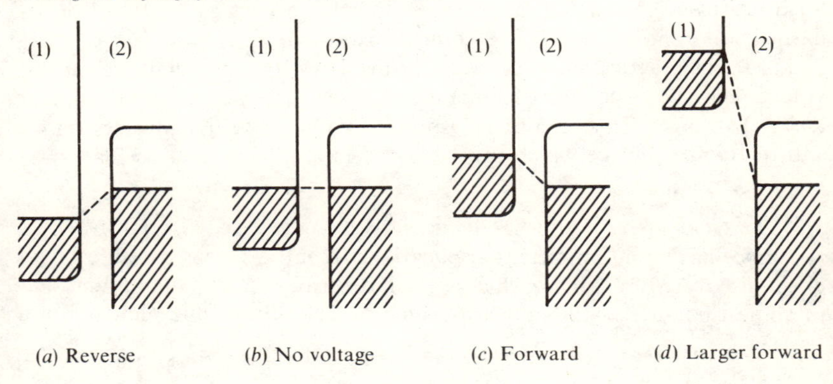

that the state 1 is occupied and that the state 2 is unoccupied. We may discuss the tunneling for a system at absolute zero in which case these probabilities of occupation are one or zero and can be read immediately from Fig. 3.17.

When no voltage is applied as in Fig. 3.17*b*, no transitions occur since every state which is occupied on the left can make only transitions to states of the same energy on the right and these are all occupied; similarly, no transitions from right to left can occur. (We could alternatively again drop the Pauli principle and find transitions to the right but also transitions to the left which exactly cancel them.)

When a small forward voltage is applied as in Fig. 3.17*c*, a small number of occupied levels on the left are raised in energy above the Fermi energy on the right. Each of these will have a tunneling probability of simply P_{12}; by summing over them we obtain a current that is proportional to the small applied voltage. As the forward voltage is increased further, ultimately the highest-energy occupied states on the left will be raised above the valence-band maximum on the right as in Fig. 3.17*d* and again no tunneling from these levels will be possible; there are no levels on the right into which they can tunnel. The current must reach a maximum and ultimately drop to zero as the conduction-band edge on the left passes the valence-band edge on the right. If a reverse voltage is applied, as in Fig. 3.17*a*, there will be a reverse current that is again initially proportional to the voltage. In this case, however, the reverse current can increase indefinitely as the reverse voltage is increased. The reverse current will not remain linear in voltage indefinitely because of variations in the tunneling matrix element and in the densities of states. The resulting current-voltage characteristic that we obtain is shown schematically in Fig. 3.18. Actually there are additional processes that may take electrons across the tunneling region at high enough applied forward voltages. In particular, an electron may lose energy during the tunneling process by giving it to the lattice in the form of lattice vibrations. Such processes become important at high forward voltages and lead to the dashed curve shown in Fig. 3.18 at high forward voltages.

From a practical point of view the most important aspect of the current voltage characteristic of the tunnel diode is the region of negative slope at forward voltages. This region corresponds to a negative dynamic resistance and therefore a device which can be useful in amplifiers and oscillators.

3.3 The Gunn effect Another semiconductor device based upon negative resistance has found important applications recently. This is based upon an effect discovered by Gunn[1] which we will describe briefly. Gunn found that in some three-five semiconductors, such as gallium arsenide and indium

[1] J. B. Gunn, *Solid-State Communications*, **1**:88 (1963).

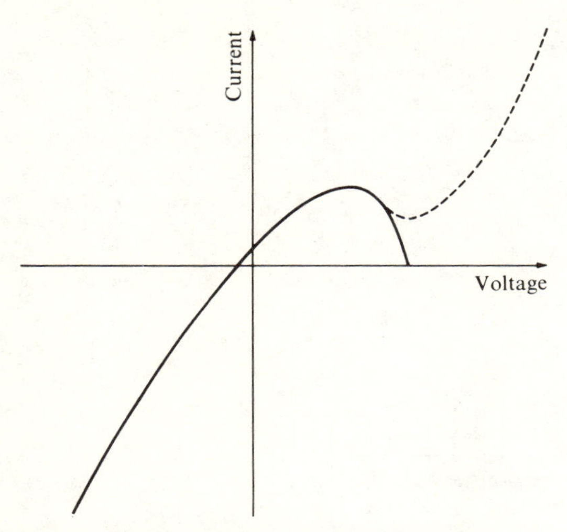

Fig. 3.18 A schematic representation of the current-voltage characteristic for a tunnel diode. Forward voltage is to the right. The solid curve is deduced from consideration of Fig. 3.17. Actually additional processes become important at high forward voltages deflecting the curve as shown by the dashed line.

phosphide, the application of a high dc electric field could cause the generation of microwave-frequency oscillations. This provides the possibility then for simple and direct production of microwaves.

This effect is now thought to arise from energy-band structures such as shown in Fig. 3.19. The conduction-band minimum has high curvature corresponding to a very low effective mass and high electrical conductivity. There exist also local minima at higher energy which are describable by a much higher effective mass.

At low applied fields the electron-drift velocity will be small and all electrons will remain in the neighborhood of the low-mass minimum, thus the conductivity will be high. However, as the fields are increased and the drift velocities are increased, it becomes possible for electrons to be transferred to the local minima. This could occur either from the direct acceleration over the local maximum between, but it seems more likely that the transfer occurs by scattering of electrons which have sufficient energy that they may scatter to the high-mass minima. If this occurs the effect of the high electric field is to transfer electrons to states of low conductivity. Thus the current may decrease as the electric field is increased in much the same way as we have indicated in Fig. 3.18 for the tunnel diode.

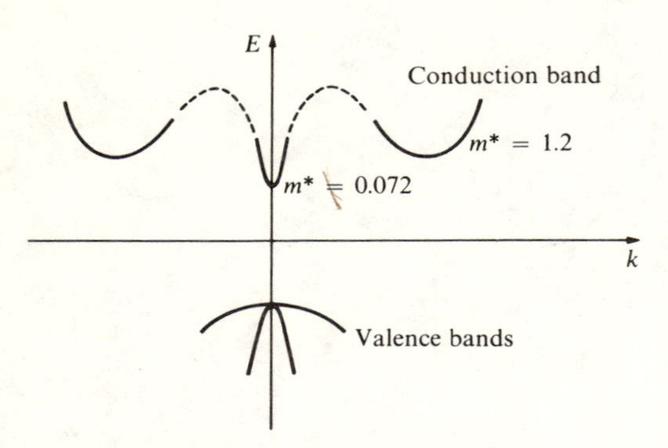

Fig. 3.19 The band structure of GaAs [after H. Ehrenreich, Phys. Rev., *120:1951(1960)]. A conduction band of this form can give rise to Gunn oscillations. At sufficiently high currents electrons may be transferred to the secondary high-mass minima.*

This mechanism for negative resistance has an important distinction from that in the tunnel diode. There we found negative resistance through a very thin sheet corresponding to the tunnel junction. In the Gunn effect we have found a *bulk* negative dynamic conductivity of the semiconductor. If the conductivity were very slightly negative, a microwave propagating through the specimen would be amplified. When the effect is large, as it is in the Gunn diode, the crystal may divide into propagating domains of high and low conductivity and the treatment in detail is much more difficult.

4. Screening

For the most part in our discussion we have specified the potentials that are to be used in the calculation of electronic states and electron behavior. In discussing band structures we thought of superimposing simple atomic potentials. As we indicated at that time, this is only an approximate approach. To construct a more accurate potential we must do a self-consistent calculation, obtaining a potential which will give rise to electronic states which correspond to a charge density which, in turn, will lead to the beginning potential.

In some sense our division of the donor potential $-Ze/r$ by a dielectric constant includes such a self-consistent approach. The polarization of a dielectric can be determined self-consistently and leads to the dielectric constant which we used in that treatment. Similarly, in semiconductor systems, we indicated that the raising and lowering of the bands on the two sides of the *p-n* junction was consistent with the expected charge densities.

We will now explore this self-consistency more systematically. We will begin by using an approach based upon transport theory. The transport equations tell us the distribution of charge to be expected in the presence of given fields. Poisson's equation tells us the potential that will arise from a given distribution of charges; these then can be solved self-consistently. We will then proceed to quantum-mechanical treatments of the same effect. In both cases we will seek the linear response of the system to small applied fields, which corresponds in the classical case to the use of the linearized Boltzmann equation. An important feature of the linear theory is that we may make a Fourier decomposition of a completely arbitrary applied field in time and position and compute the response to each Fourier component separately. Thus a calculation of the response to a potential $V_0\, e^{i(q \cdot r - \omega t)}$ (with V_0 a constant amplitude) as a function of q and ω is a calculation for a completely general weak applied potential.

Furthermore, the linear-response problem can be stated for a general uniform system. A weak applied potential $V_0\, e^{i(q \cdot r - \omega t)}$ will give rise to a fluctuation in the electron density with the same dependence upon space and time, $n_s\, e^{i(q \cdot r - \omega t)}$. For sufficiently weak V_0, the response will be linear.

$$n_s = F(q,\omega)\, V_0 \tag{3.26}$$

The $F(q,\omega)$ is the *density-response function* and is well defined for any uniform system of electrons. Even for an exact treatment of the interaction between electrons, there exists an $F(q,\omega)$ and we will discuss its evaluation in Sec. 4.5. There we will see its relation to the exact energy of a system of interacting electrons.

For the present we note that the density fluctuation represents a nonuniform charge density which causes an additional potential seen by the electrons. It is called the *screening potential* $V_s\, e^{i(q \cdot r - \omega t)}$ and may be evaluated directly from Poisson's equation.

$$\nabla^2 V_s\, e^{i(q \cdot r - \omega t)} = -4\pi e^2 n_s\, e^{i(q \cdot r - \omega t)}$$

or

$$V_s = \frac{4\pi e^2}{q^2} n_s = \frac{4\pi e^2}{q^2} F(q,\omega)\, V_0 \tag{3.27}$$

This could be done more generally by using Maxwell's equations. However, for most purposes Poisson's equation will suffice. We will simply be computing to lowest order in the ratio of the velocity (ω/q) to the speed of light. This will be sufficient for all applications we wish to make. Thus, within the self-consistent-field approximation, the total potential seen by the electrons has an amplitude

$$V_t = V_0 + V_s = \left[1 + \frac{4\pi e^2}{q^2} F(q,\omega)\right] V_0$$

The transport calculation (both classical and quantum mechanical) is most conveniently done in terms of the total potential. That is, we will compute the linear response to the total potential

$$n_s = X(q,\omega)\, V_t \tag{3.28}$$

The $X(q,\omega)$ can be seen by comparison with Eqs. (3.26) and (3.27) to be related to the density-response function by

$$X(q,\omega) = \frac{F(q,\omega)}{(1 + 4\pi e^2 F(q,\omega)/q^2)} \tag{3.29}$$

Thus a calculation of $X(\mathbf{q},\omega)$ is also a calculation of the general linear response problem. Within the self-consistent-field framework a convenient form for the result is the *dielectric function* $\epsilon(q,\omega)$ which relates the total potential to the applied potential,

$$V_t = \frac{V_0}{\epsilon(q,\omega)}$$

The dielectric function is seen to be related to $X(q,\omega)$ by

$$\epsilon(q,\omega) = 1 - \frac{4\pi e^2}{q^2} X(q,\omega) \tag{3.30}$$

It gives us the net Hartree potential in a metal in terms of known bare potentials.

We can use the dielectric function for zero frequency to estimate the potentials that occur in metals and give rise to energy-band structure. Similarly, we may use the dielectric function to obtain the scattering potential arising from any defects in the structure and need only know the form of the potential in the absence of the electron gas. Thus the dielectric function represents the solution of the screening problem.

We will see also that the frequency-and-wavenumber-dependent conductivity may be written directly in terms of the dielectric function. Thus our calculation represents also the solution of a very general conductivity problem. Since absorption of light is ultimately the resistive loss due to a very high-frequency electric field, the dielectric function will also give us an elementary theory of optical properties.

The linear-response function can also give us information about collective vibrations of the system. We will see that a vanishing of the dielectric function leads directly to the familiar plasma oscillations of metals.

We will notice in our calculations that the amplitudes of the screening potential or of the dielectric function can be complex. If, however, we superimpose two applied potentials of the form given above in order to obtain a real potential, the corresponding responses will also be real and the complex amplitudes will simply give rise to potentials that are out of phase with the applied potential.

The dielectric function is particularly useful in the Hartree approximation where it leads to the potential that enters the one-electron Schroedinger equation. If, on the other hand, exchange is to be included (e.g., in the Hartree-Fock approximation), it enters as a nonlocal (energy-dependent) potential and cannot then be included simply by modifying the dielectric function. It is possible to define energy-dependent dielectric functions, and a dielectric function for test charges which is different from that for electrons, but it seems preferable to return to the density-response function $F(q,\omega)$ which is unique and well defined. That we will do in Sec. 4.5, but for the moment no exchange is to be included and a single dielectric function is appropriate.

4.1 Classical theory of simple metals We calculate the response of the system (i.e., the electron-density fluctuation) due to a potential, which includes both applied and screening terms, of the form $V_t e^{i(q\cdot r-\omega t)}$. The first-order distribution function will then be in the form $f_1(\mathbf{p})e^{i(q\cdot r-\omega t)}$. These may be directly substituted into the linearized Boltzmann equation (Eq. (3.16) of Sec. 2.1).

$$-i\omega f_1 + \frac{\partial f_0}{\partial E}\mathbf{v}\cdot(-i\mathbf{q}V_t) + i\mathbf{q}\cdot\mathbf{v}f_1 = -\frac{f_1}{\tau} + \frac{\delta_n f_0}{\tau}$$

where we have noted $\mathbf{F} = -\nabla V_t e^{i(q\cdot r-\omega t)} = -i\mathbf{q}V_t e^{i(q\cdot r-\omega t)}$. We have also cancelled the phase factors $e^{i(q\cdot r-\omega t)}$ after substitution. We may solve immediately for f_1,

$$f_1(\mathbf{p}) = \frac{i\mathbf{q}\cdot\mathbf{v}\tau V_t(\partial f_0/\partial E) + \delta_n f_0}{1 - i\omega\tau + i\mathbf{q}\cdot\mathbf{v}\tau} \tag{3.31}$$

We wish to compute the electron-density fluctuation $n_s e^{i(q\cdot r-\omega t)}$ which is given by

$$n_s = \frac{2}{h^3}\int d^3p f_1(\mathbf{p}) \tag{3.32}$$

This in turn will be needed to determine $\delta_n f_0$, but we may proceed with the integration of Eq. (3.32) without knowing the magnitude of $\delta_n f_0$. We first take an angular average of Eq. (3.31), which is all that is needed in Eq. (3.32). We write $\mathbf{q}\cdot\mathbf{v} = qv\cos\theta$ and note that the angular integrals are of the form $\int x\,dx/(a+bx)$ and $\int dx/(a+bx)$. We obtain the average over angle,

$$\bar{f}_1(p) = V_t \frac{\partial f_0}{\partial E}\left[1 - \frac{1 - i\omega\tau}{2iqv\tau} \ln\left(\frac{1 - i\omega\tau + iqv\tau}{1 - i\omega\tau - iqv\tau}\right)\right] + \frac{\delta_n f_0}{2iqv\tau} \ln\left(\frac{1 - i\omega\tau + iqv\tau}{1 - i\omega\tau - iqv\tau}\right) \quad (3.33)$$

Now we must be more explicit about $\delta_n f_0$. We note that f_0 depends on the density through ξ.

$$\delta_n f_0(E) = -\frac{\partial f_0}{\partial E} \frac{d\xi}{dn} \delta n(\mathbf{r})$$

where $\delta n(\mathbf{r})$ is the local density fluctuation, $n_s e^{i(\mathbf{q}\cdot\mathbf{r} - \omega t)}$, and $d\xi/dn$ is $1/n(\xi)$, the reciprocal of the density of states per unit energy at the Fermi energy. Thus all terms in Eq. (3.33) are proportional to $\partial f_0/\partial E$. Up to this point the analysis is applicable to nondegenerate, as well as to degenerate, electron gases. We now restrict consideration to simple metals, substitute Eq. (3.33) in Eq. (3.32), and use the delta-function-like nature of $-\partial f_0/\partial E$. We note further that $n(\xi) = dn/d\xi = 8\pi p^2/(h^3 v)$ where p and v are evaluated at the Fermi energy. We obtain

$$n_s = -n(\xi)\, V_t\left[1 - \frac{1 - i\omega\tau}{2iqv\tau} \ln\left(\frac{1 - i\omega\tau + iqv\tau}{1 - i\omega\tau - iqv\tau}\right)\right] + \frac{n_s}{2iqv\tau} \times \ln\left(\frac{1 - i\omega\tau + iqv\tau}{1 - i\omega\tau - iqv\tau}\right)$$

We may solve for n_s and obtain by comparison with *Eq.* (3.28)

$$X(q,\omega) = -n(\xi) \frac{1 - \dfrac{1 - i\omega\tau}{2iqv\tau} \ln\left(\dfrac{1 - i\omega\tau + iqv\tau}{1 - i\omega\tau - iqv\tau}\right)}{1 - \dfrac{1}{2iqv\tau} \ln\left(\dfrac{1 - i\omega\tau + iqv\tau}{1 - i\omega\tau - iqv\tau}\right)} \quad (3.34)$$

We have accomplished our goal of obtaining the electron-density distribution in terms of the total potential that is present. From Eq. (3.30) we may immediately write the dielectric function,

$$\epsilon(q,\omega) = 1 + \frac{4\pi e^2}{q^2} n(\xi) \frac{1 - \dfrac{1 - i\omega\tau}{2iqv\tau} \ln\left(\dfrac{1 - i\omega\tau + iqv\tau}{1 - i\omega\tau - iqv\tau}\right)}{1 - \dfrac{1}{2iqv\tau} \ln\left(\dfrac{1 - i\omega\tau + iqv\tau}{1 - i\omega\tau - iqv\tau}\right)}$$

The $\epsilon(q,\omega)$ is called the *frequency-and-wavenumber-dependent dielectric function.* Note that it plays the same role as the usual dielectric *constant* in insulators which also relates applied fields to net fields. Here we have obtained the dielectric *function* for a semiclassical free-electron gas.

4.2 Limits and applications of the dielectric function The simplest application of the dielectric function is the calculation of screening. By using the classical transport equation we have restricted ourselves to long wavelength variations of the potentials and therefore to small values of q. This limitation will be removed in the next section but it is of some interest before that to see the form of the classical result.

We note that if ω is taken equal to zero, $X(q,0)$ is seen from Eq. (3.34) to be simply equal to $-n(\xi)$. This is a very plausible result physically and corresponds to the *linearized Fermi-Thomas approximation.* We may understand this approximation by considering a very long-wavelength static potential as illustrated in Fig. 3.20. (This must be of significantly longer wavelength than that corresponding to the electrons.) Since the wavelength is long we should be able to describe the occupation of states locally by integrating the density of states $2/h^3$ over the volume of momentum space corresponding to occupied states. The thermodynamic Fermi energy, computed at this location, will be the sum of the net potential at that point and the Fermi kinetic energy $p_F{}^2/2m$. Thus since the potential is different at different points and since the thermodynamic Fermi energy must be the same everywhere in the system, the Fermi momentum must fluctuate up and down. Where the potential is low the Fermi kinetic energy must be high and there must consequently be a higher density of electrons. Clearly the excess number of electrons will be simply equal to the density of states per

Fig. 3.20 *In the Fermi-Thomas approximation the Fermi kinetic energy $p_F{}^2/2m$ is taken to be a well-defined function of position and equal to the difference between the thermodynamic Fermi energy ξ and the total potential energy $V_t(\mathbf{r})$ at that position. The local electron density at each point is then computed from the local Fermi momentum using the same formula as for a uniform electron gas.*

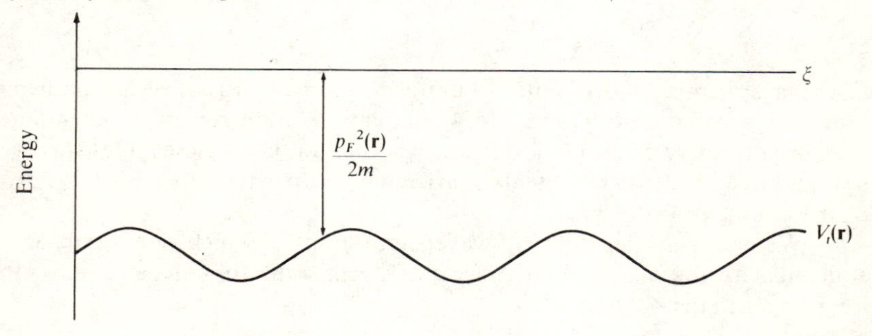

unit energy at the Fermi energy times the local increase in Fermi kinetic energy, which in turn is equal to the local decrease in potential. This leads directly to Eq. (3.28) with

$$X(q,0) = -n(\xi)$$

This argument will break down when q becomes greater than or of the order of the Fermi wavenumber, and we will see from our quantum-mechanical calculation that this formula becomes inaccurate in that limit.

It is interesting to apply this simplest approximation to screening of a point defect. In particular we consider the screening of a point coulomb charge of magnitude Ze and use the wavenumber-dependent dielectric function. The Fourier transform of this potential is simply $4\pi Ze^2/(\Omega q^2)$. (Note that a convergence factor $e^{-\mu r}$ is required in this calculation which has subsequently been taken equal to unity.) This corresponds to the applied potential. We need now divide by the dielectric function which becomes simply $\epsilon(q,0) = 1 + 4\pi e^2 n(\xi)/q^2$. Thus the net potential, which is equal to the applied potential divided by the dielectric function, has a Fourier transform given by $4\pi Ze^2/\Omega(q^2 + \kappa^2)$ where κ is the *Fermi-Thomas screening parameter*, $\kappa = \sqrt{4\pi e^2 n(\xi)}$. By transforming back to real space we see that the screened potential has become simply $Ze^2\, e^{-\kappa r}/r$. The effect of the screening in this approximation has been simply to introduce a damping of the long-range coulomb potential as shown in Fig. 3.21.

In many applications of static screening it is simpler to use the linearized Fermi-Thomas method directly without first making a Fourier expansion. Since the proportionality constant relating n_s and V_t is independent of q, we may write the response directly in terms of the position-dependent potential,

$$n_s(\mathbf{r}) = -n(\xi)\, V_t(\mathbf{r}) = -n(\xi)[V_0(\mathbf{r}) + V_s(\mathbf{r})]$$

In addition we may write $n_s(\mathbf{r})$ as $(-4\pi e^2)^{-1}\nabla^2 V_s(\mathbf{r})$ using Poisson's equation in order to obtain a differential equation in V_s,

$$\frac{1}{4\pi e^2 n(\xi)}\nabla^2 V_s(\mathbf{r}) - V_s(\mathbf{r}) = V_0(\mathbf{r})$$

Given a system which is initially uniform we may introduce an applied potential distribution $V_0(\mathbf{r})$ and solve this equation for the screening potential. Similarly, we may construct a differential equation in the screening density $n_s(\mathbf{r})$ and treat directly the potentials arising from an externally added charge distribution (see Prob. 3.8).

We next consider the long-wavelength limit (in which our results are equivalent to the quantum treatment) and retain the time dependence. At long wavelengths, we find

$$\epsilon(q,\omega) \to 1 + \frac{4\pi i N e^2 \tau}{m\omega(1 - i\omega\tau)} \tag{3.35}$$

If the scattering time is sufficiently long this becomes

$$\epsilon(\omega) \to 1 - \frac{4\pi N e^2}{m\omega^2} \tag{3.36}$$

which is plotted in Fig. 3.22. Note that ϵ has a zero at a frequency ω_p given by

$$\omega_p{}^2 = \frac{4\pi N e^2}{m} \tag{3.37}$$

This means that an electron gas has an indefinitely large response to applied fields at this frequency. In other words, there exist self-sustaining oscillations of the system. These are just the long-wavelength plasma oscillations of the system, and ω_p is the *plasma frequency*. In general, zeros in the dielectric function correspond to excited states of the system. The energy of these oscillations, $\hbar\omega_p$, is ordinarily several volts.

Fig. 3.21 *The screening of an attractive coulomb potential in the linearized Fermi-Thomas approximation. The κ is the screening parameter.*

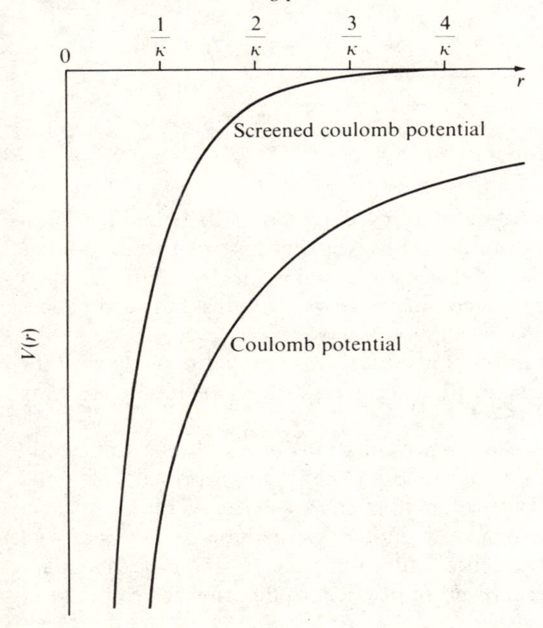

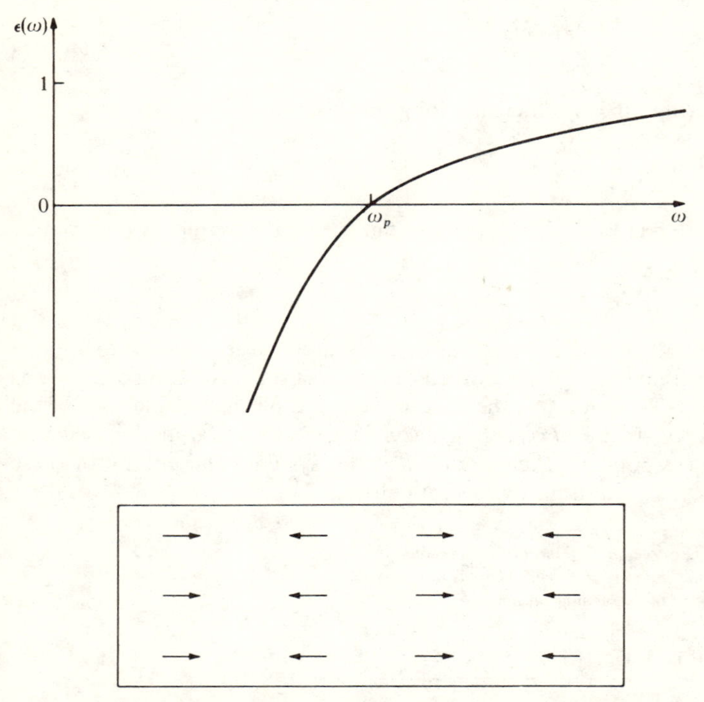

Fig. 3.22 The dielectric function for an electron gas at long wavelengths and long scattering times passes through zero at the frequency ω_p of self-sustaining plasma oscillations such as those indicated schematically above.

Physically these plasma oscillations correspond to soundlike compression waves in the electron gas. However, because of the long-range nature of the coulomb potential which sustains the oscillations, their frequency does not approach zero at long wavelengths but approaches this finite frequency. Plasma oscillations may be observed in metals by firing high-energy electrons at thin foils. Electrons interact strongly with the plasma modes and the characteristic energy loss is observable by studying the transmitted electron beam.

At the frequency of the plasma oscillations $\omega\tau$ is always sufficiently large that the step from Eq. (3.35) to Eq. (3.36) is appropriate. The small imaginary part of the dielectric function causes a decay of the plasma oscillation, or equivalently a breadth in energy of these excited states of the system. At much lower frequencies (for example, $\omega \lesssim 10^{12}$) the imaginary part dominates and it is perhaps more physical to describe the response in terms of a conductivity, which we will see is directly related to the dielectric function.

The conductivity $\sigma(\mathbf{q},\omega)$ is defined by

$$\mathbf{j}(\mathbf{q},\omega) = \sigma(\mathbf{q},\omega)\,\mathscr{E}(\mathbf{q},\omega) \tag{3.38}$$

Here $\mathbf{j}(\mathbf{q},\omega)$ and $\mathscr{E}(\mathbf{q},\omega)$ are the amplitudes of periodic current densities and electric fields. Strictly speaking $\sigma(\mathbf{q},\omega)$ is a tensor, but for the electron gas we have been discussing and for $\mathscr{E}\|\mathbf{q}$ it becomes a scalar function of q and ω.

We may write $\mathbf{j}$ and $\mathscr{E}$ in terms of the screening potential and total potential in order to obtain the relation between $\sigma(q,\omega)$ and $\epsilon(\mathrm{q},\omega)$. This argument is not restricted to any limiting values of $\mathbf{q}$ and ω. The continuity equation is

$$\nabla\cdot\mathbf{j} + \frac{\partial\rho}{\partial t} = 0$$

or

$$qj(\mathbf{q},\omega) = (-e)\omega n_s(\mathbf{q},\omega)$$

The electric field is given by

$$(-e)\mathscr{E} = -\nabla V_t$$

or

$$-e\mathscr{E}(\mathbf{q},\omega) = -i\mathbf{q}V_t$$

Substituting these in Eq. (3.38), we have

$$n_s = \left[-\frac{iq^2\sigma(q,\omega)}{\omega e^2}\right]V_t$$

The expression in square brackets may be identified with $X(q,\omega)$ from Eq. (3.28). We have

$$\sigma(q,\omega) = \frac{i\omega e^2}{q^2}X(q,\omega)$$

By comparison with Eq. (3.30), we may note further that

$$\sigma(q,\omega) = \frac{i\omega}{4\pi}[1 - \epsilon(q,\omega)] \tag{3.39}$$

This result is quite general and depends only on the definitions of $\epsilon(\mathbf{q},\omega)$ and $\sigma(q,\omega)$.

It is interesting to consider the long-wavelength form. From Eq. (3.35) and Eq. (3.39) we have

$$\sigma(q,\omega) \rightarrow \frac{Ne^2\tau}{m(1 - i\omega\tau)}$$

At low frequencies it approaches the dc conductivity we computed before. At high frequencies it approaches $iNe^2/m\omega$ which is just the classical response of N free charges with charge to mass ratio e/m. The real part of the conductivity for arbitrary frequency is

$$\text{Re } \sigma(q,\omega) \to \frac{Ne^2\tau}{m(1 + \omega^2\tau^2)}$$

which is plotted in Fig. 3.23. We may compute the absorption of light from the energy loss $\mathbf{j} \cdot \mathscr{E}$, which involves the real part of the conductivity. (Since we have let $q \to 0$, the conductivity is the same for transverse fields as for longitudinal fields.) The corresponding energy loss varying with frequency as $(1 + \omega\tau)^{-2}$ corresponds to the *Drude theory* of optical absorption and accounts for much of the experimental absorption of metals. We will return to other contributions later.

4.3 Quantum theory of screening The problem that we have just studied classically will now be redone in a quantum-mechanical framework. This will eliminate the restriction to small q required at the beginning of the classical treatment. We are asking for the response of a quantum-mechanical system to a periodic applied field. This calculation is most directly done using a density-matrix approach, which parallels rather closely the classical treat-

Fig. 3.23 The principal optical absorption in simple metals is proportional to the real part of the frequency-dependent conductivity at long wavelengths shown here.

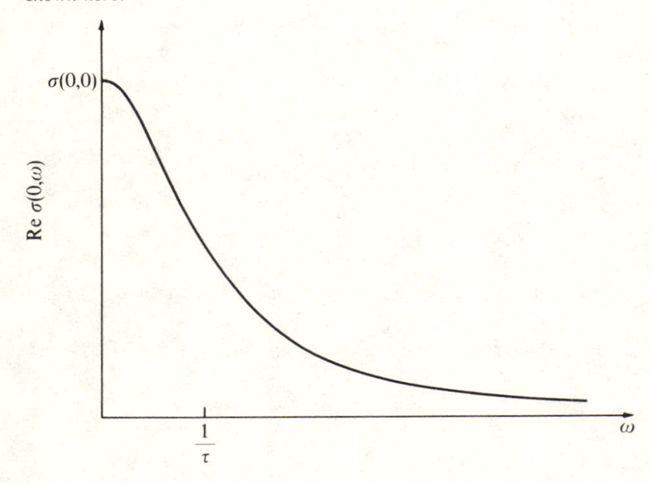

ment. We will in fact see that in the classical limit appropriate matrix elements of the density matrix become the Fourier components of the distribution function which appeared in the classical calculations. Making this identification it will become very clear what approximation has been made in the use of semiclassical methods and the Boltzmann equation.

We will begin by describing an N-electron system in terms of a single Slater determinant, $\Psi_N(\mathbf{r}_1,\mathbf{r}_2, \ldots, \mathbf{r}_N,t)$, and will later generalize the description. We will wish to evaluate one-electron operators which may be written in the form

$$O = \sum_i o(\mathbf{r}_i)$$

The corresponding expectation value may be written

$$\langle O \rangle = \sum_i \int d\mathbf{r}_1\, d\mathbf{r}_2 \ldots d\mathbf{r}_N\, \Psi_N^* o(\mathbf{r}_i)\, \Psi_N$$

To evaluate this we expand each of the Slater determinants in the ith term in the form

$$\Psi_N = \frac{1}{\sqrt{N}} \left[\psi_1(\mathbf{r}_i)\Psi_{N-1}^{(1)} - \psi_2(\mathbf{r}_i)\Psi_{N-1}^{(2)} + \cdots \right]$$

where the $\psi_j(\mathbf{r}_i)$ are the orthonormal one-electron functions entering the Slater determinant. Note that we have not assumed that these are energy eigenstates. A factor of $N^{-1/2}$ has been taken out in order that the remaining Slater determinants $\Psi_{N-1}^{(j)}$ of $N - 1$ electrons are orthonormal. (See Eq. (2.13) in Sec. 3.2 of Chap. II.) For each $\mathbf{r}_i$ we may integrate out the remaining coordinates to obtain

$$\langle O \rangle = \frac{1}{N} \sum_i \int d\mathbf{r}_i \left[\psi_1^*(\mathbf{r}_i) o(\mathbf{r}_i) \psi_1(\mathbf{r}_i) + \psi_2^*(\mathbf{r}_i) o(\mathbf{r}_i) \psi_2(\mathbf{r}_i) + \cdots \right]$$

But $\mathbf{r}_i$ has become a dummy variable and there are N identical terms canceling the N^{-1} in front. Thus we see the expectation value of any one-electron operator may be obtained from a single function of two coordinates, *the one-electron density matrix*, given by

$$\rho(\mathbf{r},\mathbf{r}',t) = \sum_n \psi_n(\mathbf{r},t)\psi_n^*(\mathbf{r}',t) \tag{3.40}$$

where the sum is over occupied states. In terms of the density matrix the expectation value of the one-particle operator is given by

$$\langle O \rangle = \int d\mathbf{r}\, [o(\mathbf{r})\rho(\mathbf{r},\mathbf{r}',t)]_{\mathbf{r}'=\mathbf{r}} \tag{3.41}$$

It is necessary to include a dependence on two spatial variables since o operates only on a single wavefunction; the expectation value cannot be evaluated without this separation. This still is a very much simpler description of the state of the system than the Slater determinant. This simplification has been accomplished by limiting our consideration to the evaluation of one-electron operators but in that context it is entirely equivalent.

In addition the use of a density matrix allows a greater generality than the use of Slater determinants. We may, for example, be interested in an electron gas interacting with an environment. The wavefunction of the system involves the coordinates not only of the electron gas but also all of the coordinates describing the environment. Because of the interaction there does not exist a wavefunction for the electron gas by itself. Nevertheless we may integrate out all of the coordinates describing the environment in order to obtain a density matrix describing the electron gas, or a one-particle density matrix. We may similarly see an immediate generalization of the one-electron density matrix that we have defined here. The system may be describable by a linear combination of Slater determinants; then the density matrix becomes a sum of the density matrices and cross terms derived from all of these determinants with appropriate weighting coefficients. This leads to the *statistical one-electron density matrix*, which we will use in our subsequent discussion. The general form, in terms of the orthonormal basis set, ψ_n, is

$$\rho(\mathbf{r},\mathbf{r}',t) = \sum_{n,n'} \psi_n(\mathbf{r},t) f(n,n') \psi_{n'}^*(\mathbf{r}',t) \tag{3.42}$$

Then $f(n,n)$ is the probability that the one-electron state $\psi_n(\mathbf{r},t)$ is occupied. We could alternatively absorb the time dependence in the f rather than in the ψ, but the form we have used is more convenient.

The density matrix may also be immediately extended for the evaluation of two-electron operators. Proceeding just as above we would integrate out all coordinates but two. The expectation value of the two-electron operator would be written

$$\langle O(\mathbf{r}_1,\mathbf{r}_2)\rangle = \int d\mathbf{r}_1\, d\mathbf{r}_2 \left[o(\mathbf{r}_1,\mathbf{r}_2)\rho(\mathbf{r}_1,\mathbf{r}_2,\mathbf{r}_1',\mathbf{r}_2')\right]_{\substack{\mathbf{r}_1'=\mathbf{r}_1 \\ \mathbf{r}_2'=\mathbf{r}_2}} \tag{3.43}$$

This would be necessary for a correct evaluation of the electron-electron interaction; however, within the self-consistent-field approximation which we use, one-electron operators suffice.

We now have a method for obtaining the expectation values we will need in terms of the density matrix. We require only the method for obtaining the density matrix itself. Thus we will need an equation giving the time dependence, the counterpart of the Schroedinger equation, and a method for specifying the initial conditions. The time dependence itself is obtained directly from Eq. (3.42) since we know the time dependence of each of the one-electron functions if the Hamiltonian can be written as a one-electron operator,

$$\sum_i H(\mathbf{r}_i,t)$$

That is

$$i\hbar\frac{\partial\psi_n(\mathbf{r},t)}{\partial t} = H(\mathbf{r},t)\psi_n(\mathbf{r},t)$$

Thus we obtain the time dependence immediately

$$\frac{\partial\rho(\mathbf{r},\mathbf{r}',t)}{\partial t} = \sum_{n,n'}\left\{\left[\frac{1}{i\hbar}H\psi_n(\mathbf{r},t)\right]\psi_{n'}^*(\mathbf{r}',t) + \psi_n(\mathbf{r},t)\left[\frac{1}{i\hbar}H\psi_{n'}(\mathbf{r}',t)\right]^*\right\}f(n,n')$$

which is conventionally written

$$i\hbar\frac{\partial\rho(\mathbf{r},\mathbf{r}',t)}{\partial t} = H(\mathbf{r},t)\rho(\mathbf{r},\mathbf{r}',t) - \rho(\mathbf{r},\mathbf{r}',t)H(\mathbf{r}',t) \tag{3.44}$$

This is the *Liouville equation.* The operator $H(\mathbf{r}',t)$ of course operates to the left. This gives us the required time dependence.

The initial conditions are customarily set by imagining a system in equilibrium in the distant past. At that past time we may evaluate the one-electron eigenstates $\psi_n(\mathbf{r},-\infty)$. The probability of occupation of any state is then simply given by the Fermi distribution function which we write $f_0(n)$. Thus the density matrix in the distant past is obtained by taking $f(n,n')$ equal to $f_0(n)\,\delta_{n,n'}$ in Eq. (3.42) and any perturbation which disrupts equilibrium, such as an applied electric field, is introduced slowly (adiabatically). A static potential $V(\mathbf{r})$ is introduced in the form

$$V(\mathbf{r},t) = V(\mathbf{r})e^{\alpha t}$$

where α is taken to be extremely small and properties are evaluated at the time $t = 0$. This adiabatic turning-on of the potential avoids any transient effects and makes the problem well defined.

Note that we have taken the density matrix as diagonal (no terms with $n' \neq n$) in the distant *past*. This is a very subtle point. A diagonal density matrix corresponds to a very low-entropy system. By picking this form in the distant past we have properly ordered the evolution of time. It might at first seem plausible to make the alternative approach of turning off the potential adiabatically and assuming that the density matrix is diagonal at infinite positive times. However, such a calculation would lead, for example, to currents flowing against applied fields rather than with them. By asserting that the final state of the system is of low entropy we would effectively reverse time. The adiabatic turning-on which we are using here is the appropriate approach.

For our purposes it will be convenient to make a Fourier expansion of the density matrix with respect to both spatial coordinates. Then the density matrix may be written

$$\rho(\mathbf{r},\mathbf{r}',t) = \frac{1}{\Omega}\sum_{k,k'} e^{ik\cdot r}\,\rho_{k,k'}(t)e^{-ik'\cdot r'} \equiv \sum_{k,k'} |\mathbf{k}\rangle \rho_{k,k'}(t)\langle\mathbf{k}'|$$

where

$$\rho_{kk'}(t) = \frac{1}{\Omega}\int d\mathbf{r}\,d\mathbf{r}'\,e^{-ik\cdot r}\,\rho(\mathbf{r},\mathbf{r}',t)\,e^{ik'\cdot r'} \equiv \langle\mathbf{k}|\rho|\mathbf{k}'\rangle$$

We may immediately rewrite Eqs. (3.41) and (3.44) to obtain

$$\langle O\rangle = \sum_{kk'} o_{kk'}\,\rho_{k'k} = \mathrm{Tr}(o\rho) \tag{3.45}$$

$$i\hbar\frac{\partial}{\partial t}\rho_{k'k} = \sum_{k''}\left[H_{k'k''}\,\rho_{k''k} - \rho_{k'k''}\,H_{k''k}\right] \tag{3.46}$$

We see here that $\rho_{k'k}$ is the matrix element of a matrix with as many rows and columns as there are wavenumbers. The expectation value is obtained by taking the matrix product of the matrix $o_{kk'} = \langle\mathbf{k}|o|\mathbf{k}'\rangle$ with the density matrix and evaluating its trace. Further we see that the time dependence is given by the commutator of the Hamiltonian matrix and the density matrix.

Given Eqs. (3.45) and (3.46) we may very directly compute the response of an electron gas to an applied potential in the self-consistent-field approximation. We follow the procedure first given by Ehrenreich and Cohen.[1] We will begin at negative infinite times with an electron gas in an equilibrium configuration. We will then apply a potential varying in space and time as $e^{i(q\cdot r-\omega t)+\alpha t}$ and compute the density matrix at time $t = 0$. In terms of this density matrix we will compute the electron density, which gives directly the $X(q,\omega)$ determining the dielectric function.

The first step is to linearize the Liouville equation as we linearized the Boltzmann equation. The Hamiltonian contains a zero-order term H_0 which is just the kinetic energy and a first-order term H_1 which contains the applied potential. The density matrix then contains a zero-order term ρ_0, the equilibrium distribution, and a first-order term, the linear response. Higher-order terms will be dropped. Substituting these in the Liouville equation, Eq. (3.46), we obtain a zero-order and a first-order equation

$$i\hbar\frac{\partial\rho_0}{\partial t} = H_0\rho_0 - \rho_0 H_0$$

[1] H. Ehrenreich and M. H. Cohen, *Phys. Rev.*, **115**:786 (1959).

and

$$i\hbar\frac{\partial\rho_1}{\partial t} = H_1\rho_0 - \rho_0 H_1 + H_0\rho_1 - \rho_1 H_0 \tag{3.47}$$

where each ρH and $H\rho$ is a matrix product. Equation (3.47) is the *linearized Liouville equation.*

We consider the $\mathbf{k}'\mathbf{k}$ matrix element of Eq. (3.47). Noting that $\rho_{0k'k} = f_0(\epsilon_k)\delta_{kk'}$ and $H_{0k'k} = \epsilon_k\delta_{kk'}$, where ϵ_k is the kinetic energy, $\hbar^2k^2/2m$, the matrix products can be directly evaluated

$$i\hbar\frac{\partial}{\partial t}\rho_{1k'k} = [f_0(\epsilon_k) - f_0(\epsilon_{k'})]H_{1k'k} + \rho_{1k'k}(\epsilon_{k'} - \epsilon_k) \tag{3.48}$$

But $H_1 = V_t e^{i(q\cdot r - \omega t) + \alpha t}$ and ρ_1 will have the same time dependence. Thus $\partial/\partial t$ can be replaced by $-i\omega + \alpha$ and we may solve for $\rho_{1k'k}$.

$$\rho_{1k'k} = \frac{f_0(\epsilon_{k'}) - f_0(\epsilon_k)}{\epsilon_{k'} - \epsilon_k - \hbar\omega - i\hbar\alpha} H_{1k'k} \tag{3.49}$$

Finally, we note that $H_{1k'k} = V_t e^{-i\omega t + \alpha t}\delta_{k',k+q}$, leading to

$$\rho_{1k+q,k} = \frac{f_0(\epsilon_{k+q}) - f_0(\epsilon_k)}{\epsilon_{k+q} - \epsilon_k - \hbar\omega - i\hbar\alpha} V_t e^{-i\omega t + \alpha t} \tag{3.50}$$

and all other first-order matrix elements vanish.

It is interesting before proceeding to make a comparison with the classical calculation in which we required q to be small. We expand both the numerator and the denominator in Eq. (3.50) for small q.

$$\begin{aligned}\rho_{1k+q,k} &\approx \frac{(\partial f_0/\partial E)(\hbar^2\mathbf{k}/m)\cdot\mathbf{q}}{(\hbar^2\mathbf{k}/m)\cdot\mathbf{q} - \hbar\omega - i\hbar\alpha} V_t e^{-i\omega t + \alpha t} \\ &= \frac{(\partial f_0/\partial E)(i\mathbf{q}\cdot\mathbf{v}\tau)}{1 + i\mathbf{q}\cdot\mathbf{v}\tau - i\omega\tau} V_t e^{-i\omega t}\end{aligned}$$

where in the last step we replaced α by $1/\tau$ and dropped the factor $e^{\alpha t}$. This replacement would have occurred naturally in the derivation if we had introduced a scattering term $-i\hbar\rho_1/\tau$ in the Liouville equation, Eq. (3.48), rather than turning on the perturbation adiabatically. (This is an alternative method for making the solution well defined.) By comparison with Eq. (3.31) of Sec. 4.1, we see that in the long-wavelength limit $\rho_{1k+q,k}$ becomes the corresponding Fourier component $f_1(\mathbf{p})$ of the distribution function as q approaches zero if we neglect the scattering-term correction $\delta_n f_0/\tau$. When we see shortly that the screening distribution is given by

$$n(\mathbf{r},t) = \Omega^{-1}\sum_k \rho_{k+q,k}\, e^{iq\cdot r}$$

we will have completed the demonstration that the approximation made in using the Boltzmann equation is exactly the neglect of terms of order $\hbar q/\hbar k = \hbar q/p$. Our classical treatment had the redeeming feature of a more careful treatment of scattering.

The evaluation of the first-order density matrix constitutes the solution of the transport problem. We now obtain the electron-density fluctuations in terms of the first-order density matrix.

If the electron density at any time is written as a Fourier expansion,

$$n(\mathbf{r},t) = \sum_q n_q(t) e^{iq\cdot r}$$

then n_q is given by

$$n_q(t) = \frac{1}{\Omega}\int n(\mathbf{r},t)e^{-iq\cdot r}\,d\tau = \int \sum_n \psi_n^*(\mathbf{r},t)\psi_n(\mathbf{r},t)\frac{e^{-iq\cdot r}}{\Omega}\,d\tau$$

Thus by comparison with Eqs. (3.40) and (3.41) we see that $e^{-iq\cdot r}/\Omega$ is the one-electron operator giving the amplitude of the density fluctuation n_q. Its matrix elements with respect to plane-wave states may be obtained immediately, $\langle \mathbf{k}|e^{-iq\cdot r}/\Omega|\mathbf{k}'\rangle = \Omega^{-1}\,\delta_{k',k+q}$. From Eq. (3.45) we evaluate the expectation value of n_q and thus $n(\mathbf{r},t)$.

$$n(\mathbf{r},t) = \frac{1}{\Omega}\sum_{k,q}\rho_{k+q,k}e^{iq\cdot r} = \frac{1}{\Omega}\sum_{k,q}\frac{[f_0(\epsilon_{k+q}) - f_0(\epsilon_k)]\,V_t e^{i(q\cdot r-\omega t)+\alpha t}}{\epsilon_{k+q} - \epsilon_k - \hbar\omega - i\hbar\alpha} \tag{3.51}$$

which has the same time and spatial dependence as the perturbing potential. But we may now let α become very small and, comparing Eq. (3.51) with $n(\mathbf{r},t) = n_s e^{i(q\cdot r-\omega t)}$, we obtain immediately n_s, which in turn is set equal to $X(q,\omega)\,V_t$. This yields

$$X(q,\omega) = \frac{1}{\Omega}\sum_k \frac{f_0(\epsilon_{k+q}) - f_0(\epsilon_k)}{\epsilon_{k+q} - \epsilon_k - \hbar\omega - i\hbar\alpha}$$

and a dielectric function given by

$$\epsilon(q,\omega) = 1 - \frac{4\pi e^2}{q^2\Omega}\sum_k \frac{f_0(\epsilon_{k+q}) - f_0(\epsilon_k)}{\epsilon_{k+q} - \epsilon_k - \hbar\omega - i\hbar\alpha}$$

This is the *Lindhard formula* for the Hartree dielectric function of a free-electron gas. It is to be evaluated with α approaching zero. It is an exact expression for the dielectric function in the Hartree approximation.

We may rewrite this dielectric function in a form which is more convenient for many purposes by changing variables in the term in $f_0(\epsilon_{k+q})$ to $\kappa = k + q$. We then note that κ may be replaced by $-\kappa$ since we sum over all κ. Finally we note that both the distribution function and the energy are independent of the sign of the wavenumber. Thus we obtain the form

$$\epsilon(q,\omega) = 1 + \frac{4\pi e^2}{q^2\Omega} \sum_k f_0(\epsilon_k)\left(\frac{1}{\epsilon_{k+q} - \epsilon_k + \hbar\omega + i\hbar\alpha} + \frac{1}{\epsilon_{k+q} - \epsilon_k - \hbar\omega - i\hbar\alpha}\right) \quad (3.52)$$

We may set ω equal to zero in Eq. (3.52) to obtain the static dielectric function. By rationalizing the denominators and letting α approach zero, we see that α has the effect of requiring principal values in the sum over **k**. That sum may be replaced by an integral and evaluated explicitly at zero temperature. The integrals are essentially the same as in Sec. 4.1. We obtain

$$X(q) = \frac{-n(\xi)}{2}\left(\frac{1-\eta^2}{2\eta}\ln\left|\frac{1+\eta}{1-\eta}\right| + 1\right) \quad (3.53)$$

where $\eta = q/2k_F$ and $n(\xi)$ is the density of states at the Fermi energy, $mk_F/\pi^2\hbar^2$. Using this $X(q)$,

$$\epsilon(q) = 1 - 4\pi e^2 \frac{X(q)}{q^2}$$

is the *static Hartree dielectric function* and is shown in Fig. 3.24.

At long wavelengths, small η, $X(q)$ approaches the classical value of $-n(\xi)$ as we know it must. At short wavelengths, large η, $X(q)$ approaches zero as $1/q^2$; the electrons do not respond appreciably to perturbations with wavelengths much less than those of the electrons. This decay of the response function at short wavelengths is a quantum-mechanical effect that we missed in the classical treatment.

The dielectric function that we obtain using this quantum-mechanical treatment is appropriate for approximating the screening of any weak potential in the free-electron gas. The situation is slightly more complicated when a nonlocal pseudopotential is to be screened and we will return to that problem in Sec. 4.4.

It is interesting to note that the static Hartree dielectric function has a logarithmic singularity of the form $x \ln x$ at $q = 2k_F$. The function is continuous but its first derivative becomes infinite logarithmically at this point. This is indeed a very weak singularity as seen in the inset in Fig. 3.24 but becomes important in many properties of metals. We will notice that it gives rise to a similar singularity in the dispersion curves (frequency as a function of wavenumber) for lattice vibrations in crystals. These are the so-called *Kohn anomalies*.

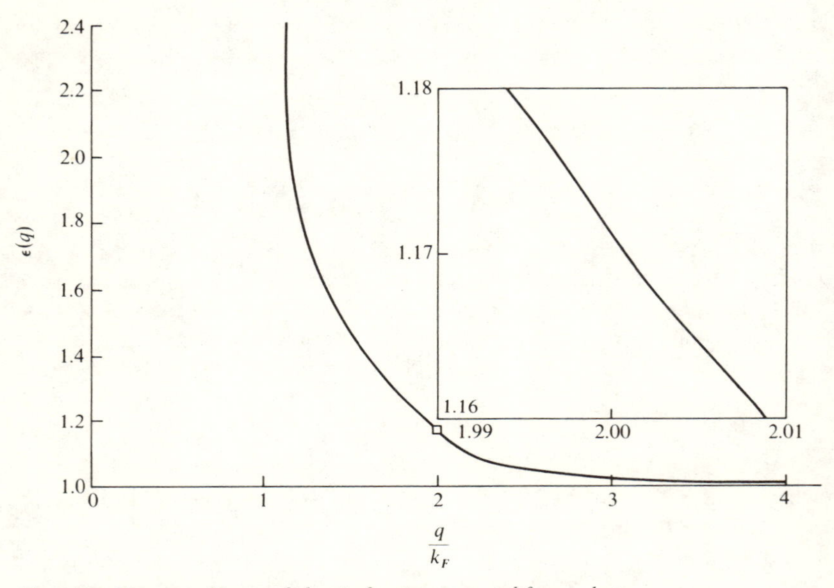

Fig. 3.24 The static Hartree dielectric function computed for an electron density equal to that for the valence electrons in aluminum. The region near the singularity at $q = 2k_F$ is expanded by a factor of 50 and plotted in the inset.

It will also tend to be important in any properties which are obtained from Fourier transforms of the dielectric function. One such case of particular importance is the screening field arising from a defect in a metal. We found that classically the screening led to an exponential decay of the potential at large distances. However, we found in our discussion of phase shifts in Sec. 8.4 of Chap. II that the assumption of a localized scattering potential led in quantum theory to Friedel oscillations in the charge density at large distances. Use of the dielectric function enables us to handle this same problem self-consistently. This could not be done in the phase-shift framework because the net potential is long range. However, it will also restrict the treatment to weak potentials whereas phase-shift analysis can be exact. In this analysis, the Friedel oscillations are seen to arise from the logarithmic singularity of the dielectric function.

Here we will outline the calculation; the algebra is somewhat complicated if given in detail. We consider the screening of a spherically symmetric potential $V^0(r)$ for which we can write the Fourier expansion

$$V^0(r) = \sum_q V_q^{\,0}\, e^{iq\cdot r} = \sum_q V_q^{\,0}\, \frac{\sin qr}{qr}$$

As we have indicated before, the screened potential takes the form

$$V(r) = \sum_q \frac{V_q^{\,0}}{\epsilon(q)} \frac{\sin qr}{qr}$$

where $\epsilon(q)$ is the static Hartree dielectric function.

We next convert the sum to an integral and seek an asymptotic expansion for the potential. This can be done most easily by successive integration by parts such that each succeeding term is of one higher order in $1/r$. Converting the sum to an integral, we write

$$rV(r) = \frac{4\pi\Omega}{(2\pi)^3} \int dq \frac{qV_q^{\,0}}{\epsilon(q)} \sin qr$$

After two integrations by parts we have:

$$rV(r) = \frac{4\pi\Omega}{(2\pi)^3} \left\{ -\frac{\cos qr}{r} \frac{qV_q^{\,0}}{\epsilon(q)} \Bigg| + \frac{\sin qr}{r^2} \frac{d}{dq}\left[\frac{qV_q^{\,0}}{\epsilon(q)}\right] \Bigg| - \frac{1}{r^2} \int dq \sin qr \frac{d^2}{dq^2}\left[\frac{qV_q^{\,0}}{\epsilon(q)}\right] \right\}$$

Since $\epsilon(q)$ is singular at $q = 2k_F$, the surface terms should be taken at $2k_F \pm \epsilon$ as well as at $q = 0$ and $q = \infty$. However, they can be seen to cancel for the two terms given. In addition, we expect $V_q^{\,0}$ and its derivatives to go to zero at large q so we need retain only the final integral.

$$V(r) = -\frac{\Omega}{2\pi^2 r^3} \int_0^\infty dq \sin qr \frac{d^2}{dq^2}\left[\frac{qV_q^{\,0}}{\epsilon(q)}\right] \tag{3.54}$$

Such integrals tend to go to zero as r becomes infinite because of the rapid oscillations of $\sin qr$. However, $(d^2/dq^2)\epsilon(q)$ will be seen to have an infinite discontinuity at $q = 2k_F$ giving a nonvanishing integral. The derivative $d^2\epsilon/dq^2$ can be obtained from Eq. (3.53), and the most singular term, $-me^2/[4\pi\hbar^2 k_F^{\,2}(2k_F - q)]$, is to be retained. The most singular term in $\dfrac{d^2}{dq^2}\left(\dfrac{qV_q^{\,0}}{\epsilon(q)}\right)$ is

$$-\frac{2k_F\, V_{2k_F}^0}{\epsilon(2k_F)^2}\left(\frac{d^2\epsilon}{dq^2}\right)$$

All other terms are slowly varying functions of q and give contributions to the integral of Eq. (3.54) which go to zero at large r. Equation (3.54) becomes

$$V(r) \approx -\frac{\Omega m e^2 V^0_{2k_F}}{4\pi^3\hbar^2 k_F \epsilon(2k_F)^2 r^3} \int_0^\infty \frac{\sin qr\, dq}{2k_F - q}$$

Finally, we change variables to $x = (q - 2k_F)r$ and can extend the limits of integration to plus and minus infinity. We expand $\sin qr = \sin x \cos 2k_F r + \cos x \sin 2k_F r$, and note that only the first contributes. Using the relation $\int_{-\infty}^{\infty} (\sin x/x)\, dx = \pi$, we obtain

$$V(r) \approx \frac{\Omega m e^2 V^0_{2k_F}}{4\pi^2\hbar^2 k_F \epsilon(2k_F)^2} \frac{\cos 2k_F r}{r^3} \tag{3.55}$$

This can be directly related to the result, Eq. (2.55) of Sec. 8.4 in Chap. II, of the phase-shift calculation. From that expression for the charge density we obtain the potential $V(r) = 4\pi e^2 n(r)/q^2$. The phase shifts are related to the backscattering form factor $\langle \mathbf{k} + \mathbf{q}|w|\mathbf{k}\rangle$ (for small δ_l) through Eq. (2.58) of the same section. Finally we identify $\langle \mathbf{k} + \mathbf{q}|w|\mathbf{k}\rangle$ with the Fourier component of the screened potential, $NV^0_{2k_F}/\epsilon(2k_F)$, in Eq. (3.55) above. We then see that Eq. (3.55) contains an additional factor of $1/\epsilon(2k_F)$. Thus, even if a screened potential $V^0_{2k_F}/\epsilon(2k_F)$ is used in the phase-shift calculation, an error of a factor of $1/\epsilon(2k_F)$ is made in assuming that the potential is localized. This correction is an additional screening of the oscillations themselves.

Such a screened coulomb potential is sketched in Fig. 3.25 for parameters appropriate to aluminum. At small distances the potential varies as $1/r$ as for the unscreened potential. This unscreened $1/r$ potential is greatly reduced at large distances as in Fermi-Thomas screening but there is a residual oscillating term.

The corresponding oscillations in the charge density may be seen in properties that depend upon local electron density. For example, nuclear-magnetic-resonance experiments show a shift in the resonant frequency which depends upon the electron density at the nucleus. A broadening of these resonances in alloys may be caused by oscillations of this type around impurity atoms. A similar broadening occurs in very tiny particles due to Friedel-like oscillations that originate at the surfaces of the crystal.

Of course these oscillations do not continue to infinite distance. We expect an exponential decay of these oscillations of the form $e^{-r/l}$, where l is the mean free path of the electrons since the oscillations arise from freely propagating electrons. In addition the singularity in the dielectric

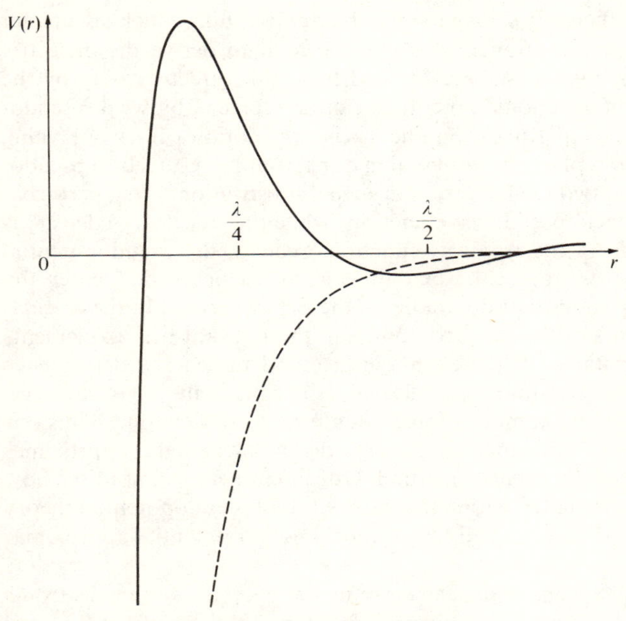

Fig. 3.25 A screened coulomb potential showing Friedel oscillations. The λ is the electron wavelength at the Fermi energy. On this scale and for parameters appropriate for a simple metal the unscreened coulomb potential would be out of the figure to the bottom while linearized Fermi-Thomas screening would give the potential shown by the broken line.

function arises from the sharpness of the Fermi surface. At finite temperatures we expect the singularity to disappear and a damping of the Friedel oscillations to occur which is presumably also exponential and probably has a form like $e^{-k_F r KT/E_F}$.

These oscillations also occur in the screening field of individual atoms in a perfect crystal and will be seen to be very important in the study of atomic properties.

4.4 Screening of pseudopotentials and of hybridization Our treatment of screening up to this point has envisaged the response of free electrons to weak potentials. If the corresponding perturbations are ionic potentials, which they frequently will be, then we have seen that the potentials are not weak.

They are in fact sufficiently strong to introduce phase shifts which are greater than π, and the perturbation theory that we used to derive the dielectric function becomes invalid. We are rescued from this difficulty only by the knowledge that the true ionic potentials can be replaced by weak pseudopotentials for which perturbation theory is appropriate. It is not valid, however, to simply replace the potential in our screening results by a pseudopotential; there are two related errors in such a substitution. First, perturbation theory leads us to pseudowavefunctions while the true charge density is obtainable only from the *true* wavefunction. Second, the pseudopotential must be regarded as nonlocal when the matrix elements that enter the calculation couple states that do not lie on the Fermi surface. In the calculation of the real part of the dielectric function the relevant matrix elements do in fact couple states off the Fermi surface and the energy dependence of the pseudopotential cannot be neglected. It is consistent, as we shall see, to neglect both of these complications though not to neglect one. Thus our simple discussions of screening can be regarded as meaningful approximations within the pseudopotential method. Our intent here is first to see how screening may be calculated within the framework of pseudopotential theory and second to investigate the validity of a local pseudopotential approximation.

As long as we were interested only in the energies of the electronic states we were able to use the pseudopotential directly; there was no need to worry that the corresponding pseudowavefunctions differed from the true wavefunctions. Whenever we consider properties that depend upon the wavefunctions themselves we must proceed more carefully. The appropriate procedure in such cases is to first write the property in question in terms of the true wavefunctions and true energies. We may then substitute for the true wavefunction in terms of the pseudowavefunction, including the orthogonalization term. Systematically keeping contributions to a given order we may then ordinarily obtain our results in terms again of the pseudopotential itself. The results will be different than if we were to replace the system by a nearly-free-electron gas and the true potential by a pseudopotential.

In the screening problem we are seeking Fourier components of the charge density from which we can compute screening fields. The first step in the problem is trivial. We may directly write the electron density in terms of a sum over occupied states,

$$n(\mathbf{r}) = \sum_k \psi_k^*(\mathbf{r})\psi_k(\mathbf{r}) \tag{3.56}$$

where the ψ_k are the true wavefunctions for the occupied states. We are assuming a static Hamiltonian; clearly a time dependence can be introduced and will not differ in any important way from that in the free-electron approximation. We may also make a Fourier expansion of the charge density,

$$n(\mathbf{r}) = \sum_q n_q e^{iq\cdot r} \tag{3.57}$$

The n_q are just the Fourier components of the screening electron density which enter a screening calculation. Equations (3.56) and (3.57) together lead directly to an expression for the n_q,

$$n_q = \frac{1}{\Omega}\sum_k \int \psi_k^*(\mathbf{r})e^{-iq\cdot r}\psi_k(\mathbf{r})\,d\tau \tag{3.58}$$

This gives rigorous specification of the screening density in terms of the true wavefunctions. We now turn to pseudopotential theory for an evaluation of the electronic eigenstates.

In Sec. 5 of Chap. II we defined a pseudowavefunction φ from which the true wavefunction could be evaluated,

$$\psi_k = (1 - P)\varphi_k \tag{3.59}$$

where we have added a subscript to φ indicating the state to which it corresponds. The projection operator P is given by a sum over core states,

$$P = \sum_{t,j} |t,j\rangle\langle t,j| \tag{3.60}$$

The pseudowavefunction itself is to be computed from the pseudopotential equation,

$$-\frac{\hbar^2}{2m}\nabla^2\varphi_k + W\varphi_k = E_k\varphi_k \tag{3.61}$$

We may think of the charge density as having two contributions. We could define a "pseudo-charge density" from $\sum_k \varphi_k^*\varphi_k$ which would include a uniform term as well as screening terms. In addition, the orthogonalization of Eq. (3.59) will "punch a hole" in the pseudo-charge density around each ion; this positive charge density is called the *orthogonalization hole*. In the standard pseudopotential calculations (e.g., Harrison)[1] this contribution is included as an additional term in the ion potential. Here we will include it in the screening calculation. The introduction of an orthogonalization hole requires a renormalization of the wavefunction since we calculate normalized pseudowavefunctions but require normalized true wavefunctions. However, the renormalization will not affect the screening density to the order we compute.

[1] W. A. Harrison, "Pseudopotentials in the Theory of Metals," Benjamin, New York, 1966.

There are many valid forms for the pseudopotential. One that will be useful here may be obtained from Eqs. (2.25) and (2.26) of Sec. 5 of Chap. II. It is defined in terms of its plane-wave matrix elements,

$$\langle \mathbf{k} + \mathbf{q}'|W|\mathbf{k}\rangle = \langle \mathbf{k} + \mathbf{q}'|V|\mathbf{k}\rangle + \sum_{t,j} (E_k - E_t)\langle \mathbf{k} + \mathbf{q}'|t,j\rangle\langle t,j|\mathbf{k}\rangle \tag{3.62}$$

where $E_k = \hbar^2 k^2/2m + \langle \mathbf{k}|W|\mathbf{k}\rangle$ and for our purposes may be thought of as the zero-order energy. We have used a $\mathbf{q}'$ in order to avoid confusion with the $\mathbf{q}$ of Eq. (3.58). If this pseudopotential is used in Eq. (3.61) and the pseudo-wavefunction calculated exactly, it will lead through Eq. (3.59) to the exact true wavefunction. Here instead we calculate the pseudowavefunction by perturbation theory. We obtain to first order in the pseudopotential

$$\varphi_k = |\mathbf{k}\rangle + \sum_{q'} \frac{|\mathbf{k} + \mathbf{q}'\rangle\langle \mathbf{k} + \mathbf{q}'|W|\mathbf{k}\rangle}{E_k - E_{k+q'}} \tag{3.63}$$

This first-order expression is to be used with Eq. (3.59) to obtain the true wavefunction to first order in the pseudopotential.

At this point the question of orders becomes tricky but also very crucial. The matrix elements of W in Eq. (3.62) are to be regarded as first order. However, we notice that the second term contains a factor of the zero-order energy E_k. Thus different matrix elements of the pseudopotential will differ from each other by a zero-order energy difference times $\langle \mathbf{k} + \mathbf{q}|P|\mathbf{k}\rangle$. To be consistent we must regard the plane-wave matrix elements of the projection operator as first-order quantities. This is also numerically consistent since such matrix elements ordinarily are of the order of 0.1. However, there are difficulties in associating an order with a projection operator since the square of a projection operator is equal to the projection operator itself, $P^2 = P$. This may be seen directly from the definition, Eq. (3.60).

$$PP = \sum_{\alpha,t,\alpha',t'} |\alpha,t\rangle\langle\alpha,t|\alpha',t'\rangle\langle\alpha',t'| = \sum_{\alpha,t} |\alpha,t\rangle\langle\alpha,t| = P$$

Here we have made use of the orthonormality of the core states. Similarly, in terms of matrix elements, we see that

$$\sum_{k''} \langle \mathbf{k}'|P|\mathbf{k}''\rangle\langle \mathbf{k}''|P|\mathbf{k}\rangle = \langle \mathbf{k}'|PP|\mathbf{k}\rangle = \langle \mathbf{k}'|P|\mathbf{k}\rangle \tag{3.64}$$

In this argument we have used the completeness relation

$$\sum_{k''} |\mathbf{k}''\rangle\langle \mathbf{k}''| = 1 \tag{3.65}$$

if the sum is over all $\mathbf{k}''$.

The contraction made in Eq. (3.64) could not have been made if the initial sum contained an energy denominator such as $E_k - E_{k''}$. The point is that the cores are strongly localized and the sum in Eq. (3.64) converges slowly. Thus small cores, leading to small $\langle \mathbf{k}'|P|\mathbf{k}\rangle$, imply many terms in Eq. (3.64); this is the numerical basis for the equivalence of the initial and contracted expressions. An energy denominator cuts off the sum so that the order of magnitude of $\sum_{k''} \langle \mathbf{k}'|P|\mathbf{k}''\rangle\langle \mathbf{k}''|P|\mathbf{k}\rangle/(E_{k''} - E_k)$ corresponds to $1/E_F$ times a second-order quantity. The order of each term must be considered separately and this is not as systematic as we might like.

We are now in a position to evaluate the wavefunction in terms of the first-order pseudowavefunction of Eq. (3.63). Keeping for the moment all four terms we obtain

$$\psi_k = |\mathbf{k}\rangle - P|\mathbf{k}\rangle + \sum_{q'} \frac{|\mathbf{k} + \mathbf{q}'\rangle\langle \mathbf{k} + \mathbf{q}'|W|\mathbf{k}\rangle}{E_k - E_{k+q'}} - \sum_{q'} \frac{P|\mathbf{k} + \mathbf{q}'\rangle\langle \mathbf{k} + \mathbf{q}'|W|\mathbf{k}\rangle}{E_k - E_{k+q'}}$$

Because of the dependence of the energy denominator upon $\mathbf{k} + \mathbf{q}'$, the final sum will converge rapidly and is a second-order term. We may drop it in our evaluation of electron density to first order in the pseudopotential. Keeping only the first three terms we may also evaluate ψ_k^*. (In making this evaluation it is important to notice the nonhermiticity in the pseudopotential which is apparent in Eq. (3.62), i.e., because of the dependence of E_k in the second term upon the wavenumber of the right-hand state $\langle \mathbf{k} + \mathbf{q}|W|\mathbf{k}\rangle^* \neq \langle \mathbf{k}|W|\mathbf{k} + \mathbf{q}\rangle$.) We obtain

$$\psi^* = \langle \mathbf{k}| - \langle \mathbf{k}|P + \sum_{q''} \frac{\langle \mathbf{k} + \mathbf{q}''|W|\mathbf{k}\rangle^*}{E_k - E_{k+q''}} \langle \mathbf{k} + \mathbf{q}''|$$

We may now immediately obtain the n_q of Eq. (3.58), noting for example $\langle \mathbf{k}|e^{-i\mathbf{q}\cdot\mathbf{r}}P|\mathbf{k}\rangle = \langle \mathbf{k} + \mathbf{q}|P|\mathbf{k}\rangle$,

$$n_q = \frac{1}{\Omega} \sum_k \Bigg(-\langle \mathbf{k} + \mathbf{q}|P|\mathbf{k}\rangle - \langle \mathbf{k}|P|\mathbf{k} - \mathbf{q}\rangle + \langle \mathbf{k}|Pe^{-i\mathbf{q}\cdot\mathbf{r}}P|\mathbf{k}\rangle + \frac{\langle \mathbf{k} + \mathbf{q}|W|\mathbf{k}\rangle}{E_k - E_{k+q}} + \frac{\langle \mathbf{k} - \mathbf{q}|W|\mathbf{k}\rangle^*}{E_k - E_{k-q}} + \text{second-order terms}\Bigg) \tag{3.66}$$

The third term cannot be contracted, as in Eq. (3.64), yet we can see it to be of first order by expanding with the use of Eq. (3.65).

$$\langle \mathbf{k}|Pe^{-iq\cdot r}P|\mathbf{k}\rangle = \sum_{q'q''} \langle \mathbf{k}|P|\mathbf{k} + \mathbf{q}'\rangle\langle \mathbf{k} + \mathbf{q}'|e^{-iq\cdot r}|\mathbf{k} + \mathbf{q}''\rangle \langle \mathbf{k} + \mathbf{q}''|P|\mathbf{k}\rangle \tag{3.67}$$

$$= \sum_{q'} \langle \mathbf{k}|P|\mathbf{k} + \mathbf{q}'\rangle\langle \mathbf{k} + \mathbf{q}' + \mathbf{q}|P|\mathbf{k}\rangle$$

There is no factor truncating the sum so we expect a magnitude of first order. Here we treat this term approximately. We note that k will always be less than k_F and the q of interest will be of that order of magnitude (though q' in the sum will become large). Thus if the cores are very small, the error will not be large in replacing $\mathbf{k}$ by $\mathbf{k} - \mathbf{q}$ in the second factor in Eq. (3.67). Then the contraction can be made

$$\langle \mathbf{k}|Pe^{-iq\cdot r}P|\mathbf{k}\rangle \approx \langle \mathbf{k}|P|\mathbf{k} - \mathbf{q}\rangle \tag{3.68}$$

This is an approximate treatment of the distribution of charge associated with the orthogonalization terms which is valid for sufficiently small cores.

We are led from Eqs. (3.66) and (3.68) to

$$n_q \approx \frac{1}{\Omega}\sum_k \left(-\langle \mathbf{k} + \mathbf{q}|P|\mathbf{k}\rangle + \frac{\langle \mathbf{k} + \mathbf{q}|W|\mathbf{k}\rangle}{E_k - E_{k+q}} + \frac{\langle \mathbf{k} - \mathbf{q}|W|\mathbf{k}\rangle^*}{E_k - E_{k-q}} \right) \tag{3.69}$$

Since the sum includes a $-\mathbf{k}$ for every $\mathbf{k}$, the $\mathbf{k}$ in the last term of Eq. (3.69) may be replaced by $-\mathbf{k}$ without error, and noting

$$\frac{\langle -\mathbf{k} - \mathbf{q}|W|-\mathbf{k}\rangle^*}{E_{-k} - E_{-k-q}} = \frac{\langle \mathbf{k} + \mathbf{q}|W|\mathbf{k}\rangle}{E_k - E_{k+q}} \tag{3.70}$$

we see that the contribution of the last two terms in Eq. (3.69) is identical. The first term in Eq. (3.69) clearly corresponds to the Fourier components of the orthogonalization hole.

For our purposes here, it is more convenient to combine the first two terms over a common denominator and write (using also Eq. (3.70))

$$n_q = \frac{1}{\Omega}\sum_k \frac{\langle \mathbf{k} + \mathbf{q}|W^+|\mathbf{k}\rangle + \langle \mathbf{k} + \mathbf{q}|W|\mathbf{k}\rangle}{E_k - E_{k+q}} \tag{3.71}$$

where the *conjugate pseudopotential* W^+ is defined by

$$\langle \mathbf{k} + \mathbf{q}|W^+|\mathbf{k}\rangle = \langle \mathbf{k} + \mathbf{q}|W|\mathbf{k}\rangle + (E_{k+q} - E_k)\langle \mathbf{k} + \mathbf{q}|P|\mathbf{k}\rangle$$

$$= \langle \mathbf{k} + \mathbf{q}|V|\mathbf{k}\rangle + \sum_{t,j} (E_{k+q} - E_t)\langle \mathbf{k} + \mathbf{q}|t,j\rangle\langle t,j|\mathbf{k}\rangle$$

We see by comparison with Eq. (3.62) that the conjugate pseudopotential is the same as the usual pseudopotential except that the energy E_{k+q} in the final term is determined from the left-hand wavenumber rather than the right-hand one.

In the screening of simple potentials there is no distinction between direct and conjugate potentials so in the expression corresponding to Eq. (3.71) the two terms in the numerator are identical and in fact can be taken out of the sum. Here, because of the k dependence of the pseudopotential, the matrix elements cannot be taken outside of the sum in Eq. (3.71) and that sum must be obtained numerically. This sum and the corresponding screening potential will lead to a screened form factor that approaches $-\frac{2}{3} E_F$ as $q \to 0$ just as does $-4\pi Z e^2/[q^2 \Omega_0 \epsilon(q)]$.

At this point we can make a local approximation to the pseudopotential. Then the local pseudopotential is identified with $(\langle \mathbf{k} + \mathbf{q} | W^+ | \mathbf{k} \rangle + \langle \mathbf{k} + \mathbf{q} | W | \mathbf{k} \rangle)/2$ (which for two states $|\mathbf{k}\rangle$ and $|\mathbf{k} + \mathbf{q}\rangle$ of the same energy is simply $\langle \mathbf{k} + \mathbf{q} | W | \mathbf{k} \rangle$) and the screened pseudopotential is obtained from the unscreened pseudopotential simply by dividing by the Hartree dielectric function. It is important in this approach not to include the potential due to the orthogonalization hole in the unscreened potential; this is inconsistent with the treatment of the pseudopotential as k-independent. Including it in the bare potential will lead to an erroneous limit at small q.

The calculation of screening in transition metals is further complicated by hybridization. We saw in Eq. (2.84) of Sec. 9 of Chap. II that the true wavefunction could be written in terms of the pseudowavefunction obtained in perturbation theory by

$$|\psi_k\rangle = |\varphi_k\rangle - \sum_\alpha |\alpha\rangle\langle\alpha|\mathbf{k}\rangle - \sum_d |d\rangle\langle d|\mathbf{k}\rangle + \sum_d \frac{|d\rangle\langle d|\Delta|\mathbf{k}\rangle}{E_d - E_k}$$

The first three terms are of just the form we obtain for simple metals, with the d states regarded as core states, and the corresponding contributions to the screening field are obtained in precisely the same way. However, the final term arising from hybridization is new and gives an additional term in the screening potential. Furthermore, we noted that it is necessary to recompute the d-like states treating the admixture of plane waves to atomic d states in perturbation theory. These changes also contribute to the screening density $n(\mathbf{r})$.

We can see that if the unscreened potential is taken as a superposition of free-ion potentials (including atomic d states in the ions) we will obtain first a contribution to n_q identical to that given in Eq. (3.71) but with the W representing the transition-metal pseudopotential of Eq. (2.80) of Sec. 9 of Chap. II,

$$\langle \mathbf{k} + \mathbf{q}|W|\mathbf{k}\rangle = \langle \mathbf{k} + \mathbf{q}|V|\mathbf{k}\rangle + \sum_{\alpha} (E_k - E_\alpha)\langle \mathbf{k} + \mathbf{q}|\alpha\rangle\langle\alpha|\mathbf{k}\rangle$$
$$+ \sum_d (E_k - E_d)\langle \mathbf{k} + \mathbf{q}|d\rangle\langle d|\mathbf{k}\rangle + \sum_d (\langle \mathbf{k} + \mathbf{q}|d\rangle\langle d|\Delta|\mathbf{k}\rangle$$
$$+ \langle \mathbf{k} + \mathbf{q}|\Delta|d\rangle\langle d|\mathbf{k}\rangle) \quad (3.72)$$

and with $\langle \mathbf{k} + \mathbf{q}|W^+|\mathbf{k}\rangle$ the same as in Eq. (3.72) except for the replacement of E_k by E_{k+q}. In addition we obtain hybridization terms due to the admixture of atomic d states in the k-like states and due to the admixture of plane waves in the d-like states.

It is not surprising that there is much cancellation between these last two contributions. This perhaps is seen most easily in terms of Slater determinants. We first imagine states determined without hybridization. Then an electron density could be determined from a Slater determinant containing k-like and d-like states. If we now introduce hybridization, some d-like wavefunction is added to each k-like state (and vice versa). This corresponds, however, to the addition of constant multiples of one row of the determinant to other rows and does not change the value of the determinant nor the resulting electron density. Thus if a d-like state is occupied, the only contribution to the screening density comes from admixture of plane waves corresponding to unoccupied k-like states; all other corrections are cancelled. The resulting contribution to the screening density (again making the small-core approximation that led to Eq. (3.68) and the corresponding approximation for d states) is found to be[1]

$$n_q^{\text{hyb}} = \sum_{\substack{d \\ k>k_F}} \left[\frac{2\langle \mathbf{k} + \mathbf{q}|\Delta|d\rangle\langle d|\Delta|\mathbf{k}\rangle}{(E_d - E_k)(E_k - E_{k+q})} - \frac{\langle \mathbf{k}|\Delta|d\rangle\langle d|\Delta|\mathbf{k}\rangle}{(E_d - E_k)^2}\right] \quad (3.73)$$

This result is plausible with respect to the range of summation, the quadratic dependence on Δ, and the energy denominators, but it would have been difficult to guess the detailed form. The same analysis for a system with unoccupied d-like states plausibly enough leads to the same form as Eq. (3.73) but with a change in sign and with the summation over $k<k_F$ rather than $k>k_F$. It is striking that in both cases the formulas avoid the singular sum through resonance and therefore are performed only over regions of energy where perturbation theory is expected to be accurate.

The electron density given in Eq. (3.73) gives an additional potential which in principle should be added to the V entering the pseudopotential. However, it is more convenient to include this extra potential and its screening with the hybridization. This can be accomplished simply by dividing Eq. (3.73) by the Hartree dielectric function; the resulting screening term

[1] W. A. Harrison, *Phys. Rev.*, **181**:1036 (1969).

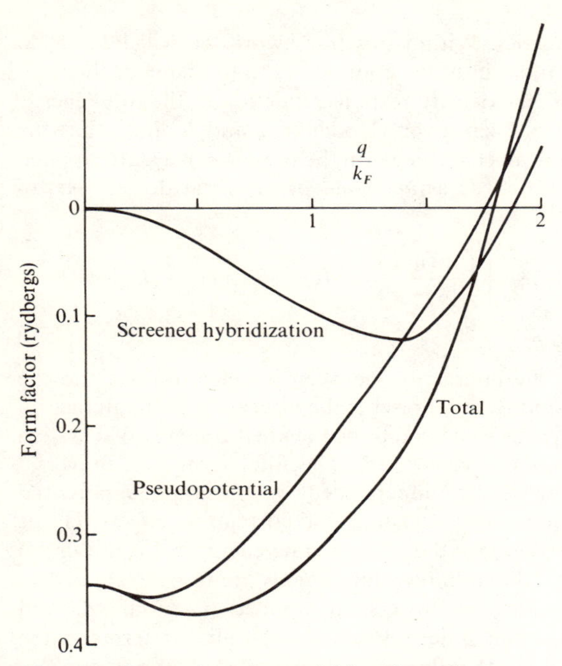

***Fig. 3.26** Computed pseudopotential form factor for copper, screened hybridization form factor for copper, and the sum of the two.* [*After W. A. Harrison,* Phys. Rev., ***181**:1036(1969)*].

may be added to the hybridization term to give the *screened hybridization* shown in Fig. 3.26 for copper. Note that screening has had the effect of canceling entirely the hybridization form factor at long wavelengths. This screened hybridization form factor may be added to the screened pseudopotential form factor, also shown in Fig. 3.26, to give the total form factor for copper which was shown in Fig. 2.46 and which enters directly the calculation of electronic properties.

4.5 The inclusion of exchange and correlation The self-consistent-field theory that we have been describing has been extended to include exchange and correlation effects by Kohn and Sham.[1] The formulation, which is exact, will lead us to a density-response function in terms of parameters that are obtainable by field-theoretic techniques, at least for the limiting

[1] W. Kohn and L. J. Sham, *Phys. Rev.*, **140**:A1133 (1965).

case of high electron densities. Within this framework we will be able to indicate also some of the more intuitive approaches to the same problem.

We begin by relating the density response function to the total energy of the system. A proof has been given (Hohenberg and Kohn)[1] that the total ground-state energy of an electron gas in the presence of a static applied potential $V^0(\mathbf{r})$ can be written as a *functional* of the local density $n(\mathbf{r})$ of electrons. We write

$$E = \int V^0(\mathbf{r})n(\mathbf{r})\,d\tau + \frac{e^2}{2}\iint \frac{n(\mathbf{r})n(\mathbf{r}')}{|\mathbf{r} - \mathbf{r}'|}\,d\tau\,d\tau' + T[n(\mathbf{r})] + E_{xc}[n(\mathbf{r})] \tag{3.74}$$

The first term represents the interaction between the electron gas and the applied potential, the second term represents the direct coulomb interaction between electrons, and $T[n(\mathbf{r})]$ is defined to be the kinetic energy of a system of noninteracting electrons of density $n(\mathbf{r})$. The final term is a universal functional of the electron density, independent of the applied potential except through $n(\mathbf{r})$, which includes exchange and correlation effects. This is regarded as an exact expression for the ground-state energy but is not useful of course until we see how the unknown functionals are to be evaluated.

Again, we will be interested in a small applied potential and will seek the linear response of the system. Because of nonlinear terms in the density in Eq. (3.74) it will be desirable to write a real applied potential

$$V^0(\mathbf{r}) = V_q^{\,0} e^{iq\cdot r} + V_q^{\,0*} e^{-iq\cdot r} \tag{3.75}$$

Then the density of the system will contain a linear response and may be written

$$n(\mathbf{r}) = n_0 + n_q e^{iq\cdot r} + n_q^* e^{-iq\cdot r} \tag{3.76}$$

where n_0 is the uniform density in the absence of the applied potential. This form is to be substituted into Eq. (3.74). The evaluation of the first term is immediate. In performing the integrations in the second term we change variables to $\mathbf{r} - \mathbf{r}'$ and $\mathbf{r}'$. In the final two terms functional derivatives can be performed and the functional written as a power-series expansion in the n_q. The lowest-order terms must be quadratic since neither can depend upon the sign of n_q. We obtain

$$E = E_0 + \Omega(n_q^* V_q^{\,0} + n_q V_q^{\,0*}) + \Omega\left(\frac{4\pi e^2}{q^2}\right) n_q^* n_q + 2\Omega T_q n_q^* n_q + 2\Omega E_q n_q^* n_q \tag{3.77}$$

[1] P. Hohenberg and W. Kohn, *Phys. Rev.*, **136**: B864 (1964).

where T_q and E_q are the functional derivatives of the corresponding terms. They are simple functions of q. This expression is exact to second order in V_q^0.

We may now immediately find the density fluctuation in terms of the applied potential. We note that in equilibrium the total energy is stationary with respect to variations of n_q; that is, $\partial E/\partial n_q = \partial E/\partial n_q^* = 0$. We perform the differentiations with respect to n_q^* and solve for n_q.

$$n_q = F(q)V_q^0$$

where

$$F(q) = -\frac{1}{4\pi e^2/q^2 + 2T_q + 2E_q} \tag{3.78}$$

We have now found the density-response function in terms of the parameters T_q and E_q which are parameters describing the electron gas.

We may obtain T_q immediately by noting that, if we repeat the same analysis without the contributions of exchange and correlation, Eq. (3.78) with E_q set equal to zero may be identified with the density-response function in the Hartree approximation, and by comparison with Eq. (3.29) (Sec. 4) and Eq. (3.53) (Sec. 4.3), we obtain

$$T_q = \left[\frac{1-\eta^2}{2\eta}\ln\left|\frac{1+\eta}{1-\eta}\right| + 1\right]^{-1}\frac{1}{n(\xi)} \tag{3.79}$$

where η is again $q/2k_F$ and $n(\xi)$ is the density of states at the Fermi energy. Thus we have recovered the response function in the Hartree approximation but we may now improve upon that with an estimate of E_q. Kohn and Sham have argued that if the perturbing potential is sufficiently slowly varying (i.e., if q is sufficiently small) the correct exchange and correlation energy can be evaluated locally as for a free-electron gas.

$$E_{xc}[n(\mathbf{r})] \approx \int n(\mathbf{r})\epsilon_{xc}(n(\mathbf{r}))\,d\tau$$

where $\epsilon_{xc}(n)$ is the exchange and correlation energy per electron in a uniform electron gas at the density n. By expanding $n\epsilon_{xc}$ around n_0 we may evaluate the integral and identify E_q. We obtain

$$\lim_{q\to 0} E_q = \frac{1}{2}\frac{d^2}{dn^2}[n\epsilon_{xc}(n)]\Big|_{n=n_0}$$

The energy ϵ_{xc} has been evaluated for a high-density electron gas.[1] Though the density of an electron gas in a metal is not sufficiently high to

[1] M. Gell-Mann and K. A. Brueckner, *Phys. Rev.*, **106**:364 (1957).

justify this limit, it may be used as an approximation. In particular for exchange alone we obtain simply (see Sec. 1 of Chap. V)

$$\lim_{q\to 0} E_q\,(\text{exchange}) = -\frac{\pi e^2}{2k_F{}^2} \tag{3.80}$$

Thus use of Eqs. (3.79) and (3.80) in (3.78) leads to the density-response function including exchange for long-wavelength perturbations.

Kohn and Sham have made an identification of this approach with the self-consistent-field method. We note that if we redo the derivation in Eqs. (3.74)–(3.77) in the Hartree approximation, but replace the coulomb potential seen by an electron due to the other electrons by an effective potential,

$$e\varphi(\mathbf{r}) = e^2 \int \frac{n(\mathbf{r}')\,d\tau'}{|\mathbf{r}-\mathbf{r}'|} \to V_{\text{eff}}(\mathbf{r}) = e^2 \int \frac{n(\mathbf{r}')\,d\tau}{|\mathbf{r}-\mathbf{r}'|} + 2E_{q\to 0}n(\mathbf{r}) \tag{3.81}$$

We are led (to within a constant) to precisely Eq. (3.77) from which we obtained the response function for small q. Thus by introducing an effective potential of the form of Eq. (3.81) we may proceed in a self-consistent-field approximation. We may compute one-electron eigenstates which will lead to the correct density fluctuations and to the correct change in energy of the system due to the introduction of a perturbing potential.

However, we do not have the counterpart of Koopmans' theorem which would allow us to attach meaning to the individual one-electron eigenvalues. Thus the meaning of a band calculation based upon a self-consistent field for exchange is very unclear and this has led to a degree of confusion. Recently band calculations have been performed using the potential obtained from Eqs. (3.80) and (3.81). This is called the *Kohn-Sham exchange potential.* Much earlier Slater[1] proposed an exchange potential which was larger than that given here by a factor of $^3/_2$. In the framework of the density-response function which we have given here, the Slater treatment would give an inaccurate density distribution; however, when the total energy is computed in either method, Eq. (3.77) is to be used with E_q given by the Kohn-Sham value, though with slightly different n_q. Thus the Slater method requires an additional correction (like the correction for electron-electron interactions counted twice) when the total energy is computed from the Hartree-like parameters. Slater's choice is equal to the true exchange potential (which is energy dependent) averaged over occupied states, whereas Kohn and Sham's corresponds to that evaluated at the Fermi energy. Thus one may argue that the Hartree parameters obtained with the Slater method give a better overall description of the true band structure whereas

[1] J. C. Slater, *Phys. Rev.*, **81**:385 (1951).

Kohn and Sham's is more appropriate for the bands near the Fermi energy. In either case the interpretation of the Hartree-like parameters as one-electron energies is on a weak footing.

The effective potential for exchange and correlation also has meaning in connection with atomic systems. The original Kohn-Sham treatment was not based upon linear-response theory as was the discussion here (nor in fact was the Slater method). However, the approximation of slow variation, small q, was made which led after minimization of the energy to an exchange potential proportional to $[n(\mathbf{r})]^{1/3}$, where $n(\mathbf{r})$ is the total density. With this exchange potential, the ground-state energy of the atom can be calculated just as simply as in the Hartree method, but it includes exchange. The only approximation is the assumption of slow variation of the density with position. This approximation, however, is a serious one in a free atom. Kohn and Sham also extended the treatment to excited states, and in particular to the electronic specific heat of a free-electron gas. This required the introduction of additional parameters and the results have not proven useful to date.

Kohn and Sham noted that a more accurate calculation could be made within the framework of the Hartree-Fock formalism rather than the Hartree formalism given above. They noted that $E_{xc}[n(\mathbf{r})]$ could be divided into an exchange term and a correlation term. The correlation term could be expanded for small fluctuations as we expanded the exchange *and* correlation terms above. This leads to an effective potential for correlation which can be incorporated in the Hartree-Fock method. Within this method both the kinetic energy and the exchange energy can be computed exactly. Only the correlation energy need be expanded. Ma and Brueckner[1] have recently used this approach in an attempt to improve upon calculated energies in free atoms. They made an expansion of E_q (correlation) in q. They obtained the zero-order term from the earlier treatment of the uniform free-electron gas and computed the q^2 term by field-theoretic methods for a high-density electron gas. This did not improve agreement with experiment in the free atom, presumably because the density fluctuations are so great.

However, we can include the q^2 term which they found in Eq. (3.78) if we also include the E_q (exchange) term which is quadratic in q. I. Ortenburger (private communication) has obtained this exchange contribution to E_q, which when added to the $q = 0$ term of Eq. (3.80) gives

$$E_q(\text{exchange}) = -\frac{\pi e^2}{2k_F{}^2} - 0.154\,\frac{e^2}{k_F{}^2}\left(\frac{q}{k_F}\right)^2 + \cdots \tag{3.82}$$

[1] S. Ma and K. A. Brueckner, *Phys. Rev.*, **165**: 18 (1968).

The contribution of correlation given by Ma and Brueckner may be written in our notation as

$$E_q(\text{correlation}) = -0.154\,\frac{me^4}{\hbar^2 k_F{}^3} + 0.338\,\frac{e^2}{k_F{}^2}\left(\frac{q}{k_F}\right)^2 + \cdots \tag{3.83}$$

If the contribution, Eqs. (3.82) and (3.83) and also Eq. (3.79), is used in Eq. (3.78), we obtain the density-response function, to order q^2 in the denominator. The result contains exactly the leading terms in an expansion in e^2; i.e., in the high-density limit.

In the spirit of the self-consistent-field expansion for describing exchange and correlation this would seem to be the most accurate expression that can presently be used. However, neither the high-density expansion nor the expansion in q/k_F is appropriate in a metal and it may well be preferable to use the lowest-order approximation—the Hartree approximation—rather than the expansions of Eqs. (3.82) and (3.83).

Returning again to the exact E_q, which could be used if it were known, there are two ways to define a dielectric function for the system. Perhaps the simplest is to seek an $\epsilon(q)$ such that $V_q{}^0/\epsilon(q)$ is the total potential, including the effective potential for exchange and correlation, seen by an electron within the self-consistent-field approximation. Using a little algebra this dielectric function is found to be

$$\epsilon(q)_{xc} = 1 - \left(\frac{4\pi e^2}{q^2} + 2E_q\right)X(q) \tag{3.84}$$

where X is given in Eq. (3.53). This expression is not very different from the Hartree dielectric function. If exchange alone is included according to Eq. (3.80) or (3.82), we may see that the correction due to E_q is small for small q but becomes appreciable when $q \approx 2k_F$. However, when q is that large the true dielectric function must be near unity in any case. Thus the inclusion of the self-consistent field for exchange does not greatly modify the energy-band calculation and, due to the uncertainty in the meaning of the Hartree-like parameters, the changes are not clearly improvements.

We can instead define a dielectric function which is the potential seen by a test charge in the electron gas in the presence of a weak applied potential. This then includes only the direct coulomb potential and not the effective potential for exchange and correlation. Such a dielectric constant differs very markedly from the Hartree dielectric function but such a potential does not enter the band structure in any case. It is perhaps the similarity between Eq. (3.84) and the Hartree dielectric function that makes ordinary energy-band calculations as accurate as they are.

5. *Optical Properties*

We will begin by discussing the optical properties of metals. Our later discussion of interband optical absorption will be applicable to semiconductors and insulators as well.

5.1 The penetration of light in a metal The response of a metal to light waves is a complicated transport problem, largely due to the complexity of the geometry. Even with a plane-polarized wave reflecting from a semi-infinite metal, the form of the fields within the metal is not simple and a correct calculation of the self-consistent response is complicated. We will use this simple geometry, illustrated in Fig. 3.27, and discuss the fields and currents within the metal.

In a simple metal we may assume the response is describable by a conductivity which can be frequency dependent but is independent of wavenumber. In making this assumption we are asserting that the fields do not vary appreciably over an electronic-mean-free path. We will need at the end to examine our result and see whether it is in fact true. Under this assumption the current at a point may be written

$$\mathbf{j} = \sigma \mathscr{E}$$

The currents and fields will vary with a frequency ω, and σ may be a function of that frequency. This represents an approximate solution of the transport problem. We need also Maxwell's equation. For the transverse electric field envisaged there will be no charge accumulation and Poisson's equation is inadequate. If there is no charge accumulation, $\nabla \cdot \mathscr{E} = 0$. Maxwell's equation leads us to a vector potential $\mathbf{A}$ satisfying

$$\nabla^2 \mathbf{A} - \frac{1}{c^2} \frac{\partial^2 \mathbf{A}}{\partial t^2} = -\frac{4\pi \mathbf{j}}{c}$$

Fig. 3.27 A schematic sketch of the spatial variation in electric field at one instant as light is reflected from the surface of a metal.

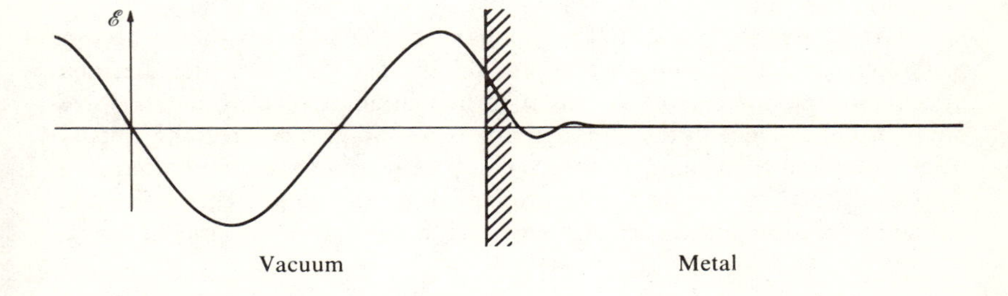

or, noting $\mathscr{E} = -(1/c)\,\partial\mathbf{A}/\partial t$,

$$\nabla^2\mathscr{E} - \frac{1}{c^2}\frac{\partial^2\mathscr{E}}{\partial t^2} = \frac{4\pi}{c^2}\frac{\partial\mathbf{j}}{\partial t}$$

We may assume for the moment a spatial and time variation given by $e^{i(\mathbf{q}\cdot\mathbf{r}-\omega t)}$ and substitute for $\mathbf{j}$ using the transport equation. We obtain

$$-q^2\mathscr{E} + \frac{\omega^2}{c^2}\mathscr{E} = -\frac{4\pi i\omega}{c^2}\sigma\mathscr{E}$$

There is no solution for both q and ω real (unless the conductivity is purely imaginary). This of course simply reflects the fact that a medium with losses will not transmit the light without attenuation. Our reflecting light has given us a fixed real frequency and therefore we find a solution only for complex wavenumbers. We may solve immediately for q

$$q = \frac{\omega}{c}\left(1 + \frac{4\pi i\sigma}{\omega}\right)^{1/2} \tag{3.85}$$

Since q is complex we have found that our wave shows an exponential decay into the metal. The expression $1 + 4\pi i\sigma/\omega$ is, as we have shown in Eq. (3.39), simply the complex dielectric function, but in this case it describes the response to transverse electric fields. Its square root, which enters the above equation directly, is called the *complex refractive index.*

In a metal σ will be much greater than ω at optical frequencies and only the second term in the square root of Eq. (3.85) is of importance. Thus the real and the imaginary part of q are both equal to $\sqrt{2\pi\omega\sigma}/c$. The fields die within the metal over a distance of the order of the reciprocal of the real part of q. This reciprocal is called the *classical skin depth*, $\delta = c/\sqrt{2\pi\omega\sigma}$.

We may return now to the question as to whether our assumption of slow variation over a mean-free path is appropriate. In ordinary metals at room temperature the mean-free path will be small compared to the classical skin depth. Thus in most circumstances our assumption is self-consistent and our treatment reasonable.

At low temperatures, on the other hand, the mean-free path may become very long in comparison to the skin depth. The problem then becomes much more complicated. We might at first attempt to introduce a wavenumber dependence in the conductivity as we did in our screening calculation. We might then continue this to complex wavenumbers and attempt to apply it directly to the attenuating wave within the metal. This is in fact a possible solution for the electromagnetic wave within the metal but an additional assumption has been made. For a propagating plane wave it is legitimate to assume that the currents and the fields will have the same dependence on position. For the attenuating wave at the surface of the metal this ordinarily will no longer be true. The presence of the surface will in general modify the form of the currents and fields and it becomes necessary to return to the Boltzmann equation in solving the transport problem. We may note in particular that the field is confined to a very thin layer on the surface. Then those electrons which happen to run almost parallel to the surface will tend to dominate the conductivity. This gives rise to a modification in the fields at the surface which is called the *anomalous skin effect*.

Pippard[1] has given an intuitive treatment of this situation, defining a region on the Fermi surface for which the electrons are effective in contributing to the current and neglecting the contribution of other electrons. This is the so-called *ineffectiveness concept*. Analytical treatments of the same effect have also been carried out[2] for ellipsoidal Fermi surfaces but not for the general case. It turns out that the response of the system gives a measure of the curvature of the Fermi surface in the effective zone since this curvature determines the number of electrons which are effective. The anomalous skin effect is interesting in its own right but is not a subject that we will pursue.

In the absence of the anomalous skin effect the optical properties divide themselves neatly into parts. Given the frequency-dependent conductivity we may match the propagating waves in space to the attenuating waves within the material. We may obtain directly a reflection coefficient in terms of the complex conductivity. Similarly, we may compute the transmission of a thin metallic film. The measurement of these two quantities enables us to determine both the real and imaginary parts of the conductivity. These aspects of the problem are part of the classical theory of optics.

The second aspect of the problem is the determination of the complex conductivity in terms of the properties of real materials. This is in the realm of solid state physics and we wish to pursue it further.

5.2 The optical conductivity In general the absorption of radiant energy is simply the joule heat, which is the dot product of the current and the local electric field. When we can define a local conductivity as we have done here, the current is simply given by σ times the electric field. The joule heat depends upon only the component of the current in phase with the electric field and therefore only upon the real part of the electrical conductivity.

[1] A. B. Pippard, *Proc. Roy. Soc. (London)*, **A191**:385 (1947).

[2] G. E. H. Reuter and E. H. Sondheimer, *Proc. Roy. Soc. (London)*, **A195**:336 (1948).

The form which we have obtained for the real part of the conductivity for a free-electron gas in Sec. 4.2 is

$$\mathrm{Re}[\sigma(0,\omega)] = \frac{\sigma_0}{1 + \omega^2\tau^2}$$

Use of this form corresponds to the Drude theory of optical absorption and gives a good account of the absorption in simple metals at long wavelengths.

This result of course is applicable only to the response of a free-electron gas. In real metals and in insulating materials we are interested in the response of a more complicated system and a more general formulation is required.

We will first carry out the calculation for a system of noninteracting electrons, but allowing the one-electron states to be arbitrary. This allows a compact description of the state of the system and, more importantly, a simple statement of the field present. We will introduce a classical electric field $\mathscr{E} = \mathscr{E}_0 e^{i(q \cdot r - \omega t)}$, with $\mathscr{E}_0$ constant and perpendicular to $\mathbf{q}$, and seek a current response of the same form. The difficulty in an inhomogeneous system, e.g., a periodic lattice, is that an applied field of this form will cause responses with the same time dependence, but with the spatial periodicity of the system. Such currents do not enter the absorption of energy directly since the spatial average of the corresponding contributions to $\mathscr{E} \cdot \mathbf{j}$ are zero. For interacting electrons, however, the charge densities from these currents cause potentials which, in addition to the applied field, couple electronic states and therefore affect the absorption. This effect tends to be small as we shall see in Sec. 5.4. Furthermore, it should be pointed out that this subtle complication is very easy to overlook and in fact was for many years—and still frequently is. One may be continually on the watch for such overlooked subtleties in the theory of solids—particularly when an apparently well-founded theory disagrees with experiment.

For noninteracting electrons we need not distinguish between the applied and total electric field. The simplest way to add the electromagnetic field is to replace the momentum operator in the Hamiltonian by $\mathbf{p} - (-e)\mathbf{A}/c$ where $\mathbf{A}$ is the vector potential, $\mathbf{A}_0 e^{i(q \cdot r - \omega t)}$, and $\mathscr{E} = -(1/c)\,\partial\mathbf{A}/\partial t$. We have, as before, written the charge on the electron as $-e$. For the situations of interest $\mathbf{q}$ will always be so small that we may neglect it altogether. We may write out the kinetic-energy term in the Hamiltonian and obtain the perturbing term of interest.

$$H_1 = -\frac{ie\hbar}{mc}\mathbf{A} \cdot \boldsymbol{\nabla} = -\frac{e\hbar}{m\omega}\mathscr{E} \cdot \boldsymbol{\nabla} \tag{3.86}$$

Given this form for the perturbation we may seek the response of the system using the density matrix, but now based upon the eigenstates ψ_k rather than plane waves.

$$\rho_{k'k}(t) = \int d\mathbf{r}\, d\mathbf{r}'\, \psi_{k'}^*(\mathbf{r})\rho(\mathbf{r},\mathbf{r}',t)\psi_k(\mathbf{r}')$$

and

$$H_{1k'k} = -\frac{e\hbar}{m\omega}\mathscr{E}\cdot\int d\mathbf{r}\psi_{k'}^*(\mathbf{r})\nabla\psi_k(\mathbf{r})$$

The matrix element $H_{1k'k}$, without the factor $\mathscr{E}$, is called the *oscillator strength* for the transition from ψ_k to $\psi_{k'}$. Use of general ψ_k enables us to treat insulators and real metals as well as free-electron gases. The derivation through Eq. (3.49) of Sec. 4 remains appropriate and gives a first-order density matrix

$$\rho_{1k'k} = \frac{f_0(\epsilon_{k'}) - f_0(\epsilon_k)}{\epsilon_{k'} - \epsilon_k - \hbar\omega - i\hbar\alpha} H_{1k'k}$$

The ϵ_k are the eigenvalues for ψ_k. The H_1 now corresponds to a uniform applied field, though the zero-order system need not be uniform. We seek the spatial average of the current density in order to obtain the conductivity. To lowest order in the vector potential the current-density operator is $-\frac{1}{\Omega}\frac{e\hbar}{im}\nabla$ and there is no current in the zero-order state so we may obtain the current density from Eq. (3.45)

$$\begin{aligned}\langle \mathbf{j}\rangle &= \sum_{k,k'} \rho_{k'k}\left\langle \psi_k \left| -\frac{e\hbar}{im\Omega}\nabla \right| \psi_{k'}\right\rangle \\ &= -\frac{ie^2\hbar^2}{m^2\omega\Omega}\sum_{k,k'}\frac{f_0(\epsilon_{k'}) - f_0(\epsilon_k)}{\epsilon_{k'} - \epsilon_k - \hbar\omega - i\hbar\alpha}\langle\psi_k|\nabla|\psi_{k'}\rangle\langle\psi_{k'}|\nabla|\psi_k\rangle\cdot\mathscr{E}\end{aligned}$$

We may immediately extract the conductivity tensor σ_{ij} such that

$$\langle j\rangle_i = \sigma_{ij}\mathscr{E}_j$$

We seek the real part of the conductivity as $\alpha \to 0$, thus the imaginary part of the final sum in the equation for $\langle \mathbf{j}\rangle$. In evaluating the real part of such sums (for $\omega = 0$) we found that the α had the effect of requiring principal values where the denominator vanished [Eq. (3.53)]. In evaluating the imaginary part we now find that the energy denominator behaves as a delta function. Note that

$$\mathrm{Im}\left[\frac{f_0(\epsilon_{k'}) - f_0(\epsilon_k)}{\epsilon_{k'} - \epsilon_k - \hbar\omega - i\hbar\alpha}\right] = i\hbar\alpha\frac{f_0(\epsilon_{k'}) - f_0(\epsilon_k)}{(\epsilon_{k'} - \epsilon_k - \hbar\omega)^2 + \hbar^2\alpha^2}$$

is strongly peaked at $\epsilon_{k'} - \epsilon_k - \hbar\omega = 0$. Furthermore,

$$\int_a^b \frac{\hbar\alpha\, dx}{x^2 + (\hbar\alpha)^2} = \arctan \frac{x}{\hbar\alpha}\bigg|_a^b \to \pi$$

if $b/\hbar\alpha$ is large and $a/\hbar\alpha$ is large and negative. Thus when $\hbar\alpha$ becomes small the entire contribution to an integral over energy comes from the region where the energy denominator vanishes and is equal to π times the remaining factors in the integrand evaluated there. We see that the real part of the conductivity is given by

$$\lim_{\alpha\to 0} \operatorname{Re} \sigma_{ij} = \frac{\pi e^2\hbar^2}{m^2\omega\Omega} \sum_{k',k} [f_0(\epsilon_{k'}) - f_0(\epsilon_k)]\langle\psi_k|\nabla_i|\psi_{k'}\rangle\langle\psi_{k'}|\nabla_j|\psi_k\rangle \times \delta(\epsilon_{k'} - \epsilon_k - \hbar\omega)$$

This result can be expressed in more conventional form by writing

$$f_0(\epsilon_{k'}) - f_0(\epsilon_k) = f_0(\epsilon_{k'})[(1 - f_0(\epsilon_k)] - f_0(\epsilon_k)[1 - f_0(\epsilon_{k'})]$$

Since we sum over all **k** and **k**′ we may interchange **k** and **k**′ in the sum over the first term in this expression. We obtain

$$\lim_{\alpha\to 0} \operatorname{Re} \sigma_{ij} = \frac{\pi e^2\hbar^2}{m^2\omega\Omega} \sum_{k',k} f_0(\epsilon_k)[1 - f_0(\epsilon_{k'})] \times [-\langle\psi_k|\nabla_i|\psi_{k'}\rangle\langle\psi_{k'}|\nabla_j|\psi_k\rangle] \cdot [\delta(\epsilon_{k'} - \epsilon_k - \hbar\omega) - \delta(\epsilon_{k'} - \epsilon_k + \hbar\omega)] \quad (3.87)$$

This is called the *Kubo-Greenwood formula*[1] for the real part of the conductivity, since it is derivable as a one-electron approximation to an exact formula due to Kubo and Greenwood.

This form is physically plausible. The real part of the conductivity represents transitions and energy absorption and the result may be understood in those terms. A contribution arises when two states **k** and **k**′ are coupled by a matrix element of the perturbation [Eq. (3.86)]. There is only a contribution when the state **k** is occupied and the state **k**′ unoccupied. The contribution is positive if $\epsilon_{k'} = \epsilon_k + \hbar\omega$, corresponding to the absorption of a photon, and negative if $\epsilon_{k'} = \epsilon_k - \hbar\omega$, corresponding to the emission of a photon. (Note that the diagonal components of $-\langle\psi_k|\nabla|\psi_{k'}\rangle\langle\psi_{k'}|\nabla|\psi_k\rangle$ can be seen to be positive by performing an integration by parts in either matrix element.)

[1] R. Kubo, *J. Phys. Soc. Japan*, **12**:570 (1957); D. A. Greenwood, *Proc. Phys. Soc. (London)*, **A71**:585 (1958).

This is a very general form for the conductivity and is applicable to a wide variety of systems. It is even applicable to the calculation of light absorption by atomic systems, though in that case, with a discrete eigenvalue spectrum, it is necessary to introduce a range of frequencies of light in order to make use of the energy delta function. It is mainly of interest in solid state systems.

5.3 Simple metals It is well known that a free electron cannot absorb a single photon; it is not possible in such an absorption to conserve both momentum and energy. This is true for the classical equations for low-energy electrons and light and remains true in the relativistic equations. It is reflected in Eq. (3.87) in the fact that $\langle\psi_{k'}|\nabla|\psi_k\rangle$ vanishes for $\mathbf{k}' \neq \mathbf{k}$ if the states are plane waves. The Drude absorption given earlier was made possible by scattering centers which may take up momentum in the absorption. Similarly, with a periodic crystal potential the lattice may take up momentum and allow absorption. In both cases the extra potential must be represented by a pseudopotential if perturbation theory is to be used, as it ordinarily is. So we turn to that problem next.

As we indicated in our discussion of screening of pseudopotentials, the proper approach is to obtain an expression for the property of interest (in this case conductivity) in terms of wavefunctions and energies, and then use pseudopotential theory to evaluate these wavefunctions and energies. Here Eq. (3.87) gives the conductivity; we seek first an evaluation of the matrix elements. The analysis follows closely that for the screening calculation. To first order in the pseudopotential (recall the projection operator is of order W) we have

$$\psi_k = (1 - P)\varphi_k = |\mathbf{k}\rangle - P|\mathbf{k}\rangle + \sum_q \frac{|\mathbf{k} + \mathbf{q}\rangle\langle\mathbf{k} + \mathbf{q}|W|\mathbf{k}\rangle}{\epsilon_k - \epsilon_{k+q}}$$

and

$$\psi_{k'}^* = \varphi_{k'}^*(1 - P) = \langle\mathbf{k}'| - \langle\mathbf{k}'|P + \sum_{q'} \frac{\langle\mathbf{k}' + \mathbf{q}'|W|\mathbf{k}'\rangle^*\langle\mathbf{k}' + \mathbf{q}'|}{\epsilon_{k'} - \epsilon_{k'+q'}}$$

We may then evaluate $\langle\psi_{k'}|\nabla|\psi_k\rangle$ to first order in W. Terms of the form PW are second order and are dropped. As in the screening calculation the order of the P^2 term is ambiguous. It is given by

$$\langle\mathbf{k}'|P\nabla P|\mathbf{k}\rangle = \sum_{\alpha,\beta} \langle\mathbf{k}'|\alpha\rangle\langle\alpha|\nabla|\beta\rangle\langle\beta|\mathbf{k}\rangle$$

The matrix element $\langle\alpha|\nabla|\beta\rangle$ may be rewritten using an identity from Schiff,[1]

[1] L. I. Schiff, "Quantum Mechanics," p. 247, 1st edition (p. 404 in 3rd edition) McGraw-Hill, New York, 1949.

$$\langle \mathbf{k}'|P\nabla P|\mathbf{k}\rangle = \sum_{\alpha,\beta} -\frac{m}{\hbar}(\epsilon_\alpha - \epsilon_\beta)\langle\alpha|\mathbf{r}|\beta\rangle\langle\mathbf{k}'|\alpha\rangle\langle\beta|\mathbf{k}\rangle$$

This term would vanish if all core states were s states. In any case it becomes small in a small-core limit since $\langle\alpha|\mathbf{r}|\beta\rangle$ is of the order of the core radius. Thus with some justification this term can be regarded as of higher order and discarded. We are left with

$$\langle\psi_{k'}|\nabla|\psi_k\rangle = i\mathbf{k}'\left(\frac{\langle\mathbf{k}'|W|\mathbf{k}\rangle}{\epsilon_k - \epsilon_{k'}} - \langle\mathbf{k}'|P|\mathbf{k}\rangle\right) - i\mathbf{k}\left(\frac{\langle\mathbf{k}|W|\mathbf{k}'\rangle^*}{\epsilon_k - \epsilon_{k'}} + \langle\mathbf{k}'|P|\mathbf{k}\rangle\right)$$

By comparison with Eq. (3.71) of Sec. 4 we see that the first expression in parentheses is of the form of the conjugate pseudopotential,

$$\frac{\langle\mathbf{k}'|W|\mathbf{k}\rangle}{\epsilon_k - \epsilon_{k'}} - \langle\mathbf{k}'|P|\mathbf{k}\rangle = \frac{\langle\mathbf{k}'|W^+|\mathbf{k}\rangle}{\epsilon_k - \epsilon_{k'}}$$

The second expression in parentheses may be similarly contracted using the form Eq. (3.62) for W,

$$\frac{\langle\mathbf{k}|W|\mathbf{k}'\rangle^*}{\epsilon_k - \epsilon_{k'}} + \langle\mathbf{k}'|P|\mathbf{k}\rangle = \frac{\langle\mathbf{k}'|W|\mathbf{k}\rangle}{\epsilon_k - \epsilon_{k'}}$$

Finally we note that contributions to the optical conductivity of Eq. (3.87) occur only where $|\epsilon_{k'} - \epsilon_k| = \hbar\omega$ (in evaluating the matrix element we do not distinguish between the zero-order energy difference and the true difference in eigenvalues), so we may write

$$|\langle\psi_{k'}|\nabla|\psi_k\rangle| = \frac{1}{\hbar\omega}|\mathbf{k}'\langle\mathbf{k}'|W^+|\mathbf{k}\rangle - \mathbf{k}\langle\mathbf{k}'|W|\mathbf{k}\rangle| \tag{3.88}$$

If W were a simple potential, there would be no distinction between W and W^+ and the expression would be simply

$$|\langle\psi_{k'}|\nabla|\psi_k\rangle| \cong \left|\frac{(\mathbf{k}' - \mathbf{k})}{\hbar\omega}\langle\mathbf{k}'|W|\mathbf{k}\rangle\right| \tag{3.89}$$

This is the form which arises in the nearly-free-electron approximation.[1] However, the correct form, Eq. (3.88), can modify the results by as much as a factor of 2† and improves agreement with experiment.

Equations (3.87) and (3.88) properly include the presence of a static pseudopotential in the optical conductivity. They may be used for a disordered structure where the pseudopotential gives rise to scattering; this

[1] P. N. Butcher, *Proc. Phys. Soc. (London)*, **A64**:765 (1951).

†A. O. E. Animalu, *Phys. Rev.*, **163**:557 (1967).

will lead to the Drude result. They may also be used to find the effect of the periodic potential; this will correspond to interband absorption which occurs in semiconductors and insulators as well as in metals.

We first consider briefly a scattering pseudopotential in order to make an identification between the Kubo-Greenwood formula and the transport approach. This will require an identification of the scattering rate in terms of the parameters in Eqs. (3.87) and (3.88). We may readily make the identification if $\hbar\omega$ is sufficiently low that both initial and final states lie near the Fermi surface. (Effects which are left out by taking this limit are corrections to the relaxation-time approximation of transport theory.) Then the difference between W and W^+ may be neglected and Eq. (3.89) for the matrix elements becomes appropriate. The difference in wavenumbers appears squared in the conductivity and for an isotropic system may be written $(k_x - k_x')^2 = \frac{1}{3}|\mathbf{k} - \mathbf{k}'|^2 = 2k^2(1 - \cos\theta)/3$, if the frequency is sufficiently low that we may neglect the difference in the magnitude of $\mathbf{k}$ and of $\mathbf{k}'$. We have written θ as the angle between $\mathbf{k}$ and $\mathbf{k}'$.

The sum over $\mathbf{k}'$ is now of just the form we gave for the relaxation time for conductivity including even the factor $(1 - \cos\theta)$. It differs only in that the energy delta function contains an $\hbar\omega$ which we neglect in comparison to Fermi energies. Note that the neglect of the energy difference between the initial and final state, which we have made twice, becomes appropriate as $\hbar$ approaches zero and therefore is in keeping with the classical treatment that led to the Drude formula.

We must retain that energy difference, however, when we consider the effects of the distribution function. At absolute zero $f_k(1 - f_{k'})$ will be nonzero only over an energy range equal to $\hbar\omega$. Thus in our summation over $\mathbf{k}$ we will obtain a factor $n(E)\hbar\omega$. For a free-electron gas this is $3N\hbar\omega/2E_F$. Collecting all of these factors together we obtain

$$\sigma(\omega) = \frac{Ne^2}{m\omega^2\tau} \tag{3.90}$$

We recall that the Drude form was $Ne^2\tau/(1 + \omega^2\tau^2)m$. But here the calculation is only to lowest order in the scattering rate $1/\tau$, and the Drude form goes to Eq. (3.90) to lowest order in $1/\tau$. This is a complicated way to obtain the Drude result, but indicates the general applicability of Eq. (3.87), clarifies the approximations made in a scattering-time approximation, and establishes the relation between the different approaches used.

5.4 Interband absorption The pseudopotential in a perfect crystal can also give rise to absorption. This fact forms the basis, in the alkali metals, for a theory of interband absorption.

The matrix elements in Eq. (3.88) vanish for a perfect crystal unless $\mathbf{k}' - \mathbf{k}$ is a lattice wavenumber. Thus transitions occur only between states which differ by a lattice wavenumber. Our description here is based upon an extended-zone scheme in which the wavenumbers indexing the states run over all wavenumber space. In other systems it will be more convenient to use the reduced-zone scheme, in which case the restriction to changes by a lattice wavenumber becomes a condition that the transitions occur between states of the same wavenumber *within the reduced zone*, but between different bands. Even in the alkalis, it may be easier to understand the absorption using the reduced zone as indicated in Fig. 3.28. This may be viewed as a one-dimensional analog, or a plot of the bands along a line perpendicular to the nearest zone face in an alkali metal. The change in wavenumber by a lattice wavenumber in the extended-zone scheme becomes a "vertical" transition in the reduced zone.

The Fermi surface lies within the first zone in the alkali metals, as shown, so a minimum energy absorption, near that shown, occurs. This is called an *absorption edge.* We note that the bands near the edge are close to free-electron bands; the distortion due to finite pseudopotentials is restricted primarily to the region of the zone faces. Thus the free-electron

Fig. 3.28 *Interband optical absorption in an alkali metal can be viewed in the extended-zone scheme as a simultaneous photon absorption and Bragg reflection. In the reduced zone this becomes a vertical transition between bands.*

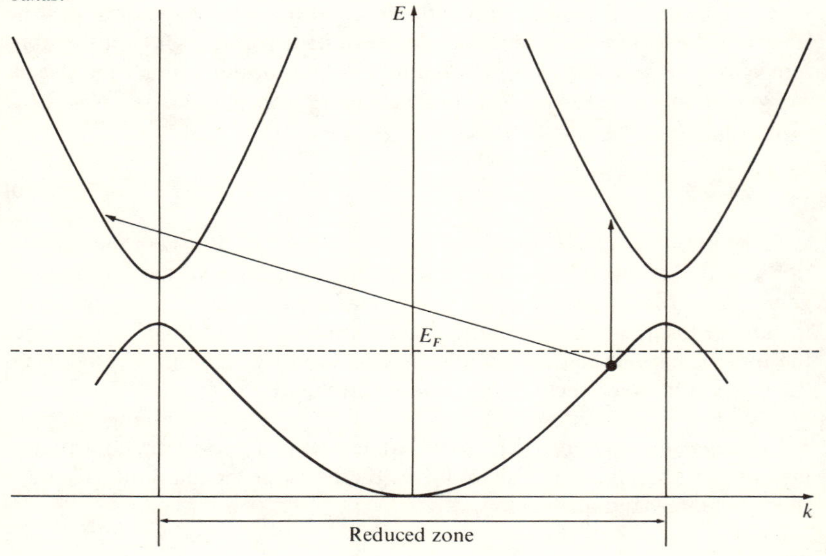

energies may be used in Eq. (3.87) for the conductivity. This is confirmed experimentally; the interband absorption edges do occur close to the thresholds predicted by free-electron theory. These edges show up as extra absorption at energies of a few volts where the Drude absorption is becoming small.

The intensity of the absorption may be computed using the Kubo-Greenwood formula, Eq. (3.87). The addition of a Brillouin zone face, corresponding to a lattice wavenumber **q**, gives an additional contribution to the conductivity by coupling the occupied state ψ_k with an unoccupied state ψ_{k+q}.

The geometry of the summation over states is complicated and will be simplified here by use of the approximate form, Eq. (3.89), for the oscillator strengths. It should be noted, however, that an error of the order of a factor of 2 or more may be made by using Eq. (3.89) rather than the correct Eq. (3.88).† Taking a coordinate system with z axis along the electric-field vector of the light, we see that only the component σ_{33} of the conductivity will enter absorption; combining Eqs. (3.87) and (3.89) we obtain

$$\operatorname{Re}\sigma_{33} = \sum_q \frac{\pi e^2 q^2 \cos^2\varphi}{m^2\omega^3\Omega} \sum_k f_0(\epsilon_k)[1 - f_0(\epsilon_{k+q})] \times |\langle \mathbf{k} + \mathbf{q}|W|\mathbf{k}\rangle|^2 \, [\delta(\epsilon_{k+q} - \epsilon_k - \hbar\omega) - \delta(\epsilon_{k+q} - \epsilon_k + \hbar\omega)]$$

where φ is the angle between **q** and the electric-field vector. Thus we have a direct superposition of the effects of each lattice wavenumber.

We focus attention on a single lattice wavenumber. At zero temperature $f_0(\epsilon_k)$ will be nonzero only for $k < k_F$, and in an alkali metal all coupled states ϵ_{k+q} will lie above the Fermi energy, so $[1 - f_0(\epsilon_{k+q})] = 1$. Further, taking $\omega > 0$, only the first delta function can contribute. Finally, we make a local approximation to the pseudopotential, so $\langle \mathbf{k} + \mathbf{q}|W|\mathbf{k}\rangle$ is taken as independent of **k**. This is consistent with the use of Eq. (3.89) for the matrix elements, but it is only semiquantitatively valid. We are left with

$$\operatorname{Re}\sigma_{33} = \sum_q \frac{\pi e^2 q^2 \cos^2\varphi}{m^2\omega^3\Omega} |\langle \mathbf{k} + \mathbf{q}|W|\mathbf{k}\rangle|^2 \sum_{k<k_F} \delta(\epsilon_{k+q} - \epsilon_k - \hbar\omega)$$

The summation over **k** can now be performed. Using zero-order energies in the delta function, which we have seen is reasonable for the monovalent alkali metals, the summation becomes

$$\sum \equiv \sum_{k<k_F} \delta(\epsilon_{k+q} - \epsilon_k - \hbar\omega)$$

$$= \frac{2\Omega}{(2\pi)^3} \int 2\pi\, \delta\left[\frac{\hbar^2}{2m}(2\,kq\cos\theta + q^2) - \hbar\omega\right] \times \sin\theta\, d\theta k^2\, dk$$

† *Ibid.*

where θ is the angle between $\mathbf{k}$ and $\mathbf{q}$. The integration runs over occupied states and will be nonzero only if

$$\hbar\omega \geq \frac{\hbar^2}{2m}(q^2 - 2k_F q) \tag{3.91}$$

since otherwise the delta function lies outside the realm of the integration. This gives the energy of the absorption edge which might be directly obtained by consideration of Fig. 3.28.

We may perform the angular integration first. This angular integration, for fixed k, will be nonzero only if the region of integration includes the point where the argument of the delta function is zero; i.e.,

$$k > \frac{q}{2} - \frac{m\omega}{\hbar q} \tag{3.92}$$

and

$$k > -\frac{q}{2} + \frac{m\omega}{\hbar q} \tag{3.93}$$

Equation (3.93) will be satisfied except for very high-energy photons, so near the absorption edge we need not apply that condition. However, Eq. (3.92) restricts the contributing region of the k integration for energies near the absorption edge. Thus, after performing the angular integration, we obtain

$$\sum = \frac{\Omega m}{2\pi^2\hbar^2 q}\int_{q/2 - m\omega/\hbar q}^{k_F} k\,dk = \frac{\Omega m}{4\pi^2\hbar^2 q}\left[k_F{}^2 - \left(\frac{q}{2} - \frac{m\omega}{\hbar q}\right)^2\right] \tag{3.94}$$

if Eq. (3.91) is satisfied, i.e., if the Σ of Eq. (3.94) is positive, and the conductivity becomes

$$\operatorname{Re}\sigma_{33} = \sum_{[q]}\frac{e^2 q}{4\pi m\hbar^2\omega^3}|\langle\mathbf{k} + \mathbf{q}|W|\mathbf{k}\rangle|^2\cos^2\varphi\left[k_F{}^2 - \left(\frac{q}{2} - \frac{m\omega}{\hbar q}\right)^2\right] \tag{3.95}$$

where the sum is over all $\mathbf{q}$ satisfying Eq. (3.91). The contribution due to a single lattice wavenumber is as shown in Fig. 3.29 and is to be added to the contribution of the Drude term.

This result is only semiquantitative because of our approximate treatment of the pseudopotential. In addition it is in error because it is based upon a formula for noninteracting electrons. Hopfield[1] has pointed out that within the framework of a weak lattice potential the effects of the

[1] J. J. Hopfield, *Phys. Rev.*, **139**: A419 (1965).

electron-electron interaction can be included by a simple modification of Eq. (3.95). We have implicitly included some contribution of the electron-electron interaction in our calculation if we use a screened pseudopotential. For a local pseudopotential we could write the matrix element in Eq. (3.95) in terms of the unscreened pseudopotential, W^0, by using the static dielectric function,

$$|\langle \mathbf{k} + \mathbf{q}|W|\mathbf{k}\rangle|^2 \rightarrow \frac{|\langle \mathbf{k} + \mathbf{q}|W^0|\mathbf{k}\rangle|^2}{|\epsilon(q,0)|^2} \tag{3.96}$$

Hopfield has shown that if electron-electron interactions are included exactly, the same form, Eq. (3.95), is obtained but with Eq. (3.96) replaced by

$$|\langle \mathbf{k} + \mathbf{q}|W|\mathbf{k}\rangle|^2 \rightarrow \frac{|\langle \mathbf{k} + \mathbf{q}|W^0|\mathbf{k}\rangle|^2}{|\epsilon(q,\omega)|^2}$$

The screening of the pseudopotential for this calculation is as if the pseudopotential oscillated in time with the light frequency. This correction to the interband absorption is small since the dielectric function is near unity (usually within 25 percent) for the q of interest in any case.

Equation (3.95) gives a qualitative description of interband absorption edges in the solid alkali metal, with discrepancies in the magnitude of as much as a factor of 4. Use of a nonlocal pseudopotential reduces the discrepancy by a factor of 2. The remainder of the error may well arise from the inherent arbitrariness in the pseudopotential or the inaccuracy in computing σ_{33} only to lowest order in the pseudopotential; as we indicated in Sec. 5.1 of Chap. II, these are equivalent sources of error.

Fig. 3.29 A plot of the real part of the optical conductivity arising from interband absorption near a zone face in an alkali metal. Contributions from each zone face could be added to the Drude contribution to obtain a total optical conductivity. The ω_0 is the frequency of the absorption edge.

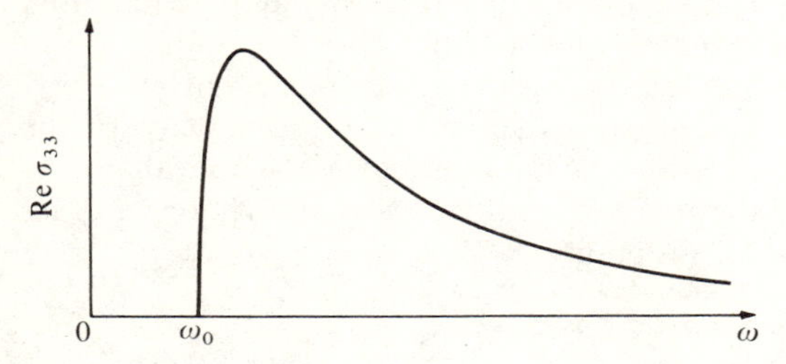

Equation (3.95) is also directly applicable to liquid metals.[1] The sum over lattice wavenumbers becomes an integration over wavenumber q with $|\langle \mathbf{k} + \mathbf{q}|W|\mathbf{k}\rangle|^2$ going to $S^*(q)S(q)|\langle \mathbf{k} + \mathbf{q}|w|\mathbf{k}\rangle|^2$. The counterpart of interband absorption in the solid arises from the peaks in $S^*(q)S(q)$ in the liquid, and the calculation includes automatically the Drude contribution with a scattering rate arising from the disorder in the liquid.

Interband absorption takes a very much different form in the polyvalent metals. Transitions arising from Brillouin zone faces that do not intersect the Fermi surface are describable by Eq. (3.95), but those arising from Brillouin zone faces that intersect the Fermi surface require special treatment.

Here again the absorption does not ordinarily arise at symmetry points. The geometry is a little difficult to sort out. We may consider the simplest case where a free-electron-like sphere overlaps a Brillouin zone face. This is illustrated in Fig. 3.30. We have first shown the resulting Fermi surface in the extended-zone scheme and have then gone to the reduced Brillouin zone. Interband absorption can occur only where states are occupied in the lower band and unoccupied in the upper band. Since both bands are occupied at L no absorption can occur at that point. Absorption, however, can occur at the zone face in the region indicated in the figure. Looking down on the zone face we see that this forms a circular ribbon on the zone face. Absorption can also occur further into the zone but the bands are separating rapidly in that region and the absorption edge will occur at the band gap evaluated at the ribbon. In our studies of the band structure of simple metals we saw that the band gap in this region is given by just twice the appropriate OPW form factor (for simple metals with one atom per primitive cell). Thus we expect absorption edges to occur at energies equal to twice the form factors for the Brillouin zone faces that intersect the Fermi surface. The form of the edge is readily obtained[2] and is given by

$$\operatorname{Im} \sigma_{33} = \frac{A}{(\hbar\omega - 2|\langle \mathbf{k} + \mathbf{q}|W|\mathbf{k}\rangle|)^{1/2}} \qquad \text{for } \hbar\omega > 2|\langle \mathbf{k} + \mathbf{q}|W|\mathbf{k}\rangle|$$

and A is a constant. In contrast to the monovalent metals, here the position of the absorption edge is sensitive to the value of the pseudopotential itself and the calculation is correspondingly less reliable.

Interband absorption of this type should occur generally in polyvalent metals and should give direct experimental measurements of the OPW form factors. This picture seems to be confirmed in the optical properties of aluminum but the interpretation of the optical absorption in other simple metals is incomplete at this point.

[1] Animalu, *op. cit.*

[2] W. A. Harrison, *Phys. Rev.*, **147**:467 (1966).

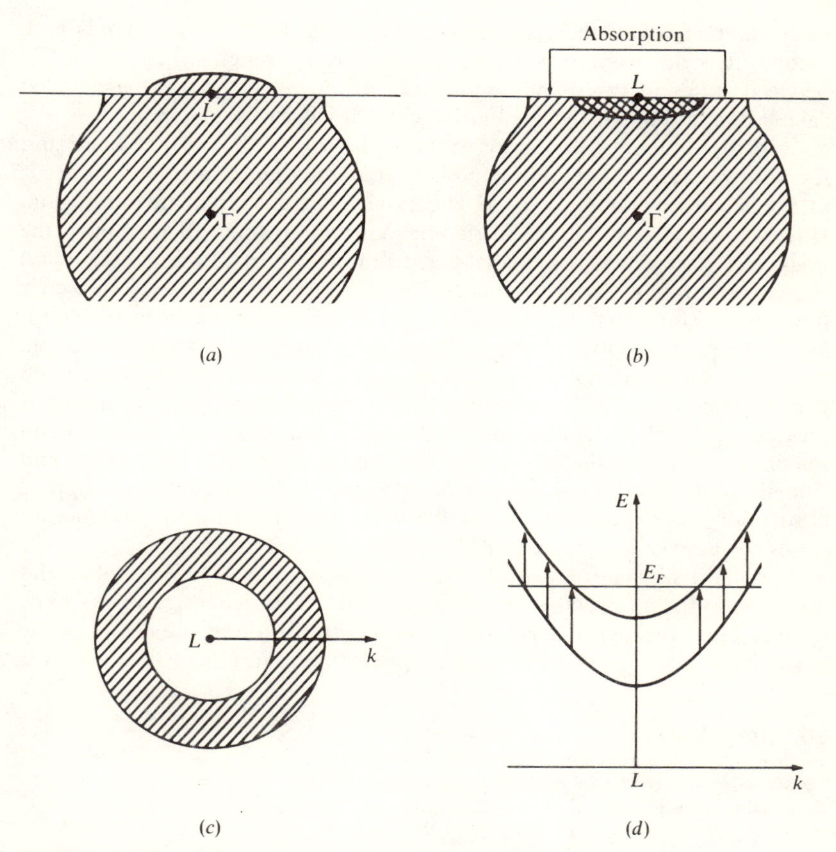

Fig. 3.30 (*a*) *Occupied states in a polyvalent metal where the Fermi surface overlaps a Bragg plane. Interband optical absorption occurs with electron transfer between states of the same wavenumber in the reduced zone* (*b*), *and can occur only at regions where states are occupied in one band but not the other.* (*c*) *Intersection of the Fermi surface with the Bragg plane, showing the region of minimum energy absorption.* (*d*) *Energy bands plotted along the plane and the allowed transitions.*

In a semiconductor with small numbers of carriers (or in an insulator), each band is either full or empty and only interband absorption is possible. The concepts we have used in interband absorption in metals, and also the Kubo-Greenwood formula, Eq. (3.87), remains valid. It is natural to describe the bands in a reduced-zone scheme, and symmetry requires that transitions be vertical, as illustrated in Fig. 3.31. Here, however, the calculation is much more difficult. First, there is no simple analytic form for the bands and, second, it is not easy to compute the oscillator strengths, $\langle \psi_k^{(n)} | \nabla | \psi_k^{(m)} \rangle$, between states in two bands, numbered n and m. Although the detailed calculation is difficult the qualitative behavior is easily understood.

If the difference in energy between the top of the valence band and the bottom of the conduction band is E_g, certainly no absorption can occur at frequencies below $\omega_g = E_g/\hbar$. The crystal is transparent in this region. Had we calculated the full dielectric constant (rather than simply the imaginary part which is proportional to the real part of the conductivity) the corresponding *virtual* transitions would have led to a real dielectric function different from unity and therefore a refractive index different from 1. We find refraction of the light at these low frequencies but no absorption.

Let us now extend the frequencies above that associated with the band gap. Consider a semiconductor in which the conduction-band minimum is located at a different point in the Brillouin zone than the valence-band maximum. This was the case in Fig. 3.31 and is the case in germanium and silicon. With $\hbar\omega$ slightly greater than E_g we will still obtain no optical absorption by vertical transitions. We will return later to possible mechanisms for nonvertical transitions.

At still higher frequencies direct absorption becomes possible and gives rise to strong structure in the absorptivity of semiconductors. We may, for example, consider absorption by vertical transition at the center of the zone. Here we note that the matrix element of Eq. (3.87), the oscillator

Fig. 3.31 Schematic energy bands in a semiconductor showing a vertical transition. Note that all vertical transitions require greater energy than the band gap E_g for these bands.

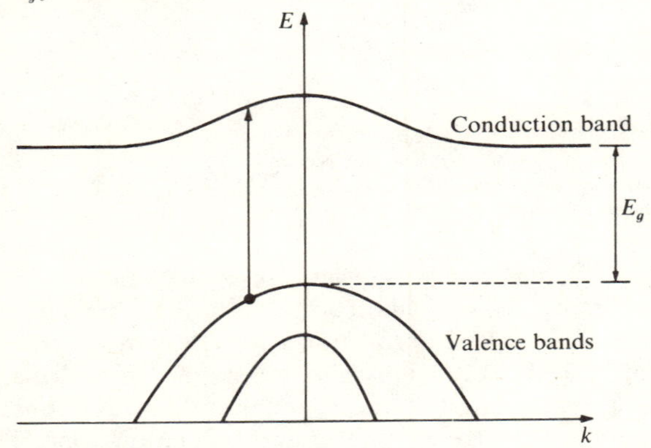

strength, is of the same form as the matrix elements that entered $\mathbf{k} \cdot \mathbf{p}$ perturbation theory. By knowing the symmetry of the wavefunctions at Γ in the different bands we may determine whether such a transition is forbidden or allowed. If it is allowed we may expect an absorption edge to occur at the corresponding energy or perhaps a singularity of some other type in the absorptivity. If, for example, the energy gap decreases as we move away from Γ, there should be a sharp drop in the absorption at the energy corresponding to the band gap at Γ. In any case, singularities are expected in the optical conductivity which reflect the band gaps at symmetry points, so optical studies would seem an ideal way of studying band structures in semiconductors. Unfortunately, the predominant structure in many cases does not arise from energy bands near the symmetry points (e.g., Kane)[1] but from large regions of wavenumber space where the bands are nearly parallel. This was also the case in our discussion of interband absorption in the polyvalent metals, and the tendencies to parallel bands in semiconductors may ultimately have the same origin. Experimental studies of optical absorption in semiconductors unquestionably give information about the energy bands, but the interpretation is difficult and often ambiguous.

In our discussion of interband absorption, we have taken the transitions to be vertical, as required by translational symmetry. This assumption will break down if an electron creates a lattice vibration in the course of the transition. Such a transition is called *indirect* and may be viewed as a second-order process in which an electron makes a virtual vertical transition and then emits a phonon (quantized lattice vibration) taking it to an energy conserving state. This is illustrated in Fig. 3.32. Phonon energies are generally less than 0.1 ev, small compared to the photon energies, so the primary role of the phonon is in transferring momentum. Indirect transitions are allowed at all energies above the gap E_g. Because the matrix element for phonon emission (or absorption) depends upon temperature, the strength of indirect absorption edges depends upon temperature and can be made quite small at low temperatures.

Static imperfections can also take up momentum and relax the requirement of vertical transitions. In fact, the transitions that we obtained from the Kubo-Greenwood formula when we went to the Drude formula correspond to such nonvertical (but in that case *intraband*) transitions. In addition, optical absorption frequently occurs near a crystalline surface and we might expect the surface also to take up momentum during the transition. It may also be that nonvertical transitions occur because of electron-electron interactions which are beyond the one-electron approximation we are utilizing here. In any of these cases which do not involve lattice vibrations, transitions are called *nondirect*. In spite of all these possibilities, in almost all

[1] E. O. Kane, *Phys. Rev.*, **146**:558 (1966).

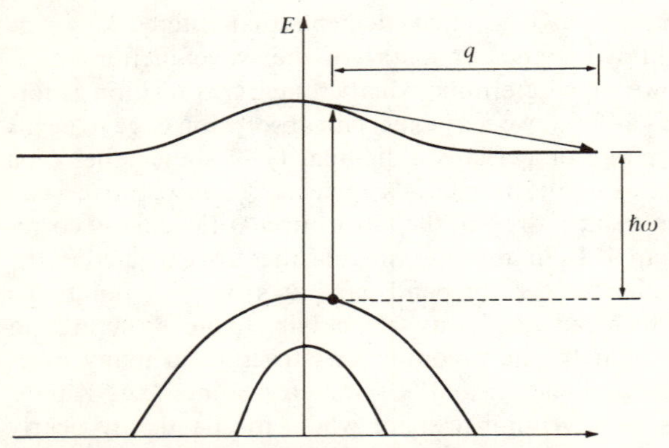

Fig. 3.32 An indirect transition shown as a virtual vertical transition followed by a phonon emission. The ħω is the energy of the absorbed photon (if we neglect the small energy of the phonon), and q is the wavenumber of the emitted phonon.

cases the observed interband optical absorption is found to be consistent with the assumption of vertical transitions.

Interband absorption occurs also in the transition metals and in the noble metals due to the presence of the *d* bands. In many cases these bands are quite flat and can give rise to sharp absorption. Because of this flatness much of the interpretation of the experimental absorption spectra could be equally well made on the basis of vertical transitions or nondirect transitions. In the noble metals there is strong absorption which is believed to come from transitions from the *d* band to the conduction band. These absorptions in the blue are responsible for the reddish color of both copper and gold.

5.5 Photoelectric emission The interpretation of optical absorption is made difficult by the fact that one has no way of deciding which portion of the band structure has caused an observed absorption edge unless the band structure is very simple or very accurately and completely known. Studies of photoelectric emission greatly narrow down the possibilities.[1]

If the energy of the excited electron is greater than the energy of an electron at rest in the vacuum outside the specimen, there is some chance that the electron will emerge from the specimen. This is simply photoelectric emission. By measuring the energy of such an electron, one learns the final-state energy ϵ_k', in addition to the energy of the transition $\hbar\omega =$

[1] See, e.g., F. Abeles (ed.), "Optical Properties and Electronic Structure of Metals and Alloys," C. N. Bergland, p. 285; W. E. Spicer, p. 296. North-Holland, Amsterdam, 1966.

$\epsilon_{k'} - \epsilon_k$. This aids greatly the association of absorption edges with specific features of the band structure and, by providing a second variable, $\epsilon_{k'}$, it allows much more detailed information than that provided by optical properties alone.

An additional uncertainty, the probability of emission of an excited electron as a function of **k** and band index, has been introduced. Even ignoring this complication, the interpretation of electron-emission data is difficult without simplifying assumptions. The most extreme simplification is to assume that the oscillator strengths $\langle\psi_{k'}|\nabla|\psi_k\rangle$ are completely independent of **k** and **k**′. This assumes, among other things, the dominance of nondirect transitions. If this simplification is made, Eq. (3.87) for the conductivity becomes

$$\text{Re}\,\sigma = \frac{A}{\omega}\sum_{k,k'}\delta(\epsilon_{k'} - \epsilon_k - \hbar\omega)$$

where A is a constant. As we indicated following Eq. (3.87), the conductivity can be thought of as a superposition of transitions or absorption processes. Thus we may define a partial conductivity $\sigma(\epsilon',\omega)$ such that $\sigma(\epsilon',\omega)\,d\epsilon'$ is the contribution to the conductivity (at frequency ω) arising from electrons excited to energies in the range $d\epsilon'$. Clearly

$$\sigma(\epsilon',\omega) = \frac{A}{\omega}n(\epsilon')\sum_k\delta(\epsilon_{k'} - \epsilon_k - \hbar\omega)$$

where $n(\epsilon')$ is the density of states per unit energy in the solid. But $\sum_k\delta(\epsilon_{k'} - \epsilon_k - \hbar\omega) = \int n(\epsilon)\,\delta(\epsilon - \epsilon' + \hbar\omega)\,d\epsilon = n(\epsilon' - \hbar\omega)$. Thus

$$\sigma(\epsilon',\omega) = \frac{A}{\omega}\,n(\epsilon')n(\epsilon' - \hbar\omega)$$

For this very crude approximation the partial conductivity $\sigma(\epsilon',\omega)$, which is proportional to the number of electrons emitted with energy ϵ' by light of frequency ω, depends directly upon the joint density of initial and final states, $n(\epsilon')n(\epsilon' - \hbar\omega)$. Experimental emission densities as a function of ϵ' and ω can be interpreted to give separately the density of occupied states [from $n(\epsilon' - \hbar\omega)$] and the density of unoccupied states [from $n(\epsilon')$].

However, the assumption of constant oscillator strengths is so arbitrary and implausible that the $n(\epsilon)$ so determined are generally called *optical densities of states*, serving as a reminder that they may bear very little relation to the true densities of states.

5.6 Color centers and the Franck-Condon principle Our discussion has centered upon the absorption in perfect crystals. In metals the effects arising from the conduction electrons and in some cases the d-band states tend to swamp any effects arising from defects or impurities. In insulators, on the other hand, which are transparent in some wavelength regions, the impurities and defects may make themselves strongly felt.

The elementary theory of absorption by impurities in insulators is quite simple. The electronic states associated with these impurities are very much the same as those in the free atoms though the states may be split and shifted by the presence of the crystalline field. Thus the absorption by electrons in impurity states can be treated just as the corresponding absorption in free atoms. Impurities such as gold in an otherwise transparent crystal lead to sharp absorption lines and bright colors. We will return to a discussion of impurity states when we describe lasers.

Optical absorption arising from defects such as vacancies in insulators has received extensive study. It provided one of the earliest tools for the study of crystalline defects. The most familiar such defect is the F center in alkali halides. This center is a halide vacancy. The missing negative ion leaves a positive potential sufficient to bind an electron. The halide vacancy with its bound electron constitutes the F center. The first excited state of the F center lies a volt or two above the ground state and causes an easily observable absorption band at that energy. Similarly divacancies and other defects cause characteristic absorption bands. The study of these defects, using optical properties and other probes, is a large field in its own right (e.g., Brown)[1] but not one which we will pursue further.

F centers do, however, illustrate nicely a feature of electron-lattice interaction of such conceptual importance in solids that we should discuss it. Though the absorption by an F center occurs at perhaps 2 volts, the subsequent emission occurs at only 1 volt. Reemission at a lower energy is called *fluorescence* and the shift in frequency is called the *Stokes shift*. It is understandable in terms of the distortion of the lattice near the center. We will first give a qualitative description of the effect and then examine the picture more closely.

We characterize the distortion of the lattice by a single parameter x, which might be the magnitude of radial displacements. For the moment we let this be a classical variable. We could imagine computing the lowest-lying electronic state as a function of the lattice-distortion parameter x. We would obtain a total energy (including the elastic energy of distortion) as a function of x as shown in Fig. 3.33. The ground state of the system then corresponds to the lowest-lying electronic state and x with the value x_0 at which the total energy is a minimum. We could similarly construct the first excited electronic state as a function of x. Because of the interaction between the electron and its neighbors we would expect the minimum in the

[1] F. C. Brown, "The Physics of Solids," Benjamin, New York, 1967.

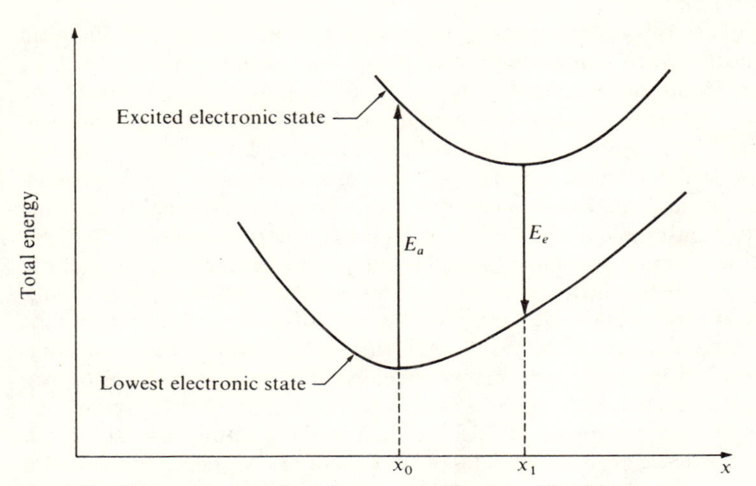

Fig. 3.33 *The total energy of an F center as a function of local lattice distortion x. If the lattice relaxes after excitation at energy E_a, the energy of emission E_e will be lower than E_a.*

energy for the excited state to occur at a different value of x, again as shown in Fig. 3.33.

We now place the system in the ground state with $x = x_0$ and introduce a radiation field. The Franck-Condon principle states, in the framework given here, that the electronic transition will occur vertically in the figure. The idea is that the electron is so light and fast that the transition will occur before the ions have had a chance to move or to change their velocity. Thus a photon energy E_a, shown in the figure, is required for the transition. However, if the excited state is of sufficiently long life, we may expect the lattice to readjust such that $x = x_1$ before emission occurs. The emission will again be vertical and the emitted photon energy will be E_e which can be seen to always be less than E_a.

This properly accounts for the Stokes shift. However, we relied on the Franck-Condon principle and in attempting to justify it the question immediately arises: *How fast does the transition occur*? This same question arises in the Mössbauer effect, in tunneling in polarizable media, and in many other phenomena. However, the question as stated is not meaningful since it does not have a well-defined experimental answer, even in principle. It can be made well defined in terms of an experiment and of course optical absorption is a suitable experiment. Thus the question "How fast does the transition occur?" can be answered only by solving the whole problem in the first place.

We do this by describing the electron and the lattice quantum mechanically when computing the transition probability. We will pick as simple a Hamiltonian as possible which still contains the physics of the problem. The Hamiltonian will contain H_{el} which includes the electron kinetic energy and the potential in the undistorted lattice ($x = 0$). It will contain a lattice Hamiltonian H_{lat} which consists of elastic energy, proportional to x^2, and the kinetic energy of the lattice, proportional to $\dot{x}^2$; this is simply a harmonic-oscillator Hamiltonian in the coordinate x.

Now we must also introduce a coupling between the electron and the distortion since this is the origin of the Stokes shift. It must depend upon x and it must depend upon the energy (or the state) of the electron. The simplest way to introduce such a term is in the form $-\mu H_{el}x$, where μ is a constant. This form is somewhat arbitrary but will be seen to lead to total energy curves such as shown in Fig. 3.33.

Finally, we introduce the electric field of the exciting light $H_{\mathscr{E}}$ which depends on the electron coordinates (or momenta) but we neglect interaction of the light with the lattice distortion. (It will not in fact depend upon x if the contemplated distortion has inversion symmetry.) To do the transition-rate problem correctly we would need to introduce a range of light frequencies to absorb the energy delta function in the Golden Rule. However, we will seek only correction factors here and therefore need not worry explicitly about this.

Thus the postulated Hamiltonian is

$$H = H_{el} + H_{lat} - \mu H_{el}x + H_{\mathscr{E}}$$

We first consider the Hamiltonian without the electric field to find the eigenstates. We will then introduce $H_{\mathscr{E}}$ as a perturbation in order to compute the transitions between eigenstates. In the first step we will see that the eigenstates without the electric field can be represented in terms of Fig. 3.33.

The eigenstates of the system are functions of the electron position $\mathbf{r}$ and the distortion parameter x. However, for the simple Hamiltonian we have chosen they can be separated,

$$\Psi(\mathbf{r},x) = \psi(\mathbf{r})\varphi(x) \tag{3.97}$$

since by operating on Eq. (3.97) with the Hamiltonian $H' = H - H_{\mathscr{E}}$, we obtain

$$H'\Psi(\mathbf{r},x) = E_{el}\psi(\mathbf{r})\varphi(x) + \psi(\mathbf{r})H_{lat}\varphi(x) - \mu E_{el}\psi(\mathbf{r})x\varphi(x) \tag{3.98}$$

where E_{el} is the electronic eigenvalue $H_{el}\psi(\mathbf{r}) = E_{el}\psi(\mathbf{r})$. But the right side of Eq. (3.98) can be written $(E_{el} + E_{lat})\psi(\mathbf{r})\varphi(x)$ if

$$(H_{lat} - \mu E_{el}x)\varphi(x) = E_{lat}\varphi(x) \tag{3.99}$$

Eq. (3.99), which does not contain the electronic coordinates, can be solved immediately. The H_{lat} is a harmonic-oscillator Hamiltonian containing a potential energy which may be written $\kappa x^2/2$. The additional term $-\mu E_{\text{el}}x$ simply shifts the origin of the potential energy expression,

$$\frac{\kappa x^2}{2} - \mu E_{\text{el}} x = \frac{\kappa(x - \mu E_{\text{el}}/\kappa)^2}{2} - \frac{(\mu E_{\text{el}})^2}{2\kappa} \tag{3.100}$$

Thus the eigenstates of Eq. (3.99) are harmonic-oscillator wavefunctions centered at $x = \mu E_{\text{el}}/\kappa$ and the eigenvalues are $\hbar\omega_{\text{lat}}(n + \frac{1}{2}) - (\mu E_{\text{el}})^2/2\kappa$ where ω_{lat} is the classical frequency of the harmonic oscillator and n is the quantum number. We have found eigenstates of H',

$$H'\psi(\mathbf{r})\varphi_n\left(x - \frac{\mu E_{\text{el}}}{\kappa}\right) = \left[E_{\text{el}} - \frac{(\mu E_{\text{el}})^2}{2\kappa} + \hbar\omega_{\text{lat}}(n + \tfrac{1}{2})\right] \psi(\mathbf{r})\varphi_n\left(x - \frac{\mu E_{\text{el}}}{\kappa}\right)$$

We may now make an identification with Fig. 3.33. Let the lowest two eigenvalues E_{el} be E_0 and E_1. From the potential, Eq. (3.100), we see that the minimum total energies for these two states are at $x_0 = \mu E_0/\kappa$ and $x_1 = \mu E_1/\kappa$. In the quantum-mechanical description we have replaced the classical coordinate x by a dependence of the wavefunction on x. The system is the same; only the description has changed.

The optical absorption represents transitions in which the electron state changes from ψ_0 to ψ_1. We will take the ground state of the system $\psi_0(\mathbf{r})\varphi_0(x - x_0)$ as the initial state and consider transitions to all excited eigenstates $\psi_1(\mathbf{r})\varphi_n(x - x_1)$. It is useful first to perform the calculation for the case in which the electron-lattice coupling constant μ is zero. Then $x_0 = x_1$ and the total energies are $E_{\text{el}} + \hbar\omega_{\text{lat}}(n + \frac{1}{2})$. The matrix element for a transition is

$$\langle \psi_1(\mathbf{r})\varphi_n(x) | H_{\mathscr{E}} | \psi_0(\mathbf{r})\varphi_0(x) \rangle$$

The $H_{\mathscr{E}}$ is a function only of the electron coordinates so we may integrate out the coordinate x directly, $\langle \varphi_n(x) | \varphi_0(x) \rangle = \delta_{n0}$ by the orthonormality of the harmonic-oscillator states, so transitions only occur to final states in which the lattice is in its ground state. This would of course not have been the case if we had allowed $H_{\mathscr{E}}$ to depend upon x. Then direct absorption of light by the lattice would have been possible. However, we are not interested at the moment in this contribution to the absorption, so the corresponding term in the Hamiltonian was omitted. Here we are left with the matrix element $\langle \psi_1(\mathbf{r}) | H_{\mathscr{E}} | \psi_0(\mathbf{r}) \rangle$ just as if the possibility of distortion had been eliminated. Furthermore the energy delta function becomes

$\delta(E_1 - E_0 - \hbar\omega)$. The proper result has been obtained in the absence of coupling. We now seek the changes arising from the coupling.

Again the initial state is the ground state, which is $\psi_0(\mathbf{r})\varphi_0(x - x_0)$ and the final states are $\psi_1(\mathbf{r})\varphi_n(x - x_1)$. We may again evaluate the matrix element, performing the integration over x first.

$$\langle\psi_1(\mathbf{r})\varphi_n(x - x_1)|H_{\mathscr{E}}|\psi_0(\mathbf{r})\varphi_0(x - x_0)\rangle$$
$$= \langle\psi_1(\mathbf{r})|H_{\mathscr{E}}|\psi_0(\mathbf{r})\rangle\langle\varphi_n(x - x_1)|\varphi_0(x - x_0)\rangle$$

The first factor is the same as before, but now the x integration does not give a delta function. Transitions are possible to final states with the lattice in any vibrational state. The energy delta function becomes $\delta(E_1' - E_0' + \hbar\omega_{\text{lat}}n - \hbar\omega)$ where E_0' and E_1' are the electronic energies shifted by the polaronlike contribution, $E_i' = E_i - (\mu E_i)^2/2\kappa$. If we compute the absorption as a function of the frequency of the applied electric field, we find absorption peaks not only at the "no-phonon" energy, $\hbar\omega = E_1' - E_0'$, but also at a series of energies greater than this by $n\hbar\omega_{\text{lat}}$. The relative amplitude of each peak is given by $|\langle\varphi_n(x - x_1)|\varphi_0(x - x_0)\rangle|^2$. To see any relation to the qualitative picture given earlier we must consider these relative amplitudes.

We consider first a transition with $n = 0$. This is a transition from the ground state to the lowest-lying excited state and corresponds in the classical picture to the arrow in Fig. 3.34. Such a transition which we ruled out before is now allowed but its probability is reduced by the square of the overlap of the shifted harmonic-oscillator functions. The physical interpretation of this reduction is very clear. The spread of the harmonic-oscillator functions represents the zero-point fluctuation of the lattice. The transition can occur only if a zero-point fluctuation provides the distortion required to generate the final state with its appropriate lattice distortion. If the required distortions are small or the zero-point excursions large, the transition probability will be near that for the system without electron-lattice distortion. Correspondingly, the probability of transition to a state of $n \neq 0$ will be small because of the near orthogonality of the corresponding oscillator states. In the context of slow and fast transitions discussed earlier, we would say that the transition is slow.

It is interesting to note that in general the sum of squared oscillator strengths over all final states does not change when the final states are modified since, with the initial state, they form a complete set. This is readily demonstrated by summing over all possible final states Ψ_f different from the initial state Ψ_0

$$\sum_f{}' \langle\Psi_0|H_{\mathscr{E}}|\Psi_f\rangle\langle\Psi_f|H_{\mathscr{E}}|\Psi_0\rangle = \langle\Psi_0|H_{\mathscr{E}}\,H_{\mathscr{E}}|\Psi_0\rangle$$
$$- \langle\Psi_0|H_{\mathscr{E}}|\Psi_0\rangle\langle\Psi_0|H_{\mathscr{E}}|\Psi_0\rangle$$

where we have used the completeness relation $\sum_f |\Psi_f\rangle\langle\Psi_f| = 1$. For our case $\langle\Psi_0|H_{\mathscr{E}}|\Psi_0\rangle = 0$ and $\langle\Psi_0|H_{\mathscr{E}}{}^2|\Psi_0\rangle$ is independent of the coupling constant. Such a sum rule can be very useful and powerful. We can say here that if the $n = 0$ transition probability is near that without coupling, the sum rule is "exhausted" and very little oscillator strength is available for $n \neq 0$ transitions.

We may consider the other extreme in which the difference in distortion $x_1 - x_0$ is so large that the necessary zero-point fluctuations are very unlikely. This corresponds more closely to the F-center problem. Then the $n = 0$ transition becomes very unlikely and the sum rule requires that

Fig. 3.34 The classical energy versus distortion is shown at top with a schematic representation of the $n = 0$ transition. The corresponding harmonic-oscillator wavefunctions are shown below. For the choice of parameters here the squared overlap $|\langle\varphi_0(x - x_0)|\varphi_0(x - x_1)\rangle|^2 = 0.2$ so the distortion reduces the probability of the corresponding transition by a factor of 0.2.

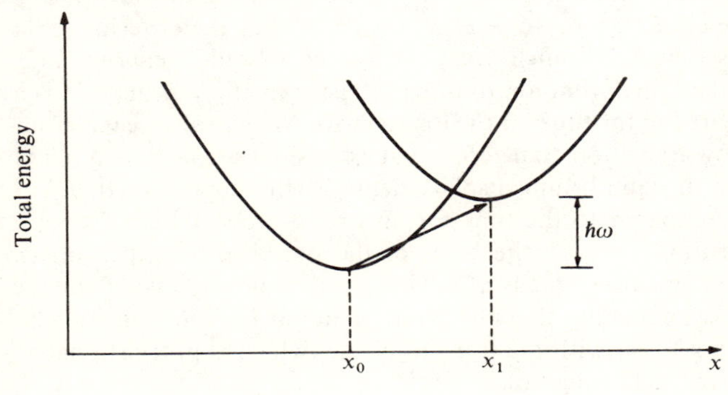

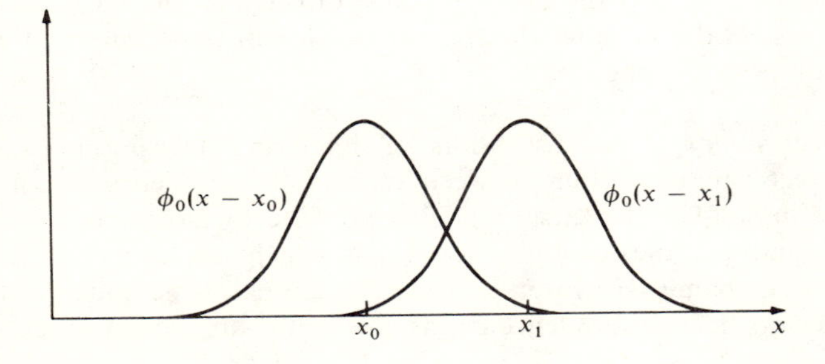

$n \neq 0$ transitions occur. If $x_1 - x_0$ is large, we note that not only is the $\langle \varphi_n(x - x_1)|\varphi_0(x - x_0)\rangle$ overlap small for $n = 0$, but also for small n. Thus the transitions must take place for large n, suggesting that for this case we will obtain the classical limit obtained earlier; i.e., the electronic transition will be "fast" in comparison to the motion of the lattice. We can see this in detail by considering again the overlaps. For large n the envelope of the harmonic-oscillator function is given by the square root of the classical probability distribution for an oscillator with that energy. We see then that the overlap will be large only for n corresponding to a classical amplitude equal to $x_1 - x_0$. This is illustrated in Fig. 3.35. Such states have energy corresponding to the vertical transition E_a of Fig. 3.33. Thus as expected we return to the classical Fig. 3.33 and transitions which are fast in the context of optical absorption. The criterion for this is that the zero-point excursions are small in comparison to the required distortions. Physically the transition occurs before the ions can move and they are left vibrating after the transition.

An alternative statement of the criterion for fast or slow transitions can be made in terms of the energies or times. We obtained the classical result, fast transitions, when a large-n final state occurred, and this $n\hbar\omega_{\text{lat}}$ was equal to the elastic energy $\Delta = \kappa(x_1 - x_0)^2/2$. Thus the criterion for a fast transition is that Δ be much greater than the vibration energy $\hbar\omega_{\text{lat}}$. If we call $1/\omega_{\text{lat}}$ the time for an ion to move, a fast transition means $\hbar/\Delta$ is a short time compared to the time for the ion to move. We can also see that our criterion for a slow transition, that zero-point excursions are large compared to the difference in equilibrium displacements, is the condition that $\hbar/\Delta$ be a long time compared to the time for an ion to move. Thus, if we are to define a transition time on the basis of the polarization experiments, we find that that transition time is $\hbar/\Delta$. The same result obtains, of course, for emission. It is interesting that the "time required for the transition to occur" has nothing to do with the transition rate (which is proportional to $\sigma(\omega)$ and the electric field squared).

If we view the polarizable environment as a device to measure the transition time, we find that the result depends directly on the measuring device. In the case of the F center the transition is found to be fast and the Franck-Condon principle applies.

5.7 X-ray spectroscopy Our discussion of interband transitions was restricted to electronic transitions between various of the valence or conduction bands in solids. Of course it is also possible to have transitions between these bands and the deep-lying core levels, which may or may not be described as energy bands. However, because the core levels lie quite deep, frequently 100 volts below the valence bands, the "light" absorbed in such a

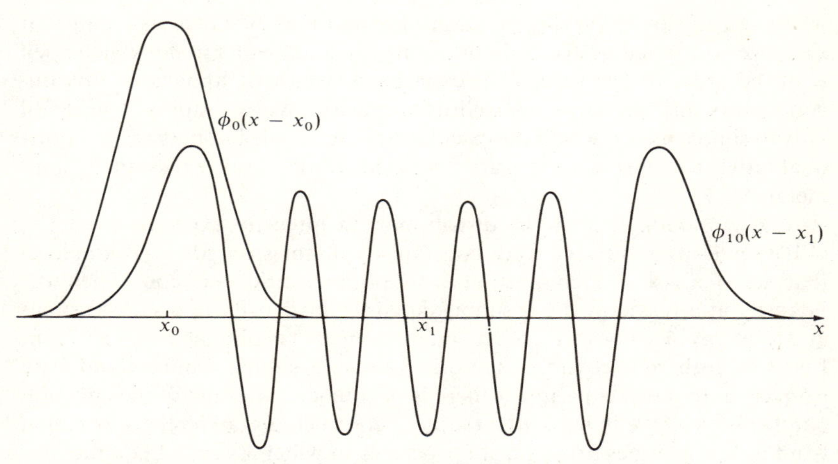

Fig. 3.35 *Harmonic-oscillator functions which become appropriate when the difference between equilibrium distortions becomes large compared to the zero-point excursions. The corresponding electronic transition would leave the lattice with 10 quanta of vibrational energy.*

transition is in fact an x-ray. These are soft x-rays, however, with wavelengths of the order of 100 Å, so the assumption of long wavelengths on an atomic scale remains appropriate and we may use again Eq. (3.87) for the optical conductivity.

For the case of x-rays the optical conductivity can be simplified because of our knowledge of the core states. We consider a particular core state with a known wavefunction ψ_{core} which is the same as in the free atom. (We could alternatively write tight-binding linear combinations of the core states, but this would not affect the results and it is perhaps simpler to think in terms of a core eigenstate localized on a single atom.) We describe x-ray emission since this gives information about the occupied valence states and is frequently of more interest. Thus we may imagine the initial state of the system in which a core state has been emptied (by, e.g., a previous irradiation by x-rays) and consider the subsequent emission. Thus the final one-electron state will be ψ_{core} and we are to sum over occupied initial states which may become empty during the transition. The emitted x-ray intensity due to such events is proportional to the corresponding optical conductivity and we rewrite Eq. (3.87) in terms of the intensity

$$I(\hbar\omega) = A \sum_{\mathbf{k}} |\langle\psi_{\text{core}}|\nabla|\psi_{\mathbf{k}}\rangle|^2 \, \delta(\epsilon_{\text{core}} + \hbar\omega - \epsilon_{\mathbf{k}}) \tag{3.101}$$

where A is a slowly varying function of ω and may be taken as a constant since we will be interested only in a range of a few volts in $\hbar\omega$ which itself is of the order of 100 volts. This expression is directly applicable to semiconductors and insulators as well as to metals. We will apply it in detail only to simple metals where the calculations are simple but may readily draw qualitative conclusions afterward concerning soft x-ray emission in nonmetals.

Let us look first at the distribution in intensity expected from Eq. (3.101) in terms of the very crude constant-oscillator-strength approximation that we discussed in connection with photoelectric emission. Then the intensity at any energy $\hbar\omega$ is simply equal to the density of occupied states at an energy $\hbar\omega$ above ϵ_{core}. An intensity may be obtained as shown in Fig. 3.36. If there were more than one core state, similar bands would arise from each. In a real experiment there is of course smoothing of the emission band at both edges. In particular there is a conspicuous tail on the absorption band at low energies arising from processes in which a second conduction-band electron is excited as the first drops into the core state. This is called the *Auger tail* since the mechanism is the counterpart of Auger processes in atoms.

In addition there are important changes in the shape of the emission band due to variations in the oscillator strength. In the framework of pseudopotential theory it is easy to see the nature of these distortions as we include the effects of the pseudopotential to successively higher orders.

In zero order the pseudowavefunctions become plane waves, but of course the wavefunction entering the matrix element in Eq. (3.101) becomes an orthogonalized plane wave rather than a plane wave. In the context of our discussion of interband optical transitions this is a first-order effect, and we include it in our first approximation. We may write out the matrix element of interest

Fig. 3.36 X-ray intensity in a simple metal, assuming constant oscillator strengths. The dashed line represents schematically the corresponding measured emission band.

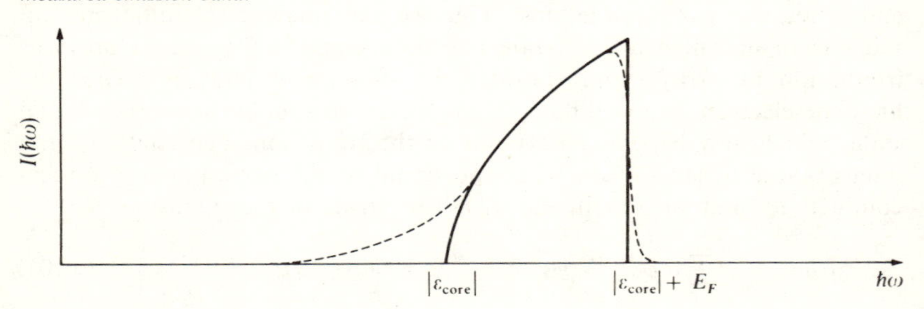

$$\langle\psi_{\text{core}}|\nabla|(1-P)|\mathbf{k}\rangle = \langle\psi_{\text{core}}|\nabla|\mathbf{k}\rangle - \sum_{t,j}\langle\psi_{\text{core}}|\nabla|t,j\rangle\langle t,j|\mathbf{k}\rangle$$
$$= i\mathbf{k}\,\langle\psi_{\text{core}}|\mathbf{k}\rangle - \sum_{t}\langle\psi_{\text{core}}|\nabla|t\rangle\langle t|\mathbf{k}\rangle \qquad (3.102)$$

In the final expression we have neglected any overlap of core functions on different atoms.

It is useful at this stage to distinguish the emission into core s states from the emission into core p states. For an s state $\langle\psi_{\text{core}}|\mathbf{k}\rangle$ is approximately independent of $\mathbf{k}$. In addition the second term can be seen to vary as an odd power of k. Thus the oscillator strength will tend to be proportional to k, as is the density of states per unit energy, so that the intensity given by Eq. (3.101) will tend to vary as k^3 or as $E^{3/2}$ rather than as $E^{1/2}$ as indicated in Fig. 3.36. (Here $E = \hbar^2k^2/2m = \hbar\omega - |\epsilon_{\text{core}}|$.) This result was obtained much earlier by writing the valence-band state as a Bloch function with a periodic factor independent of k. Detailed numerical calculation with Eq. (3.102) gives a result that rises less rapidly than the $3/2$ power of energy but not conspicuously so. Such curves computed with appropriate geometrical sums over states are shown for sodium in Fig. 3.37.

If the core state is a p state then $\langle\psi_{\text{core}}|\mathbf{k}\rangle$ tends to be proportional

Fig. 3.37 Computed x-ray emission intensity (in arbitrary units) for sodium, assuming single-OPW conduction-band states. The E is measured from the emission edge for each type of transition though the photon energy at the edge differs for different transitions. Curves at energies above $E = E_F$ are observable only through x-ray absorption.

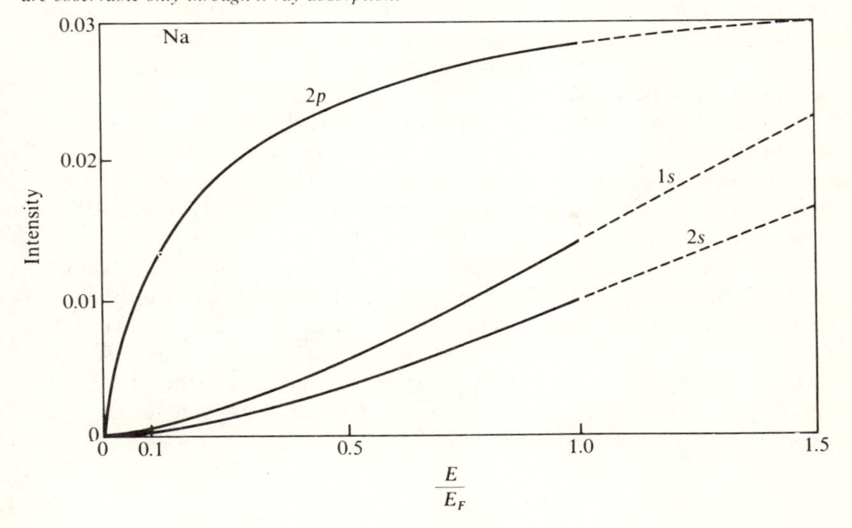

to k and the first term in the oscillator strength varies as k^2. If we omitted the term from orthogonalization we would deduce an intensity varying with the $\frac{5}{2}$ power of energy. However, whenever there is a core p state into which electrons can drop there will also be a core s state and the corresponding contribution to the second term in Eq. (3.102) will be independent of k. Thus at small wavenumbers (for these purposes the Fermi wavenumber is small in the simple metals) the intensity in Eq. (3.101) tends to rise as $E^{1/2}$, as illustrated in Fig. 3.36. The result of a more careful calculation for sodium is shown also in Fig. 3.37. If we attempted to interpret soft x-ray emission with the constant-oscillator-strength model we would have deduced a different density of states in measurements of soft x-ray emission into p states than in experiments with soft x-ray emission into s states.

There is an additional first-order correction to the intensity in Eq. (3.101) which arises from the use of first-order pseudowavefunctions rather than plane-wave pseudowavefunctions. In evaluating this first-order correction we can again, as in the calculation of optical properties, use the pseudowavefunction directly for the matrix elements. In the framework of the perturbation theory it is proper to add the effects due to each Brillouin zone plane. Each of these corrections can be calculated taking appropriate angular averages. There is a divergence in the correction factor which arises because of vanishing energy denominators. However, the corrections should be reliable away from this singularity. In Fig. 3.38 we have shown the result of such a calculation for the effects of the [111] Brillouin zone planes in aluminum, assuming a pseudopotential form factor independent of $\mathbf{k}$. The corrections for each Brillouin zone face are found to be small enough to justify the use of perturbation theory (corrections of the order of 3 percent for each plane). However, when the effects are added for all eight planes they become quite appreciable.

There is also a first-order correction to the density of states since the first-order energy contains a term $\langle \mathbf{k}|w|\mathbf{k}\rangle$ which is energy dependent. This, however, is a rather small correction and does not show any structure. Thus to first order in the pseudopotential the *structure* arises entirely from modifications in the oscillator strength.

Structure in the density of states arises only in second order. Again, we can superimpose the shift in the density of states due to the introduction of each Brillouin zone plane. This can be done using Eq. (2.31) of Sec. 5.7 of Chap. II, which gives the component κ_t of the wavenumber transverse to the lattice wavenumber in terms of the component κ_z parallel to the lattice wavenumber at fixed energy. (See Fig. 3.39.) Atomic units were used in this expression which may be rewritten as

$$\kappa_t^{\ 2} = 2E - \left(\frac{q}{2}\right)^2 - \kappa_z^{\ 2} \mp \sqrt{(q\kappa_z)^2 + 4w^2} \tag{3.103}$$

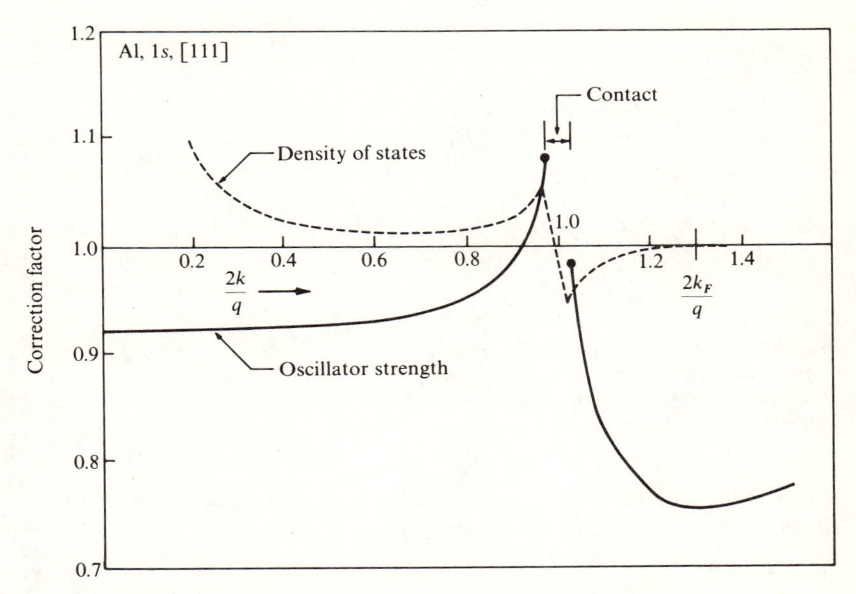

Fig. 3.38 First-order and second-order correction factors to the 1s *absorption intensity in aluminum which arises from the eight [111] Brillouin zone faces. The solid curve is the first-order correction and may be attributed to variations in oscillator strength. The dashed curve is the second-order correction and may be associated with density-of-states variations. Energy,* $h^2k^2/2m$ *measured from the band minimum, increases to the right;* q *is the [111] lattice wavenumber.*

The minus or plus appropriately allows two solutions for some ranges of κ_z. The total volume of a constant-energy surface for fixed E is obtained from the integral

$$\Omega(E) = \pi \int d\kappa_z \kappa_t^{\,2} \tag{3.104}$$

This integral is somewhat awkward but the derivative of $\Omega(E)$ with respect to E is quite simple.

We consider first the integral of Eq. (3.104) for an energy sufficiently low that the Fermi surface does not reach the zone face. Then the plus obtains in Eq. (3.103) over the entire range. The lower and upper limits on κ_z in Eq. (3.104) are written κ_z^- and κ_z^+, respectively, and are obtained by setting $\kappa_t = 0$ in Eq. (3.103). Differentiating the integral for $\Omega(E)$ with respect to energy we obtain derivatives of the limits with respect to energy, but since the integrand vanishes at both limits these do not contribute. We also obtain

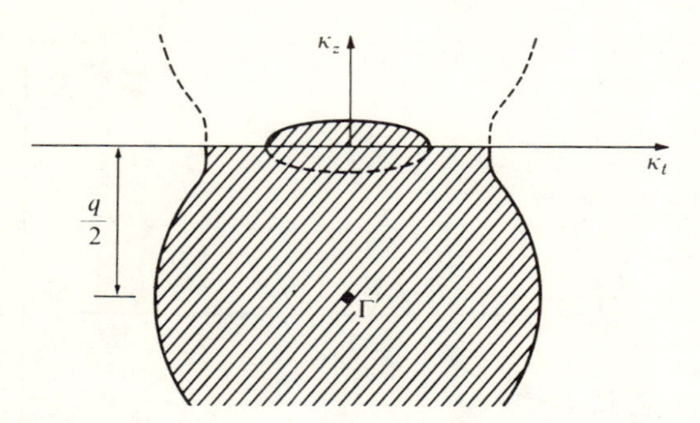

Fig. 3.39 *Coordinate system used in studying constant-energy surfaces near a Brillouin zone face. Solid lines represent surfaces which are used in the calculation of volume.*

the derivative of the integrand with respect to energy, but only the first term in Eq. (3.103) contributes and we see that

$$\frac{d\Omega(E)}{dE} = 2\pi(\kappa_z{}^+ - \kappa_z{}^-)$$

This simple result obtains also when there is contact but no overlap of the zone face since in that case the upper limit becomes $q/2$ and is explicitly independent of energy. Finally, it remains appropriate when there is overlap at the Brillouin zone face, though in that case we must be careful to use the correct signs in evaluating the limits from Eq. (3.103) (again, see Fig. 3.39). By using these exact expressions (for a single Bragg plane and energy-independent pseudopotentials) we avoid any divergence near the zone face and can plot the entire curve as shown in Fig. 3.38. As expected the structure is somewhat weaker than that due to the oscillator strengths since it arises in higher order. Of course, the singularities in the intensity versus energy due to the density of states corrections are real. The point which we wish to stress is that the oscillator-strength effects are larger.

There is a second important distinction that may be made between the first- and second-order corrections. The matrix element of the pseudopotential corresponding to a particular Brillouin zone face may be either positive or negative, depending upon whether the odd or the even state lies lowest at the Brillouin zone face. Wilson,[1] thinking in terms of matrix elements of the true potential, thought naturally that the *s* state would always lie lower and considered only negative matrix elements. In most simple metals the pseudo-

[1] A. H. Wilson, "Theory of Metals," p. 140, Cambridge, London, 1936.

potential matrix elements are in fact found to be positive. Modifications in oscillator-strength corrections, being proportional to the matrix elements of the pseudopotential, change sign with a change in the sign of the pseudopotential. The density-of-states corrections, on the other hand, being proportional to the square of the pseudopotential matrix elements, are independent of the sign of those matrix elements.

These results for the simple metals reinforce misgivings we might have for any system in assuming constant oscillator strengths and interpreting experiments directly in terms of the density of states. In addition it now appears that structure in the spectra may arise from purely many-body effects (See Sec. 5.8) so that particular caution is required in drawing conclusions about electronic structure from the data.

In spite of the difficulty in interpreting x-ray spectra as well as any other optical properties, they can be important tools for studying electronic structure: first, because they give information about the states far from the Fermi energy and, second, because they are applicable to alloys in which the short mean-free paths prevent the use of Fermi surface studies.

In alloys soft x-ray emission arises for transitions into core states of both constituents. It is possible to proceed quite far in the theoretical study of dilute alloys using the concepts that we have used in our considerations of impurity states.[1] Somewhat surprisingly the experiments in most cases give results that are very similar to the results that would be obtained for a system made up of the two components, each in its pure state. This is understandable in the noble and transition metals but is difficult to understand in the simple metals. It may well be a consequence of the fact that the experiments inevitably are performed with concentrated alloys.

5.8 Many-body effects Our discussion of optical properties has been based primarily on the self-consistent-field approximation. We noted, however, that direct use of the Kubo-Greenwood formula with a model of noninteracting electrons leads to an error (even with the inclusion of static screening of the pseudopotential). The calculation instead of the response in the presence of the three perturbations (light, unscreened pseudopotential, and electron-electron interaction) leads to the replacement of the static dielectric function by a frequency-dependent dielectric function. In terms of the processes taking place during absorption (or in terms of perturbation theory) the more refined calculation corresponds to the addition of processes in which, for example, an electron absorbs a photon, collides with a second electron, scatters from the lattice, and collides again with the second electron. The unsettling feature of this result is that this more complicated process,

[1]For a review, see D. J. Fabian (ed.), "Soft X-Ray Band Spectra," W. A. Harrison, p. 227. Academic Press, New York, 1968.

which must be considered higher order in the sense of perturbation theory, nevertheless leads to a correction of the same order in the pseudopotential as the process involving noninteracting electrons. In this case the effect turned out to be small but we cannot be sure that that will be the case in all other processes we might consider. This dilemma has recently been partially resolved, at least for the case of soft x-ray spectra, by work of Nozières and coworkers.[1] Although that work was based upon many-body techniques which we have not discussed, the central results are understandable in the context we have developed here. A more extensive discussion from a point of view similar to ours has been given by Friedel.[2]

One error entailed in the neglect of the electron-electron interaction in treating soft x-ray absorption was described many years ago by Friedel[3] in terms of *Fermi-gas projections*. In calculating the absorption as in Eq. (3.87) we have written a matrix element of ∇ between one-electron states. In fact, of course, even in the self-consistent-field approximation, we should use the many-electron states, product states, or Slater determinants. The assumption required in using the one-electron expression in Eq. (3.87) is that all of the other occupied states (aside from the conduction-band state which is to be emptied and the core state which is to be filled) are the same before and after the transition and the overlap between these other states (the Fermi gas projection) simply gives a factor of unity.

Since in the transition a localized core state becomes filled there will be some change in the remaining states. In particular, in the presence of the core hole, there will be a nonzero phase shift in all of the conduction-band states before transition and no phase shift afterward. When no bound state is formed we may show that typically the overlap for each conduction-band state differs from unity by a term of the order of the reciprocal of the number of atoms, $1/N$. However, the product of approximately N such overlaps enters the Fermi gas projection and it is not clear whether the result is near one or not. Friedel suggested that it was but was unable to evaluate the overlap explicitly. Very recently Anderson[4] addressed this problem in a different context and found that the total overlap is given approximately by[5]

$$\frac{1}{N^{\pi^{-2}\sum_l (2l+1)\delta_l^2}}$$

[1] B. Roulet, J. Gavoret, and P. Nozières, *Phys. Rev.*, **178**:1072 (1969); P. Nozières, J. Gavoret, and B. Roulet, *Phys. Rev.*, **178**:1084 (1969); P. Nozières and C. T. de Dominicis, *Phys. Rev.*, **178**:1097 (1969).

[2] J. Friedel, *Comments on Solid State Physics*, **2**:21 (1969).

[3] J. Friedel, *Phil. Mag.*, **43**:1115 (1952).

[4] P. W. Anderson, *Phys. Rev. Letters*, **18**:1049 (1967).

[5] P. W. Anderson, private communication. The form given in the preceding reference is in error. Note that in the correct form a factor $N^{-1/2}$ is introduced for each bound state (π phase shift) formed; that is, $(N^{-1/2})^{2(2l+1)}$.

The δ_l are characteristic phase shifts for the various angular-momentum quantum numbers.

This is a most remarkable result. Clearly the exponent is positive and nonzero. Thus the Fermi gas projections are of the order of one over the number of atoms present raised to a nonzero power. In the limit of a large number of particles, which we ordinarily consider in solid state physics, the expression approaches zero. We would conclude that these Fermi gas projections profoundly modify the transition probabilities. Of course the result of a full calculation of the intensity for a large system will not depend upon the size of the system as suggested by this expression. This simply reflects the failure of the one-electron approximation.

It is interesting to note the similarity between the Fermi gas projections and the lattice distortion effects that we discussed in Sec. 5.6 of this chapter. In both cases a one-electron matrix element is reduced by a factor which is the overlap of the initial and final wavefunctions of the rest of the system. Here, as in the case of lattice distortion, there is a conservation of oscillator strength which requires that any suppression of the direct transition be compensated by the matrix element to other excited states of the system. In the case of Fermi gas projections the excited states are electronic; i.e., there are many possibilities for optical transitions in which additional electrons are excited from their unperturbed states. It is such additional excitations (which Friedel calls *shake-off* electrons) which cause the Auger tail at the low-energy end of the emission spectrum, and the recent work has demonstrated that they can cause important effects in both the emission and absorption spectrum near the threshold; that is, at x-ray energies near $|\varepsilon_{\text{core}}| + E_F$. (See Fig. 3.36.)

The probability of a particular shake-off excitation can in principle be computed just as we computed the probability of vibrational excitations in Sec. 5.6. Consider states of a particular angular momentum. As an electron in the state of wavenumber k_0 drops into the core state, the electrons in all other occupied states k_i may end up in the corresponding states k_i'. (The primed states are defined to have the same number of nodes as the unprimed states of the same index, but the latter are phase-shifted with respect to the former.) As indicated above, intensity is reduced by the overlap factor which we denote schematically as $\Pi_i\langle k_i|k_i'\rangle^2$. Instead, one of the occupied states k_1 may become unoccupied and a different state k_2' may become occupied during the transition. The probability of this occurring is small, of order $\langle k_1|k_2'\rangle^2 \sim N^{-1}$ times the probability of no shake-off electron; here again N is a large number approaching infinity as the system becomes infinite (but for a particular angular momentum, N is proportional to the cube root of the number of atoms). However, for a given state k_0 being emptied and a given x-ray energy being emitted we must sum over a set of k_1-to-k_2' shake-off transitions of equal energy change, compensating for the factor N^{-1}. Furthermore, the overlap $\langle k_1|k_2'\rangle$ diverges as $(k_1 - k_2)^{-1}$ when k_2 approaches k_1, leading to a singularity in the intensity curve at the threshold. Nozières found that the intensity as a function of energy ϵ measured from threshold had terms of the form ϵ^β, where β depended upon the phase shifts and could be of either sign. Thus these effects can either depress the intensity at the threshold or lead to peaks.

The analysis has not been extended beyond the behavior precisely at threshold nor has it been made quantitative for real systems as yet. Nevertheless, it seems likely that these effects are responsible for much of the observed structure in soft x-ray spectra in metals which had previously been interpreted in terms of densities of states. It is reasonable to expect important extensions of this work in the future.

5.9 Lasers Transitions among impurity levels in insulators have become important in recent years in the construction of lasers. The variety of host materials, as well as of impurity atoms, has allowed great diversity in solid state lasers. The theory is much the same as for gaseous lasers and masers. However, it is a sufficiently important solid state system that we should give some description of the phenomenon.

In order to obtain laser action we must arrange a system of electron levels and external radiation such that for some pair of levels, the level of higher energy has a higher probability of occupation than the level of lower energy. To obtain such a system we imagine a rare-earth atom dissolved in an insulator. The ground state of the impurity will include electrons in the f states ($l = 3$). We denote the ground state by F_0. There will also be excited states which consist again of f states but with differing total quantum numbers. We denote two of these excited states by F_1 and F_2 in order of increasing energy. Finally, there will be excited configurations corresponding to, for example, an electron transferred from an f state to a d state. We call such a state D, and let its energy be greater than that of F_2. With these four levels we can understand the laser action.

At any given temperature the occupation of these four levels will be determined by the equilibrium distribution function. For example, the probability of the state D being occupied on any impurity is $f_D \approx e^{-E_D/KT}$, where E_D is the energy of the state D measured from the ground state. Let us now introduce external radiation with frequency $\hbar\omega_p = E_D$. According to the semiclassical Eq. (3.87) transitions will occur from the F_0 to D at a rate proportional to the intensity of the light and to the probability factor $f(F_0)[1 - f(D)]$. Similarly, any atoms in the state D will be stimulated to drop to the state F_0 at a rate proportional to $f(D)[1 - f(F_0)]$. These expressions would suffice for a discussion of the light applied to or emitted by the laser where the quantum numbers of the radiation field are large, and a classical treatment of the radiation field is applicable. They would, however, yield incorrect answers if applied to thermal radiation where the quantum numbers are small. Thus it is preferable to convert to expressions based upon quantized fields before proceeding. Then, of course, the intensity of light is replaced by the number of photons n_p with energy $\hbar\omega_p$, the probability of absorption of a photon is proportional to n_p and the probability of emission is proportional to $n_p + 1$. (The addition of 1 to n_p gives rise to spontaneous emission.) We then see that the *net* rate of transitions from F_0 to D is given by

$$R_{F_0D} \propto f(F_0)[1 - f(D)]n_p - f(D)[1 - f(F_0)](n_p + 1)$$
$$= [f(F_0) - f(D)]n_p - f(D)[1 - f(F_0)]$$

If n_p corresponded to thermal radiation, $n_p = [e^{\hbar\omega_p/KT} - 1]^{-1}$, rather than external radiation, the system would be in equilibrium. Substituting this n_p and Fermi distributions for the f, we obtain $R_{F_0D} = 0$. If n_p is fixed externally we may seek the steady state by setting $R_{F_0D} = 0$ and solve for $f(D)$ in terms of $f(F_0)$. We obtain

$$f(D) = \frac{n_p f(F_0)}{1 - f(F_0) + n_p}$$

Thus, as we might expect, at sufficiently high intensities (n_p sufficiently large) the probability of occupation of the two states approaches equality. The absorption at this frequency is said to be *saturated*. This condition was obtained by requiring only that $R_{F_0D} = 0$ and remains valid even if other transitions in the system are occurring. Note that use of semiclassical equations corresponds to infinite n_p and therefore complete saturation.

Now let us introduce the possibility of transitions between D and F_2. These may be radiative; they could instead involve the absorption and emission of phonons—vibrational quanta. In any case, if we have not introduced the quanta externally, the corresponding density of quanta will be determined by the temperature of the lattice. By solving the steady-state equations for this transition just as above we find that $f(F_2)$ and $f(D)$ are related by the equilibrium condition for the lattice temperature,

$$f(F_2) = \frac{f_0(F_2) f(D)}{f_0(D)}$$

where the f_0 are equilibrium distribution functions. The probability of occupation of F_2 will be greater than that of D since it is of lower energy. Similarly, if we introduce transitions between F_0 and F_1 we find that $f(F_1) < f(F_0)$ since F_1 is of higher energy.

For the system postulated, the light of frequency ω_p has *inverted the population* of F_2 with respect to F_1.

$$f(F_2) > f(D) \approx f(F_0) > f(F_1)$$

The occupation of F_2 is higher than that of F_1 though its energy is higher. The light has *pumped* electrons from F_1 to F_2. This population inversion is the essence of laser action.

Of course transitions are also possible between any pair of states, not just the particular pairs chosen here. However, if the transitions given here have the greatest rates, the inversion of the populations will occur.

We now allow a weak radiative transition between F_2 and F_1—it should be sufficiently weak not to destroy the population inversion. If then a photon of energy $\hbar\omega_l = E_{F_2} - E_{F_1}$ enters the system, it will be absorbed with a probability proportional to $f(F_1)[1 - f(F_2)]$ but will stimulate the emission of another photon with a probability proportional to $f(F_2)[1 - f(F_1)]$. But since $f(F_2) > f(F_1)$ the emission is more likely. Light of frequency ω_l is amplified rather than attenuated by the medium. This amplification of course transfers electrons from F_2 to F_1, but they will then return to the ground state and are again pumped up to D and then to F_2. This pumping of electrons around the circuit is illustrated in the energy-level diagram shown in Fig. 3.40. The device simply converts energy at the pumping frequency ω_p to energy at a laser frequency ω_l.

The important feature of the laser is that the stimulated emission is into precisely the same mode as that which stimulated it. Thus inverted populations give immense quantum numbers in a very few modes, which corresponds to very intense *coherent* light. In contrast, if a large number of impurities spontaneously emitted photons the light would emerge in all directions and with a range of frequencies. The stimulated emission, which enters the name **l**ight **a**mplification by **s**timulated **e**mission of **r**adiation, is the essential feature.

A laser is constructed with parallel reflecting ends which "tune" the laser to the desired mode. With one mirror having partial transmission, the very intense parallel laser light emerges. The laser described above

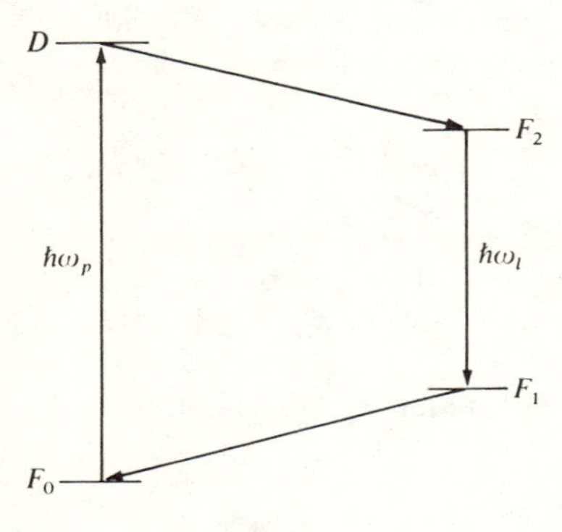

Fig. 3.40 *Level diagram for a four-level laser. Electrons are pumped by an external source of light from the ground state F_0 to an excited state D. They then transfer to the metastable level F_2, emit photons of energy $\hbar\omega_l$, and return to the ground state by way of F_1.*

operates continuously. In many cases the pumping radiation is introduced in a pulse and a pulse of laser light emerges as the energy is converted to the lower frequency. Clearly a number of transition rates and parameters need to be suitably matched if the device is to work. Clearly also there is considerable freedom to vary the system. For example, it is possible to let the levels F_0 and F_1 become the same, forming a three-level laser. The most familiar laser system is made of Al_2O_3 with chromium impurities, the ruby laser. In many cases rare-earth impurities in insulators are used, as in the example we took.

6. Landau Theory of Fermi Liquids[1]

It is desirable before leaving a discussion of the effects of the electron-electron interaction to discuss briefly the remarkable insight into the problem represented by Landau's theory. It is a tribute to the extent of the insight to note that the theory was formulated in order to study the properties of liquid helium three. It is equally applicable to the theory of the electron gas. It also represents phenomenology at its best. To be sure, there are parameters of the system which must be postulated and which cannot be determined within the framework of the theory, in contrast to a microscopic theory such as many-body perturbation theory. Yet in this case it has turned out that microscopic theory has been unable to provide realistic determinations of these parameters and therefore ultimately requires the same dependence upon experiment.

We will speak specifically about a gas of electrons in a uniform positive background and we will neglect effects of electron spin. There are three assumptions which enter the theory, all of which Landau regarded as self-evident. We enumerate these first.

1. There is a one-to-one correspondence between the states of the interacting system and of the noninteracting system. The latter are of course the plane-wave states that we have been discussing in the self-consistent-field approximation. The former are the so-called *quasiparticle states*. There is in particular a low-lying quasiparticle state corresponding to each of the low-lying excitations of the ideal noninteracting Fermi gas.
2. The state of the system is entirely described by the distribution function $n(\mathbf{p},\mathbf{x})$, a function of momentum $\mathbf{p}$ and position $\mathbf{x}$, such that

$$\frac{n(\mathbf{p},\mathbf{x})\, d^3p\, d^3x}{h^3}$$

is equal to the number of quasiparticles in a differential region of momentum

[1] L. D. Landau, *Soviet Phys. JETP*, **3**:920 (1957).

and real space. We have used the notation of Landau. This n corresponds to the distribution function f which we used in our transport calculations. Note that this entails a semiclassical approximation but for a very wide range of problems this is a very good approximation. For the properties of the uniform electron gas this is of course no approximation at all.
3. Since the electrons themselves had a charge $-e$, the quasiparticles will respond to fields as with a charge $-e$.

It is surprising how rich the consequences of these assumptions are. Since the state of the system is entirely describable in terms of n it follows that the total energy of the system must be a *functional* of $n(\mathbf{p},\mathbf{x})$. Note that this is not the assumption that the total energy is a function of the magnitude of n alone as would be true for a noninteracting gas. In this case the energy associated with a particular quasiparticle will depend upon the occupation of other quasiparticle states.

We will wish to make small variations $\delta n(\mathbf{p},\mathbf{x})$ in the distribution function from a reference distribution function $n(\mathbf{p},\mathbf{x})$ which may or may not be the equilibrium distribution function. Then we may write the corresponding variation in energy in the form

$$\delta E = \frac{1}{h^3}\int E[\mathbf{p},\mathbf{x};n(\mathbf{p},\mathbf{x})]\,\delta n(\mathbf{p},\mathbf{x})\,d^3p\,d^3x$$

where $E(\mathbf{p},\mathbf{x};n)$ is the functional derivative of the total energy with respect to $n(\mathbf{p},\mathbf{x})$. It is the increment in energy due to the addition of a quasiparticle at the position $\mathbf{x}$ with the momentum $\mathbf{p}$ to a system which has a distribution $n(\mathbf{p},\mathbf{x})$.

The $E(\mathbf{p},\mathbf{x};n)$ is called the *quasiparticle energy*. It may be regarded as the Hamiltonian of the added particle and therefore the quasiparticle velocity is equal to $\partial E(\mathbf{p},\mathbf{x};n)/\partial\mathbf{p}$. This is not regarded as an additional assumption. It follows further that the current density is given by

$$\mathbf{j}(\mathbf{x}) = \frac{1}{h^3}\int(-e)n(\mathbf{p},\mathbf{x})\,\frac{\partial E(\mathbf{p},\mathbf{x};n)}{\partial\mathbf{p}}\,d^3p \tag{3.105}$$

Since quasiparticles respond to fields as with a charge $-e$, fields may be included in the theory by modifying the Hamiltonian for added quasiparticles in the form

$$E(\mathbf{p},\mathbf{x};n) \rightarrow E\left(\mathbf{p} + \frac{e\mathbf{A}}{c},\mathbf{x};n\right) - e\varphi$$

where $\mathbf{A}$ and φ are the vector and scalar potentials.

It follows in addition that quasiparticles will be conserved since charge must be conserved. Then $n(\mathbf{p},\mathbf{x})$ satisfies a continuity equation and we may derive the Boltzmann equation as in the classical case.

$$\frac{\partial n}{\partial t} + \frac{\partial n}{\partial \mathbf{x}} \cdot \frac{\partial E}{\partial \mathbf{p}} - \frac{\partial n}{\partial \mathbf{p}} \frac{\partial E}{\partial \mathbf{x}} + \left.\frac{\partial n}{\partial t}\right|_{\text{collisions}} = 0$$

The $E(\mathbf{p},\mathbf{x};n)$ enters the theory much as a usual Hamiltonian. However, we should note that the total energy is not directly calculable in terms of $E(\mathbf{p},\mathbf{x};n)$ in the sense which it is in the case of noninteracting particles. That is,

$$E_{\text{total}} \neq \frac{1}{h^3} \int E(\mathbf{p},\mathbf{x};n) n(\mathbf{p},\mathbf{x}) \, d^3p \, d^3x$$

However, we can compute the energy required to add particles one at a time. This is all that is needed since we will always wish to compute first-order distribution functions as we did with the Boltzmann equation and for these small corrections we wish the functional derivative evaluated for the equilibrium distribution.

We can immediately compute the equilibrium distribution itself, which is obtained using the usual variational argument of statistical mechanics. In this calculation we make variational excursions from the equilibrium distribution, and the $E(\mathbf{p},\mathbf{x};n)$ which enters the Fermi distribution is that evaluated at equilibrium. The only assumption that is made is again that $n(\mathbf{p},\mathbf{x})$ describes the system completely.

In many cases we will wish to compute the first-order change in $E(\mathbf{p},\mathbf{x};n)$ due to a first-order change in the distribution function. This would, for example, enter the second-order shift in the energy due to a first-order shift in the distribution function. We write the result immediately

$$\delta E(\mathbf{p},\mathbf{x};n) = \frac{1}{h^3} \int f(\mathbf{p},\mathbf{p}',\mathbf{x},\mathbf{x}';n) \, \delta n(\mathbf{p}',\mathbf{x}') \, d^3p' \, d^3x'$$

where $f(\mathbf{p},\mathbf{p}',\mathbf{x},\mathbf{x}';n)$ is the functional derivative of $E(\mathbf{p},\mathbf{x};n)$ with respect to $n(\mathbf{p},\mathbf{x})$. (Note that this is the second functional derivative of the total energy with respect to $n(\mathbf{p},\mathbf{x})$, evaluated at $n(\mathbf{p},\mathbf{x})$.) The f is called the *quasiparticle interaction* since it is the change in energy of a particle $(\mathbf{p},\mathbf{x})$ due to the addition of a particle $(\mathbf{p}',\mathbf{x}')$. It is of course a functional of $n(\mathbf{p},\mathbf{x})$. The expressions $E(\mathbf{p},\mathbf{x};n_0)$ and $f(\mathbf{p},\mathbf{p}',\mathbf{x},\mathbf{x}';n_0)$ are the two phenomenological functions which enter the theory and upon which the calculation of properties is based; note that both are evaluated at the equilibrium distribution.

At this point there is the danger of simply using the formalism to justify a one-electron description of properties. It is not uncommon to suggest that the many-body effects simply modify the $E(\mathbf{p})$ from what we would find from an energy-band calculation, to ignore the interaction between quasiparticles, and to proceed in the manner that we have used before. The danger in this approach can be seen by examining the current density as we have given it above in Eq. (3.105). We might write the distribution function as an equilibrium zero-order term plus a first-order term and seek to compute the first-order current density. In the classical treatment we could simply insert the first-order distribution in Eq. (3.105) and use the zero-order $\partial E/\partial \mathbf{p}$ since no current was carried by the zero-order distribution. In the theory of Fermi liquids, however, we note that there is a first-order shift in $E(\mathbf{p},\mathbf{x};n)$ due to the first-order term in the distribution function. This can be combined with the zero-order distribution function to give a first-order term in the current. This second term is directly proportional to the quasiparticle interaction. The result of this calculation is obtained rather directly and is given by

$$j = -\frac{e}{h^3}\int \delta n(\mathbf{p},\mathbf{x})\left[\frac{\partial E(\mathbf{p},\mathbf{x};n_0)}{\partial \mathbf{p}} - \frac{1}{h^3}\int \frac{\partial n_0(\mathbf{p}',\mathbf{x}')}{\partial \mathbf{p}'} f(\mathbf{p}',\mathbf{p},\mathbf{x}',\mathbf{x};n_0)\, d^3p'\, d^3x'\right] d^3p$$

where the zeros refer to evaluations at equilibrium. (In this derivation we have done a partial integration on $\mathbf{p}$ and interchanged dummy indices.) The point of course is that the excitation of one quasiparticle modifies the velocity of all other quasiparticles and gives a second contribution to the current in addition to that arising from the velocity of the excited quasiparticle. This is illustrated in Fig. 3.41.

We have seen that the change in current due to the excitation of a quasiparticle differs from the noninteracting value both through a modification of the quasiparticle velocity and through the quasiparticle interaction. The relative magnitude of these two contributions will depend upon the circumstances in question. We may notice, in particular, that if we simply shift the entire equilibrium distribution by a fixed drift velocity, the total current flowing must simply be $-e$ times the density of particles present times the drift velocity. This follows from Galilean invariance; the shift is equivalent to a coordinate transformation to a moving reference frame. The current can also be calculated in the laboratory frame. There are then shifts in the current both due to the quasiparticle excitation and due to the shift in the current carried by other quasiparticles. Thus in this case the many-body effects cancel out completely. This point can be made formally and leads to a relation between the quasiparticle velocity and the quasiparticle interactions (see Prob. 3.14).

From arguments such as that given above we may show that if a

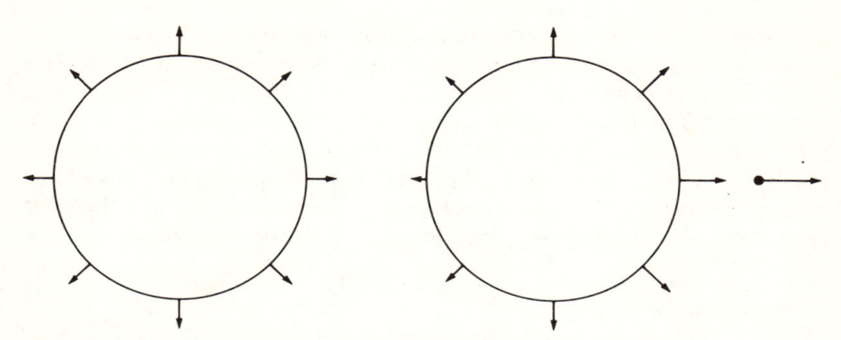

Fig. 3.41 A schematic representation of the contributions to the current in a gas of interacting electrons. At left is the Fermi sphere in the ground state. The contribution to the current from quasiparticles of opposite wavenumbers cancels. On the right, a quasiparticle excitation is added, giving an unbalanced contribution to the current. In addition, the velocity of all other quasiparticles is modified by the excitation so that they also contribute to the current.

current is set up as a uniform drift as above and a magnetic field is then applied perpendicular to this current, the current will rotate with a cyclotron frequency ω_c equal to the value obtained for noninteracting electrons. If, on the other hand, we set up a current by exciting a few quasiparticles out of the equilibrium distribution, and again apply a magnetic field, we will find that the current will rotate with a cyclotron frequency appropriate to a Fermi velocity of $\partial E(\mathbf{p},\mathbf{x};n_0)/\partial \mathbf{p}$. It turns out that the Azbel-Kaner resonance corresponds very closely to this type of excitation. These two frequencies can be significantly different and it is absolutely necessary to make our calculations for the system envisaged in the experiment.

Similarly, we may calculate the electronic specific heat due to quasiparticle excitations. We find that the result is equivalent to that obtained for noninteracting particles but with the density of states replaced by that based upon $E(\mathbf{p},\mathbf{x};n_0)$. On the other hand, if we calculate the Pauli susceptibility, which in a noninteracting picture depends also directly upon the density of states at the Fermi energy, we find that the appropriate density of states is not given by the quasiparticle density of states but by the value closer to that for noninteracting particles.

Any measurement of a property which depends upon the quasiparticle parameters gives information about them. Thus the specific heat masses given in Sec. 5.6 of Chap. II give a measure of the average quasiparticle velocity. The shifts from values obtained from energy-band calculations for that case, however, are thought to arise largely from the electron-phonon interaction rather than the electron-electron interaction. Recently there has been success also in obtaining information about the quasiparticle interactions from experiments on the propagation of high-frequency electromagnetic waves in metals.[1]

The Fermi liquid theory enables us to tell unambiguously which properties are affected by the interaction between electrons and which will not be. It also enables us to relate the one-electron-like parameters associated with one property with those of another. To a large extent the field-theoretic approaches to the many-body problem do very little more for us.

7. *Amorphous Semiconductors*[2]

We return to the description of amorphous semiconductors which was mentioned briefly in Sec. 10.2 of Chap. II. We will describe a model of the electronic structure and then see how the transport and optical properties can be understood in terms of it.

We imagine first a solid semiconductor with each atom surrounded tetrahedrally by four neighbors. We then begin to introduce disorder but retain a predominantly tetrahedral coordination. With disorder there will inevitably be atoms without tetrahedral coordination and we may imagine unfilled, or "dangling," bonds giving rise to states, or traps, in the forbidden band. Similarly, such errors in the coordination may cause localized states to rise into the gap from the valence band. The energies of such states will depend upon the details of the local structure and therefore will be distributed throughout the gap, though predominantly in the neighborhood of the valence- and conduction-band edges. Those lying nearest to the center of the forbidden region will be expected to be well localized; those near the band edges will be spread out. As we increase the disorder we may expect ultimately to obtain a system such as shown schematically in Fig. 3.42.

We may expect enough electrons to be present to fill the states to an energy near the center of what was the forbidden gap. At that energy the density of levels, all well localized, may be fairly low but it is not zero. Thus even if impurities are added, the Fermi level will be shifted only slightly and will remain in a region where the states are all localized. This accounts for the fact that amorphous semiconductors appear to be intrinsic even when they are impure.

We may next understand the conductivity of such a system. Conduction can occur only by electrons in the propagating states well above the Fermi energy or by holes well below. The occupation of these levels is determined

[1] For a recent reference, see P. M. Platzman, W. M. Walsh, Jr., and E-Ni Foo, *Phys. Rev.*, **172**:689 (1968).

[2] M. H. Cohen, H. Fritzsche, and S. R. Ovshinsky, *Phys. Rev. Letters*, **22**:1065 (1969).

Fig. 3.42 A schematic energy-level diagram for an amorphous semiconductor. The energy gap of the crystalline semiconductor is replaced by an energy range in which the levels have low density and are localized, the localization being represented here by the shortness of the lines. At low temperatures the states below the Fermi energy ξ are occupied; those above, unoccupied. Conduction is possible only by electrons or holes that are thermally excited into propagating levels; i.e., levels that lie outside the mobility gap.

by the Fermi function just as if there were not a large number of localized traps and the conductivity is therefore very much the same as in the intrinsic crystalline semiconductor with an energy gap given by the mobility gap shown in Fig. 3.42. At reasonable temperatures this is a very low conductivity.

Although we have seen that it is not possible to raise the conductivity by doping, it may be possible to raise the conductivity by injecting a large number of electrons (or holes); i.e., by applying a large voltage to metal contacts on the amorphous semiconductor. If the voltage is subsequently reduced, these electrons may drop from conducting states into the higher-lying traps and subsequently be easily reexcited into conducting states. Such a nonequilibrium situation could give an occupation of states in the neighborhood of the top of the mobility gap very much as if the Fermi energy had been raised to that region. This is in fact the interpretation which has been given to the *Ovshinsky effect*, which is the observed increase in conductivity of amorphous semiconductors following a pulsed voltage.

We may consider one further property of the amorphous semiconductors, the optical absorption. Since we have levels distributed at all energies we might not expect transparency at low frequencies nor the absorption edge which we have discussed in the crystalline semiconductors. However, it is found experimentally that the optical properties are very similar to those of the crystalline semiconductors. This also may be understood in terms of the model. We note that though there are occupied states just below the

Fermi energy and unoccupied states just above, both are strongly localized and are ordinarily not close to each other in the crystal. Thus the oscillator strength for absorptions between such levels will be zero simply because the initial and final wavefunctions do not overlap each other. We are in fact not likely to find unoccupied states overlapping a given localized state below the Fermi energy except for unoccupied states with sufficiently high energy that they are delocalized, i.e., near the top of the mobility gap. Similarly, we expect no excitation into localized unoccupied states except from occupied states near the bottom of mobility gap. Thus in both cases there will tend to be very little absorption except at frequencies greater than or equal to half of the mobility gap. The fact that the observed absorption edges are quite sharp suggests that the transition with changing energy from localized to propagating states is quite abrupt at the top and the bottom of the mobility gap.

Though this picture of the electronic structure of amorphous semiconductors is very plausible, it is also very much different from the picture presented for crystalline semiconductors. It is surprising that it leads to such similar properties. It is the possible technological importance of the Ovshinsky effect as a switching device that has drawn so much attention to the problem recently.

Problems

3.1 *a.* Consider a semiconductor with band gap Δ and parabolic electron and hole bands corresponding to effective masses m_e and m_h respectively. Making suitable expansions of the Fermi function, obtain the Fermi energy as a function of temperature. Note the direction of the shift with T if $m_h > m_e$.

b. Add a small atom fraction of donor atoms. Note the low-T and high-T limits of the Fermi energy (for $m_h = m_e$). (Don't worry about excited bound states of the donors.) You may be able to see through this without calculation.

3.2 In indium antimonide the electron effective mass is about 0.01 m. Assuming 10^{18} donors per cubic centimeter and a donor-level binding energy of 0.01 ev, find the position of the Fermi level with respect to the conduction-band edge at room temperature.

3.3 *a.* Consider an electron gas in equilibrium for $t < 0$. Let a uniform electric field be introduced at $t = 0$ and held constant. Using the linearized Boltzmann equation in the collision-time approximation, determine the current at all $t > 0$. Note the value of $d\mathbf{j}/dt$ at $t = 0$.

b. Using the same approximation, find the current as a function of time in the presence of a spatially uniform time-varying field $\mathscr{E}\cos\omega t$.

c. Since the current is proportional to $\mathscr{E}$, we may define a complex current due to a field $\mathscr{E}e^{-i\omega t}$ and a complex proportionality constant $\sigma(\omega)$. What is the meaning of the phase of $\sigma(\omega)$? What are the high- and low-frequency limits?

3.4 Show from the steady-state Boltzmann equation in the collision-time approximation for a uniform system *a.* that

$$j = \frac{Ne^2\tau}{m}\mathscr{E}$$

exactly as long as $E = p^2/2m$

or

b. by writing $f = f_0 + f_1 + f_2 + f_3 + \cdots$ in increasing order of $\mathscr{E}$, evaluate f_0, f_1, f_2, and f_3 and show that only f_1 contributes to the current.

3.5 Imagine two one-dimensional solids. We construct states with vanishing-boundary conditions on the surface so the states are of the form of sines rather than complex exponentials, and the number of states in a wavenumber range $d\kappa$ is $(2L/\pi)\,d\kappa$ with the 2 for spin, L is the length of each solid, and κ is always positive. Let the bands measured from the Fermi energy be as below.

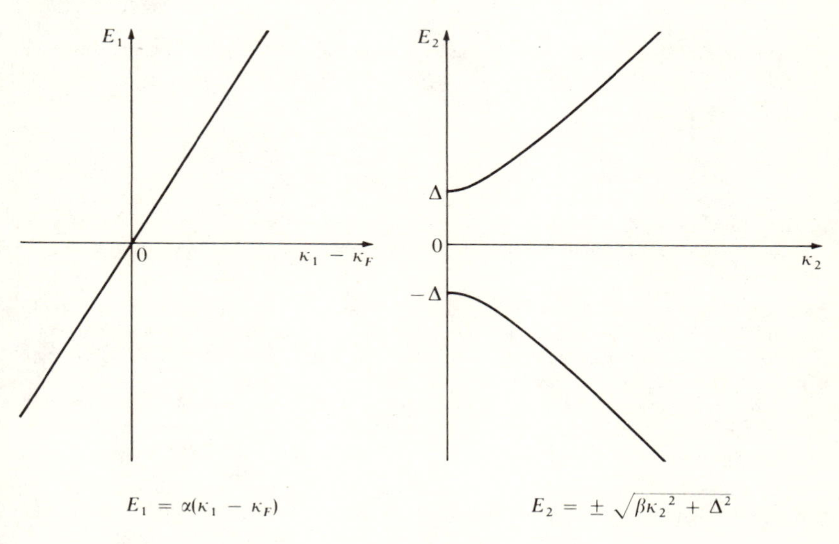

$$E_1 = \alpha(\kappa_1 - \kappa_F) \qquad\qquad E_2 = \pm\sqrt{\beta\kappa_2{}^2 + \Delta^2}$$

Evaluate the tunneling current between the two as a function of applied voltage

$$J = \sum_{\kappa_1} eP_{1,2}(\kappa_1)[f(\kappa_1) - f(\kappa_2)]$$

where the $f(\kappa)$ are Fermi distributions at $T = 0$. Assume the tunneling matrix element $T_{\kappa_1\kappa_2}$ is a constant for electrons of the same spin. Sketch the result.

3.6 Consider two identical one-dimensional metals with parabolic bands at zero temperature.

a. Compute the tunneling current at all voltages assuming the transmission coefficient P is independent of energy.

b. By noting the low-voltage and high-voltage limit, sketch the same current versus voltage when the tunneling matrix element is assumed independent of energy.

(The factors of 2 are tricky, but may be discarded for the problem.)

3.7 Consider a *p-n* junction, shown on p. 362. Fields vary only in the x direction though the

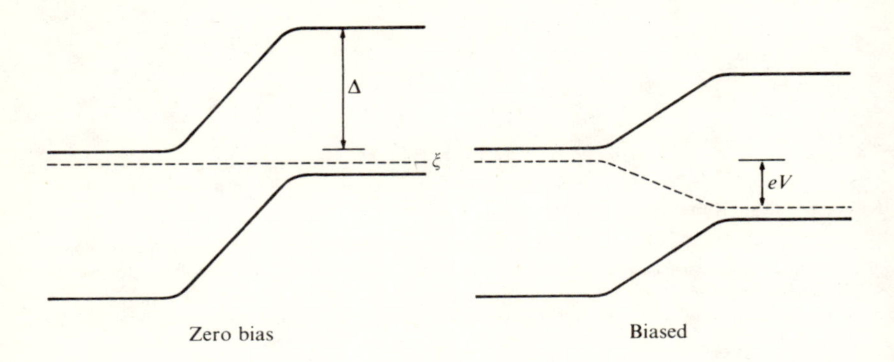

Zero bias Biased

system is three dimensional. The bands on either side are given by $E = p^2/2m +$ band minimum. Consider only electrons (not holes). If the density of electrons on the left is

$$n_0 = \frac{2}{h^3}\iiint Ae^{-p^2/2mKT}\,dp_x\,dp_y\,dp_z$$

a. What is the density on the right?

b. Compute the current as a function of V. Note an electron can cross to the right only if $p_x{}^2/2m > \Delta - eV$.

c. Sketch the result and compute the ratio of current with $V = 0.1$ ev to that with $V = -0.1$ ev at room temperature.

3.8 Consider a metal with ion density $n_0 - \delta n$ for $x < 0$, and ion density $n_0 + \delta n$ (both constant) for $x > 0$. Using the linearized (around n_0 with δn first order) Fermi-Thomas approximation, find the potential self-consistently for all x and sketch the result. Note at $x = 0$ there is no surface charge, so dV/dx as well as V is continuous.

The result corresponds to charge neutrality at large distances from the interface but nonzero fields locally. It also describes qualitatively the surface potentials at a metal-vacuum interface, but is quantitatively weak since the corresponding δn is not small compared to n_0.

3.9 *a*. Consider a conductor in a uniform magnetic field H_A in the z direction. For an electric field with low frequency and long wavelength, the electronic current and the field are related by

$$\mathscr{E} = \frac{1}{\sigma}\mathbf{j}_1 - R\mathbf{j}_1 \times \mathbf{H}$$

where $\mathscr{E}$, $\mathbf{H}$, and $\mathbf{j}_1$ are functions of $\mathbf{r}$ and t.

Consider an applied transverse circularly polarized electric field propagating along $\mathbf{H}_A$.

$$\mathscr{E} = -\frac{1}{c}\frac{\partial \mathbf{A}}{\partial t}, \qquad \mathbf{A} = \frac{(\hat{x} + i\hat{y})}{\sqrt{2}}A_0 e^{i(qz - \omega t)}$$

This implies that some applied current $\mathbf{j}_0$ (perhaps an ion current due to a sound wave) must be present since Maxwell's equations lead to

$$-\nabla^2\mathbf{A} + \frac{1}{c^2}\frac{\partial^2}{\partial t^2}\mathbf{A} = \frac{4\pi\mathbf{j}}{c}$$

(if $\nabla \cdot \mathbf{A} = 0$ as it is here and no charge accumulation is present; we expect none here).

The applied field will give rise to current $\mathbf{j}_1$ which through Maxwell's equations will give rise to additional electric fields. Solve the problem self-consistently to obtain a transverse dielectric function $\epsilon(q,\omega,H_A)$ which is the ratio of applied to total fields. This need be obtained only in the limit as $\sigma \to \infty$.

b. Find low-frequency solutions ($\omega \ll qc$) of $\epsilon = 0$ corresponding to a divergent response, and therefore excited states of the system. These are called "helicons." Sketch their dispersion curve $[\omega(q)]$ and estimate their frequency at $H_A = 10{,}000$ gauss, $N = 10^{22}$ cm^{-3}, and a wavelength of 1 cm.

3.10 Consider a system with two quantum levels of interest,

$$H_0|1\rangle = E_1|1\rangle$$

$$H_0|2\rangle = E_2|2\rangle$$

The density matrix can then be expanded $\rho(\mathbf{r},\mathbf{r}',t) = |1\rangle\rho_{11}(t)\langle 1| + |1\rangle\rho_{12}(t)\langle 2| + \cdots$. Now add a perturbation H_1 coupling the states,

$$H_1|1\rangle = \lambda\ |2\rangle$$

$$H_1|2\rangle = \lambda^*|1\rangle$$

where $\lambda = \lambda_0 e^{-i\omega t}\, e^{\alpha t}$

a. Using the Liouville equation, find exact expressions for the time dependence of all four ρ_{ij} in terms of each other.

b. If at $t = -\infty$ we have $\rho_{11} = 1$, $\rho_{12} = \rho_{21} = \rho_{22} = 0$, find the probability of occupation of $|2\rangle$ (that is, ρ_{22}) at $t = 0$ to lowest order in λ_0. Sketch the result as a function of ω.

3.11 Imagine a small steplike applied potential

$$V(x) = V_0 \qquad \text{for } x > 0$$

$$= 0 \qquad \text{for } x < 0$$

in a three-dimensional electron gas. Clearly Friedel-like oscillations in electron density will arise. Obtain the asymptotic form for $n(x)/\bar{n}$ at large x.

3.12 Consider a semiconductor with N donors per unit volume, each with an electron bound to it (and no electrons in the conduction band). The optical conductivity is nonzero due to the possibility of excitation of the bound electrons into the conduction band. Compute the corresponding $Re\sigma_{ij}(\omega)$ using the Kubo-Greenwood formula.

The bound wavefunctions may be taken as $\psi_d(r) = Ae^{-\alpha r}$ (i.e., 1s functions); the conduction-band wavefunctions may be taken as plane waves orthogonalized to the donor states (which do not overlap).

The donor binding energy is $-(\hbar^2\alpha^2/2m)$. The form of the result is simpler if written in terms of ω,α and the magnitude of the wavenumber of the final state k. Thus $\hbar^2k^2/2m + \hbar^2\alpha^2/2m = \hbar\omega$. Sketch the result, $\sigma(\omega)$.

3.13 *a.* Find the reduction in intensity of the no-phonon transition for the F center described in Sec. 5.6 due to lattice polarization. Write the result in terms of the difference in equilibrium distortions $x_1 - x_0$ and the mean-square zero-point displacement. Note the ground-state harmonic-oscillator function is

$$\varphi(x - x_i) = \sqrt{\frac{\alpha}{\sqrt{\pi}}}\, e^{-\frac{1}{2}\alpha^2(x - x_i)^2}$$

and we take α the same in the initial and final states.

b. How is the reduction in intensity computed if there are two types of lattice distortion corresponding to two different vibrational normal modes?

c. Redo part *a* with $x_1 - x_0 = 0$, but with the coefficient α different before and after the transition.

3.14 In a uniform Fermi liquid the current is given by

$$\mathbf{j} = \frac{-e}{h^3}\int \delta n(\mathbf{p})\left[\frac{\partial E(\mathbf{p},n_0)}{\partial \mathbf{p}} - \frac{1}{h^3}\int \frac{\partial n_0(\mathbf{p}')}{\partial \mathbf{p}'} f(\mathbf{p}',\mathbf{p};n_0)\, d^3p'\right] d^3p$$

For an isotropic system, clearly both terms in square brackets are parallel to $\mathbf{p}$ and $\partial n_0(\mathbf{p}')/\partial \mathbf{p}'$, for example, can be written $\partial n_0(p')/\partial p' \cos\theta$ with θ the angle between $\mathbf{p}'$ and $\mathbf{p}$.

We noted that for a uniform drift $[n(\mathbf{p}) = n_0(\mathbf{p} - \delta\mathbf{p}_0)]$ in a translationally invariant system $\mathbf{j}$ must equal $(-Ne/m)\,\delta\mathbf{p}_0$ where m is the free-electron mass and N the electron density. It follows that the expression in square brackets, evaluated for $\mathbf{p}$ at the Fermi surface, must equal $\mathbf{p}/m$. We assume low temperature throughout.

If we assume the quasiparticle energy is given by

$$E(\mathbf{p};n_0) = \frac{p^2}{2m^*}$$

with m^* constant, not equal to m, and the quasiparticle interaction is given by

$$f(\mathbf{p}',\mathbf{p};n_0) = |\mathbf{p} - \mathbf{p}'|^2 \frac{A}{N}$$

with A constant (this form not intended to be realistic, only illustrative), then

a. Determine A in terms of m^*.

b. What is the current carried by the system with a single quasiparticle excitation of momentum $\mathbf{p}$?

IV

LATTICE VIBRATIONS AND ATOMIC PROPERTIES

We found in Sec. 5.3 of Chap. I on the basis of symmetry arguments alone that the vibrations of a lattice can be assigned a wavenumber just as can the electronic states of a perfectly periodic lattice. If we assign a wavenumber $\mathbf{q}$ to a given mode and number different modes of the same wavenumber by λ, then for crystals with a single atom per primitive cell the displacement of an atom with equilibrium position $\mathbf{r}_j$ is given by

$$\delta\mathbf{r}_j = \frac{\mathbf{u}_{q,\lambda}}{\sqrt{N}}\, e^{i(q\cdot r_j - \omega_{q,\lambda}t)} \tag{4.1}$$

where $\mathbf{u}_{q,\lambda}$ is the vector amplitude of the wave and $\omega_{q,\lambda}$ is the angular frequency of the wave. To obtain real displacements we may simply take the real (or imaginary) part of Eq. (4.1) or add the complex-conjugate expression. For more than one atom per primitive cell an additional $\mathbf{u}_{q,\lambda}$ must be introduced for each additional atom in the cell. The wavenumber $\mathbf{q}$ may be restricted to the first Brillouin zone, which is determined of course by the primitive cell. Thus there are as many wavenumbers in the Brillouin zone as there are primitive cells. Since there are three degrees of freedom for each atom in the primitive cell, the index λ must run over a number of integers equal to three times the number of atoms in the primitive cell. We will simplify our notation by considering crystals with one atom per cell.

1. Calculation with Force Constants[1]

The calculation of the modes and frequencies for a given lattice is a very old problem. The traditional approach has been to note that the total energy of the system may be written as a function of the positions of all of the atoms, $W(\mathbf{r}_1,\mathbf{r}_2,\ldots)$. The vibrations of the lattice involve small excursions from the equilibrium position so we may expand this energy in terms of these displacements.

$$E_{\text{tot}} = W(\mathbf{r}_1,\mathbf{r}_2,\ldots) + \sum_i \frac{\partial W}{\partial \mathbf{r}_i}\,\delta\mathbf{r}_i + \frac{1}{2}\sum_{i,j}\delta\mathbf{r}_i\,\frac{\partial^2 W}{\partial\mathbf{r}_i\partial\mathbf{r}_j}\,\delta\mathbf{r}_j + \cdots \tag{4.2}$$

where the $\mathbf{r}_i$ now refer to the equilibrium positions and all W are evaluated for these positions. The $\delta\mathbf{r}_i$ are the displacements from equilibrium. The equilibrium energy itself does not interest us. Furthermore, in equilibrium the first-order term must be equal to zero. Thus for the calculation only the second-order term is needed. The $\partial^2 W/\partial\mathbf{r}_i\partial\mathbf{r}_j$ is called the *interaction matrix*.

We may now directly compute the force on the atom which at equilibrium would lie at the position $\mathbf{r}_i$.

$$\mathbf{F}_i = -\frac{\partial E_{\text{tot}}}{\partial \mathbf{r}_i} = -\frac{1}{2}\sum_j\left(\frac{\partial^2 W}{\partial\mathbf{r}_i\partial\mathbf{r}_j}\,\delta\mathbf{r}_j + \delta\mathbf{r}_j\,\frac{\partial^2 W}{\partial\mathbf{r}_j\partial\mathbf{r}_i}\right)$$

$$= -\sum_j \frac{\partial^2 W}{\partial\mathbf{r}_i\partial\mathbf{r}_j}\,\delta\mathbf{r}_j \tag{4.3}$$

Note the term $j = i$ is included. Furthermore, the acceleration of the particle at $\mathbf{r}_i$ is simply equal to this force divided by the atomic mass M. Substituting the known form of the solution from Eq. (4.1) we obtain directly

$$-M\omega_{q,\lambda}{}^2\,\mathbf{u}_{q,\lambda} = -\sum_j \frac{\partial^2 W}{\partial\mathbf{r}_i\partial\mathbf{r}_j}\,\mathbf{u}_{q,\lambda}\,e^{iq\cdot(r_j - r_i)} \tag{4.4}$$

The translational symmetry of the lattice with one atom per primitive cell guarantees that $\partial^2 W/\partial\mathbf{r}_i\partial\mathbf{r}_j$ will depend only upon the difference $\mathbf{r}_j - \mathbf{r}_i$. (For crystals with more than one atom per primitive cell we would require separate equations for each of the atoms in the primitive cell.) For crystals with one atom per primitive cell we have three simultaneous equations corresponding to the three components of Eq. (4.4). If we know $\partial^2 W/\partial\mathbf{r}_i\partial\mathbf{r}_j$,

[1] For representative studies using this approach, see R. F. Wallis (ed.), "Lattice Dynamics," Pergamon, New York, 1965.

we can in principle solve the three simultaneous equations to obtain the eigenfrequencies and the relative magnitudes of the components of the amplitude vectors for each of the modes.

All of this formulation is quite general. The difficulty lies in determining the interaction coefficients $\partial^2 W/\partial \mathbf{r}_i \partial \mathbf{r}_j$. The traditional approach to this problem is a very plausible one. It is noted that these interaction constants represent a coupling between pairs of atoms. It has been assumed that the coupling between near neighbors is very much larger than that between more distant neighbors. Thus, if we think of Eq. (4.4) as representing point particles coupled by springs, we expect the spring constants to be much larger for the nearest neighbors. The corresponding spring constants or coupling coefficients are taken as disposable parameters. As many neighboring interactions are included as can be determined from known parameters of the system, such as the elastic constants.

The general structure of the interaction matrix allows nine independent coupling constants between any two atoms. This number of course can be drastically reduced by symmetry arguments since the interaction matrix must go into itself under any of the point-group symmetry operations of the crystal. Thus the number of independent constants describing the interaction between nearest neighbors is frequently only two. Two or three more may be required to describe the interaction with next-nearest neighbors and so forth. It may be convenient to think of these interactions as arising from elastic *bonds* between atoms. We can then speak of bond-stretching and bond-bending force constants. Once a few of these are determined, by fitting experimental parameters and assuming that all other force constants are negligible, it is possible to compute directly the entire vibration spectrum for the lattice.

This general approach is quite rigorous. In carrying it out the only approximation is the restriction of the number of force constants included. Nonetheless the approach has turned out to be quite unsuccessful for quantitative description of the vibration spectrum of solids, largely because of inherent long-range forces. In metals such long-range forces arise from an interaction between ions modulated by the conduction electrons. The origin of these forces may be seen from the Friedel oscillations in the screening field that we described. Such fluctuating charge densities arising from one atom must inevitably lead to fluctuating forces on atoms at large distances away. In ionic crystals the coulomb interaction between atoms extends of course to large distances. Even in semiconductors the forces turn out to be long range and seem to arise from dipole or quadripole moments induced on the atoms during distortion.

The failure of this approach has become clear as an increasing amount of information about the vibration spectra has been obtained. Initial fits based upon elastic constants provided calculated vibration spectra. Subsequent measurement of specific modes gave additional information about the spectra which turned out typically to be inconsistent with the earlier calculations. These were then corrected by the addition of other force constants so that then all known information was accounted for. This, however, required sizable changes in all of the previously determined force constants and gave little indication that the force constants were dropping rapidly with distance. This process has continued with the inclusion of interaction out to as many as seven sets of neighbors. Success has been achieved only when long-range forces have been introduced. One method that has met with particular success is the *shell model* in which distortions of the atom and resulting long-range forces are introduced phenomenologically.[1]

In spite of these difficulties with the force-constant approach as a quantitative method, it remains a valuable model for qualitative studies of the nature of lattice vibrations. Therefore it will be desirable to begin our discussion by the application of the force-constant model to a very simple case.

1.1 Application to the simple cubic structure We consider a simple cubic solid and begin with a discussion of $\partial^2 W/\partial \mathbf{r}_i \partial \mathbf{r}_j$. We look first at the interaction between nearest neighbors. See Fig. 4.1. We expect a radial interaction between these atoms; i.e., we expect the energy to change as the distance between near neighbors is changed. Each atom has six nearest neighbors so the same spring constant, which we shall call κ_1, is associated with each of these interactions. We might seek then an interaction constant relating the lateral displacement of neighboring atoms. Such a displacement has the effect of changing the orientation of the bond (or the difference in position vector) between atoms. Such a term corresponding to an increase in energy quadratic in the rotation of the bond is clearly unphysical since it would lead to an increase in energy if the entire crystal were rotated in space. These lateral constants must therefore be set equal to zero unless we are to include interactions with more distant neighbors which cancel this rotational energy of the entire crystal. Thus, if we restrict our attention to nearest-neighbor interactions only, they must be radial and describable by the single force constant κ_1. This, however, is not sufficient since these forces offer no resistance to a shear of the lattice and the stability of the lattice is not maintained. We must therefore extend the interaction to the next-nearest neighbors, which lie in [110] directions in a simple cubic lattice.

These can include a radial interaction between next-nearest neighbors

[1] *Ibid.*

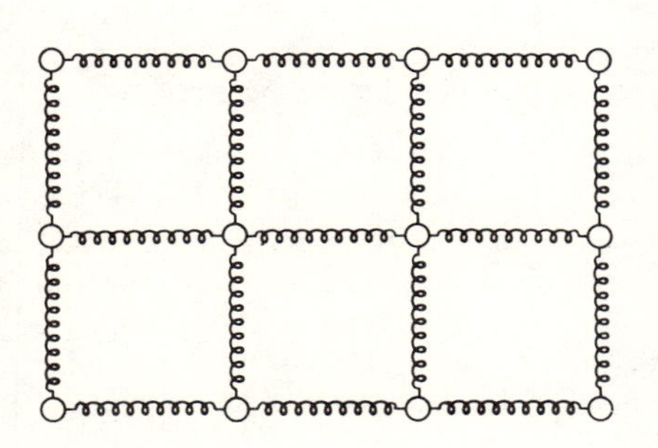

Fig. 4.1 The interaction between nearest neighbors may be thought of as springs connecting them. Purely radial interactions correspond to springs free to pivot at the atoms and would not stabilize a simple cubic structure. Rigid connections at the atoms would give resistance also to a change in angle between adjacent springs and stabilize the structure. Such bond-bending interactions couple next-nearest as well as nearest neighbors, since three atoms are required to define the angle between springs.

with a new force constant κ_2. These two radial interactions alone are sufficient to stabilize the lattice. However, we might further introduce lateral interaction constants for nearest and next-nearest neighbors. Such interactions would correspond physically to a change in energy when the angle between two adjacent bonds is changed. Such interactions are frequently called *bond-bending forces.* If we apply again the requirement of the rotational invariance of the total energy, we are reduced to three independent constants describing the interactions between nearest and next-nearest neighbors. These could be fit to the three independent elastic constants of a simple cubic system and the entire vibration spectrum computed.

For simplicity we will consider the propagation of sound along a [100] direction and will discuss only the longitudinal mode. The form of the modes that may be propagated in this direction is clear physically and may be verified by symmetry arguments. There will be a longitudinal mode in which the displacements of the atoms are parallel to the direction of propagation. In addition there will be two transverse modes in which the displacements are perpendicular to the propagation, and these two modes must have the same frequency at the same wavenumber. The geometry for the longitudinal mode is shown in Fig. 4.2. The displacements of every atom in a given plane perpendicular to $\mathbf{q}$, the wavenumber of the mode, will be the same and, even with nearest- and next-nearest-neighbor interactions, the force on a given atom will depend only upon the relative displacements of the first plane to the right and the first plane to the left. Thus we can construct a composite force constant for the interactions between planes and the problem has become essentially one-dimensional, equivalent to the elementary one-dimensional chain discussed in most elementary solid state texts. We will not deduce here (see Prob. 4.1) the composite force constant in terms of those defined above but will simply define the constant κ which describes the change in energy, per atom in a plane transverse to the direction of propagation, due to the change in distance between two neighboring planes. One contribution to κ will be the κ_1 defined above and the two constants become identical as the interactions between next-nearest neighbors become small. The elastic energy of a single line of atoms parallel to $\mathbf{q}$ will be given by

$$\sum_i \tfrac{1}{2}\kappa(\delta r_i - \delta r_{i+1})^2 = \kappa \sum_i (\delta r_i{}^2 - \delta r_i \delta r_{i+1})$$

This corresponds to a one-dimensional chain with energy describable by Eq. (4.2) and with all interaction constants $\partial^2 W/\partial r_i \partial r_i = 2\kappa$ and $\partial^2 W/\partial r_i \partial r_{i+1} = \partial^2 W/\partial r_i \partial r_{i-1} = \kappa$ and all other $\partial^2 W/\partial r_i \partial r_j$ equal to zero.

These force constants may be directly substituted into the secular equation, Eq. (4.4), writing the nearest-neighbor distance as a. We obtain

$$-M\omega^2 = -\kappa(2 - e^{iqa} - e^{-iqa})$$

which may be directly solved for the frequency

$$\omega = \sqrt{\frac{\kappa}{M}}\, 2 \sin \frac{qa}{2} \tag{4.5}$$

The same analysis would yield precisely the same form for a transverse mode propagating in the same direction but with a modified (and ordinarily weaker) composite interaction constant κ'. The corresponding dispersion curves are shown in Fig. 4.3.

Note that at long wavelengths, small q, the frequency is linear in the wavenumber giving a well-defined speed of longitudinal sound equal to $d\omega/dq = a\sqrt{\kappa/M}$ and transverse sound $a\sqrt{\kappa'/M}$. At large wavenumbers the dispersion curve approaches the Brillouin zone face horizontally just as the

Fig. 4.2 *A schematic display of the displacements in a part of the lattice for a longitudinal wave propagating to the right.*

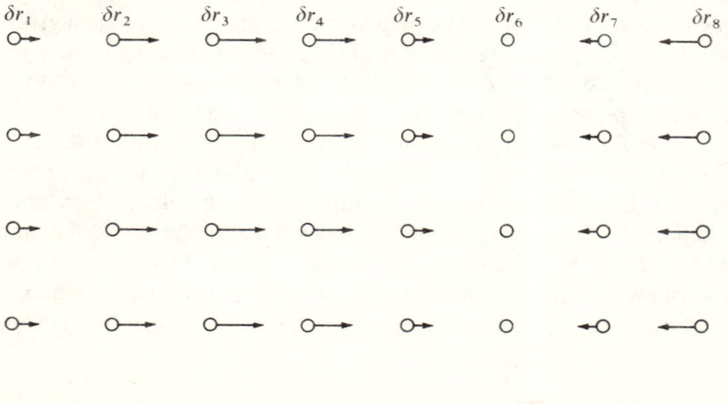

δr_1 δr_2 δr_3 δr_4 δr_5 δr_6 δr_7 δr_8

energy bands for electrons ordinarily do. We may note further that the choice of a wavenumber outside the Brillouin zone leads by Eq. (4.1) to displacements identical to those of the reduced wavenumber within the Brillouin zone. Our dispersion curves give a complete set of vibrations propagating in the [100] direction for the cubic crystal under consideration.

We should note that if we had chosen an arbitrary direction for the propagation of the sound rather than a symmetry direction, we would not have been able to determine by symmetry the polarization of the individual modes. Displacements in a given direction would have given rise to forces in the other directions. It would therefore have been necessary to simultaneously consider the displacements in all three directions; we would have obtained three coupled equations which could have then been solved for the frequencies and the directions of polarization. The modes would not have been purely transverse nor purely longitudinal though they might well be approximately so. The degeneracy between the two transverse modes would have been split but otherwise the nature of the dispersion curves would have been similar. We might further note that had we included an interaction constant between third-nearest neighbors we would have obtained, in addition to the terms in ω^2 obtained here, terms with an exponential e^{2iqa} and the corresponding higher Fourier components in the dispersion curve. The analogy between this force-constant treatment of the vibration spectrum and the tight-binding treatment of electronic states is clearly quite close.

Fig. 4.3 The vibration spectrum for modes propagating in a **[100]** *direction in a simple cubic crystal with nearest and next-nearest neighbor interactions only. The transverse modes are doubly degenerate.*

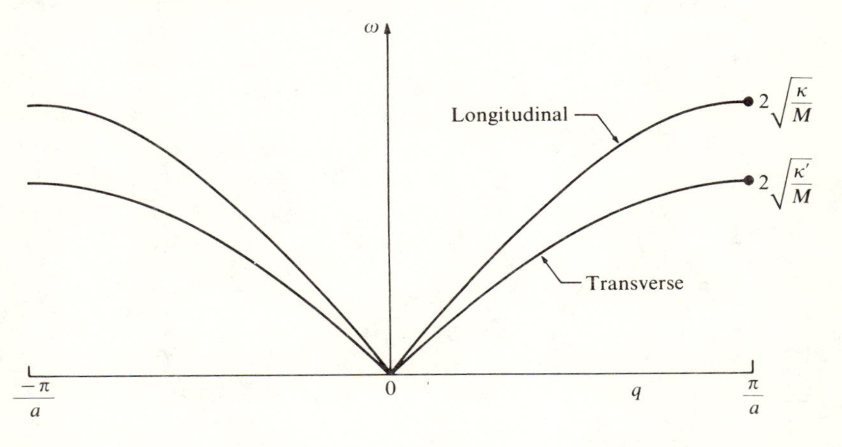

1.2 Two atoms per primitive cell We may generalize the discussion by considering the propagation of lattice vibrations in crystals with more than one atom per primitive cell. One such generalization of the above model can be obtained by modifying the mass of alternate atoms in the simple cubic crystal described. We may do this such that the mass of nearest neighbors differs but the mass of next-nearest neighbors is the same, in the manner of the sodium chloride structure. An even simpler generalization may be made by modifying masses on alternate planes as shown in Fig. 4.4. Then for propagation in a [100] direction the problem remains one-dimensional and can be easily solved. We will carry out the analysis in analogy with that given in Sec. 1.1.

The atom planes are again indexed by j, and the mass of an atom in the jth plane is M_1 if j is odd and M_2 if j is even. For a longitudinal mode, displacements δr_j in the x direction may be written

$$\delta r_j = \begin{cases} \dfrac{u_1}{\sqrt{N}} e^{iqaj} e^{-i\omega t} & \text{for } j \text{ odd} \\[2ex] \dfrac{u_2}{\sqrt{N}} e^{iqaj} e^{-i\omega t} & \text{for } j \text{ even} \end{cases} \tag{4.6}$$

Interplanar force constants may again be defined and the force on the jth ion [Eq. (4.3)] becomes

$$F_j = \kappa[(\delta r_{j+1} - \delta r_j) + (\delta r_{j-1} - \delta r_j)]$$

Fig. 4.4 A simple cubic array in which masses of the atoms on alternate y–z planes are different.

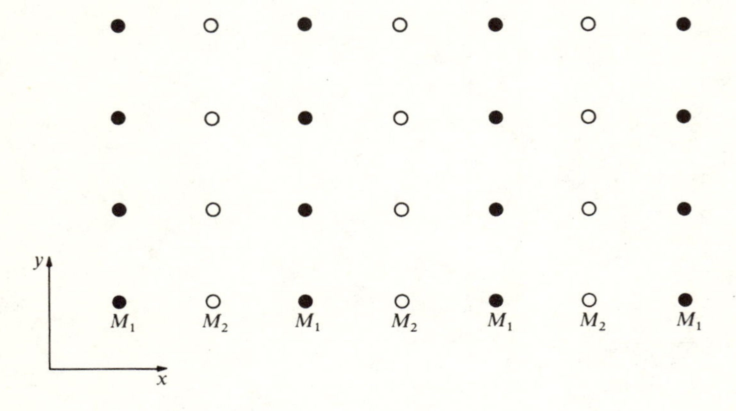

or

$$F_j = \begin{cases} \dfrac{\kappa}{\sqrt{N}}(u_2 e^{iqa} + u_2 e^{-iqa} - 2u_1)e^{iqaj}e^{-i\omega t} & \text{for } j \text{ odd} \\ \dfrac{\kappa}{\sqrt{N}}(u_1 e^{iqa} + u_1 e^{-iqa} - 2u_2)e^{iqaj}e^{-i\omega t} & \text{for } j \text{ even} \end{cases}$$

Then substitution for δr_j in $F_j = M_j \delta \ddot{r}_j$ gives

$$\begin{aligned} 2\kappa(u_2 \cos qa - u_1) &= -M_1\omega^2 u_1 \\ 2\kappa(u_1 \cos qa - u_2) &= -M_2\omega^2 u_2 \end{aligned} \tag{4.7}$$

We have cancelled the factor $e^{iqaj}e^{-i\omega t}$ after substitution. These may be solved for ω^2 to give

$$\omega^2 = \frac{2\kappa}{M_1 M_2}\left[\frac{M_1 + M_2}{2} \pm \sqrt{\frac{(M_1 - M_2)^2}{4} + M_1 M_2 \cos^2 qa}\right] \tag{4.8}$$

We note first the solution for $M_1 = M_2$, which must be equivalent to that given before.

$$\omega^2 \to \frac{2\kappa}{M}(1 \pm \cos qa)$$

or

$$\omega \to 2\sqrt{\frac{\kappa}{M}} \times \begin{bmatrix} \cos \dfrac{qa}{2} & \text{for } + \\ \sin \dfrac{qa}{2} & \text{for } - \end{bmatrix}$$

which are plotted in Fig. 4.5. The sine solution is that obtained before, Eq. (4.5), and the cosine solution may be seen to be a duplication of the same solutions. Since in Eq. (4.6) only odd integers j appear in e^{iqaj}, changing q by π/a changes every δr_j for j odd by a factor $e^{i\pi} = -1$, and that factor can be absorbed in u_1. Thus each point in the cosine curve represents a mode already represented by the sine curve and we may discard the cosine curve. This is an *extended-zone representation* in the sense we used in describing electron states.

We could instead represent all modes in the *reduced zone*, $-\pi/2a < q \leq \pi/2a$, and include both curves. In the reduced zone, with $M_2 = M_1$, we find that $u_1 = u_2$ for the lower band, and $u_1 = -u_2$ for the upper band. This is an awkward description when $M_1 = M_2$, but is equivalent to that given earlier. It is based upon a primitive cell with dimensions $2a$, a, a,

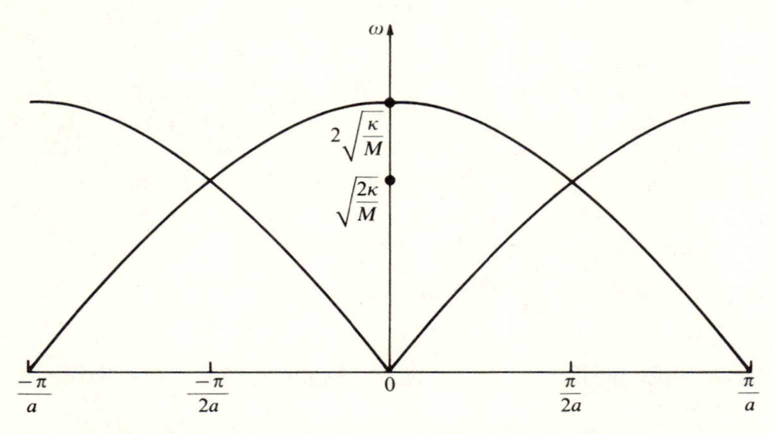

Fig. 4.5 *Solutions of ω versus q for the crystal of Fig. 4.4 for longitudinal modes propagating in the* **[100]** *direction when $M_1 = M_2$.*

rather than a, a, a which was appropriate when alternate atoms were not distinguished.

Let us now let M_1 differ slightly from M_2. We see from Eq. (4.8) that the larger frequency is slightly raised and the lower frequency slightly lowered and they are never equal. The curves are shown in Fig. 4.6. Just as in the nearly-free-electron approximation, gaps are opened up at the faces of the

Fig. 4.6 *Solutions of ω versus q for the crystal of Fig. 4.4 for longitudinal modes propagating in the* **[100]** *direction with $M_1 > M_2$. The reduced zone is $-\pi/2a < q \leq \pi/2a$.*

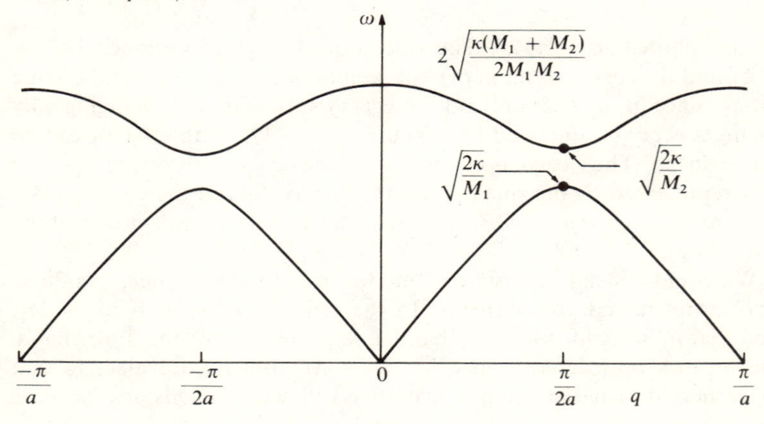

smaller Brillouin zone based upon the larger $(2a,a,a)$ primitive cell. The upper band is called an *optical-mode band* and the lower band the *acoustical-mode band* for reasons which may be seen by examining the form of the modes at small q. Substituting the limiting frequencies in Eq. (4.7) we see that for the acoustical modes $u_1 = u_2$; both types of atoms move together in a long-wavelength pressure wave. For the optical mode at small q we find $u_1 = -M_2u_2/M_1$; the two types of atoms vibrate against each other. If there is a charge difference on the two atoms, this will cause a polarization wave of wavenumber q. Optical modes occur at infrared frequencies and can be excited with infrared radiation. We will return briefly to the question later. We may also obtain the form of the modes at the zone face $q = \pi/2a$. These are shown in Fig. 4.7. Note that in each mode only atoms of one type move. The mode in which the heavier atoms move will have the lower frequency.

This same splitting into acoustical and optical branches occurs for any crystal with two atoms per primitive cell. It also occurs for the transverse as well as the longitudinal modes. Thus, for a general crystal with two atoms per primitive cell, we have six modes, as shown in Fig. 4.8.

The structure chosen for our treatment was selected for mathematical simplicity. A more typical crystal with two atoms per primitive cell is the sodium chloride structure shown in Fig. 4.9. Again in this case the Cl^- and the neighboring Na^+ ions will move in phase with each other in the long-wavelength acoustical modes. Again in long-wavelength optical modes the Na^+ sublattice moves together as does the Cl^- sublattice. Within each primitive cell the vibration is very much the same as that of a molecular vibration of the two atoms present. For [100] propagation of optical modes in the sodium chloride structure these "molecular" vibrations represent a bond-stretching vibration and two rotational modes. For additional atoms

Fig. 4.7 *A sketch of the displacement amplitudes (or plucking patterns) for longitudinal modes of wavenumber $q = \pi/2a$ for the crystal of Fig. 4.4. The filled circles represent heavier masses than do the empty circles.*

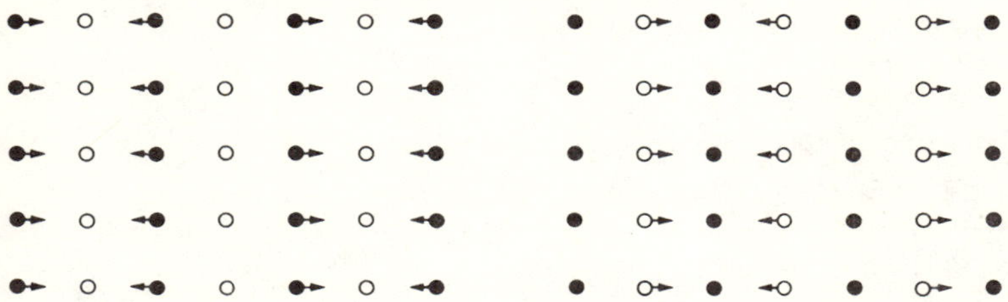

Acoustic modes Optical modes

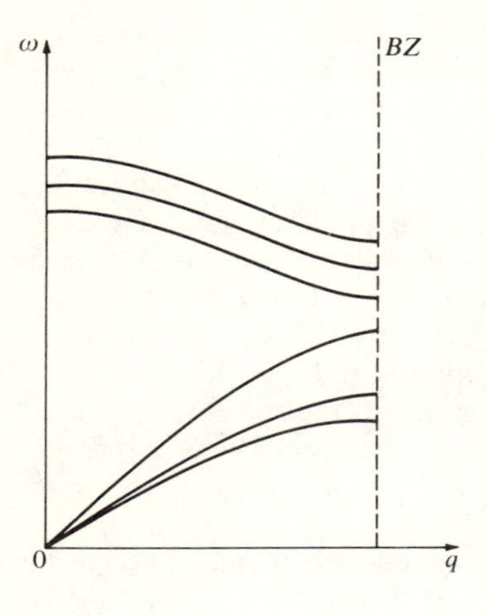

Fig. 4.8 Optical and acoustical modes for a general crystal (no degeneracies) with two atoms per primitive cell. They may be labeled by their form at small q as longitudinal acoustical, longitudinal optical, transverse acoustical, or transverse optical. The Brillouin zone face is labeled BZ.

per primitive cell we would introduce additional molecular vibrations corresponding to different optical modes.

In ionic crystals the two atoms in the primitive cell will have different charge. One of the optical modes corresponds to modulation along a direction parallel to the displacement of ions. This will be characteristic of the longitudinal mode. This modulated polarization will lead to charge accumulation and an increase in the frequency due to the same electrostatic resistance to deformation which gave rise to plasma oscillations in an electron gas. The two transverse optical modes will be degenerate at lower frequency since they do not give rise to charge accumulation. We may note that this distinction between longitudinal and transverse modes breaks down for the wavenumber $q \equiv 0$. Because of the long-range nature of the coulomb interaction the dispersion curves behave peculiarly here. To

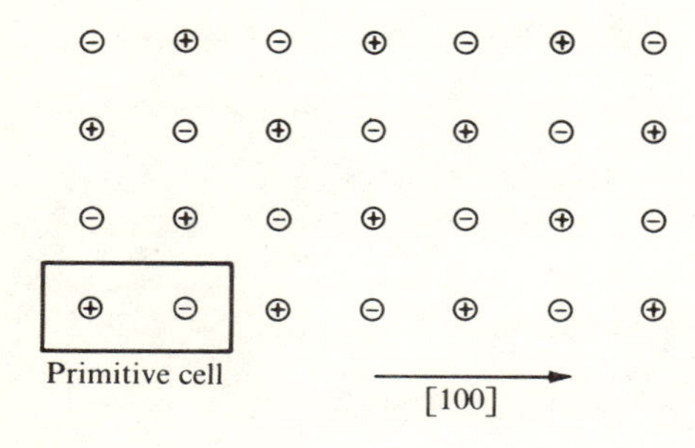

Fig. 4.9 One plane of ions in a sodium chloride crystal, showing a primitive cell.

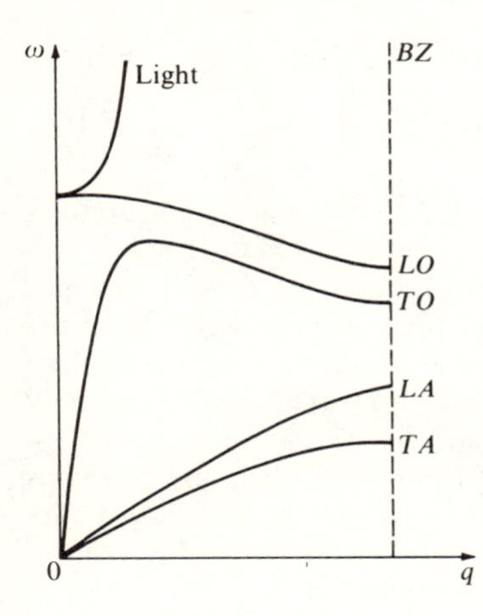

Fig. 4.10 Dispersion curves for the lattice vibrations in sodium chloride and for light, which couples with the transverse optical modes. The various modes have been labeled TA for transverse acoustic, etc. The modes TA, TO, and light are doubly degenerate for the direction of propagation shown. The slope of the light curve (the speed of light) has been greatly reduced for illustrative purposes. The light curve would reach the Brillouin zone face at x-ray frequencies and would be split, as are energy bands and vibration spectra, due to the diffraction by the lattice.

understand the behavior we consider the force-constant equations with coulomb interactions omitted. These lead to vibrational modes as for nonionic crystals. We separately consider Maxwell's equations which lead to light waves. Finally we introduce the coupling. We may either describe the interaction between light and the optical modes of a polar crystal in terms of absorption of the light or we may think of the light and the optical vibration modes as mixtures of composite excitations of the system which include both the light and the vibrations. From this last point of view the coupling gives rise to a mixing just as in the mixing between conduction band and atomic d states in the transition metals. The velocity of light is of course very much greater than the velocity of sound so the dispersion curves for the entire system will look schematically as shown in Fig. 4.10. Note that as q goes to zero all three optical modes are degenerate but at larger q the waves are split as we have indicated.

2. *Phonons and the Lattice Specific Heat*[1]

In constructing the vibration spectrum for a set of N atoms we have obtained $3N$ independent harmonic oscillators corresponding to the $3N$ modes of the system. At finite temperature these modes will be excited thermally and the total vibrational energy present will constitute the thermal energy. If the system behaves classically we expect an energy KT on the average to be associated with each degree of freedom. Thus the derivative of the

[1]For a more complete discussion, see F. Seitz, "Modern Theory of Solids," pp. 99ff., McGraw-Hill, New York, 1940.

thermal energy with respect to temperature, the specific heat, is independent of the frequencies of the oscillators and is equal simply to

$$C_V = 3NK$$

This is called the law of *Dulong and Petit*. This law is obeyed to a good approximation in many solids at high temperatures. However, at low temperatures the specific heat drops toward zero. The resolution of this discrepancy came only with the advent of quantum mechanics. The construction of the normal modes of the system may be regarded as a canonical transformation on the Hamiltonian of the system of interacting atoms. Thus we may directly apply quantum mechanics to the resulting system of harmonic oscillators. A given mode with classical frequency ω can be excited only in integral steps of the vibrational quantum $\hbar\omega$. The energy in a particular mode is given by $E_q = (n_q + \frac{1}{2})\hbar\omega_q$ where the integer n_q describes the degree of excitation of the system. We ordinarily describe such an excitation by saying there are n_q *phonons* of wavenumber $\mathbf{q}$ present. The vibrational energy of the system contains the energy of the phonons present as well as the zero-point energy $\frac{1}{2}\hbar\omega_q$ for each mode. This will be sufficient for our present purpose. At a later stage we will carry out the quantization in more detail.

From statistical mechanics we may compute the average excitation of each oscillator at a given temperature T. The result is

$$n_q = \frac{1}{e^{\hbar\omega_q/KT} - 1}$$

This corresponds equivalently, of course, to a distribution of phonons that satisfy Bose-Einstein statistics.

Using this distribution and a known or calculated distribution of vibration frequencies, we may directly compute, now quantum mechanically, the thermal energy of the system.

If we calculate or measure the distribution of modes as a function of frequency in a solid we find that it is rather strongly peaked at a frequency of the order of the frequency at a Brillouin zone face. We might then, as a first approximation, replace the vibration spectrum by a set of $3N$ oscillators at a single frequency, called the *Einstein frequency* since this approximation was originally proposed by Einstein. Then the total thermal energy of the system is given by

$$E_{\text{tot}} = \frac{3N\hbar\omega_E}{e^{\hbar\omega_E/KT} - 1}$$

The specific heat is thus given by

$$C_V = \frac{dE_{\text{tot}}}{dT} = \frac{3N(\hbar\omega_E)^2\, e^{\hbar\omega_E/KT}}{KT^2(e^{\hbar\omega_E/KT} - 1)^2}$$

which is shown in Fig. 4.11. We see immediately that at high temperatures this expression approaches $3NK$, the Dulong-Petit value, as it should. At low temperatures, on the other hand, the specific heat drops exponentially to zero.

This result is qualitatively correct but the form at low temperatures is quite incorrect. The error arises from the model since it implies that at low temperatures the thermal energy KT is much less than the vibrational quantum $\hbar\omega$ for all modes. Thus we find that the occupation of every mode drops exponentially at low temperature. In fact there are of course modes at arbitrarily low frequencies and these become important at low temperatures.

We may see very crudely what the effect of these low-frequency modes must be. At a particular temperature T we expect essentially classical occupation of states which have frequency ω less than the critical value given by

$$\hbar\omega = \hbar v_s q = KT$$

where v_s is the speed of sound. The number of such modes will be proportional to $4\pi q^3/3$ and thus to T^3. The classical energy in these states then will be proportional to T^4 leading to a specific heat proportional to T^3. This is the familiar T^3 law of specific heats observed in many solids.

An approximate calculation of the lattice specific heat which gives the

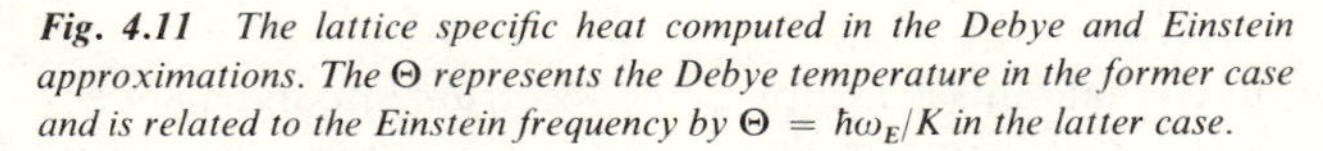

Fig. 4.11 *The lattice specific heat computed in the Debye and Einstein approximations. The* Θ *represents the Debye temperature in the former case and is related to the Einstein frequency by* $\Theta = \hbar\omega_E/K$ *in the latter case.*

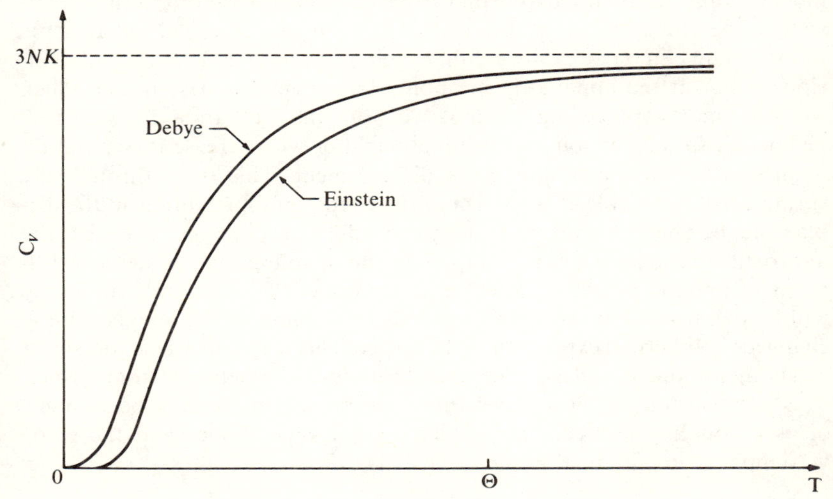

proper behavior in both low- and high-temperature limits has been given by Debye. Focusing attention on the long-wavelength modes, we may describe them approximately by a longitudinal and a transverse velocity. As a still simpler approximation we may regard these modes as degenerate with an average velocity v_s. Then the number of modes in a frequency range $d\omega = v_s\, dq$ is given by $3[\Omega/(2\pi)^3]4\pi q^2\, dq$. Then the sum over modes using the Bose-Einstein distribution is replaced by an integral over wavenumber space. In order to retain the correct number of modes the integral is cut off at a Debye frequency ω_D chosen such that the number of modes within the corresponding sphere in **q** space is equal to the number of modes in the zone. This has the effect of replacing the Brillouin zone by a sphere of equal volume. A Debye temperature Θ_D is associated with this cut-off frequency, $\Theta_D = \hbar\omega_D/K$. The resulting specific heat may be calculated directly and is found to be given by

$$C_V = 9NK\left(\frac{T}{\Theta_D}\right)^3 \int_0^{\Theta_D/T} \frac{x^4 e^x}{(e^x - 1)^2}\, dx$$

and is shown in Fig. 4.11. This leads to the Dulong-Petit value at high frequencies and at low frequencies approaches

$$C_V = \frac{12\pi^4}{5} NK\left(\frac{T}{\Theta_D}\right)^3$$

It should be noted that our separation of the spectrum into independent modes is only approximate. We may expect higher-order terms in the expansion of the energy given by Eq. (4.2). These are the so-called anharmonic terms. We may in fact readily see that such terms *must* arise in three dimensions. Even in the apparently harmonic force-constant description which we gave in the preceding section, we note that an atom moving perpendicular to the line joining it with a neighbor will increase its separation in proportion to the square of its displacement. This then, through the spring constant, will lead to a term in the energy of fourth order in the displacement. These anharmonic terms produce a coupling between the approximate modes we have computed. The coupling may be described in terms of phonon-phonon scattering. A second effect of the anharmonicity will be to cause a change in the equilibrium volume as the temperature is changed; i.e., thermal expansion of the lattice. This may either be understood as a nonparabolicity and asymmetry of the interactions between atoms, which causes expansion as the temperature is increased, or it may be thought of as a coupling between the ordinary vibrational modes and the zero-wavenumber longitudinal mode.

3. *Localized Modes*

The method of classical Green's functions has been used very effectively in recent years for the treatment of vibrations in lattices with defects.[1] The basic difficulty in treating vibrations in crystals with defects arises from the loss of translational periodicity which allowed us in perfect crystals to construct the form of the modes by symmetry. When we introduce a single defect, however, the remainder of the lattice remains perfect and it turns out to be possible to write the effect of this perfect portion of the lattice in terms of a Green's function and then focus our attention on the defect itself, reducing the problem essentially to that of the "molecular" vibrations of the defect.

We will first state the normal mode problem mathematically in a convenient form. We will note how this is solved for the perfect crystal by diagonalization of a matrix, and then how it may be solved in terms of Green's functions for the perfect crystal. Finally we will consider a crystal with a defect, the simplest defect being the modification of the mass of a single atom.

We return to Eq. (4.3) giving the force on the ith atom in terms of the displacements $\delta\mathbf{r}_j$ of all of the atoms. In a normal mode, every atom will oscillate with the same angular frequency ω so we may write each displacement as $\delta\mathbf{r}_j e^{-i\omega t}$ where the $\delta\mathbf{r}_j$ are complex amplitudes, but are independent of time. We set this force equal to the mass of the ith ion, M_i, times its acceleration, $-\omega^2\,\delta\mathbf{r}_i e^{-i\omega t}$. We obtain

$$\mathbf{F}_i = -\sum_j \frac{\partial^2 W}{\partial\mathbf{r}_i\,\partial\mathbf{r}_j}\,\delta\mathbf{r}_j e^{-i\omega t} = -M_i\omega^2\,\delta\mathbf{r}_i e^{-i\omega t}$$

which may be rewritten as

$$\sum_j \left(\frac{W_{ij}}{M_i} - \omega^2\,\delta_{ij}\right)\delta\mathbf{r}_j = 0 \tag{4.9}$$

where $W_{ij} = \partial^2 W/\partial\mathbf{r}_i\,\partial\mathbf{r}_j$.

There are three components of $\delta\mathbf{r}_j$ for each of the N atoms present, so there are $3N$ solutions. If we number these by $\mathbf{q}$, these $3N$ solutions are

$$\delta\mathbf{r}_j = \mathbf{S}_{jq} \qquad \text{with} \qquad \omega^2 = {\omega_q}^2$$

for $3N$ values of $\mathbf{q}$. Here we have let $\mathbf{S}_{jq}$ and $\mathbf{q}$ be vectors, each having three

[1] For reference, see R. J. Elliot, in W. H. Stevenson (ed.), "Phonons," Oliver & Boyd, Edinburgh, 1966.

components, rather than introduce extra indices. By substitution into Eq. (4.9) we obtain

$$\sum_j \left(\frac{W_{ij}}{M_i} - \omega_q^{\,2}\,\delta_{ij} \right) \mathbf{S}_{jq} = 0 \tag{4.10}$$

For a perfect crystal with a single atom per primitive cell, the $\mathbf{q}$ can be taken to be wavenumbers for plane waves satisfying periodic-boundary conditions, and we know by symmetry that we can write

$$\mathbf{S}_{jq} = \frac{\mathbf{S}(\mathbf{q}) e^{iq \cdot r_j}}{\sqrt{N}} \tag{4.11}$$

where $\mathbf{S}(\mathbf{q})$ is a unit vector with three components and $\mathbf{q}$ runs over three Brillouin zones. We are using the extended-zone representation again to avoid extra indices and we have selected normalization factors so that

$$\sum_j \mathbf{S}^*_{jq'} \mathbf{S}_{jq} = \mathbf{S}^*(\mathbf{q}) \mathbf{S}(\mathbf{q})\, \delta_{q'q}$$

Then the matrix $\mathbf{S}_{jq}$ is unitary, which follows directly from this equation and its inverse is $\mathbf{S}_{qj}^{\;-1} = \mathbf{S}^*_{jq}$.

Even if the crystal is not perfect there exist normal modes and they yield an $\mathbf{S}_{jq}$ which is unitary and has an inverse. (The fact that $\mathbf{S}$ is unitary is not so essential as the fact that it has an inverse. Furthermore, our notation is clearer if we write the inverse as $\tilde{\mathbf{S}}^{-1}$, where the tilde indicates a matrix.) We may now rewrite Eq. (4.10) explicitly as a diagonalization by operating on the left with the inverse of $\tilde{\mathbf{S}}$.

$$\sum_{ij} \mathbf{S}_{q'i}^{-1} \frac{W_{ij}}{M_i} \mathbf{S}_{jq} - \omega_q^{\,2}\,\delta_{qq'}\, \mathbf{S}^*(\mathbf{q}) \mathbf{S}(\mathbf{q}) = 0 \tag{4.12}$$

Thus the matrix $\tilde{\mathbf{S}}^{-1}(\tilde{W}/M)\tilde{\mathbf{S}}$ is diagonal and its diagonal elements are the squared eigenfrequencies. In solving Eq. (4.9) or (4.10) we are doing a very general problem. For example, in exactly the same way we can discuss impurity electronic states in a tight-binding approximation in which case the W_{ij} become matrix elements of the Hamiltonian based upon atomic states and ω^2 becomes E. For definiteness we discuss the evaluation only in terms of localized vibrational modes.

Rather than describe the calculation of the ω_q as a diagonalization of $\tilde{W}/M$, we may describe it in terms of the Green's function defined by the equation

$$\sum_j \left(\frac{W_{ij}}{M_i} - \omega^2\,\delta_{ij} \right) G_{jk} = \delta_{ik}$$

or written in matrix notation

$$\left(\frac{\tilde{W}}{M} - \omega^2\tilde{1}\right)\tilde{G} = \tilde{1} \tag{4.13}$$

Thus $\tilde{G}$ is the inverse of $\tilde{W}/M - \omega^2\tilde{1}$ (if one exists). We may obtain $\tilde{G}$ in terms of $\tilde{S}$ from Eq. (4.13) by multiplying on the left by $\tilde{S}^{-1}$ and on the right by $\tilde{S}$. We obtain

$$\tilde{S}^{-1}\left(\frac{\tilde{W}}{M} - \omega^2\tilde{1}\right)\tilde{G}\tilde{S} = \tilde{S}^{-1}\tilde{1}\tilde{S}$$

or

$$\left(\tilde{S}^{-1}\frac{\tilde{W}}{M}\tilde{S} - \omega^2\right)\tilde{S}^{-1}\tilde{G}\tilde{S} = \tilde{1}$$

Written again in terms of components we obtain, using also Eq. (4.12),

$$({\omega_q}^2 - \omega^2)\mathbf{S}_{q'j}^{-1}G_{jk}\mathbf{S}_{kq} = \delta_{q'q}$$

Thus

$$\mathbf{S}_{q'j}^{-1}G_{jk}\mathbf{S}_{kq} = \frac{\delta_{q'q}}{{\omega_q}^2 - \omega^2}$$

or

$$G_{jk} = \sum_q \frac{\mathbf{S}_{jq}\mathbf{S}_{qk}^{-1}}{{\omega_q}^2 - \omega^2} \tag{4.14}$$

As in the case of the one-electron Green's function there are poles in the Green's function at the eigenvalues. If we knew G_{jk} explicitly as a function of ω^2 for an imperfect crystal, we could seek poles to obtain the eigenvalues. To obtain the Green's function from Eq. (4.14), however, we must first solve the eigenvalue problem so nothing is gained; it is simply an alternative way of solving the problem. Here we will obtain the Green's function by solving the eigenvalue problem for the perfect crystal and use it to obtain eigenvalues in the presence of a defect. For a perfect crystal Eqs. (4.11) and (4.14) may be combined to give

$$G_{jk}{}^0 = \sum_q \frac{\mathbf{S}(\mathbf{q})\mathbf{S}^*(\mathbf{q})e^{i\mathbf{q}\cdot(\mathbf{r}_j - \mathbf{r}_k)}}{({\omega_q}^2 - \omega^2)N} \tag{4.15}$$

where the $\mathbf{S}(\mathbf{q})$ again are unit vectors in the direction of polarization of the mode of wavenumber $\mathbf{q}$. The superscript zero indicates the solution without a defect.

We now return to Eq. (4.9) which is to be solved for a crystal with a defect. We write W_{ij}/M_i as W_{ij}/M_i for the perfect crystal plus the change

$C_{ij} = \delta(W_{ij}/M_i)$ due to the defect. If there are changes in mass as well as in force constants it is convenient to multiply Eq. (4.9) through by the unperturbed mass M_i before introducing the δW_{ij} and δM_i. We divide afterward again by M_i. We then see that C_{ij} may be written as

$$C_{ij} = \frac{\delta W_{ij}}{M_i} - \omega^2 \frac{\delta M_i}{M_i} \delta_{ij}$$

Multiplying on the left by $\tilde{G}^0$, we see

$$\sum_{ij} G_{ki}{}^0\left(\frac{W_{ij}}{M_i} - \omega^2\delta_{ij}\right)\delta\mathbf{r}_j + \sum_{ij} G_{ki}{}^0 C_{ij}\delta\mathbf{r}_j = 0$$

or

$$\delta\mathbf{r}_k + \sum_{ijq} \frac{\mathbf{S}(\mathbf{q})e^{iq\cdot(r_k - r_i)}}{(\omega_q{}^2 - \omega^2)N} C_{ij}\mathbf{S}^*(\mathbf{q}) \cdot \delta\mathbf{r}_j = 0 \tag{4.16}$$

In the final equation we have noted that $G_{ki}{}^0$ is the inverse of $W_{ij}/M_i - \omega^2\delta_{ij}$ in evaluating the first term and have substituted for $G_{ki}{}^0$ from Eq. (4.15) in the second.

Solution of Eq. (4.16) again requires the diagonalization of very large matrix. That matrix $\tilde{M}$ is given by

$$M_{kj}(\omega^2) = \sum_{iq} \frac{\mathbf{S}(\mathbf{q})e^{iq\cdot(r_k - r_i)}}{(\omega_q{}^2 - \omega^2)N} C_{ij}\mathbf{S}^*(\mathbf{q}) \tag{4.17}$$

in terms of which Eq. (4.16) is

$$\sum_j [M_{kj}(\omega^2) + \delta_{kj}]\delta\mathbf{r}_j = 0 \tag{4.18}$$

However, the advantage is in the simple form which M_{kj} takes when C_{ij} arises from a localized defect. Thus, if C_{ij} is nonzero only for a small range of i and j (i.e., only for a few atoms), then M_{kj} is nonzero only in a few columns. If, for example, the defect is the modification of a single mass, say the nth mass, then M_{kj} is nonzero only for $j = n$. The matrix $M_{kj}(\omega^2) + \delta_{kj}$ takes the form

$$\tilde{M} + \tilde{1} = \left|\begin{matrix} 1 & 0 & 0 & 0 & 0 & M_{1n} & 0 & \ldots \\ 0 & 1 & 0 & 0 & 0 & M_{2n} & 0 & \ldots \\ 0 & 0 & 1 & 0 & 0 & \ldots & \ldots & \ldots \\ 0 & 0 & 0 & 1 & 0 & \ldots & \ldots & \ldots \\ 0 & \ldots & \ldots & \ldots & 1 & M_{(n-1)n} & \ldots & \ldots \\ \ldots & \ldots & \ldots & \ldots & \ldots & 1 + M_{nn} & 0 & \ldots \\ \ldots & \ldots & \ldots & \ldots & \ldots & M_{(n+1)n} & 1 & \ldots \\ \ldots & \ldots & \ldots & \ldots & \ldots & \ldots & \ldots & 1 \\ \ldots & \ldots & \ldots & \ldots & \ldots & \ldots & 0 & \ldots \end{matrix}\right| \tag{4.19}$$

and the solution of Eq. (4.18) requires that the determinant of Eq. (4.19) vanish. The determinant may be expanded trivially so that the condition $\mathrm{Det}(\tilde{M} + \tilde{1}) = 0$ becomes $1 + M_{nn} = 0$. Evaluating M_{nn} from Eq. (4.17) with $C_{nn} = -\omega^2 \delta M/M$, we obtain

$$1 - \frac{\delta M}{NM} \sum_q \mathbf{S}(\mathbf{q})\mathbf{S}^*(\mathbf{q}) \frac{\omega^2}{\omega_q^{\,2} - \omega^2} = 0 \tag{4.20}$$

If the C_{ij} had, for example, been nonzero for six values of i and of j, the determinantal equation would have contained a 6-by-6 matrix. In any case, this approach reduces the problem to a calculation with a number of degrees of freedom comparable to the number of force constants or masses changed, which was the goal of the approach. Furthermore, within the framework of the force-constant model the approach is exact; there is no restriction to small changes in parameters.

We must finally come to grips with the vector notation we have been using. The $\mathbf{S}(\mathbf{q})$ is a unit vector in the direction of polarization for the mode $\mathbf{q}$. Thus $\mathbf{S}(\mathbf{q})\mathbf{S}^*(\mathbf{q})$ is a projection operator selecting the component of any vector it operates upon (specifically $\delta\mathbf{r}_n$) parallel to $\mathbf{S}(\mathbf{q})$. Thus $\mathbf{S}(\mathbf{q})\mathbf{S}^*(\mathbf{q})$ is a matrix with elements in cartesian coordinates of $S_i(\mathbf{q})S_j^*(\mathbf{q})$ where, for example, S_i is the component of $\mathbf{S}$ in the ith direction. Similarly, the 1 in Eq. (4.20) is δ_{ij}. Written in components Eq. (4.20) becomes

$$\frac{\delta M}{NM} \sum_q S_i(\mathbf{q})S_j(\mathbf{q}) \frac{\omega^2}{\omega_q^{\,2} - \omega^2} = \delta_{ij}$$

For an exact solution we need to know the frequencies and polarizations of each of the normal modes of the perfect crystal. The solution then will lead to the frequencies of all of the normal modes of the crystal with the defect. For a defect with cubic symmetry in a cubic crystal it is clear that the result of the summation over $\mathbf{q}$ must have cubic symmetry; i.e., it must be a multiple of the unit matrix δ_{ij}. For all ω and for each i it must be equal to one-third of the value obtained by replacing $S_i(\mathbf{q})S_j(\mathbf{q})$ by 1. Thus for the cubic case we may write

$$\frac{\delta M}{3MN} \omega^2 \sum_q \frac{1}{\omega_q^{\,2} - \omega^2} = 1 \tag{4.21}$$

The nature of the solutions of Eq. (4.21) may be easily seen by sketching the left side, which we call Σ, as a function of ω, noting that there is a pole at each eigenvalue ω_q of the perfect crystal. We note also that for ω above the highest ω_q every term in the sum is negative. Figure 4.12 gives the plot in the upper curve when $\delta M > 0$. We show only $\omega > 0$ since Eq. (4.21) is clearly even in ω. Note that $\Sigma = 1$ at a value slightly below each of the unperturbed mode frequencies ω_q. An increase in the mass of one atom slightly depresses each mode, but never further than the frequency difference to the next lowest mode. This is reminiscent of our finding with phase shifts that a perturbing local potential slightly shifts each energy eigenvalue. In

Fig. 4.12 *A schematic plot of the summation Σ as a function of frequency ω, shown above for a heavy impurity and below for a light impurity. The plot is made with only 11 distinct frequencies for the 3N modes of the perfect crystal. Solutions with the impurity are obtained where $\Sigma = 1$. Note that for a light impurity the highest frequency mode ω_{3N} is pulled out of the band to the frequency ω'_{3N}. It is a local mode.*

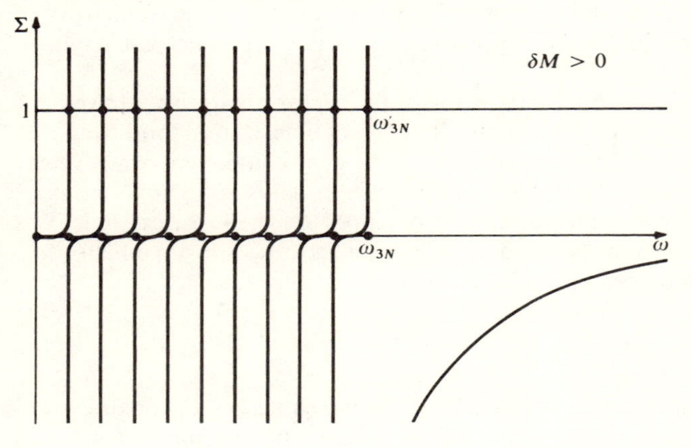

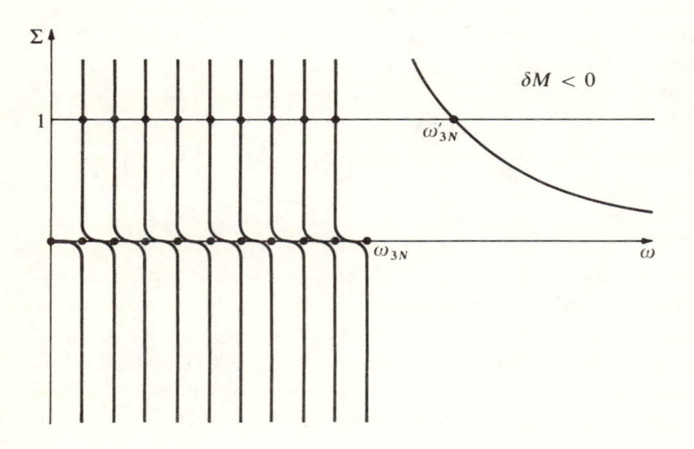

this case, however, no local mode can be pulled out of the bottom of the band since the band extends to zero frequency. The $\omega = 0$ mode (uniform translation of the lattice) is not obtained explicitly since Σ is not defined at $\omega = 0$.

In the lower portion of the figure we have plotted Σ for the case $\delta M < 0$. We see that each frequency is raised slightly by the lighter mass. Again, for all but the uppermost mode, the shifts are restricted to be very tiny, of order $\omega_{3N}/3N$. The shift of the uppermost mode, however, may be very large. For a light defect a localized mode can be pulled out of the continuum. It is plausible physically that a light mass can generate a local mode while a heavy mass cannot. The light mass can vibrate by itself, leaving the medium largely undisturbed, while the heavy mass will inevitably carry the medium with it. We may note also that for the cubic defect we expect the mode pulled out to be triply degenerate, with amplitudes in three perpendicular directions.

The frequency of the local mode may be obtained from Eq. (4.21). Noting that for $\omega > \omega_{3N}$ there are no singularities in the summand, we may replace the sum over $\mathbf{q}$ by an integration over ω, introducing the density of modes $n(\omega)$. Then Eq. (4.21) becomes

$$\frac{\delta M \omega^2}{M} \frac{1}{3N} \int \frac{d\omega_q n(\omega_q)}{\omega_q^{\,2} - \omega^2} = 1$$

The integration may be performed if the distribution of frequencies $n(\omega_q)$ has been calculated or measured. If we assume a Debye spectrum, then $n(\omega_q)/3N = 3\omega_q^{\,2}/\omega_D^{\,3}$ where ω_D is the cutoff frequency and the integration can be performed analytically. We obtain the condition

$$\frac{3\delta M}{M}\left(\frac{\omega}{\omega_D}\right)^2\left(1 - \frac{\omega}{2\omega_D}\, ln\left|\frac{\omega/\omega_D + 1}{\omega/\omega_D - 1}\right|\right) = 1 \tag{4.22}$$

A plot of $(\omega - \omega_D)/\omega_D$ as a function of $-\delta M/M$ is given in Fig. 4.13.

It is also possible to evaluate the squared amplitude of the local mode at the defect atom relative to the squared amplitude of displacement averaged over the crystals. For a truly localized mode, the amplitudes decay exponentially with distance and this ratio diverges for an infinite crystal. For the Debye spectrum we find a local mode is always produced, but this is due to the assumed discontinuity in $n(\omega)$. For a realistic $n(\omega)$, a finite δM is required to obtain an exponentially decaying mode. This is also in direct analogy with the localization of electronic states by an attractive local potential.

If the same treatment is applied to a crystal with optical modes separated in frequency from the acoustical modes, the curves analogous to Fig. 4.12 show immediately that a light defect may pull a local mode from the top of the optical branch as well as from the top of the acoustical branch.

Similarly, a heavy defect may pull a local mode from the *bottom* of the optical band.

It is not apparent from Fig. 4.12 how a resonant vibrational state can arise, though the analogy with the phase-shift treatment of electronic states would suggest the possibility. We have sketched Σ as a tangent function of the frequency and this is appropriate when δM is sufficiently small. Then Σ crosses zero midway between unperturbed values. For larger values of δM we may estimate the value of Σ at such midpoints by replacing the sum by an integral and obtaining the principal value. Again using the Debye approximation for the frequencies we find that these midpoint values of Σ grow quadratically with ω so for sufficiently large masses the upper curve of Fig. 4.12 is deformed as shown in Fig. 4.14. Clearly a resonant mode has been inserted and the neighboring modes displaced away from it just as in electron resonances. The resonant frequency may be obtained from Eq. (4.22) since that came from the evaluation of precisely the integral we need here, though it was obtained in connection with localized modes. We see from Eq. (4.22) that if δM is positive and large compared to M the resonance will occur at a low frequency given approximately by $\omega \sim \omega_D\sqrt{M/(3\,\delta M)}$.

We have completed the calculation here for a defect in which only the mass is changed since the mathematics is slightly simpler for this case. It has been natural for the theory to focus upon such mass defects not only because

Fig. 4.13 The relative frequency shift of the local mode as a function of the relative lowering of mass computed using the Debye spectrum (after R. J. Elliott, in W. H. Stevenson (ed.), "Phonons," Oliver & Boyd, Edinburgh, 1966).

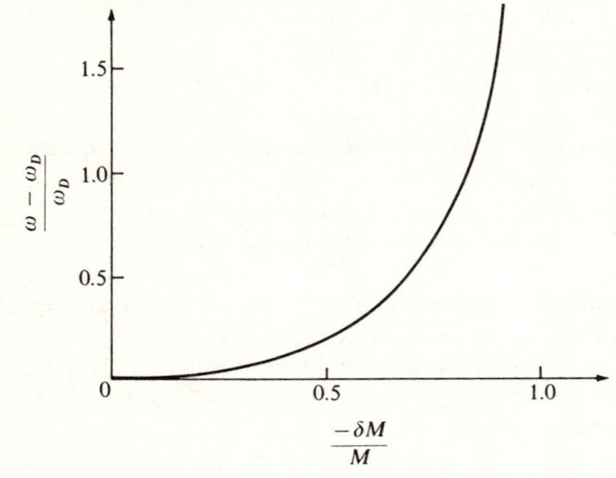

the mathematics is slightly simpler but also because we know the mass difference for any given defect, whereas it would be difficult to estimate changes in force constants due to an impurity. One of the most interesting cases, however, is a very low-lying resonant mode due to lithium substituted for potassium in KBr.† Since the impurity is light we might expect a truly localized mode. The appearance of a resonance indicates that the coupling of the impurity to the lattice must be very weak; experiments indicate a coupling of only 0.6 percent of that in the host lattice. The small lithium ion is virtually rattling around in the large hole left by the potassium ion. It seems likely that in many cases the effects arising from changes in coupling may be large compared to the effects of mass difference.

4. *Electron-Phonon Interactions*

It is clear that the vibration of the lattice must modify the electron behavior in a solid. In metals, for example, a longitudinal vibration of the ions will give rise to charge accumulation and, when properly screened, a potential with spatial dependence of the same form as that of the lattice vibration. This potential of course enters the electron Hamiltonian and gives a coupling between the electrons and the vibration. It would be possible to solve the electron-phonon problem completely in order to obtain eigenstates of the

†A. J. Sievers and S. Takeno, *Phys. Rev.*, **140**:A1030 (1965).

Fig. 4.14 *A plot of Σ as in Fig. 4.12 for the case when the impurity is very heavy in comparison to the host atoms.*

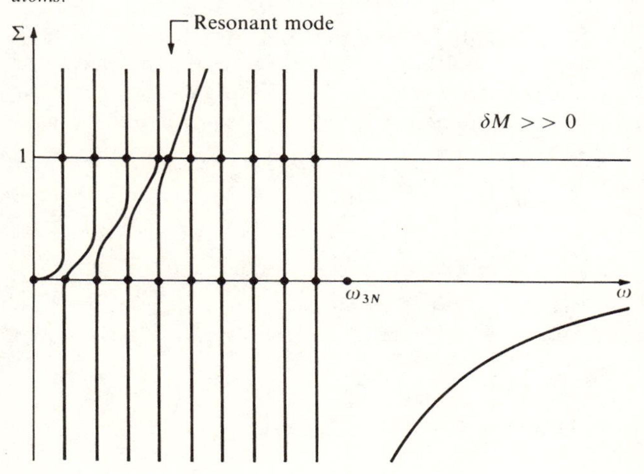

electron-phonon system. We have already partially made such a solution when we included electron screening in the vibration mode itself. This removes a portion of the electron-phonon interaction and leaves us with a residual screened field. Our construction of polarons in ionic crystals was another case in which some of the interaction between the electrons and the lattice was incorporated in our description of the electronic state. It is not ordinarily propitious to carry this solution for eigenstates all the way. Generally it is much more convenient to obtain approximate eigenstates both for the electrons and the lattice vibrations and to consider the residual interaction as a perturbation which we call the *electron-phonon interaction.* The electron-phonon interaction is not uniquely defined; it depends on the extent to which the initial coupling has already been incorporated in what we call the electrons or the phonons. However, the procedure in any given system is sufficiently standardized that there is ordinarily no ambiguity.

4.1 Classical theory *Ionic crystals.* The strongest and perhaps the simplest electron-phonon interaction occurs between longitudinal optical modes and electrons in ionic crystals. In an ionic crystal such as sodium chloride the positive and negative ions are oppositely displaced in such modes at long wavelengths. For simplicity of notation we neglect the difference in magnitude of the two since only the net separation will enter.

$$\delta \mathbf{r}_i^+ = \frac{\mathbf{u}}{\sqrt{N}} e^{i(q \cdot r_i^+ - \omega t)}$$

$$\delta \mathbf{r}_i^- = -\frac{\mathbf{u}}{\sqrt{N}} e^{i(q \cdot r_i^- - \omega t)}$$

For longitudinal modes $\mathbf{u}$ is parallel to $\mathbf{q}$. At long wavelengths we may neglect the difference in position of the two ions in a primitive cell and write the dipole moment in the cell as $2Ze\mathbf{u}N^{-1/2}e^{i(q \cdot r_i - \omega t)}$ with Ze the charge on each ion. Thus at long wavelengths the dipole moment per unit volume (the polarization) is given by

$$\mathbf{P} = \frac{2Ze}{\Omega_c} \frac{\mathbf{u}}{\sqrt{N}} e^{i(q \cdot r - \omega t)}$$

where Ω_c is the cell volume. The local charge density is simply the negative of the divergence of the polarization. From the charge density we may compute the electrostatic potential energy using Poisson's equation. We obtain an electron potential

$$V(\mathbf{r}) = \frac{8\pi Ze^2}{q^2 \Omega_c \sqrt{N}} i\mathbf{q} \cdot \mathbf{u} e^{i(q \cdot r - \omega t)}$$

The $V(\mathbf{r})$ is called the electron-phonon interaction for this case. Note that it diverges as q approaches zero. This is a strong coupling case and it is this divergence which gave rise to the strong polaron effects in ionic crystals discussed earlier.

Such a divergence does not occur for transverse optical modes in which there is no charge accumulation. Similarly, in long-wavelength acoustic modes the neighboring charges tend to move in phase and the strong electrostatic coupling does not occur. Only in piezoelectric crystals where uniform or slowly varying strains give rise to electric polarization does the electrostatic coupling arise for acoustical modes. In that case the electron-phonon interaction is calculable from the piezoelectric constant.[1] Its divergence is weaker at long wavelengths than that of the optical mode.

Semiconductors In nonpolar semiconductors such as germanium or silicon there is no polar effect and we must look more closely to obtain the electron-phonon interaction. In this problem also, attention is ordinarily focused upon the long-wavelength vibration modes. Electrons and holes that are present tend to be concentrated in small regions of wavenumber space at the band extrema and the scattering of electrons occurs predominantly with small-wavenumber phonons. Consider first the interaction with longitudinal modes which is ordinarily described in terms of a *deformation potential.*

We envisage an energy-band structure with valence-band maxima and conduction-band minima as shown in Fig. 4.15. The bands at the left may have been obtained from a band calculation for the crystal at the normal atomic volume. If we were to expand the crystal corresponding to a fractional change in volume given by the dilatation Δ, we would obtain slightly different bands as shown to the right of the figure. It should be pointed out that in fact an energy-band calculation would give only the change in the band *gap*, i.e., the difference between the two shifts, since the zero of energy is always somewhat arbitrary in an energy-band calculation. However, the shifts themselves could in principle be calculated by introducing not a uniform dilatation but a spatially varying deformation, and obtaining a self-consistent solution of the problem. For our purposes this is not an important point since the actual magnitudes of the shifts are ordinarily treated as disposable parameters and fit to experiment. The shifts in the energy of the band extrema will in general be linear in the dilatation Δ. The proportionality constant is called the *deformation-potential constant.*

$$\delta V_c = D_c \Delta \qquad \delta V_v = D_v \Delta$$

[1] H. J. G. Meijer and D. Polder, *Physica*, **19**:355 (1953); W. A. Harrison, *Phys. Rev.*, **100**:903 (1956).

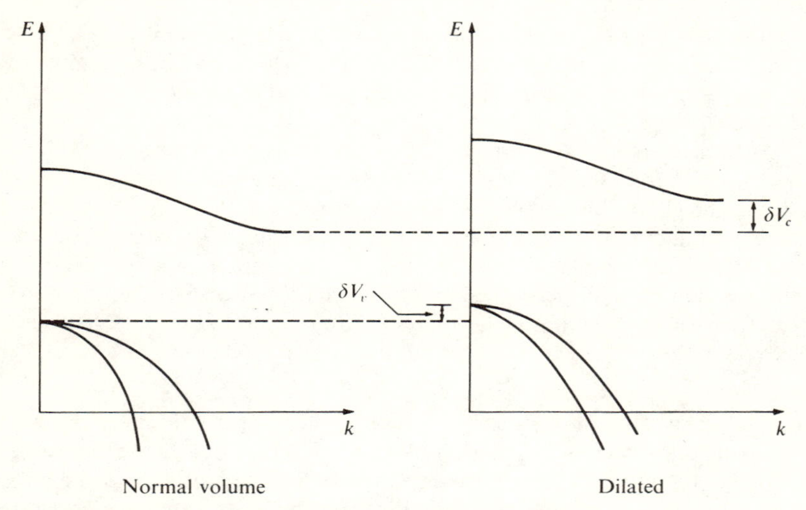

Fig. 4.15 The deformation of the valence and conduction bands in a semiconductor due to a uniform dilation of the crystal.

Thus an electron at the conduction-band minimum, for example, will have its energy shifted by an amount proportional to the dilatation Δ. It is then assumed that if the dilatation varies with position there is an effective potential seen by the electron which also varies with position.

$$\delta V_c(\mathbf{r}) = D_c\,\Delta(\mathbf{r}) \qquad \delta V_v(\mathbf{r}) = D_v\,\Delta(\mathbf{r})$$

The effective potentials would of course be slightly different for electrons away from the band extrema. However, these constants are evaluated at the band minimum and used for all electrons near the minimum.

We would expect intuitively that the magnitude of the deformation-potential constants should be of the order of the band gaps. This turns out to be the case. Given these constants we have directly the electron-phonon interaction just as in the polar crystals. In the case of polar crystals $i\mathbf{q}\cdot(\mathbf{u}/\sqrt{N})e^{i(q\cdot r-\omega t)}$ is the dilatation, but instead of the constant D we obtained $8\pi Ze^2/(\Omega_c q^2)$. In the case of nonpolar crystals we do not have the long-wavelength divergence. This same deformation-potential interaction occurs also in the polar crystals. However, it is ordinarily dominated by the much stronger polar interaction and need not be considered.

This analysis may be directly extended to shear strains and therefore to the interaction with transverse modes and optical modes. In general the strain is written

$$\epsilon_{ij} = \frac{1}{2}\left(\frac{\partial \delta_i}{\partial x_j} + \frac{\partial \delta_j}{\partial x_i}\right)$$

where $\partial\delta_i/\partial x_j$ is the change of displacement in the ith direction with distance in the jth direction. Then the deformation-potential constant is replaced by a tensor and we have

$$\delta V(\mathbf{k},\mathbf{r}) = \sum_{ij} D_{ij}(\mathbf{k})\epsilon_{ij}(\mathbf{r})$$

For the present we will focus on the simpler case, that of longitudinal acoustical waves.

Because of the low frequencies of the acoustical modes it is possible to correctly compute their contributions to the electron scattering by conceptually freezing the atoms at their positions in the deformed crystal and computing the electron scattering associated with the corresponding distortions just as we calculated the scattering by defects in crystals. (We will see how the motion of the lattice is properly included when we quantize the lattice vibrations.) An electron may be scattered from the state of wavenumber $\mathbf{k}$ to the state of wavenumber $\mathbf{k}'$ by the Fourier component of wavenumber $\mathbf{q} = \mathbf{k}' - \mathbf{k}$ of the deformation potential. This in turn will arise only from the lattice vibrations of wavenumber $\pm\,\mathbf{q}$ if we have written displacements in the form $(\mathbf{u}/\sqrt{N})e^{i(q\cdot r-\omega t)} + (\mathbf{u}^*/\sqrt{N})e^{-i(q\cdot r-\omega t)}$.

The scattering rate between these two states for the electron will be proportional to the square of the deformation potential and therefore proportional to the square of the dilatation. The energy in the mode will also be proportional to the square of the dilatation, and at high temperatures will be proportional to KT. There is also a dependence of the scattering rate on the energy of the electron through the density of states. Since the average energy of electrons varies also with temperature, this leads to additional temperature dependence. The calculation leads finally (see Sec. 4.3) to an electron mobility which varies as $T^{-3/2}$.

In addition there may be scattering processes in which electrons are scattered between the different degenerate conduction-band minima. Such processes, which are called *intervalley scattering*, may be handled in essentially the same way but require a generalization of the deformation-potential tensor defined above. The necessary generalization was described in Sec. 6.1 of Chap. II.

In electron scattering by optical modes the frequency of the modes themselves clearly cannot be neglected. The emission of an optical phonon causes a change in electron energy equal to $\hbar\omega$ for the optical mode. This will ordinarily be of the order of or larger than room temperature. Such processes become important only when the electrons acquire large energies (become hot). In that case optical-mode scattering can become the predominant mechanism for energy loss of the electrons since scattering by acoustical modes involves very small changes in energy, and in fact changes which we have neglected in the above calculation by freezing the lattice.

Simple metals In metals it becomes necessary to consider the interaction with short-wavelength lattice vibrations since we will be interested in scattering that takes electrons across the Fermi surface and therefore involves very large changes in wavenumber. Fortunately, in the simple metals the pseudopotential method provides a clear approach to the treatment of electron-phonon interaction.

We have noted that the ions in the metal interact with the electrons through a pseudopotential, the matrix elements of which may be written

$$\langle \mathbf{k} + \mathbf{q}|W|\mathbf{k}\rangle = S(\mathbf{q})\langle \mathbf{k} + \mathbf{q}|w|\mathbf{k}\rangle$$

where $\langle \mathbf{k} + \mathbf{q}|w|\mathbf{k}\rangle$ is the form factor which depends only on the properties of an individual ion. All of the information about the ion configuration is contained in the geometrical structure factor $S(\mathbf{q})$. The electron-phonon interaction is describable completely in terms of the additional matrix elements which occur due to the displacements of the ions from their equilibrium positions and this in turn enters only through the structure factor.

We will again simplify the problem by freezing the lattice, or equivalently considering the instantaneous positions of the ions. We will also for simplicity write our equations for crystals with one atom per cell. A lattice vibration of wavenumber $\mathbf{Q}$ will cause displacements from equilibrium given by

$$\delta\mathbf{r}_j = \frac{\mathbf{u}_Q}{\sqrt{N}} e^{iQ\cdot r_j} + \frac{\mathbf{u}_Q^*}{\sqrt{N}} e^{-iQ\cdot r_j}$$

We use a capital $\mathbf{Q}$ for the phonon wavevector here to distinguish it from the general wavenumber $\mathbf{q}$. We think of the amplitude vectors as constant in time but could alternatively regard them as functions of time which would of course simply give a time dependence to the matrix elements in the end. We may substitute the ion positions $\mathbf{r}_j + \delta\mathbf{r}_j$ into the structure factor and expand the exponent for small amplitudes.

$$S(\mathbf{q}) = \frac{1}{N}\sum_j e^{-iq\cdot[r_j + (u_Q/\sqrt{N})e^{iQ\cdot r_j} + (u_Q^*/\sqrt{N})e^{-iQ\cdot r_j}]}$$

$$= \frac{1}{N}\sum_j e^{-iq\cdot r_j}\left(1 - \frac{i\mathbf{q}\cdot\mathbf{u}_Q}{\sqrt{N}} e^{iQ\cdot r_j} - \frac{i\mathbf{q}\cdot\mathbf{u}_Q^*}{\sqrt{N}} e^{-iQ\cdot r_j}\right)$$

The first term is simply the structure factor that would occur in the absence of distortion. It is equal to unity for lattices with one atom per primitive cell if **q** is a lattice wavenumber and is zero otherwise. The first-order terms are the electron-phonon interaction and may be rewritten

$$\frac{-i\mathbf{q}\cdot\mathbf{u}_Q}{\sqrt{N}}\,\frac{1}{N}\sum_j e^{-i(q-Q)\cdot r_j} - \frac{i\mathbf{q}\cdot\mathbf{u}_Q^*}{\sqrt{N}}\,\frac{1}{N}\sum_j e^{-i(q+Q)\cdot r_j}$$

The sums are of course over the perfect lattice. Thus in the first term, for example, the sum will be identically zero unless $\mathbf{q} - \mathbf{Q}$ is equal to a lattice wavenumber, in which case it will be equal to N. (We will denote lattice wavenumbers by $\mathbf{q}_0$.) Similarly, the second will vanish unless $\mathbf{q} + \mathbf{Q}$ is equal to a lattice wavenumber. These new electron-phonon matrix elements have arisen at wavenumbers which are satellites to each of the lattice wavenumbers for the perfect crystal as shown in Fig. 4.16.

We will wish to consider the effects of these new matrix elements both as electron scatterers and through their shift in the energy of the electronic states. We begin with scattering and therefore with time-dependent perturbation theory. In the pseudopotential method we will take as zero-order states the plane-wave pseudowavefunctions. The single vibrational mode which we have introduced has given rise to many matrix elements and therefore couples any single zero-order state with many others.

Fig. 4.16 Additional nonzero structure factors introduced by the presence of a lattice vibration of wavenumber **Q** *are given by empty points. Filled points give the lattice wavenumbers of the undistorted lattice.*

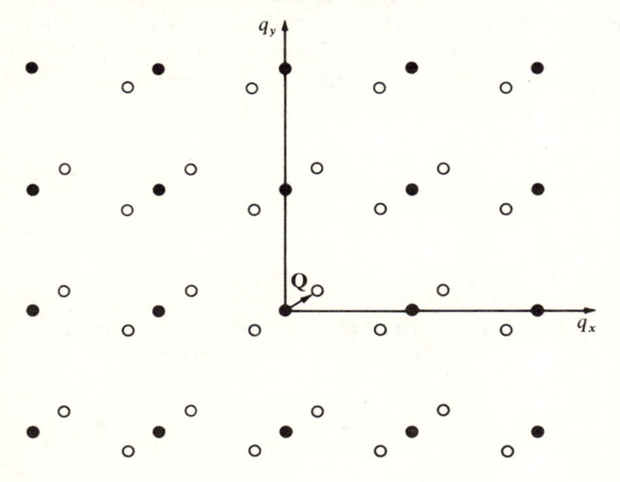

Considering first the satellites of the lattice wavenumber at the origin, $\mathbf{q}_0 = 0$, we may immediately write the electron-phonon interaction matrix element for $\mathbf{q} = \mathbf{Q}$

$$S(\mathbf{Q})\langle\mathbf{k} + \mathbf{Q}|w|\mathbf{k}\rangle = \frac{-i\mathbf{Q}\cdot\mathbf{u}_Q}{\sqrt{N}}\langle\mathbf{k} + \mathbf{Q}|w|\mathbf{k}\rangle$$

Note that if the vibrational modes are purely longitudinal and purely transverse, only the longitudinal mode will give nonvanishing coupling. This result is formally the same as that of deformation-potential theory but with the deformation-potential constant now replaced by the negative of the form factor. The leading factor is minus the dilatation. In this case the deformation-potential constant depends upon the wavenumber of the phonon involved. At long wavelengths the form factor approaches $-\frac{2}{3}E_F$. Thus the longitudinal deformation-potential constant approaches $\frac{2}{3}E_F$ at long wavelengths. Scattering in metals which arises from these satellites to the origin in wavenumber space is called *normal scattering*.

We may also have scattering from the satellites to neighboring lattice wavenumbers. Such events are called *Umklapp scattering*. We may see how such satellites can give rise to scattering between two states on the Fermi surface in Fig. 4.17. The electron-phonon interaction matrix element is not so simple in this case but may be written immediately

$$S(\mathbf{q})\langle\mathbf{k} + \mathbf{q}|w|\mathbf{k}\rangle = -i(\mathbf{q}_0 + \mathbf{Q})\cdot\frac{\mathbf{u}_Q}{\sqrt{N}}\langle\mathbf{k} + \mathbf{q}_0 + \mathbf{Q}|w|\mathbf{k}\rangle$$

In this case even if the modes are purely longitudinal and transverse, we obtain contributions from both. In such a scattering event the electron changes its wavenumber by the sum of the phonon wavenumber $\mathbf{Q}$ and a lattice wavenumber. Thus even with long-wavelength phonons there may be very large-angle scattering. Because of this Umklapp frequently dominates the resistivity.

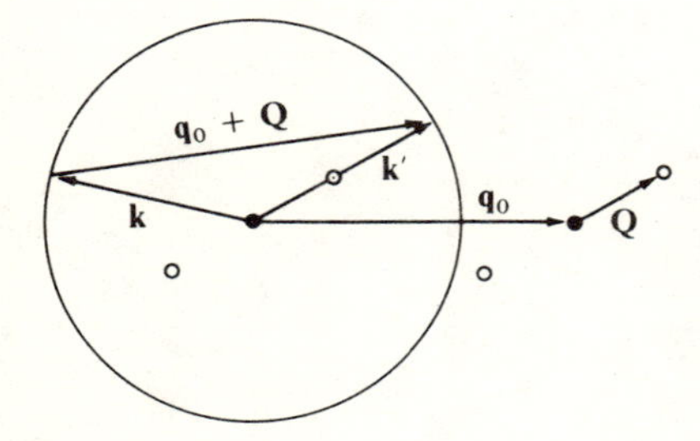

Fig. 4.17 Umklapp scattering of an electron from a state of wavenumber $\mathbf{k}$ *to one of wavenumber* $\mathbf{k}'$ *which differs from* $\mathbf{k}$ *by the sum of a lattice wavenumber* $\mathbf{q}_0$ *and the wavenumber of the lattice vibration* $\mathbf{Q}$.

It is possible to think of such scattering events as simultaneous interactions with the lattice vibration and Bragg reflections by the periodic potential of the perfect lattice. However, it is important to note that these are not second-order processes. The scattering amplitude is proportional only to the first power of the pseudopotential and there are no energy denominators.

There is no fundamental distinction between normal and Umklapp scattering. The distinction is sharp only at long wavelengths. As we increase the wavenumber of a phonon until it approaches the Brillouin zone face we continue to call the scattering normal. We may describe the electrons in a reduced-zone scheme, but the vibration in a periodic-zone scheme. Then as we continue the vibrational wavenumber beyond the Brillouin zone face the polarization of the wave varies continuously with wavenumber and similarly the electron-phonon interaction varies continuously. However, we conventionally represent such a vibration of the system in the reduced zone and would therefore conventionally represent the wavenumber of the phonon not by a $\mathbf{Q}$ lying outside the Brillouin zone, but by one of the other satellites which has now moved into the Brillouin zone. Physically there has been no discontinuity. It is simply a discontinuity arising from our convention.

It is sometimes a mathematical convenience to describe Umklapp scattering in monovalent metals by representing the vibration in the reduced-zone scheme but the electrons in a periodic-zone scheme such as is shown in Fig. 4.18. We would then say that a phonon of wavenumber $\mathbf{Q}$ (which by convention must lie within the Brillouin zone) may either scatter an electron from one state in the first Brillouin zone to another (normal scattering) or it may scatter an electron between Brillouin zones (Umklapp scattering). Since the Fermi surfaces in different Brillouin zones are simply equivalent representations of the same states, these Umklapp events may be reinterpreted in terms of scattering within the Brillouin zone. We can see here one important feature of Umklapp scattering. It is restricted to wavenumbers that are greater than or equal to the closest approach of two adjacent Fermi surfaces. Thus at sufficiently low temperatures when only the long-wavelength modes are excited, the Umklapp scattering can be frozen out. It turns out, however, that in sodium, for example, this minimum wavenumber is only of the order of 20 percent of the Brillouin zone radius and Umklapp scattering dominates even down to helium temperatures.

4.2 Second quantization In our discussion of the electron-phonon interaction in the preceding section we postulated displacements of the lattice and then computed the resulting potential seen by the electrons. This constitutes a classical treatment of the lattice vibrations. When we treated

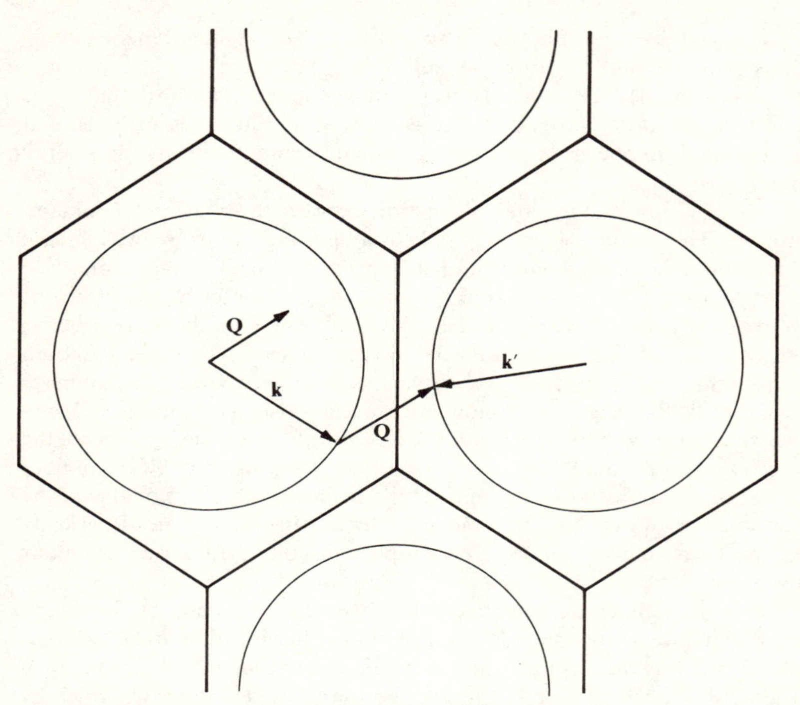

Fig. 4.18 Umklapp scattering when the electrons are represented in a periodic-zone scheme, but the vibrations are represented in a reduced-zone scheme.

the contribution to the specific heat due to lattice vibrations we found it convenient and necessary to treat within a quantum-mechanical framework the Hamiltonian obtained classically. It is also convenient and sometimes necessary to make a similar quantum-mechanical treatment of the electron-phonon interaction. In particular this will become necessary when we examine the microscopic theory of superconductivity. It is therefore convenient at this point to systematically introduce a quantum-mechanical description of both the electron and the vibrational states of the system including the electron-phonon interaction.

The essential approach is to imagine a wavefunction for the system which is then a function both of the electron coordinates and of the ion coordinates. We write the latter, of course, in terms of the vibration amplitudes, or normal coordinates. In the absence of the electron-phonon interaction the Hamiltonian would contain two terms, $\mathscr{H}_e + \mathscr{H}_\varphi$. The first term depends only upon the electron coordinates, the second only upon the vibration coordinates. Then the wavefunction for the system could be written as a product of electronic wavefunctions and vibration wavefunctions. Our approach will be to formulate these product wavefunctions and then to introduce the electron-phonon interaction, $\mathscr{H}_{e\varphi}$, which contains both electron and vibrational coordinates and which therefore couples various of these unperturbed states.

Electron states In nonrelativistic quantum mechanics the use of second quantization for electron states is purely a matter of notation. We begin by describing states in the one-electron approximation, in which case we may determine all of the one-electron states which we denote by wavenumbers $\mathbf{k}_1$, $\mathbf{k}_2$, ... These may be the solutions of the Hartree-Fock equations (2.14) described in Sec. 3.2 of Chap. II. Each state may be occupied by an electron of spin up or spin down. We let the index $\mathbf{k}_i$ specify both the wavenumber and the spin of the state in question. If we have N electrons occupying the states $\mathbf{k}_1, \mathbf{k}_2, \ldots, \mathbf{k}_N$, then we saw in Sec. 3.2 of Chap. II that the many-electron wavefunction could be written as a Slater determinant,

$$\Psi = \frac{1}{\sqrt{N!}} \begin{vmatrix} \psi_{k_1}(\mathbf{r}_1) & \psi_{k_1}(\mathbf{r}_2) & \ldots & \psi_{k_1}(\mathbf{r}_N) \\ \psi_{k_2}(\mathbf{r}_1) & \psi_{k_2}(\mathbf{r}_2) & \ldots & \ldots \\ \psi_{k_3}(\mathbf{r}_1) & \ldots & \ldots & \ldots \\ \cdots & \cdots & \cdots & \cdots \\ \psi_{k_N}(\mathbf{r}_1) & \ldots & \ldots & \psi_{k_N}(\mathbf{r}_N) \end{vmatrix}$$

This is a very cumbersome expression but all of the information contained in it may be given simply by listing the indices for the occupied states, $\mathbf{k}_1, \mathbf{k}_2, \mathbf{k}_3, \ldots, \mathbf{k}_N$ in the proper order. We may therefore use a shorthand notation for the state; the customary form is

$$\Psi = c_{k_N}{}^{+} c_{k_{N-1}}^{+} \ldots c_{k_2}{}^{+} c_{k_1}{}^{+} |0\rangle \tag{4.23}$$

We think of the state $|0\rangle$ as the vacuum with $c_{k_i}{}^{+}$ being called a *creation operator* which adds a particle in the state $\mathbf{k}_i$. Since this expression contains all of the information contained in the Slater determinant, we may of course construct one from the other.

In addition we can deduce properties of the creation operators from the known properties of the Slater determinant. For example, we know that interchanging two rows in the Slater determinant changes the sign of the wavefunction. Thus we may write immediately the *commutation relations* for the creation operators which represent the antisymmetry of the wavefunction.

$$c_{k_1}{}^+ c_{k_2}{}^+ + c_{k_2}{}^+ c_{k_1}{}^+ = 0 \tag{4.24}$$

It follows that if any state appears more than once in the expression of Eq. (4.23) for the wavefunction that the wavefunction must be zero. This is immediate if the identical operators are adjacent; if they are not, they can be brought together by commuting one to the other step by step using Eq. (4.24). This of course also follows from the Slater determinant and is simply a reflection of the Pauli principle.

We know in addition that the Slater determinant is a normalized wavefunction. This will be achieved in our second quantized notation if we define the complex-conjugate wavefunction by

$$\Psi^* = \langle 0| c_{k_1} c_{k_2} c_{k_3} \ldots c_{k_N}$$

and if we require that

$$c_{k_1} c_{k_2} + c_{k_2} c_{k_1} = 0$$

$$c_{k_1}{}^+ c_{k_2} + c_{k_2} c_{k_1}{}^+ = \delta_{k_1 k_2}$$

$$c_k|0\rangle = 0$$

$$\langle 0|c_k{}^+ = 0$$

$$\langle 0|0\rangle = 1$$

We have introduced *annihilation operators* c_{k_1}. Note that the complex-conjugate wavefunction has the ordering of the indices reversed.

We may readily verify the orthonormality of the many-particle states by following the above commutation relations. We may write out the normalization integral,

$$(\Psi, \Psi) = \langle 0| c_{k_1} c_{k_2} \ldots c_{k_N} c_{k_N}{}^+ c_{k_{N-1}}{}^+ \ldots c_{k_2}{}^+ c_{k_1}{}^+ |0\rangle$$

Note that because of the reordering of the operators in the left-hand wavefunction, annihilation and creation operators for the same state are brought together at the center. We may interchange these giving $1 - c_{k_N}{}^+ c_{k_N}$. The term with the 1 is simply the normalization integral for a state with one less electron; the effect of the c_{k_N} has simply been to annihilate the electron in the state $\mathbf{k}_N$. The remaining term may be shown to be equal to zero by commuting the c_{k_N} to the right until it reaches $|0\rangle$. We may continue to contract the integral step by step until we reach $\langle 0|0\rangle = 1$, thus demonstrating the normalization. If either the right- or the left-hand wavefunction contained a state which was not contained in the other, that operator could be commuted to the appropriate side $c_{k_i}|0\rangle$ or $\langle 0|c_{k_i}{}^+$, either of which is zero, demonstrating the orthogonality of the states.

We have simply defined a method for writing the many-electron states. We will now see how quantum-mechanical operators are written in this same framework. It is possible to show that each of the prescriptions which we will write is entirely equivalent to the corresponding prescription using Slater determinants. These demonstrations are completely straightforward; they require careful bookkeeping in using the determinants. Here we will simply state the prescription.

The states we have denoted by $\mathbf{k}_i$ are eigenstates of energy ϵ_{k_i} of the electron Hamiltonian $\mathscr{H}_e$, which we may write in second-quantized form.

$$\mathscr{H}_e = \sum_k \epsilon_k c_k^+ c_k \tag{4.25}$$

We may immediately obtain the expectation value of this Hamiltonian with respect to our many-particle eigenstates. Making use of the commutation relations as we did in the normalization integral, we may directly show that the expectation value of the Hamiltonian is simply the sum of ϵ_k over all occupied states.

We may also wish to add a potential $V(\mathbf{r})$ which is seen by each of the electrons. This takes the form

$$V(\mathbf{r}) = \sum_{k,k'} \langle \mathbf{k}' | V | \mathbf{k} \rangle c_{k'}^+ c_k \tag{4.26}$$

Here the number $\langle \mathbf{k}' | V | \mathbf{k} \rangle$ is the matrix element of the potential between the one-electron eigenstates,

$$\langle \mathbf{k}' | V | \mathbf{k} \rangle = \int \psi_{k'}^*(\mathbf{r}) V(\mathbf{r}) \psi_k(\mathbf{r})\, d^3r \tag{4.27}$$

The diagonal terms, $\mathbf{k}' = \mathbf{k}$, simply add to the zero-order energies, ϵ_k. The off-diagonal matrix elements connect many-electron wavefunctions in which one of the electrons has changed its state. The integration in Eq. (4.27) is over spin coordinates as well as spatial coordinates, but since $V(\mathbf{r})$ does not depend upon spin only states $\mathbf{k}$ and $\mathbf{k}'$ of the same spin give nonzero matrix elements, and only states of the same spin are coupled.

We may also be interested in adding to our zero-order Hamiltonian an interaction between electrons, $\frac{1}{2} \sum V(\mathbf{r}_i - \mathbf{r}_j)$. This takes the form

$$V(\mathbf{r}_1,\mathbf{r}_2) = \tfrac{1}{2} \sum_{k_1 k_2 k_3 k_4} \langle \mathbf{k}_4,\mathbf{k}_3 | V | \mathbf{k}_2,\mathbf{k}_1 \rangle c_{k_4}^+ c_{k_3}^+ c_{k_2} c_{k_1} \tag{4.28}$$

Here

$$\langle \mathbf{k}_4,\mathbf{k}_3 | V | \mathbf{k}_2,\mathbf{k}_1 \rangle = \int d^3r_1\, d^3r_2 \psi_{k_4}^*(\mathbf{r}_1) \psi_{k_3}^*(\mathbf{r}_2) V(\mathbf{r}_1 - \mathbf{r}_2) \psi_{k_2}(\mathbf{r}_2) \psi_{k_1}(\mathbf{r}_1) \tag{4.29}$$

Again the integrations are over spin coordinates as well as spatial coordinates, and nonzero matrix elements are obtained only if the spins of $\mathbf{k}_3$ and $\mathbf{k}_2$ are the same and the spins of $\mathbf{k}_4$ and $\mathbf{k}_1$ are the same. With this notation we may directly proceed with perturbation theory (see Prob. 4.5). The notation simply makes a convenient way of doing our bookkeeping.

It is interesting to see in this framework the existence of exchange as well as direct terms. The expectation value of $V(\mathbf{r}_1,\mathbf{r}_2)$ with respect to a state $c_{k_2}{}^+ c_{k_1}{}^+|0\rangle$ contains two distinct types of terms (each with an equivalent term added to it).

$$\langle 0|c_{k_1}c_{k_2}(\langle \mathbf{k}_1,\mathbf{k}_2|V|\mathbf{k}_1,\mathbf{k}_2\rangle c_{k_1}{}^+ c_{k_2}{}^+ c_{k_1}c_{k_2} + \langle \mathbf{k}_2,\mathbf{k}_1|V|\mathbf{k}_1,\mathbf{k}_2\rangle c_{k_2}{}^+ c_{k_1}{}^+ c_{k_1}c_{k_2})c_{k_2}{}^+ c_{k_1}{}^+|0\rangle \tag{4.30}$$

From the form of $\langle \mathbf{k}_1,\mathbf{k}_2|V|\mathbf{k}_1,\mathbf{k}_2\rangle$ in the first term of Eq. (4.30) we see that it is precisely what we called an exchange interaction. It is nonzero only if the spins of $\mathbf{k}_2$ and $\mathbf{k}_1$ are the same. There is an exchange interaction only between electrons of parallel spin. We see that the effect of the operators $c_{k_1}{}^+ c_{k_2}{}^+ c_{k_1}c_{k_2}$, operating on the right-hand state, has been to exchange the states $\mathbf{k}_1$ and $\mathbf{k}_2$ (or equivalently interchange the two rows in the Slater determinant). The corresponding change in sign of the matrix element appeared in our earlier discussion of the exchange interaction in Sec. 3.2 of Chap. II. Thus the term exchange interaction is a natural one in this context. The second term in Eq. (4.30) is simply the direct interaction which enters also in the Hartree approximation. It couples states of antiparallel as well as parallel spin.

It is interesting that one of the summations in Eq. (4.28) may be eliminated by noting that momentum is conserved in the collision between two electrons if the two electrons are free. Thus many of the matrix elements of Eq. (4.29) are identically zero. This may be seen mathematically by writing out Eq. (4.29) for plane-wave states.

$$\langle \mathbf{k}_4,\mathbf{k}_3|V|\mathbf{k}_2,\mathbf{k}_1\rangle = \frac{1}{\Omega^2}\int d^3r_1\, d^3r_2\, e^{i(k_1-k_4)\cdot r_1}\, V(\mathbf{r}_1 - \mathbf{r}_2)\, e^{i(k_2-k_3)\cdot r_2}$$

$$= \frac{1}{\Omega^2}\int d^3r_1\, d^3r_2 [e^{i(k_1-k_4)\cdot(r_1-r_2)}\, V(\mathbf{r}_1 - \mathbf{r}_2)] \, e^{i(k_2-k_3+k_1-k_4)\cdot r_2}$$

We may now change variables of integration to $\mathbf{r}_1 - \mathbf{r}_2$ and $\mathbf{r}_2$ and perform the integration over $\mathbf{r}_2$ first holding $\mathbf{r}_1 - \mathbf{r}_2$ fixed. We obtain zero unless $\mathbf{k}_3 - \mathbf{k}_2$ is the negative of $\mathbf{k}_4 - \mathbf{k}_1$; if it is we obtain a factor of Ω. Thus if we write $\mathbf{k}_4 - \mathbf{k}_1$ as $\mathbf{q}$, proportional to the momentum transfer, the electron-electron interaction of Eq. (4.28) becomes

$$V(\mathbf{r}_1,\mathbf{r}_2) = \tfrac{1}{2} \sum_{k_1,k_2,q} V_q c_{k_1+q}^+ c_{k_2-q}^+ c_{k_2} c_{k_1}$$

where

$$V_q = \frac{1}{\Omega} \int d^3r e^{-iq\cdot r} V(\mathbf{r})$$

This simplified form is exactly equivalent to Eq. (4.28) when the electron-electron interaction conserves momentum and is the form most usually used. This term in the Hamiltonian is associated with a "process" in which an electron in state $|\mathbf{k}_1\rangle$ interacts with an electron in state $|\mathbf{k}_2\rangle$ and exchanges a momentum $\hbar\mathbf{q}$. This process may be represented by a diagram of the form shown in Fig. 4.19.

When we discussed electron tunneling, we were interested in transitions in which an electron is removed from one part of the system and appears in another. The corresponding terms in the Hamiltonian clearly contain a product of the corresponding annihilation and creation operators.

In almost all of our use of second quantization we will wish to use the annihilation and creation operators based upon Hartree-Fock states as discussed above. However, second quantization can be formulated more generally and in fact can be used to generate the Hartree-Fock approximation itself. This formulation illuminates the self-consistent-field method and will be necessary in our treatment of superconductivity, so we introduce it here.

We have seen that a state describable by a single Slater determinant may be written as a single product $\prod_k c_k^+|0\rangle$ where the indices $\mathbf{k}$ denote the occupied one-electron states. The most general antisymmetric many-

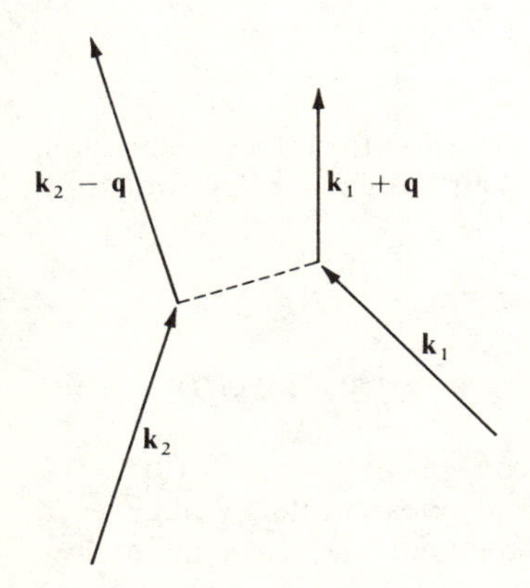

Fig. 4.19 A diagrammatic representation of a term in the electron-electron interaction, $V_q c_{k_1+q}^+ c_{k_2-q}^+ c_{k_2} c_{k_1}$. The dashed line represents the interaction between the two electrons and the arrows represent their initial and final wavenumbers in the corresponding scattering event.

electron wavefunction can be written as a linear combination of Slater determinants, a linear combination

$$\Psi = \sum_n A_n \prod_{\{k\}_n} c_k^{\ +}|0\rangle \tag{4.31}$$

where the product is over various sets $\{\mathbf{k}\}_n$ of indices. An example of such a two-electron state is

$$\Psi = \frac{1}{\sqrt{2}} c_1^{\ +} c_2^{\ +}|0\rangle + \frac{1}{\sqrt{2}} c_2^{\ +} c_3^{\ +}|0\rangle$$

which is readily seen to be antisymmetric and normalized. The indices **k** now denote any complete orthonormal set of functions of the coordinates of a single electron. These could be the eigenstates of a Hartree-Fock Hamiltonian, but no one-electron approximation is required; Eq. (4.31) is a completely general many-electron state.

The particular functions in which we have expanded are no longer central to the problem and can be eliminated by defining the electron-field operators,

$$\begin{aligned} \psi(\mathbf{r}) &= \sum_k \psi_k(\mathbf{r}) c_k \\ \psi^+(\mathbf{r}) &= \sum_k \psi_k^*(\mathbf{r}) c_k^{\ +} \end{aligned} \tag{4.32}$$

We may obtain the commutation relations for the field operators from the commutation relations for the annihilation and creation operators. We obtain

$$\begin{aligned} \psi(\mathbf{r})\psi(\mathbf{r}') + \psi(\mathbf{r}')\psi(\mathbf{r}) &= 0 \\ \psi^+(\mathbf{r})\psi^+(\mathbf{r}') + \psi^+(\mathbf{r}')\psi^+(\mathbf{r}) &= 0 \\ \psi^+(\mathbf{r})\psi(\mathbf{r}') + \psi(\mathbf{r}')\psi^+(\mathbf{r}) &= \delta(\mathbf{r} - \mathbf{r}') \end{aligned} \tag{4.33}$$

The last equation follows from the defining equation for the delta function, $f(\mathbf{r}) = \int f(\mathbf{r}')\delta(\mathbf{r} - \mathbf{r}')\, d^3r'$. We see that $\sum_k \psi_k(\mathbf{r})\psi_k^*(\mathbf{r}') = \delta(\mathbf{r} - \mathbf{r}')$ by evaluating

$$\int \sum_k \psi_k(\mathbf{r})\psi_k^*(\mathbf{r}') f(\mathbf{r}')\, d^3r' = \sum_k \left(\int \psi_k^*(\mathbf{r}') f(\mathbf{r}')\, d^3r' \right) \psi_k(\mathbf{r})$$

The integral is simply the kth coefficient in an expansion of $f(\mathbf{r})$ in the functions ψ_k.

The field operators provide a very convenient way of writing the Hamiltonian and other operators representing observables, though even within this framework we may well wish to write the states themselves in terms of the c_k and c_k^+. For example, the electron-density operator may be written $\rho(\mathbf{r}) = \psi^+(\mathbf{r})\psi(\mathbf{r})$. We see immediately, using Eq. (4.32), that if we obtain the expectation value of this operator with respect to a many-electron state, as given in Eq. (4.31), we obtain the expectation value of the electron density at the point **r**. We may also see, using the density operator, that $\psi^+(\mathbf{r})$ creates an electron at the position **r**. This may be seen by operating with $\psi^+(\mathbf{r})$ on the vacuum; operation upon the resulting state with the density operator gives, using Eq. (4.33), a delta function at **r** times the same state. The state $\psi^+(\mathbf{r})|0\rangle$ ordinarily will not be a state that we will wish to consider; the fact that it contains components of all energies is one reason. Thus while it is convenient to use the field operators for expressing operators corresponding to variables, we will ordinarily wish to use the annihilation and creation operators for specifying states.

We may write a potential-energy operator as $\psi^+(\mathbf{r})V(\mathbf{r})\psi(\mathbf{r})$. Again we see that the expectation value of this operator with respect to the many-electron state gives the expectation value of the potential energy due to electrons at the point **r**. If we wish to obtain the total potential energy we simply integrate the result over **r**. More generally if the Hamiltonian includes an electron kinetic energy and a one-electron potential $V(\mathbf{r})$ we may write the Hamiltonian operator as

$$\mathscr{H}_0 = \int d^3r\psi^+(\mathbf{r})\left[-\frac{\hbar^2\nabla^2}{2m} + V(\mathbf{r})\right]\psi(\mathbf{r}) \tag{4.34}$$

If we wish to introduce the electron-electron interaction we simply add the term

$$V_{\text{elel}} = \frac{1}{2}\int\int d^3r\, d^3r'\psi^+(\mathbf{r})\psi^+(\mathbf{r}')\frac{e^2}{|\mathbf{r}-\mathbf{r}'|}\psi(\mathbf{r}')\psi(\mathbf{r}) \tag{4.35}$$

The Hamiltonian written in terms of field operators is not usually used in the Schroedinger representation but it is not difficult to see that it may be. For a one-electron Hamiltonian, as given in Eq. (4.34), the eigenstates of this system may be written as in Eq. (4.31) with only a single contributing value of n, and the **k**'s may denote the eigenstates of the one-electron Hamiltonian. We may then see that $\mathscr{H}_0\Psi$ simply gives a constant times Ψ and the constant is equal to the sum of the eigenvalues for the occupied states. If the electron-electron interaction of Eq. (4.35) is included the eigenstate will include an infinite number of terms in the expansion in Eq. (4.31). It can be shown, though it is not so obvious, that if Ψ is an eigenstate of the Hamiltonian including the electron-electron interaction, then $\mathscr{H}_0 + V_{\text{elel}}$ operating on this state will again yield the total energy eigenvalue times the same state.

In terms of the electron-electron interaction as written in Eq. (4.35), the self-consistent-field approximation is the replacement of the term involving four field operators by terms involving only two field operators and replacing the remaining two by their expectation value. For example, the expression $\psi^+(\mathbf{r}')\psi(\mathbf{r}')$ has a nonvanishing expectation value $\langle\psi^+(\mathbf{r}')\psi(\mathbf{r}')\rangle$ for the many-electron state, which is in fact just the expectation value of the electron density at the position $\mathbf{r}'$. Note that the expectation value is obtained by eliminating annihilation and creation operators after ψ^+ and ψ are expanded as in Eq. (4.32). The dependence on position remains. Thus a term is introduced in the self-consistent-field Hamiltonian which is of the form

$$\frac{e^2}{2}\int d^3r\psi^+(\mathbf{r})\left[\int d^3r' \frac{\langle\psi^+(\mathbf{r}')\psi(\mathbf{r}')\rangle}{|\mathbf{r}-\mathbf{r}'|}\right]\psi(\mathbf{r})$$

and the expression appearing in brackets is a simple function of $\mathbf{r}$. An exactly equivalent contribution is obtained from the expectation value of $\psi^+(\mathbf{r})\psi(\mathbf{r})$. The two added together give exactly the Hartree electron-electron potential which may be added directly to the potential $V(\mathbf{r})$ in Eq. (4.34).

Similarly, we may note that $\psi^+(\mathbf{r})\psi(\mathbf{r}')$ has an expectation value which is a function of $\mathbf{r}$ and $\mathbf{r}'$. Interchanging the first two field operators in Eq. (4.35) (and picking up a minus sign thereby) we obtain a contribution to the self-consistent-field Hamiltonian given by

$$-\frac{e^2}{2}\int d^3r\, d^3r'\psi^+(\mathbf{r}')\frac{\langle\psi^+(\mathbf{r})\psi(\mathbf{r}')\rangle}{|\mathbf{r}-\mathbf{r}'|}\psi(\mathbf{r})$$

This term, in addition to an entirely equivalent term involving the expectation value of $\psi^+(\mathbf{r}')\psi(\mathbf{r})$, gives the exchange energy of the Hartree-Fock approximation.

We would expect that these are the only combinations of field operators that would have a nonvanishing expectation value. For example, the pair of field operators, $\psi(\mathbf{r}')\psi(\mathbf{r})$, when operating on the many-electron state Ψ, reduces the number of electrons by two and therefore yields a state orthogonal to Ψ and a vanishing expectation value. This proof, however, requires that the state Ψ has a well-defined number of particles. If the linear combination in Eq. (4.31) contained terms with differing numbers of electrons, it would be possible to obtain a nonvanishing expectation value of the product $\psi(\mathbf{r}')\psi(\mathbf{r})$. In our discussion of superconductivity we will see that the superconducting ground state is ordinarily written with an ill-defined number of particles and that there is a nonvanishing expectation value for such a product. Thus in a discussion of superconductivity in the self-consistent-field approximation, additional terms appear corresponding to an additional macroscopic parameter $\langle\psi(\mathbf{r})\psi(\mathbf{r})\rangle$ which enters in much the same way as the electron density does in the Hartree approximation. In normal (i.e., not superconducting) solids the self-consistent-field approximation includes only the direct and exchange terms that we have discussed before.

In this formulation, the self-consistent-field approximation is seen as the replacement of interaction terms containing four field operators by averaged terms containing only two. This replacement is essential in the definition and calculation of energy bands and in the principle approach for the treatment of cooperative phenomena.

For the present we will not use the field operators, but will describe the occupation of one-electron states using annihilation and creation operators and in addition will write one-electron operators in terms of these annihilation and creation operators.

Phonon states We now wish to make a similar set of prescriptions for dealing with the phonons and the electron-phonon interaction. In order to properly treat the phonons we must proceed systematically, first obtaining classically the Hamiltonian for the vibrating lattice. We view the use of vibration amplitudes as a transformation to normal coordinates. It is convenient again to include a normalization factor in the definition of these normal coordinates. Thus we expand the displacements of the individual ions in terms of the normal coordinates u_q.

$$\delta r_i = \frac{1}{\sqrt{N}} \sum_q u_q e^{iq \cdot r_j}$$

We have taken a system with one ion per primitive cell. Both δr_i and u_q are in fact vectors and three normal coordinates are required for each $\mathbf{q}$ in the Brillouin zone. However, these are nonessential notational complications so we will suppress the additional indices as if we had a one-dimensional system. Since the displacements vary with time, the normal coordinates must also. The normal coordinates are complex, but the displacements are real since u_{-q} is always equal to u_q^*. In terms of these normal coordinates we may write the total energy at any time.

$$E_{\text{tot}} = \sum_j \frac{M}{2} (\delta \dot{r}_j)^2 + \sum_{i,j} \tfrac{1}{2} \delta r_i \frac{\partial^2 W}{\partial r_i \, \partial r_j} \delta r_j \tag{4.36}$$

In terms of normal coordinates the second term may be written in the form

$$\frac{1}{2N} \sum_{q,q'} u_q u_{q'} \sum_{i,j} e^{iq \cdot r_i} \frac{\partial^2 W}{\partial r_i \, \partial r_j} e^{iq' \cdot r_j}$$

$$= \frac{1}{2N} \sum_{q,q'} u_q u_{q'} \sum_{i,j} \left(e^{iq \cdot (r_i - r_j)} \frac{\partial^2 W}{\partial r_i \, \partial r_j} \right) e^{i(q' + q) \cdot r_j}$$

In the final form the first exponential and the interaction matrix are both functions only of $r_i - r_j$. We may therefore first sum over r_j, holding $r_i - r_j$ fixed. The summation over the final factor yields N if $\mathbf{q}' = -\mathbf{q}$ and is zero otherwise. Thus this term in the energy becomes

$$\tfrac{1}{2} \sum_q u_q u_{-q} \lambda_q \tag{4.37}$$

where *the dynamical matrix* λ_q may be obtained by performing the sum over $r_i - r_j$.

$$\lambda_q = \sum_j e^{iq\cdot(r_i - r_j)} \frac{\partial^2 W}{\partial r_i \, \partial r_j}$$

Because we have described the mode in terms of a single parameter, u_q, the dynamical matrix contains only one element. Had we written three components, it would have been a 3-by-3 matrix. The kinetic energy term in Eq. (4.36) is directly evaluated in a similar way and we obtain

$$E_{\text{tot}} = \sum_q \left(\frac{M}{2} \dot{u}_q \dot{u}_{-q} + \frac{\lambda_q}{2} u_q u_{-q} \right)$$

Thus we have written the energy in terms of the coordinates u_q and the corresponding velocities $\dot{u}_q$. We may immediately define a Lagrangian and thus the momenta conjugate to u_q.

$$L = \sum_q \left(\frac{M}{2} \dot{u}_q \dot{u}_{-q} - \frac{\lambda_q}{2} u_q u_{-q} \right)$$

$$P_q = \frac{\partial L}{\partial \dot{u}_q} = M \dot{u}_{-q} \tag{4.38}$$

$$P_{-q} = \frac{\partial L}{\partial \dot{u}_{-q}} = M \dot{u}_q$$

Note that two terms in the sum ($\mathbf{q}$ and $-\mathbf{q}$) contribute to each derivative. The Hamiltonian in terms of these normal coordinates and canonical momenta is

$$\mathscr{H}_\varphi = \sum_q \left(\frac{1}{2M} P_q P_{-q} + \tfrac{1}{2} \lambda_q u_q u_{-q} \right) \tag{4.39}$$

Now that we have obtained the Hamiltonian classically we may move directly to quantum mechanics. Then P_q and u_q become operators with the commutation relation

$$P_q u_q - u_q P_q = \frac{\hbar}{i}$$

(In the Schroedinger representation of these operators $P_q = (\hbar/i)\partial/\partial u_q$. However only the commutation relations will be necessary here.)

It is now convenient to define two new operators, which will be seen to be the phonon annihilation and creation operators. This corresponds to a canonical transformation.

$$
\begin{aligned}
a_q^+ &= \left(\frac{M}{2\hbar\omega_q}\right)^{1/2}\left(\omega_q u_{-q} - \frac{i}{M}P_q\right) \\
a_q &= \left(\frac{M}{2\hbar\omega_q}\right)^{1/2}\left(\omega_q u_q + \frac{i}{M}P_{-q}\right)
\end{aligned}
\tag{4.40}
$$

Here ω_q is the classical frequency of the mode, $\omega_q^2 = \lambda_q/M$. The constants have been selected so that the commutation relations that may be obtained by inspection from Eqs. (4.40) and the one preceding it.

$$a_q a_{q'}^+ - a_{q'}^+ a_q = \delta_{qq'}$$

These are the commutation relations for the annihilation and creation operators for bosons.

In order to obtain the Hamiltonian in terms of these operators we write out $a_q^+ a_q$ explicitly using Eq. (4.40). By rearranging the result we obtain

$$\frac{1}{2M}P_{-q}P_q + \frac{1}{2}\lambda_q u_{-q}u_q = \hbar\omega_q\left[a_q^+ a_q + \frac{i}{2\hbar}(P_q u_q - u_{-q}P_{-q})\right]$$

The Hamiltonian is obtained by summing over all $\mathbf{q}$. In the last terms on the right we combine $\mathbf{q}$ and $-\mathbf{q}$ terms and use the commutation relations to obtain

$$\mathscr{H}_\varphi = \sum_q \hbar\omega_q(a_q^+ a_q + 1/2) \tag{4.41}$$

It follows from comparison with the usual expression for the energy of a harmonic oscillator that $a_q^+ a_q$ is the number operator for phonons in the $\mathbf{q}$th mode.

It further follows that a_q^+ has the effect of creating a phonon. We may demonstrate this by letting $|n\rangle$ be a state with n phonons in the $\mathbf{q}$th mode. We will suppress the index $\mathbf{q}$ throughout. We may then evaluate the operation of the number operator on the state $a^+|n\rangle$.

$$(a^+a)a^+|n\rangle = a^+(aa^+)|n\rangle = a^+(a^+a + 1)|n\rangle = (n + 1)a^+|n\rangle$$

Thus $a^+|n\rangle$ is an eigenstate of the number operator corresponding to $n + 1$ phonons. Similarly we may show that $a|n\rangle$ is an eigenstate of the number operator with $n - 1$ phonons.

At this stage we must be careful of the normalization. We note in particular that if the state $|n\rangle$, the n-photon state, is normalized, $a^+|n\rangle$ is not. The normalization integral becomes

$$\langle n|aa^+|n\rangle = \langle n|a^+a + 1|n\rangle = (n + 1)\langle n|n\rangle$$

We may obtain normalized states by writing

$$|n + 1\rangle = \frac{a^+}{\sqrt{n + 1}}|n\rangle$$
$$|n - 1\rangle = \frac{a}{\sqrt{n}}|n\rangle \tag{4.42}$$

consistent with all of our above notations. It then follows that $|n\rangle = (a^+)^n|0\rangle/(n!)^{1/2}$.

We can now write the eigenstates of the phonon Hamiltonian $\mathscr{H}_\varphi$ of Eq. (4.41). We write the no-phonon state as a vacuum state $|0\rangle$ and define a number n_q of phonons in the qth mode. Then the state of the lattice is

$$\Phi = \prod_q \frac{(a_q{}^+)^{n_q}}{(n_q!)^{1/2}}|0\rangle \tag{4.43}$$

and in the absence of electron-phonon coupling, the eigenstates of $\mathscr{H}_{el} + \mathscr{H}_\varphi$ are $\Phi\Psi_N$ obtained using the operations given in Eqs. (4.23) and (4.43).

Phase coherence and off-diagonal long-range order Before proceeding further with the analysis of phonons and phonon interactions we should pause to look more deeply at the meaning of some of the results in the preceding section.

In a series of steps we have eliminated the phonon amplitudes u_q in order to write the states and the Hamiltonian in terms of the annihilation and creation operators for phonons. We can now rewrite the amplitudes themselves in terms of the annihilation and creation operators using Eq. (4.40). We obtain

$$u_q = \left(\frac{\hbar}{2M\omega_q}\right)^{1/2}(a_q + a_{-q}{}^+) \tag{4.44}$$

We should note a peculiarity of this result which will arise again in our study of superconductivity but which is perhaps easier to understand here. If the state of the system has a well-defined number of phonons, the expectation value of the displacement amplitude u_q is equal to zero. This is not a statement that the displacements are as often in one direction as the other, since u_q is an amplitude. It is more precisely a statement that the phase of the amplitude is not defined. We will see that the phase of the amplitude is the conjugate variable to the number of phonons. Thus the uncertainty principle requires that if the number of phonons is well-defined the phase must be ill-defined and vice versa.

This is a very general result. It is true also for electromagnetic radiation. The phase of the electric field is conjugate to the number of photons. In the classical treatment of antenna radiation where we compute explicitly the electromagnetic fields present we are describing a state of the system with an ill-defined number of photons. In the calculation of the emission of a photon by an atom we describe a system in which the phase of the electromagnetic field is ill-defined.

We may make these considerations more precise by attempting to construct a state with a well-defined amplitude. Clearly this can be done only by taking a linear combination of states with different numbers of phonons present. We consider a particular mode of wavenumber **q** and suppress the subscripts **q**. Thus we write a state

$$\Phi = \sum_n A_n |n\rangle \tag{4.45}$$

The number of phonons present is ill-defined but has an expectation value of

$$\langle n \rangle = \sum_n A_n^* A_n n$$

We may next seek the expectation value of the amplitude u using Eq. (4.44). We obtain

$$\begin{aligned} \langle u \rangle &= \left(\frac{\hbar}{2M\omega} \right)^{1/2} \sum_{n,m} A_m^* A_n \langle m|a|n \rangle \\ &= \left(\frac{\hbar}{2M\omega} \right)^{1/2} \sum_n A_{n-1}^* A_n \sqrt{n} \end{aligned} \tag{4.46}$$

(There is also an expectation value of the amplitude for the mode of negative wavenumber but we need not consider it here.)

If the phases of the A_n vary randomly from one to the other, the phases of the many terms in Eq. (4.46) will be random and the expectation value will be very tiny. The expectation value of the amplitude will be maximum if the phases are identical or, in fact, if each phase A_{n-1} differs from the phase of A_n by the same amount, which we call φ. Thus we will obtain the maximum amplitude if we write

$$A_n = |A_n| \, e^{in\varphi} \tag{4.47}$$

We can now see that this phase coherence leads to a classical description of the problem. We may let $|A_n|$ vary smoothly with n but be sharply peaked around a value $\langle n \rangle$. In Eq. (4.46) the factor $\sqrt{n}$ may be taken outside the sum and, with $A_{n-1}^* A_n \approx |A_n|^2 e^{i\varphi}$, the sum becomes simply the normalization sum for the state and is equal to 1. We obtain

$$\langle u \rangle = \left(\frac{\hbar \langle n \rangle}{2M\omega} \right)^{1/2} e^{i\varphi}$$

We square the amplitude of both sides to obtain

$$\tfrac{1}{2} M\omega^2 \langle u \rangle^2 = \frac{\langle n \rangle \hbar\omega}{4}$$

The left side is the kinetic energy in the mode in question. If we add an equal potential energy we obtain the total energy in the mode, and if we finally add the energy in the mode of opposite wavenumber we obtain a total of $\hbar\omega\langle n \rangle$, the correct energy for the mode in the large quantum-number (classical) limit. All of the energy is in a coherent mode with well-defined phase and amplitude.

We may also see that the variables n and φ are conjugate to each other though for a rigorous demonstration we require a more careful and general definition of φ. Using the coherent form, Eq. (4.47), in the wavefunction Eq. (4.45), we see that

$$\frac{\hbar}{i} \frac{\partial}{\partial\varphi} \Phi = \hbar \sum_n n A_n |n\rangle$$

Thus the operator $(\hbar/i)\partial/\partial\varphi$ is just $\hbar$ times the number operator, just the relation (except for a factor $\hbar$) satisfied by any pair of conjugate variables in quantum mechanics. This of course can be pursued very much further but it is not necessary for our purposes.

The phase coherence giving rise to well-defined amplitudes is an exceedingly important concept in solid state physics. We have seen above its conceptual importance in seeking the relation between classical and quantum treatments of the lattice vibrations. It is also of central importance in cooperative effects, which may be seen even in the context of lattice vibrations.

Let us imagine that the lattice is unstable under a particular vibrational mode; i.e., the energy of the system drops when the lattice distorts from perfect periodicity in the form of a particular mode. Classically, of course, we will expect that deformation to arise spontaneously, and in terms of the undistorted structure a well-defined equilibrium amplitude is established. In terms of the description of the state we are using here, well-defined values of A_n are established and, more importantly, there is phase coherence between the amplitudes A_n and A_{n-1}. The term *off-diagonal long-range order*, ODLRO, is associated with such a transition. This term arises from a description of the system in terms of the density matrix, which for the phonons may be written $\rho_{nm} = A_n^* A_m$. Thus the essence of the transition is the existence of a well-defined phase for the off-diagonal matrix elements $\rho_{n-1,n}$. The value of this coherent phase becomes a thermodynamic variable of the system. It is an *order parameter* which represents long-range order in the system.

We will see that the superconducting transition may be described in a similar way. In that case the product of the electron-field operators $\psi_\uparrow(\mathbf{r})\psi_\downarrow(\mathbf{r})$, one for spin up and one for spin down, plays the same role as the amplitude u_q in the phonon system. The superconducting transition may be understood as the establishment of a macroscopic expectation value for this product which represents ODLRO for the system and a corresponding lack of definition of the number of electrons present. The fluidity in helium also represents the establishment of such ODLRO. Similarly, the formation of ferromagnetic and antiferromagnetic states represents the establishment of this long-range order. A more complete discussion of this ODLRO, particularly as applied to liquid helium, is given by Anderson.[1]

The interaction None of this formalism was necessary when we treated the specific heat and needed only the results given by Eqs. (4.25) and (4.41). The formalism is necessary only when we need the states themselves and these are required to obtain the electron-phonon interaction.

For conceptual simplicitly we will describe the electron-phonon interaction with a deformation-potential constant. Then the potential seen by an electron at the position $\mathbf{r}$ due to instantaneous displacements of the lattice corresponding to instantaneous amplitudes $\mathbf{u}_q$ (we temporarily make explicit the vector character of the amplitude) is given by

$$V(\mathbf{r}) = D \sum_q \frac{i\mathbf{q} \cdot \mathbf{u}_q}{\sqrt{N}} e^{iq \cdot r} \tag{4.48}$$

In the pseudopotential framework D is replaced by a pseudopotential form factor and $i\mathbf{q} \cdot \mathbf{u}_q/\sqrt{N}$ is replaced by the appropriate structure factor. Here we consider only coupling to the longitudinal mode and can write $i\mathbf{q} \cdot \mathbf{u}_q$ as iqu_q. In this context we also take the one-electron eigenstates to be normalized plane waves. This is the simplest case and one widely used. There is no fundamental complication in extending the treatment to include band-structure effects, but at the same time most of the physics is included in the simple model we use.

We could immediately obtain matrix elements of the interaction of Eq. (4.48) between two electronic states, but to obtain matrix elements

[1] P. W. Anderson, *Rev. Mod. Phys.*, **38**:298 (1966).

between phonon states we must first write u_q in terms of our canonical variables a_q and a_q^+ using Eq. (4.44). We obtain

$$V(\mathbf{r}) = D \sum_q \left(\frac{\hbar}{2NM\omega_q} \right)^{1/2} iq(a_q + a_{-q}^+)e^{iq \cdot r}$$

We have seen in Eq. (4.42) how to compute matrix elements of the a_q and a_q^+ between the phonon states. From Eq. (4.26) we see matrix elements of $e^{iq \cdot r}$ between electronic states are obtained by replacing $e^{iq \cdot r}$ by $\sum_k c_{k+q}^+ c_k$ if the one-electron states are normalized plane waves as we have assumed. Thus in second-quantized form the electron-phonon interaction becomes

$$\mathscr{H}_{e\varphi} = \sum_{q,k} \frac{V_q}{N^{1/2}} (a_q + a_{-q}^+) c_{k+q}^+ c_k \tag{4.49}$$

where

$$V_q = D \left(\frac{\hbar}{2M\omega_q} \right)^{1/2} iq \tag{4.50}$$

The scattering of an electron from the state $\mathbf{k}$ to the state $\mathbf{k} + \mathbf{q}$ is accompanied either by the absorption of a phonon of a wavenumber $\mathbf{q}$ or the emission of a phonon of wavenumber $-\mathbf{q}$.

We have completed the task of writing the electron-phonon Hamiltonian in second-quantized notation. In most problems we will wish to regard the noninteracting electrons and phonons as a zero-order state and to treat the electron-phonon interaction as a perturbation.

4.3 Applications Three such problems come immediately to mind. First is the resistive scattering of electrons by lattice vibrations which we discussed previously in a semiclassical framework. Second is the shift in the electron energies due to their interactions with the phonons. We discussed the existence of such an effect when we considered Fermi surfaces in metals. The third is the electron-electron interaction induced by phonons which is the origin of superconductivity. The main difficulty in each of these problems has already been overcome in obtaining the Hamiltonian in second-quantized form and we will consider each of these problems briefly.

Electron scattering Scattering is to be computed using the Golden Rule.

$$P_{12} = \frac{2\pi}{\hbar} |\langle 2|\mathscr{H}_{e\varphi}|1\rangle|^2 \, \delta(E_2 - E_1) \tag{4.51}$$

where the states $|1\rangle$ and $|2\rangle$ represent many-electron, many-phonon states.

From Eq. (4.49) we see that there are matrix elements coupling only states in which a single electron has changed its wavenumber. Therefore we consider a process in which an electron is scattered from the state $|\mathbf{k}_1\rangle$ to the state $|\mathbf{k}_2\rangle$ with all other electrons remaining in their initial states. The only contributing term in Eq. (4.49) then is for $\mathbf{k} = \mathbf{k}_1$ and $\mathbf{q} = \mathbf{k}_2 - \mathbf{k}_1$. We let the initial state contain n_q phonons of wavenumber $\mathbf{q}$ and n_{-q} phonons of wavenumber $-\mathbf{q}$. We write the corresponding phonon state $|n_q,n_{-q}\rangle$, then the phonon operators of Eq. (4.49) on this phonon state yield

$$(a_q + a_{-q}{}^+)|n_q,n_{-q}\rangle = \sqrt{n_q}\,|n_q - 1,n_{-q}\rangle + \sqrt{n_{-q} + 1}\,|n_q,n_{-q} + 1\rangle$$

Thus there are matrix elements with final states of the electron distribution specified above and with one less or one more phonon.

We consider first transitions to a state $|2\rangle$ in which there is a loss of one phonon in the $\mathbf{q}$th mode. The corresponding matrix element is

$$\langle 2|\mathscr{H}_{e\varphi}|1\rangle = \sqrt{\frac{n_q}{N}}\,V_q$$

When this is squared and inserted in Eq. (4.51) we find that the probability of the scattering occurring with the absorption of the phonon is proportional to the number of phonons n_q. We note also that the energy delta function becomes

$$\delta(E_2 - E_1) = \delta(\epsilon_{k_2} - \epsilon_{k_1} - |\hbar\omega_q|)$$

which includes both the difference in the electron energy and in phonon energy between the two states.

Similarly, when we compute the probability of emission of a phonon, we obtain the same expression with n_q replaced by $n_{-q} + 1$ and with the energy delta function replaced by $\delta(\epsilon_{k_2} - \epsilon_{k_1} + |\hbar\omega_{-q}|)$. Having evaluated the matrix elements we may directly perform the sum over final states to obtain the scattering rate.

Including both processes in which an electron makes a transition from $|\mathbf{k}\rangle$ to $|\mathbf{k} + \mathbf{q}\rangle$, we have

$$P_{k,k+q} = \frac{2\pi}{\hbar}\frac{|V_q|^2}{N}[n_q\,\delta(\epsilon_{k+q} - \epsilon_k - |\hbar\omega_q|)$$

$$+ (n_{-q} + 1)\,\delta(\epsilon_{k+q} - \epsilon_k + |\hbar\omega_{-q}|)] \quad (4.52)$$

Using the form for V_q given in Eq. (4.50) we may correctly compute the scattering rate due to the phonon field. The equation is applicable to semiconductors as well as simple metals. It is interesting to consider limiting cases.

We note first that even with no phonons present a transition may occur due to the 1 in $n_{-q} + 1$. That is, an electron can emit a phonon if there is a final state available with energy ϵ_{k+q} lower than ϵ_k by $|\hbar\omega_{-q}|$. Of course if the electron gas is at zero temperature this will never be the case, but it will for "hot" electrons.

This dependence upon the electron distribution complicates the calculation of a relaxation time for conductivity.[1] A correct expression for the conductivity can be obtained, however, by replacing $n_{-q} + 1$ by n_{-q} in Eq. (4.52). We also note that $\hbar\omega_q$ will ordinarily be much less than the electron energies involved, so that the scattering is nearly elastic, $\epsilon_{k+q} \approx \epsilon_k$. This may be seen by writing $\hbar\omega_q = \hbar q v_s$, where v_s is the velocity of sound. The electron energy may be written $\hbar^2 k^2/2m = \hbar k v_e/2$, where v_e is the electron velocity. But the electron velocity will be much greater than v_s (an electron moving at the speed of sound, of order 10^5 cm/sec, has an energy of only 10^{-5} ev) and $q \lesssim 2k$, so $\hbar q v_s$ will be much less than the typical electron energy even in a semiconductor. Since the change in electron energy is so small, for many purposes (including calculation of the conductivity) we may neglect the $\hbar\omega$ in the energy delta function. (We could not of course make this approximation if we wished to study explicitly the decay in *energy* of hot electrons.) The neglect of $\hbar\omega$ in the energy delta functions is equivalent to replacing the vibrating lattice by a statically distorted lattice. In the sense discussed in Sec. 5.6 of Chap. III, we find that the transitions are "fast" in comparison to the time required for the lattice to move.

We may also write out the leading coefficient in Eq. (4.52) using the V_q of Eq. (4.50).

$$\frac{2\pi}{\hbar}\frac{|V_q|^2}{N} = \frac{2\pi}{\hbar N}\frac{D^2\hbar q^2}{2M\omega_q} = \frac{2\pi}{\hbar N} D^2 \frac{\hbar\omega_q}{2M{v_s}^2}$$

With these approximations the scattering rate from Eq. (4.52) becomes

$$P_{k,k+q} = \frac{2\pi}{\hbar N}\frac{D^2}{M{v_s}^2} n_q \hbar\omega_q\, \delta(\epsilon_{k+q} - \epsilon_k) \tag{4.53}$$

But $n_q\hbar\omega_q$ is the thermal energy in the **q**th mode and dividing by $M{v_s}^2$ gives the square of the appropriate dilatation amplitude, allowing an entirely classical treatment of the lattice such as outlined in Sec. 4.1. For looking at various limits it is convenient to use the form from Eq. (4.53).

We consider first the high-temperature limit, in which $KT \gg \hbar\omega_q$ for the modes of interest, i.e., at temperatures above the Debye temperature. Then $n_q\hbar\omega_q$ becomes simply KT. The scattering rate for conductivity is readily evaluated. Writing Θ as the angle between **k** and **q** and noting that

[1] A. H. Wilson, "The Theory of Metals," 2d edition, p. 277, Cambridge, London, 1954.

the usual $1 - \cos\theta$ weighting factor,[1] with θ the angle between $\mathbf{k} + \mathbf{q}$ and $\mathbf{k}$, becomes $-(q/k)\cos\Theta$, we have

$$\frac{1}{\tau} = \sum_q - P_{k,k+q}\frac{q}{k}\cos\Theta$$

$$= \frac{\Omega_0}{(2\pi)^2}\frac{D^2}{\hbar}\frac{KT}{Mv_s^2}\int\left(\frac{-q}{k}\cos\Theta\right)\delta\left[\frac{\hbar^2}{2m}(2kq\cos\Theta + q^2)\right]$$

$$\sin\Theta d\Theta 2\pi q^2 dq$$

$$= \frac{\Omega_0}{4\pi}\frac{D^2}{\hbar^3}\frac{KT}{Mv_s^2}\frac{m}{k^3}\int_0^{2k} dq q^3 = \frac{\Omega_0}{\pi}\frac{D^2}{\hbar^3}\frac{KT}{Mv_s^2}mk \tag{4.54}$$

We note first that the relaxation time for scattering by lattice vibrations depends upon the energy of the electron in question through the wavenumber $\mathbf{k}$, in contrast to the assumption of an energy-independent relaxation time used with the Boltzmann equation. It is interesting that for the approximations made here the *mean-free path* $\hbar k\tau/m$ is independent of energy. This is more typically the case, but is by no means universal.

Second, we note that for simple metals, where the relaxation time is evaluated at the Fermi energy, a single scattering time τ is appropriate and it is inversely proportional to temperature. Thus we find that at high temperature the resistivity is linear in T, as observed.

Third, in a semiconductor we must recompute the conductivity by summing over occupied states with an energy-dependent scattering time. We obtain a conductivity of $Ne^2\bar{\tau}/m$ where $\bar{\tau}$ is obtained by a suitable average over the τ of Eq. (4.54). Averaging $1/k$ over a thermal distribution gives an additional factor of $(KT)^{-1/2}$ and therefore a conductivity proportional to $T^{-3/2}$ (assuming N is independent of T). This dependence is observed for n-type germanium, but in p-type germanium and in silicon the conductivity drops more steeply with temperature, presumably due to the presence of other scattering mechanisms such as optical modes or ionized impurities.[2]

The evaluation for metals at *low* temperatures is also straightforward. In that case we must write

$$n_q = \frac{1}{e^{\hbar\omega_q/KT} - 1}$$

[1] This factor was introduced in Sec. 8.6 of Chap. II, and later obtained from the Kubo-Greenwood formula in Sec. 5.3 of Chap. III.

[2] See H. Y. Fan, "Solid State Physics," vol. 1, Academic, New York, 1955.

Then in the calculation of $1/\tau$ we replace the factor KT by $n_q \hbar\omega$ in the calculation leading to Eq. (4.54). The integration over angle remains the same, but the integral over q becomes

$$\frac{1}{\tau} = \frac{\Omega_0}{4\pi} \frac{D^2}{\hbar^3} \frac{m}{M{v_s}^2 k^3} \int_0^{2k} \hbar v_s q^4 \left(\frac{1}{e^{\hbar v_s q/KT} - 1} \right) dq$$

We change variables to $x = \hbar v_s q/KT$ and at low temperatures the upper limit of $x = 2\hbar k v_s/KT$ may be taken to be infinity. We find

$$\frac{1}{\tau} = \frac{\Omega_0}{4\pi} \frac{D^2}{\hbar^3} \frac{m}{M{v_s}^2 k^3} \frac{(KT)^5}{(\hbar v_s)^4} \int_0^{\infty} \frac{x^4\, dx}{e^x - 1}$$

The interesting feature of the result is the proportionality to T^5, which is observed approximately in many metals at low temperatures. At low temperatures phonons are excited appreciably only at small q. In changing variables to x, each factor of q in the integrand has contributed a factor of T. This is in close analogy with our calculation of the lattice specific heat which was proportional to T^3 at low temperature.

Electron self-energy[1] Let us now turn to the problem of the self-energy of the electron due to vibrational modes of the system. The change in energy of the entire system due to the electron-phonon interaction may be computed by perturbation theory.

$$E = E_0 + \langle 0|\mathscr{H}_{e\varphi}|0\rangle + \sum_n \frac{\langle 0|\mathscr{H}_{e\varphi}|n\rangle\langle n|\mathscr{H}_{e\varphi}|0\rangle}{E_0 - E_n}$$

Here $|0\rangle$ represents the state of the noninteracting electron-phonon system, and in most of our discussion the ground state. These states, as well as the vacuum state, are commonly represented by $|0\rangle$. The first-order term vanishes and we focus attention on the second-order term. For simplicity we will focus our attention on states at zero temperature. Thus the state $|0\rangle$ corresponds to no phonons in any mode. Then only the a^+ enters the second matrix element and only the a enters the first. Thus

$$\mathscr{H}_{e\varphi}|0\rangle = \sum_{q,k} \frac{V_q}{N^{1/2}} a_{-q}{}^{+} c_{k+q}^{+} c_k |0\rangle$$

Each term in the summation corresponds to a different intermediate state. Each term is in fact a constant times the corresponding intermediate state

[1] For a more extensive discussion, see J. J. Quinn, in W. A. Harrison and M. B. Webb (eds.), "The Fermi Surface," p. 58, Wiley, New York, 1960.

$|n\rangle$. Thus we may combine the two matrix elements and write the energy to second order in the form

$$E = E_0 + \frac{1}{N}\sum_{q,k} \frac{\langle 0|a_{-q}c_k{}^{+}c_{k+q}a_{-q}{}^{+}c_{k+q}^{+}c_k|0\rangle\, V_q^*\, V_q}{\epsilon_k - \epsilon_{k+q} - \hbar\omega_q}$$

Each term in this perturbation expansion can be thought of as a scattering process. The term is described by saying that an electron of wavenumber $\mathbf{k}$ scatters to a state of wavenumber $\mathbf{k} + \mathbf{q}$ by the virtual emission of a phonon which is then reabsorbed as the electron returns to its initial state. This is simply a pictorial way of describing the terms in the perturbation expansion which are mathematically well defined in any case.

The phonon operators commute with the electron operators so we may bring them together and note the $a_{-q}a_{-q}{}^{+}|0\rangle = |0\rangle$ for the ground state. Similarly we may commute c_k through the $c_{k+q}c_{k+q}^{+}$ with two changes in sign. The matrix elements become simply $n_k(1 - n_{k+q})$ where the n_k are the occupation numbers for the corresponding states.

$$E = E_0 + \frac{1}{N}\sum_{q,k} \frac{V_q^* V_q\, n_k(1 - n_{k+q})}{\epsilon_k - \epsilon_{k+q} - \hbar\omega_q} \tag{4.55}$$

This contribution to the ground-state energy of the system is of little interest. However, if we construct an excited state by adding an electron with wavenumber $\mathbf{k}_0$ outside the Fermi sphere, as illustrated in Fig. 4.20, the self-energy of the system as a function of $\mathbf{k}_0$ is of interest. A shift in dE/dk_0 gives directly a shift in the density of excited states per unit energy and in the electron velocity.

Adding an electron of wavenumber $\mathbf{k}_0$ makes two changes in the interaction energy of Eq. (4.55) which must be included. First it gives additional terms with $\mathbf{k} = \mathbf{k}_0$ and with $\mathbf{k}' = \mathbf{k} + \mathbf{q}$ outside the Fermi surface. In addition we must subtract those terms in Eq. (4.55) in which the

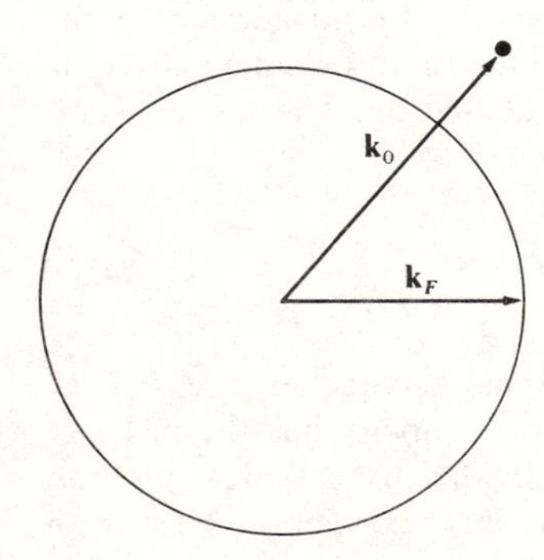

Fig. 4.20 An electron gas in the ground state with a single additional excited state occupied.

state $\mathbf{k} + \mathbf{q}$ is equal to $\mathbf{k}_0$ and the state $\mathbf{k}$ is within the Fermi surface. Thus the change in energy due to the addition of the electron is given by

$$\delta E(k_0) = \frac{\hbar^2 k_0{}^2}{2m} + \frac{1}{N} \sum_{k>k_F} \frac{V^*_{k-k_0} V_{k-k_0}}{\epsilon_{k_0} - \epsilon_k - \hbar\omega} - \frac{1}{N} \sum_{k<k_F} \frac{V^*_{k_0-k} V_{k_0-k}}{\epsilon_k - \epsilon_{k_0} - \hbar\omega} \tag{4.56}$$

where we have taken the zero-order energy to be the kinetic energy and have dropped the prime on $\mathbf{k}'$ in the first sum. This could be evaluated by replacing the summations by integrations. It turns out that the interaction terms very nearly cancel as $\mathbf{k}_0$ approaches the Fermi surface. The two integrals over the magnitude of $\mathbf{k}$ have the character, when combined, of a principal-parts integration as $\hbar\omega$ is regarded as small; the fractional correction to the Fermi energy is of the order of magnitude of a typical form factor squared divided by the Fermi energy squared, which is of order 1 percent.

However, the change in self-energy as $\mathbf{k}_0$ moves away from the Fermi surface can be seen to be quite large. This is most easily evaluated in two steps. We first take $\mathbf{k}_0$ at the Fermi surface and then increase both k_0 and k_F together by the same infinitesimal amount. The self-energy correction is very tiny when $\mathbf{k}_0$ lies on the Fermi surface as we indicated above and we may neglect the change in self-energy in this step. Next we shrink the Fermi sphere back to its initial size and must compute the change in self-energy in this second step. We see from Eq. (4.56) that a change in k_F enters only in modifying the limits of integration. For an infinitesimal change dk_F the change in self-energy becomes a surface integral over the Fermi surface times dk_F. The change in the first sum becomes

$$d\,\frac{1}{N} \sum_{k>k_F} \frac{V^*_{k-k_0} V_{k-k_0}}{\epsilon_{k_0} - \epsilon_k - \hbar\omega} = \frac{1}{N} \frac{\Omega}{(2\pi)^3} |dk_F| \int \frac{d^2k\, V^*_{k-k_0} V_{k-k_0}}{\epsilon_{k_0} - \epsilon_k - \hbar\omega}$$

All wavenumbers in the integrand now lie on the Fermi surface so the energy denominator becomes simply $-\hbar\omega$ and the matrix elements become precisely those which entered the scattering calculation. The second summation makes an equal contribution and we obtain

$$\frac{d\,\delta(E_{k_0})}{dk_0} = \frac{\hbar^2 k_0}{m} - \frac{\Omega_0}{4\pi^3} \int \frac{d^2k\, V^*_{k-k_0} V_{k-k_0}}{\hbar\omega} \tag{4.57}$$

From Eq. (4.57) we see that the fractional change in $d\,\delta E(k_0)/dk_0$ and therefore the fractional change in the Fermi velocity has the order of magnitude of a typical form factor squared divided by a typical phonon energy and the Fermi energy. This can be a very large correction, a factor of

2 for lead. We see that the shift is negative and therefore increases the density of excited states. As we indicated in Sec. 5.6 of Chap. II, it is believed to account for the large deviations of the observed specific heat masses, which we listed there, from the electronic mass. We may think of the decreased electron velocity as arising physically from a wake of virtual phonons carried by the electron.

It is interesting to note that velocity deviations occur at the Fermi surface and will move with the Fermi surface if it is distorted. For this reason the modification of the density of states does not affect the Pauli susceptibility. We may compute the self-energy shifts independently for electrons of each spin since the electron-phonon interaction does not couple states of opposite spin. Then when a magnetic field is applied, and the Fermi energies for the two spins are shifted with respect to each other, the self-energy corrections move with the respective Fermi surfaces as illustrated in Fig. 4.21. We then see that the number of electrons which shift their spin is determined by the energy bands without the self-energy correction.

The electron-electron interaction It is interesting to note that this shift in the energy of the electron states may be thought of as the result of an *effective interaction between electrons.* This point of view will be of particular value when we discuss superconductivity. In the expression above for the energy to second order we envisaged a ground state $|0\rangle$ with no phonons present. Since this state appeared on both sides of the matrix element we obtained

Fig. 4.21 *The transfer of electrons between spin bands giving Pauli paramagnetism. The dashed curves represent energy as a function of wavenumber without corrections for electron self-energy, the solid lines with these corrections. Note that the corrections do not modify the number of electrons transferred.*

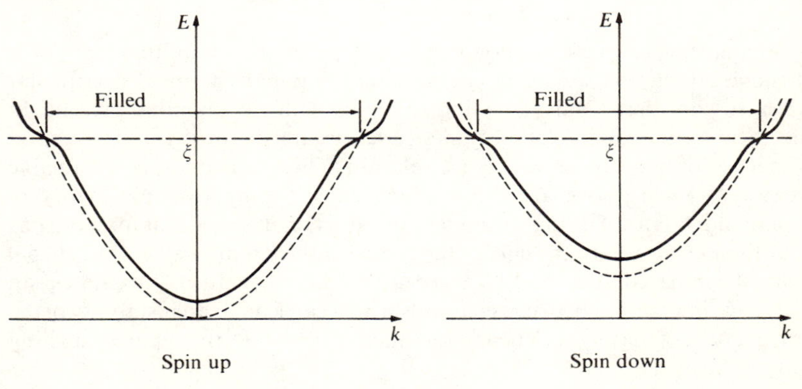

only terms with $a_{-q}a_{-q}{}^{+}|0\rangle = |0\rangle$. Thus the second-order contribution to the energy may be written

$$E^{(2)} = \frac{1}{N}\sum_{q,k}\left\langle 0\left|\frac{V_q^* V_q}{\epsilon_k - \epsilon_{k+q} - \hbar\omega_q} c_k{}^{+} c_{k+q} c_{k+q}^{+} c_k\right|0\right\rangle$$

This is the expectation value of an operator with two electron-annihilation and two electron-creation operators. The phonon annihilation and creation operators do not appear as long as we are interested in states in which no phonons are present. This is exactly the form of an operator representing the electron-electron interaction, as we indicated earlier. In this case, however, the coupling constant is given by $V_q^* V_q/(\epsilon_k - \epsilon_{k+q} - \hbar\omega_q)$. For calculating the energy to second order in the electron-phonon interaction we can simply introduce this effective interaction between electrons and compute the energy to first order in this effective interaction.

For computing the energy to second order in the electron-phonon interaction we required only the diagonal matrix element of this effective interaction between electrons. However, if we were to compute the energy to fourth order in the electron-phonon interaction, we would require second-order terms in the effective electron-electron interaction. Thus matrix elements would appear which are given by

$$\left\langle 1\left|\frac{V_q^* V_q}{\epsilon_k - \epsilon_{k+q} - \hbar\omega_q} c_{k''}^{+} c_{k''+q} c_{k+q}^{+} c_k\right|0\right\rangle$$

These are customarily written in terms of a wavenumber $\mathbf{k}' = \mathbf{k}'' + \mathbf{q}$. We may commute these operators in order to show that such terms are formally equivalent to first-order matrix elements of an effective electron-electron interaction given by

$$V_{\text{elel}} = \frac{1}{N}\sum_{q,k,k'} \frac{V_q^* V_q}{\epsilon_k - \epsilon_{k+q} - \hbar\omega_q} c_{k+q}^{+} c_{k'-q}^{+} c_{k'} c_k \tag{4.58}$$

By replacing the electron-phonon interaction in the Hamiltonian by this effective interaction we are of course discarding many terms. In particular we have considered only coupling to states with no phonons present. In the theory of superconductivity many terms are discarded and attention is focused only on those which are relevant. The terms that are obtainable from V_{elel} given above are those which enter the microscopic theory of superconductivity. The important feature, as we shall see, is that for electrons near the Fermi surface the energy difference in the denominator of Eq. (4.58) is small compared to $\hbar\omega$ and V_{elel} is predominantly negative. Thus there is an attractive interaction between electrons which can lead to an instability of the normal state of the metal. When this happens a new state, the superconducting state, is formed.

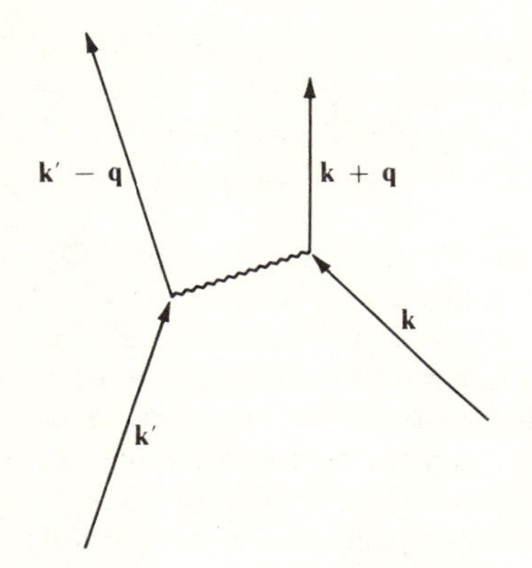

Fig. 4.22 A diagram representing the electron-electron interaction arising from the exchange of a virtual phonon. The wavy line represents a phonon of wavenumber—$\mathbf{q}$ *emitted by an electron of wavenumber* $\mathbf{k}$ *and reabsorbed by an electron of wavenumber* $\mathbf{k}'$.

We may think of the terms we have written in the electron-electron interaction as a scattering process as shown in Fig. 4.22. An electron of wavenumber $\mathbf{k}$ emits a phonon of wavenumber $-\mathbf{q}$ scattering it into a state $\mathbf{k} + \mathbf{q}$. This phonon is then absorbed by an electron of wavenumber $\mathbf{k}'$ scattering it into a state of wavenumber $\mathbf{k}' - \mathbf{q}$. The electron-electron scattering through the coulomb interaction may be described similarly as the exchange of a photon between electrons. In the theory of superconductivity we will combine this electron-electron interaction arising from the electron-phonon interaction with the coulomb interaction between electrons. We will then compute states for the corresponding system of interacting electrons. The phonons themselves will never appear in the calculation.

4.4 The Mössbauer effect[1] In recent years important studies of solids have been made using gamma rays emitted by nuclei in solids. Such studies can be useful when the natural linewidth for the emission is extremely narrow, i.e., when the emission of a gamma ray gives an extremely precise measure of the separation between the levels involved in the nuclear transition. Then if the emitting nucleus is in a different crystalline environment than a detecting nucleus the gamma-ray level separation may be sufficiently different that the gamma ray is not detected. However, by moving the detector with respect to the emitter with a velocity, perhaps that of a phonograph turntable, the gamma ray may be doppler-shifted sufficiently for the detection to take place. This velocity then gives a very precise measure of the change in the environment in the two crystals.

[1] For more detail, see C. Kittel, "Quantum Theory of Solids," p. 386, Wiley, New York, 1963, and the reprint volume by H. Frauenfelder, "The Mössbauer Effect," Benjamin, New York, 1962.

We might at first expect that such sensitive measurements would fail in any case because of the lowering of the gamma-ray energy due to the recoil of the nucleus. However, the Mössbauer effect, the possibility of recoilless emission, resolves this difficulty. It is the recoilless emission that we wish to discuss here.

Let us first imagine a free nucleus in an excited state. Let us then say that this nucleus may make a transition to the ground state with the emission of a gamma ray of wavenumber $-\mathbf{Q}$. Since total momentum must be conserved we know that the nucleus must be given a compensating momentum $\hbar\mathbf{Q}$. Such a transition may be described in ordinary time-dependent perturbation theory. The probability of the transition per unit time will be proportional to the square of the matrix element of a term in the Hamiltonian which we may write $M(\rho)e^{iQ\cdot r}$ where $\mathbf{r}$ is the center-of-mass position for the nucleus and M is a function of the internal coordinates ρ of the nucleus. We can show that the center-of-mass coordinate for the nucleus must enter in precisely this form if momentum is to be conserved by transforming the wavefunction for the nucleus to the center-of-mass and relative coordinates; then it becomes $\psi(\rho)e^{ik\cdot r}/\sqrt{\Omega}$. Indicating initial and final states with subscripts i and f, we see that the matrix element for the transition in a free nucleus becomes

$$\langle f|Me^{iQ\cdot r}|i\rangle = \frac{1}{\Omega}\int \psi_f^*(\rho)e^{-ik_f\cdot r}\,Me^{iQ\cdot r}\,\psi_i(\rho)e^{ik\cdot r}\,d\rho\,d\mathbf{r}$$

$$= \delta_{k_f - k_i - Q}\int \psi_f^* M\psi_i\,d\rho$$

Momentum is conserved and the energy delta function for the transition will require a reduction of the gamma-ray energy to compensate for the recoil energy of the nucleus.

We might now imagine holding the nucleus fixed by, e.g., embedding it in a crystalline lattice. We might then imagine that the recoil momentum is taken up by the entire lattice, with a correspondingly negligible recoil energy. It is not obvious physically that this will be the case. If, for example, we imagine the transition taking place very rapidly, recoil momentum would be transferred to the nucleus before it had sufficient time to move far enough to feel the effect of the interaction with its neighbors. If, on the other hand, we imagine the transition as being very slow, the interaction of the nucleus with this environment would hold it in its place and momentum would be transferred uniformly to the entire crystal. As in the discussion of the Franck-Condon principle in Sec. 5.6 of Chap. III, we are faced with the question of the "speed" of the transition and again the question must be resolved by carrying out a full solution of the problem. We might naively imagine that the lifetime of the state, determined by the width of the emission line, would be the relevant time. However, consideration of the discussion in Sec. 5.6 of Chap. III would correctly suggest that it is not.

We must return to the emission problem, describing the lattice as well as the emitting nucleus quantum mechanically. To be definite we assume that the vibration spectrum is that of a perfect crystal though the results do not depend upon this. We have written in Eq. (4.39) the Hamiltonian of the lattice as a sum of harmonic-oscillator Hamiltonians,

$$\mathscr{H}_\varphi = \sum_q \left(\frac{1}{2M} P_q P_{-q} + \tfrac{1}{2}\lambda_q u_q u_{-q} \right)$$

where the u_q are normal coordinates and the displacement of the jth atom is given by

$$\delta \mathbf{r}_j = \frac{1}{\sqrt{N}} \sum_q \mathbf{u}_q e^{iq \cdot r_j} \tag{4.59}$$

Here $\mathbf{u}_q$ is a vector parallel to the polarization of the mode in question. The eigenstates for the phonon system alone are given by products of harmonic-oscillator functions,

$$\Phi = \prod_q \varphi_{n_q}(u_q) \tag{4.60}$$

where n_q is the number of phonons in each mode.

We will again write the initial and final nuclear states in terms of the relative nuclear coordinates. We let the emitting nucleus correspond to $j = 0$ and take its equilibrium position $\mathbf{r}_0 = 0$. The nuclear center-of-mass coordinate $\delta \mathbf{r}_0$ enters the wavefunction of the system through Eqs. (4.59) and (4.60). Then the matrix element for a transition between the initial state $\Phi_i \Psi_i(\rho)$ and a final state $\Phi_f \Psi_f(\rho)$ will be given by

$$\langle \Phi_f \Psi_f(\rho) | M(\rho) e^{iQ \cdot \delta r_0} | \Phi_i \Psi_i(\rho) \rangle = \langle \Psi_f | M | \Psi_i \rangle \langle \Phi_f | e^{iQ \cdot \delta r_0} | \Phi_i \rangle \tag{4.61}$$

Thus the matrix element for the transition is reduced by a factor of a matrix element between and initial final phonon states of $e^{iQ \cdot \delta r_0}$.

There is again a close analogy with the theory of light emission in the F center, the overlap between shifted phonon wavefunctions being replaced by the phonon matrix element in Eq. (4.61). The completeness of the phonon wavefunctions guarantees, as it did in the case of optical transitions, that the sum of the squared phonon matrix elements over all final states will be equal to unity. Again transitions may occur in which various numbers of phonons are created or destroyed. It can in fact be shown that the *expectation value* of the energy transferred to the lattice is equal to the value for an isolated nucleus. However, there is a possibility of emission in which no phonons are created. This is recoilless emission and we wish to examine that possibility more closely.

If we specifically ask for the probability of emission without recoil, i.e., with the lattice left in the same phonon state, then both phonon states in the phonon matrix element of Eq. (4.61) become identical and the matrix element is simply the expectation value of $e^{i\mathbf{Q}\cdot\delta\mathbf{r}_0}$. These matrix elements enter many properties and have been studied in very general terms. We may obtain a result immediately for the case that the displacements $\delta\mathbf{r}_0$ are small by expanding the exponential, averaging over angle and magnitude of $\delta\mathbf{r}_0$, and reforming the exponential.

$$\langle\Phi_i|e^{i\mathbf{Q}\cdot\delta\mathbf{r}_0}|\Phi_i\rangle \approx \left\langle\Phi_i\left|1 + i\mathbf{Q}\cdot\delta\mathbf{r}_0 - \frac{(\mathbf{Q}\cdot\delta\mathbf{r}_0)^2}{2}\right|\Phi_i\right\rangle$$

$$= 1 - \frac{Q^2}{6}\langle\Phi_i|\delta r_0{}^2|\Phi_i\rangle \approx e^{-Q^2\langle\delta r_0{}^2\rangle/6}$$

The factor of $\frac{1}{3}$ came from the angular average and $\langle\delta r_0{}^2\rangle$ is the mean-square displacement of the atom. This exponential turns out to be exactly correct for a thermal distribution of phonons, including the case $T = 0$ in which each phonon state is the ground state. The probability of recoilless emission is proportional to the square of this matrix element. That square, $e^{-Q^2\langle\delta r_0{}^2\rangle/3}$, is called the *Debye-Waller factor*. Similarly, in x-ray or neutron diffraction, the intensity of a diffraction spot corresponding to diffraction without the emission of phonons is proportional to the same factor.

The Debye-Waller factor may be written in more convenient form for the ground state of the system, in which case the $\langle\delta r_0{}^2\rangle$ arises from the zero-point fluctuations. This calculation may be done in the Debye approximation by expanding δr_0 in normal modes, each of which is taken to be in the ground state. The algebra, however, is somewhat simpler in the Einstein approximation in which we imagine the emitting nucleus to be a harmonic oscillator with frequency ω_E. The expectation value of the kinetic energy is simply $\frac{1}{2}M\omega_E{}^2\langle\delta r_0{}^2\rangle$; the expectation value of the total energy is simply twice this and in the ground state is equal to $\hbar\omega_E/2$. We may immediately solve for $\langle\delta r_0{}^2\rangle$ and substitute into the Debye-Waller factor to obtain for the ground state

$$e^{-1/3\,Q^2\langle\delta r_0{}^2\rangle} = e^{-1/3\,Q^2(\hbar\omega_E/2M\omega_E{}^2)} = e^{-R/3\hbar\omega_E}$$

In the final expression we have written $\hbar^2Q^2/2M$ as the classical recoil energy R.

We see that the emission will be predominantly without recoil when the classical recoil energy is small compared to the energy of the highest-frequency phonons. This remains true in the Debye approximation.

As the temperature is increased the mean-square displacements increase and the probability of recoilless emission is reduced. Similarly, if we were to weaken the constraint on the nucleus by reducing the force constants, the appropriate Einstein frequency is reduced and the probability of recoil is increased. As we suggested at the beginning, the probability of recoilless emission does not depend at all upon the lifetime of the excited state of the nucleus. If we wish to associate a time for the decay in terms of the recoilless emission, this time becomes $\hbar/R$.

5. *Pseudopotentials and Phonon Dispersion*

In our discussion of lattice vibrations we introduced phenomenologically a set of interaction constants between the atoms at the beginning. These forces are of course calculable in principle from the Schroedinger equation for the electron-ion system. Attempts have been made to carry out this calculation for the inert gas solids, ionic crystals, simple metals, and semiconductors. The only one of these efforts that is in any sense complete is the study of the metals where pseudopotential theory provides a natural approach. We will discuss simple metals first and will formulate the problem for general properties that depend on the variation of the total energy of the system as these ions are rearranged. One such property is of course the vibration of the lattice.

We return again to the classical treatment of the ion positions. We define the positions $\mathbf{r}_j$ of the N ions present and construct the corresponding pseudopotential and compute the sum of the electronic energies by perturbation theory. This will give us one contribution to the total energy of the system which will of course depend on the positions of the ions. We will then add the direct coulomb interaction between the ions and other terms to obtain the total energy as a function of the positions. We could in principle use this energy to obtain the force constants that were introduced in our discussion of the vibration spectrum. However, it will turn out to be more convenient to use the expression obtained for the total energy directly.

5.1 The total energy Given the positions of the ions and the corresponding pseudopotential we may directly compute the energy of a state. The energy to second order in the pseudopotential is given by

$$E_k = \epsilon_k + \langle \mathbf{k}|W|\mathbf{k}\rangle + \sum_q \frac{\langle \mathbf{k}|W|\mathbf{k}+\mathbf{q}\rangle\langle \mathbf{k}+\mathbf{q}|W|\mathbf{k}\rangle}{\epsilon_k - \epsilon_{k+q}} \tag{4.62}$$

where the ϵ_k are the kinetic energies, $\hbar^2k^2/2m$. Because of the inherent nonhermiticity of the pseudopotential it is necessary to derive this expression explicitly. This form is found to be correct, rather than a form in which the matrix elements appeared as $|\langle \mathbf{k}+\mathbf{q}|W|\mathbf{k}\rangle|^2$.

We wish now to sum this result over all occupied states. In performing this sum it will be correct to second order in the pseudopotential to sum over the Fermi sphere which would exist in the absence of the pseudopotential, taking principle parts of any ill-defined integrals. This can be demonstrated explicitly.[1] The essential point is that the distortions of the Fermi surface are second order in the pseudopotential, and first-order shifts in the energies are required in rearranging the electron to the true Fermi surface. Thus the net change in energy is of third order and negligible in our second-order scheme. Thus we are to sum Eq. (4.62) over all $k < k_F$. In doing this we will factor the matrix elements of the pseudopotential into a structure factor and a form factor in the usual way.

$$E_{\text{tot}} = \sum_{k<k_F} \epsilon_k + S(0) \sum_{k<k_F} \langle \mathbf{k}|w|\mathbf{k}\rangle + \sum_{k<k_F} \sum_{q} S^*(\mathbf{q})S(\mathbf{q}) \frac{\langle \mathbf{k}|w|\mathbf{k}+\mathbf{q}\rangle\langle \mathbf{k}+\mathbf{q}|w|\mathbf{k}\rangle}{\epsilon_k - \epsilon_{k+q}}$$

The sum over kinetic energies can be directly evaluated. The structure factor in the first-order term is unity. In the second-order term we interchange the sum over $\mathbf{q}$ and that over $\mathbf{k}$. It is convenient to write the results as the energy per ion, letting Z be the number of electrons per ion. We obtain

$$E/\text{ion} = \frac{3}{5} Z \frac{\hbar^2 k_F^2}{2m} + \frac{1}{N} \sum_{k<k_F} \langle \mathbf{k}|w|\mathbf{k}\rangle + \sum_{q} S^*(\mathbf{q})S(\mathbf{q})F'(q) \tag{4.63}$$

where

$$F'(q) = \frac{1}{N} \sum_{k<k_F} \frac{\langle \mathbf{k}|w|\mathbf{k}+\mathbf{q}\rangle\langle \mathbf{k}+\mathbf{q}|w|\mathbf{k}\rangle}{\epsilon_k - \epsilon_{k+q}} \tag{4.64}$$

An absolutely essential feature of these results is the fact that the zero-order and first-order terms of Eq. (4.63) are quite independent of the detailed arrangement of the ions. They depend only upon the total volume of the system. When we compute the change in energy due to rearrangement, such as in the introduction of a lattice vibration, we will need to compute only the small second-order term. The second extremely important feature of this result is that the function $F'(q)$ is also quite independent of the configuration of the ions and again depends only upon the total volume. The calculation of $F'(q)$ turns out to be rather complex but once completed it yields a simple function of the magnitude of q which may then be used directly to compute any of the atomic properties of the simple metals. The detailed arrangement

[1] W. A. Harrison, "Pseudopotentials in the Theory of Metals," pp. 93ff., Benjamin, New York, 1966.

of the ions enters the calculation only through the simple geometrical structure factors $S(\mathbf{q})$.

This approach can be generalized to transition metals[1] using the transition-metal pseudopotentials discussed in Sec. 9 of Chap. II. We wrote in Eq. (2.85) of that section the second-order energy of the conduction-electron states. That equation plays the role which Eq. (4.62) has played here. It however contains terms arising from hybridization as well as those from the pseudopotential. Both types of matrix elements factor into a structure factor and a form factor allowing as in the simple metals the separation of the second-order structure-dependent term and the evaluation of an $F'(q)$. As we indicated in discussing transition-metal pseudopotentials, such a summation over states based upon plane waves is incomplete and must be supplemented by a sum over states based upon atomic d states. That sum, if based on direct perturbation theory, is tractable only if made over filled d bands. However, an alternative formulation based upon the one-electron Green's function (Sec. 10.3 of Chap. II) allows an approximate treatment of the total energy even when the d bands are only partially occupied.[2] For the case of full d bands the summation proceeds much the same as that required for the screening field and discussed in Sec. 4.4 of Chap. III. It leads to an $F'(q)$ given (in the notation of Chaps. II and III) by

$$F'(q) = \frac{1}{N} \sum_{k<k_F} \frac{|\langle \mathbf{k}+\mathbf{q}|w|\mathbf{k}\rangle|^2}{E_k - E_{k+q}} + \sum_{k>k_F} \frac{1}{E_k - E_{k+q}} \left(\sum_d \frac{\langle \mathbf{k}+\mathbf{q}|\Delta|d\rangle\langle d|\Delta|\mathbf{k}\rangle\langle \mathbf{k}+\mathbf{q}|w|\mathbf{k}\rangle^*}{E_d - E_k} + \text{c.c.} - \left| \sum_d \frac{\langle \mathbf{k}+\mathbf{q}|\Delta|d\rangle\langle d|\Delta|\mathbf{k}\rangle}{E_d - E_k} \right|^2 \right)$$

This reduces in essence to Eq. (4.64) if Δ is taken equal to zero. As in the screening summation the limits on the sums are such that no divergent denominators $E_d - E_k$ appear in the range of summation. Also as in the case of screening, unoccupied d bands may be included by changing the sign of the hybridization contributions and summing over $k < k_F$ rather than $k > k_F$. For our purposes here the important fact is that an $F'(q)$ can be defined for noble and transition metals as well as for simple metals though they will be more difficult to compute. Once they *have* been obtained the calculation of properties proceeds in just the same way for all metals.

This sum of the energies does not of course represent the total energy of the system. We must in particular add a direct coulomb interaction between

[1] W. A. Harrison, *Phys. Rev.*, **181**:1036 (1969).
[2] J. Moriarty, *Phys. Rev.* (in press).

the ions. In addition, in the use of the self-consistent-field approximation we count the interaction between electrons twice; this point was discussed in Sec. 3 of Chap. II. It is therefore necessary to subtract from our result this coulomb energy that includes the zero-order coulomb energy of the unperturbed states as well as that from the screening field.

The sorting out of these various terms is a truly intricate task. However, the result is quite plausible and natural. The total energy of the system can be divided into three parts. The first we call the *free-electron energy*. It contains the average kinetic energy of the electrons and some terms from $\langle \mathbf{k}|w|\mathbf{k}\rangle$. This part is independent of the configuration of the ions and depends only upon volume. It is essential that it be included if we seek the total cohesive energy of the metal, the equilibrium lattice spacing, or the compressibility. In any such calculation it is essential also to include a very careful treatment of the electron-electron interaction.

The second contribution to the energy is called the *electrostatic energy*. It is defined to be the electrostatic energy of point-positive charges located at the true positions of the ions, all embedded in a compensating uniform negative background. In the most usual treatment the magnitude of the charge on these ions is different from the true ionic charge; it includes also a correction from the charge localized at each ion due to the orthogonalization of the pseudowavefunctions to the core states. This correction to the valence charge is ordinarily of the order of 10 percent. Use of such an *effective valence* for the ion is entirely a matter of convention. If a different effective charge is used or if the true ion charge is used, this simply modifies the remaining term in the energy and the total energy is mathematically equivalent. The evaluation of this electrostatic energy is a tricky problem because of the long-range nature of the coulomb interaction. However, this term in the energy is mathematically well defined and may be evaluated by analytic methods. The most usual approach is that given originally by Ewald[1] for computing electrostatic energies of ionic crystals and extended to metals by Fuchs.[2] For many problems it is more convenient to use a modified method based on the Ewald approach.[3]

The third term in the energy is called the *band-structure energy*. It has the form of the second-order term in Eq. (4.63),

$$E_{bs} = \sum_{q} S^*(\mathbf{q})S(\mathbf{q})F(q) \tag{4.65}$$

but the function $F(q)$, which is called the *energy-wavenumber characteristic* differs from $F'(q)$ by the subtraction of screening terms.

[1] P. P. Ewald, *Ann. Physik*, **64**:253 (1921).
[2] K. Fuchs, *Proc. Roy. Soc. (London)*, **A151**:585 (1935).
[3] Harrison, "Pseudopotentials," pp. 165ff.

Evaluation of changes in the two terms that depend on configuration, the electrostatic energy and the band-structure energy, is quite straightforward for rearrangements of the ions at constant volume for any material in which $F(q)$ has been obtained.

Before applying this method to one or two simple properties, it may be desirable to notice the special form which the energy-wavenumber characteristic would take if the pseudopotential were local, i.e., if the form factors $\langle \mathbf{k} + \mathbf{q}|w|\mathbf{k}\rangle$ were independent of $\mathbf{k}$. Then the sum over occupied states can be written in terms of the Hartree dielectric function which we obtained earlier. The result we obtain is

$$F(q) = -\frac{\Omega_0 q^2}{8\pi e^2}|\langle \mathbf{k} + \mathbf{q}|w^0|\mathbf{k}\rangle|^2 \frac{\epsilon(q) - 1}{\epsilon(q)} \tag{4.66}$$

where $\langle \mathbf{k} + \mathbf{q}|w^0|\mathbf{k}\rangle$ is the unscreened form factor. Note that the dielectric function appears only once in the denominator so that the result is proportional to the product of the screened and the unscreened form factors. This difference with F' arises from the subtraction of screening terms which are counted twice in the self-consistent-field method. This local form is widely used in the treatment of atomic properties. However, it must always be regarded as an approximation.

5.2 Calculation of vibration spectra[1] We have already obtained in Sec. 4.1 the structure factors in a lattice containing a periodic distortion. We found there that structure factors, which are first order in the amplitude of the displacements, arose at satellite points to each of the lattice wavenumbers. These additional structure factors will give additional terms in the band-structure energy of Eq. (4.65) which are second order in the amplitude of the displacements and therefore are harmonic terms. In order to obtain the total change in energy to second order in the displacement, we must be sure that we have included all contributions to the energy to second order in the amplitudes. By carrying the expansion that we made in Sec. 4.1 to second order in the displacements, we find that there are also changes in the structure factor at the lattice wavenumbers which are second order in the amplitudes. When the structure factors at the lattice wavenumbers are squared it is immediately seen that these second-order shifts give second-order terms in Eq. (4.65) which must also be included. We also find that second-order structure factors are introduced at new satellites to the lattice wavenumbers but these when squared give only fourth-order contributions to the band-structure energy.

[1]For more detail, see Harrison, *op. cit.*, pp. 233ff.

Collecting the second-order contributions to the band-structure energy at each lattice wavenumber and its satellites we obtain

$$\delta E_{\text{bs}} = \sum_{\mathbf{q}_0} \big[|(\mathbf{q}_0 + \mathbf{Q}) \cdot \mathbf{u}_Q|^2 F(\mathbf{q}_0 + \mathbf{Q}) + |(\mathbf{q}_0 - \mathbf{Q}) \cdot \mathbf{u}_Q|^2 F(\mathbf{q}_0 - \mathbf{Q}) - 2|\mathbf{q}_0 \cdot \mathbf{u}_Q|^2 F(q_0) \big] \tag{4.67}$$

where again $\mathbf{Q}$ is the wavenumber of the mode and the sum is over lattice wavenumbers $\mathbf{q}_0$. The sum includes all lattice wavenumbers $\mathbf{q}_0$ but the ill-defined term $|\mathbf{q}_0 \cdot \mathbf{u}_Q|^2 F(q_0)$ is to be dropped for $\mathbf{q}_0 = 0$.

Given the energy-wavenumber characteristic we may perform a sum for any wavenumber of interest. The result is proportional to the square of the amplitude and therefore enters directly in the dynamical matrix λ_q of Eq. (4.37) in Sec. 4.2. In our construction of structure factors here we took $u_Q = u^*_{-Q}$. If both are taken real then our sum is of exactly the form entering that equation. We may also obtain the change in electrostatic energy due to the introduction of the distortion using a parallel approach but treating the long-range coulomb interaction analytically.[1]

Having obtained the dynamical matrix we may simply divide by the mass of the ions in order to obtain the square of the frequency, and the dispersion calculation is complete. The solution is this simple only if we have been able to specify the form of the mode in advance, i.e., if we know the direction of polarization. This will be the case when $\mathbf{Q}$ is in a symmetry direction in a simple structure. For arbitrary propagation directions, the energy will be a quadratic form in the three components of $\mathbf{u}_Q$ and we must solve for the three modes just as in a molecular-vibration problem. In this case the dynamical matrix λ_Q contains nine elements coupling the components of $\mathbf{u}_Q$. In a structure with two atoms per cell we require six displacement components and must solve for the six modes, three acoustical and three optical.

Such calculations to date have been moderately successful. The electrostatic contribution has been found to give a positive contribution to λ_Q and therefore to tend toward stability of the lattice. The band-structure energy ordinarily gives negative contributions to λ_Q and therefore lowers the frequencies. In the alkali metals the band-structure contribution tends to be very small and energies are given approximately by the electrostatic calculation alone. Thus small errors in $F(q)$ have been unimportant and the agreement with experiment has been good. In the metals of higher valence the band-structure contribution becomes increasingly negative and the results consequently increasingly sensitive to the details of the calculation. In aluminum, for example, the calculated frequencies[2] frequently differ by as much as a factor of 2 from experiment. However, it should be kept in mind that these

[1] *Ibid.*, pp. 247ff.

[2] *Ibid.*, p. 254.

calculations are in all cases from first principles; therefore the disagreement is not so disappointing. Furthermore, recent calculations[1] suggest that most of the discrepancy may be removed by a more complete treatment of exchange and correlations than was used earlier.

5.3 The Bohm-Staver formula[2] There are two interesting aspects of the vibration spectrum which we would like to discuss. The first concerns long-wavelength longitudinal modes. These can be studied by the approach we have described above. In particular we could focus our attention on the structure factors that occur at wavenumber satellites to $\mathbf{q}_0 = 0$; however, it may be more instructive to study these modes independently. This will lead to the Bohm-Staver formula for the speed of longitudinal sound in metals.

A long-wavelength longitudinal wave propagating through the lattice will cause a long-wavelength periodic potential. We compute this directly by again defining displacements given by

$$\delta\mathbf{r} = \frac{\mathbf{u}_Q e^{iQ\cdot r}}{\sqrt{N}}$$

where $\mathbf{r}$ is again the equilibrium position of the ion and $\boldsymbol{\delta}\mathbf{r}$ is the displacement. In longitudinal waves $\mathbf{u}_Q$ is parallel to $\mathbf{Q}$. The dilatation due to this wave is simply the divergence of $\boldsymbol{\delta}\mathbf{r}$, and the local ionic-charge density induced by this dilatation is simply the product of the dilatation and the negative of the local ionic-charge density, which is $-Ze/\Omega_0$. Again Z is the valence and Ω_0 the atomic volume. Thus there is an electrostatic-charge density due to the ionic charges with a long-wave component given by $(-iZeQu_Q/\Omega_0\sqrt{N})e^{iQ\cdot r}$, and a corresponding electrostatic potential given by $(-4\pi iZeu_Q/Q\Omega_0\sqrt{N})e^{iQ\cdot r}$. The resulting force on a given ion will be Ze times the negative of the gradient of this potential, $(-4\pi Z^2e^2u_Q/\Omega_0\sqrt{N})e^{iQ\cdot r}$. If the only forces on the ion were the forces due to this electrostatic potential arising from the ions, this force in turn would be equal to the mass of the ion M times the second derivative of u_Q with respect to time. The frequency then would be independent of wavenumber and would be the ion-plasma frequency given by

$$\omega_{\mathrm{IP}}{}^2 = \frac{4\pi Z^2e^2}{M\Omega_0}$$

This is in exact analogy to the electron-plasma frequency given in Eq. (3.37) in Sec. 4 of Chap. III except that the ratio e^2/m is replaced by Z^2e^2/M.

Of course in the metal there are other forces on the ion. In particular

[1]For example, R. W. Shaw, Jr. (to be published) in a calculation for magnesium using essentially the approach described in Sec. 4.5 of Chap. III.

[2]Reference is traditionally made to D. Bohm and T. Staver, *Phys. Rev.*, **84**:836 (1950), though the formula does not appear explicitly there.

the effect of screening will be to reduce the electrostatic potential and therefore the electrostatic force by a factor of the dielectric function $\epsilon(q,\omega)$. In addition there are forces arising from the pseudopotential but it will be interesting to find the result for the screened electrostatic interactions alone. We use here the static dielectric function which can be shown to be appropriate at the low frequencies involved. We found in Sec. 4.3 of Chap. III that at long wavelengths the dielectric function approached $4\pi e^2 n(E_F)/q^2$. Thus the frequency of the mode in the screened system is given by

$$\omega^2 = \frac{Z^2 Q^2}{M\Omega_0 n(E_F)} = \frac{2ZE_F Q^2}{3M}$$

Thus we see that with proper screening the frequency approaches zero linearly with Q; the dispersion curve corresponds to a well-defined speed of sound. This speed of sound, called the Bohm-Staver value, is $\sqrt{2ZE_F/3M}$. It may also be noticed that the Bohm-Staver speed of sound is simply $\sqrt{Zm/3M}$ times the Fermi velocity for the electrons. This gives a reasonable account of the longitudinal-sound velocity in many metals and clearly contains the physics of much of what is going on.

5.4 Kohn anomalies[1] The second aspect of the phonon-dispersion curves which we wish to discuss concerns the behavior at shorter wavelengths. We noted that the first derivative of the dielectric function became logarithmically infinite (negatively) when $q = 2k_F$. This anomaly will be reflected also in the energy-wavenumber characteristic as may be most easily seen from Eq. (4.66) giving that characteristic for a local pseudopotential. The dielectric function appears at two places in that formula but the strongest singularity arises from the numerator since the dielectric function is near unity in the range in question. Thus the infinite negative slope of $\epsilon(q)$ gives rise to an infinite positive slope in $F(q)$ at $q = 2k_F$. This remains true when the calculation is carried out for the full nonlocal pseudopotential.

Thus as we calculate the vibration spectrum and vary Q over the Brillouin zone we may see from Eq. (4.67) that there will be an infinitely rapid increase in the band-structure energy whenever $|\mathbf{q}_0 \pm \mathbf{Q}|$ (for any lattice wavenumber q_0) increases through the value $2k_F$. This infinitely rapid increase in the band-structure energy finally gives rise to an infinitely rapid increase in the frequency as the wavenumber is changed. The corresponding anomalies in the vibration spectrum are called the *Kohn anomalies*. Physically they arise from the singular change in screening as one of the relevant Fourier components of the potential moves through $2k_F$ and the effectiveness of the screening thus drops singularly.

[1] W. Kohn, *Phys. Rev. Letters*, **2**:393 (1959).

There may be many Kohn anomalies in the dispersion curves for a metal. The condition for an anomaly to occur is again that

$$|\mathbf{q}_0 \pm \mathbf{Q}| = 2k_F$$

We may obtain the corresponding wavenumber graphically by the *Kohn construction*. The condition displayed above is that $\mathbf{Q}$ lie on a sphere of radius $2k_F$ centered at a lattice wavenumber. Thus we may proceed much as in the construction of free-electron Fermi surfaces except with spheres of twice the radius. We construct such a sphere around every lattice wavenumber as indicated in Fig. 4.23 for aluminum.

It is not difficult to see that both upward and downward anomalies can occur. The frequency will increase infinitely rapidly as we move from inside to outside a sphere but will drop infinitely rapidly as we move from

Fig. 4.23 *Kohn construction for aluminum. A (110) plane in wavenumber space (containing the point* $\mathbf{k} = 0$*) is shown with the intersections with spheres at which singularities occur. Also shown is a section of the Brillouin zone, which contains the wavenumbers by which we index the lattice vibrations.*

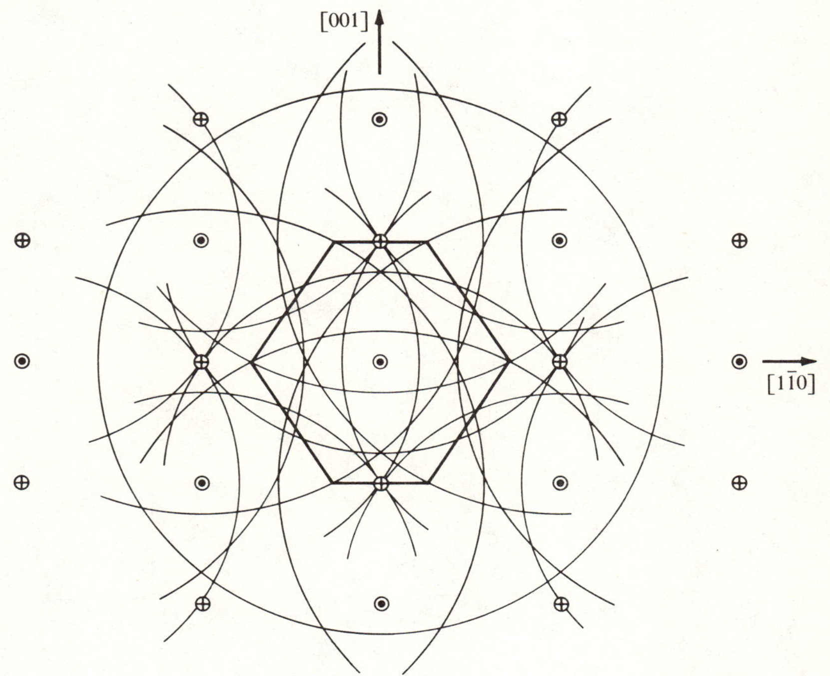

outside in. The observed vibration spectrum of aluminum is shown in Fig. 4.24 along with a calculated curve based upon a local pseudopotential with two adjustable parameters fixed to give agreement with the observed curves at two points. The anomalies in aluminum are observable in the calculated curves but are barely discernible in the experimental curves. In lead, for which Fig. 4.25 gives a similar set of curves, the anomalies are much stronger.

6. Interatomic Forces and Atomic Properties[1]

Our calculation of the total energy of the system as a function of the configuration of the ions is of course applicable to a wide range of properties in metals. The calculation of the vibration spectrum is only one such property.

[1]The material in Secs. 6.1 and 6.2 is discussed in greater detail in Harrison, "Pseudopotentials."

Fig. 4.24 *Model calculation of the vibration spectrum of aluminum, fit to experiment at the points indicated by boxes (after W. A. Harrison, "Pseudopotentials in the Theory of Metals," p. 300, Benjamin, New York, 1966). Experimental points are from J. L. Yarnell, J. L. Warren, and S. H. Koenig [in "Lattice Dynamics," R. F. Wallis (ed.), p. 57, Pergamon, New York, 1965].*

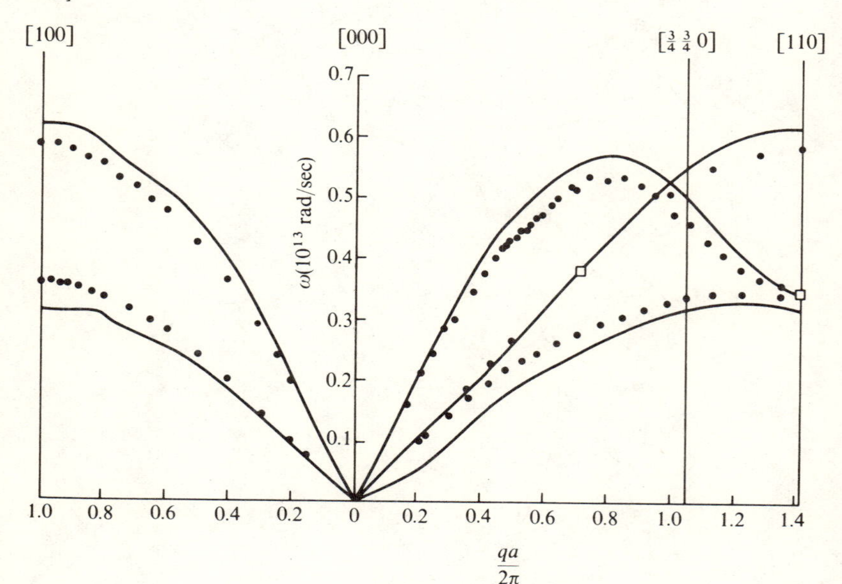

A second property that comes immediately to mind is the comparison of the energy of the crystal for different crystal structures.

6.1 Stability of metallic structures If we consider changes in crystal structure at constant volume, then only the electrostatic and the band-structure energies change. It turns out that the electrostatic energies are very nearly identical for the common metallic structures, face-centered cubic, body-centered cubic, and hexagonal close packed (with ideal axial ratio). In any case the electrostatic energy is directly calculable. The calculation of the band-structure energy for any given structure is also quite straightforward. We need simply construct a wavenumber lattice and perform the sum, Eq. (4.65), over that wavenumber lattice. The $F(q)$ drops off sufficiently rapidly at large q that we may sum by machine over several hundred lattice wavenumbers and obtain a good result. There are also short cuts which can be made if the summation is to be performed by hand so that it is important to improve the convergence.[1] This approach involves replacing the summation at large distances by an integration.

[1] *Ibid.*, pp. 191ff.

Fig. 4.25 *Model calculation as in Fig. 4.24 but for lead. Experimental points are taken from B. N. Brockhouse, T. Arase, G. Caglioti, K. R. Rao, and A. D. B. Woods* [Phys. Rev., ***128****: 1099* (*1962*)].

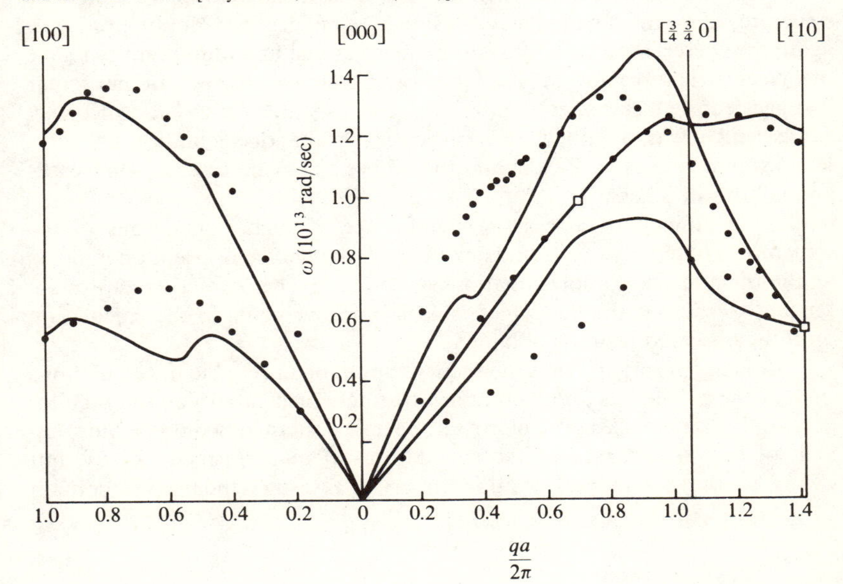

The first real attempt[1] at such comparisons of the energies of different crystal structures was made for sodium, magnesium, and aluminum following the first calculation of $F(q)$ for these metals. [Actually an earlier attempt had been made for zinc using a crude $F(q)$ computed by hand.] The energy was compared for face-centered cubic, body-centered cubic, and hexagonal close-packed structures. Since the axial ratio in the hexagonal close-packed structure is not determined by symmetry, it was necessary to calculate the energy in that structure for a number of values of axial ratio and to select the minimum energy. The results of these calculations were surprisingly good. In both sodium and magnesium the hexagonal close-packed structure was found to have the lowest energy. Both metals are known to be hexagonal close-packed at low temperatures. Aluminum was correctly found to have its lowest energy in the face-centered cubic structure. In this calculation additional information was automatically obtained for the hexagonal structures. In particular the value of the axial ratio at the minimum energy was found to be close to that observed in both sodium and magnesium. In addition the change in energy with change in axial ratio gives directly one of the elastic shear constants. The agreement with the observed elastic constant in magnesium was extremely close and the value in sodium was reasonable though the experimental value for that structure is not known.

It now appears that there was an element of luck in the treatment of the structures of these three metals. The subsequent attempt to study the structures of a wide range of metals using a crude local pseudopotential gave again the correct structure for sodium, magnesium, and aluminum but gave incorrect structures for most of the other metals studied. In particular it suggested the face-centered cubic structure as stable for zinc, an incorrect result which was obtained also earlier in the hand calculation for zinc. More recent studies by Pick[2] and others have given successes in some cases and failures in others.

It is difficult to assess the significance of such calculations of the structure. There is always some flexibility in the choice of pseudopotentials and in the numerical approximations to be made. The correct structures are known, so there is the danger of making improvements in the calculation until the observed result is obtained.

It is not surprising that the theory should be unreliable in calculations such as these. There is an inherent arbitrariness or flexibility in the pseudopotentials that are used. The electronic energies obtained would be independent of this arbitrariness if the calculations were performed exactly, but owing to the use of perturbation theory the results themselves contain

[1] W. A. Harrison, *Phys. Rev.*, **136**:A1107 (1964).
[2] R. Pick, *J. Phys. (France)*, **28**:539 (1967).

an element of flexibility. Furthermore, the screening in the electron-electron interaction is inevitably treated only approximately and it is difficult to assess the errors that are made. It seems clear nevertheless that the essential aspects of the problem can be included within the pseudopotential framework, and it is only the quantitative reliability of the results that is uncertain in any case.

Other aspects of the problem should be included in a discussion of stability of structures. The calculation described above bears on the internal energy of the system at zero temperature as a function of structure but it is of course the total free energy that is important. At finite temperatures lattice vibrations will be excited, contributing to the free energy. Even at the absolute zero of temperature the zero-point vibrations of the system should be included and will contribute to the internal energy. The inclusion of zero-point oscillations in the above calculations does not affect the results for sodium, magnesium, and aluminum, but neither are the contributions negligible. The zero-point energy is a sum over all vibrational modes of $\frac{1}{2}\hbar\omega_q$. In a very soft lattice the frequencies will be low and therefore the zero-point energy low, favoring that lattice structure. Furthermore, as the temperature is raised the entropy will increase more rapidly in a soft lattice and the free energy will drop with respect to that of the rigid lattice. In sodium, for example, the body-centered cubic structure has a transverse acoustic band of vibrations which lies at quite low frequencies. It is generally believed that the increase in entropy associated with these vibrations is responsible for the transition of sodium from the low-temperature hexagonal close-packed structure to the high-temperature body-centered cubic structure. In order to compute the transition temperature it is necessary not only to compute the difference in internal energy but also to perform the necessary sums to obtain the free energy as a function of temperature. It is clear that if this calculation had been done for sodium the result would not have agreed well with experiment since the computed difference in internal energy between hexagonal close-packed and body-centered structures differed appreciably from that deduced from experiment.

Given the energy-wavenumber characteristic for a metal there is a wide range of properties that may be explored, though there is always uncertainty as to the reliability of the results. There *is* an approach, however, which would seem to enhance the reliability. That is to first perform a pseudopotential calculation of the vibration spectrum. An adjustable parameter, or several adjustable parameters, then can be introduced in the pseudopotential and adjusted to obtain as good as possible a fit to the observed vibration spectrum. The corresponding $F(q)$ then could be used to explore other properties. The corresponding *phenomenological pseudo-*

potential would seem to correspond to a plausible treatment of the interactions but would have been adjusted in order to account accurately for some aspects of the known experimental interaction.

6.2 The effective interaction between ions The structure, the energy of formation, and the dynamics of defects in crystals provide realms of atomic properties of particular interest. If, for example, we wish to study vacancies in aluminum we might first postulate a perfect crystal with a single ion removed. The structure factors may be computed directly and the change in energy computed. Such a calculation must be performed with some care in order to keep the atomic volume for the system constant. Such a calculation was performed for aluminum and led to an energy of formation that was greater by perhaps a factor of 2 from that indicated by experiment. We note, however, that the calculation assumed that the neighboring ions remained in their original positions when the vacancy was introduced. We may expect that the neighboring ions will relax around the vacancy in a real crystal, lowering the energy slightly. This problem can also be studied within the framework of the pseudopotential theory. We simply allow relaxation of the neighboring ions, recompute the structure factor (which is a function of this distortion) and then minimize the energy with respect to distortion. This is a rather involved calculation and has not been carried through for aluminum.

We might in addition compute the energy as a function of the displacement of a neighboring ion as it is moved into the vacant site, corresponding to a motion of the vacancy from one site to the next. The maximum energy in this process would correspond to the activation energy for motion of the vacancy. Again in this calculation it would seem important to allow relaxation of the neighboring ions in the process. Any of these calculations can be performed directly with the energy-wavenumber characteristic. However, in some cases it may be simpler to rewrite the total energy in terms of an effective interaction between ions. This may be done by returning to Eq. (4.65) giving the band-structure energy and writing out the structure factors.

$$E_{bs} = \frac{1}{N^2} \sum_q \sum_{i,j} e^{i\mathbf{q}\cdot(\mathbf{r}_i - \mathbf{r}_j)} F(q)$$

$$= \frac{1}{N} \sum_{i,j} \frac{1}{N} \sum_q F(q) e^{i\mathbf{q}\cdot(\mathbf{r}_i - \mathbf{r}_j)}$$

This may be conveniently rewritten in terms of the Fourier transform of the energy-wavenumber characteristic,

$$\mathscr{V}_{bs}(r) = \frac{2}{N} \sum_q F(q) e^{i\mathbf{q}\cdot\mathbf{r}} = \frac{2\Omega_0}{(2\pi)^3} \int dq 4\pi q^2 F(q) \frac{\sin qr}{qr}$$

Then

$$E_{bs} = \frac{1}{2}\frac{1}{N}\sum_{i,j\neq i}\mathscr{V}_{bs}(\mathbf{r}_i - \mathbf{r}_j) + \frac{\Omega_0}{(2\pi)^3}\int dq 4\pi q^2 F(q)$$

where we have replaced the sum over wavenumbers by an integral. We have noted the spherical symmetry of $F(q)$ and we have written out explicitly the diagonal term, $i = j$. That term is quite independent of the detailed configuration of the ions and therefore may be added directly to the free-electron energy and neglected in any studies of rearrangement at constant volume. The E_{bs} is the band-structure energy per ion. We may multiply by the number of ions N and find that the relevant term in the band-structure energy is simply written as the total energy of ions interacting through an effective two-body central-force interaction given by $\mathscr{V}_{bs}$.

The electrostatic energy may also of course be written in precisely this form with an interaction given by $Z^{*2}e^2/|\mathbf{r}_i - \mathbf{r}_j|$ if our separation into band structure and electrostatic parts has involved the use of an effective valence Z^*. We have achieved our goal of writing the configuration-dependent energy of the crystal in terms of a two-body central-force interaction.

$$E_{tot} = \tfrac{1}{2}\sum_{j,i\neq j}\mathscr{V}(\mathbf{r}_i - \mathbf{r}_j)$$

where

$$\mathscr{V}(r) = \mathscr{V}_{bs}(r) + \frac{Z^{*2}e^2}{r}$$

The result of such a calculation of the effective interaction between ions for aluminum is shown in Fig. 4.26.[1] Several features of this result should be noted. First there is a deep minimum occurring near the nearest-neighbor distance in aluminum. We might at first think that the stability of the crystal would require that these atoms lie at the minimum. In fact, of course, we cannot even assume that the structure is stable under the interaction $\mathscr{V}(r)$ by itself. The volume-dependent terms must also be included in the calculation of the equilibrium density of the material. If they tend to expand the lattice then the effective interaction between ions must tend to contract the lattice. This is suggested by the aluminum curve.

Second we notice that at large distances the effective interaction between ions oscillates with distance. These are of course the Friedel oscillations and have the same origin as the Friedel oscillations in electron density. They are responsible for the long-range nature of the interaction between ions in a simple metal. It is interesting to note that these oscillations are of opposite sign from what we might at first expect. The oscillations in the

[1]Curves for a number of metals, computed in the same way, are given by Pick, *op. cit.*

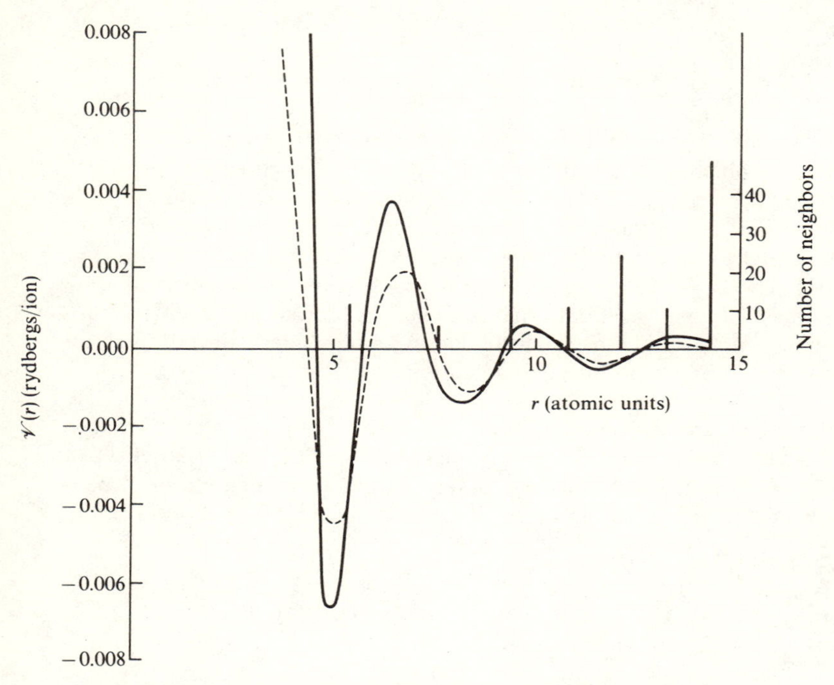

Fig. 4.26 The effective interaction between ions for aluminum at the observed volume. Also shown is the distribution of neighbors as a function of distance for the face-centered cubic structure. The dashed line is the asymptotic form given in Eq. (4.68) with $\langle -k_F|w|k_F\rangle$ taken equal to 0.1107 rydberg.

electron density will give oscillations in the electrostatic potential which will tend to favor the neighboring ions sitting at positions of maximum electron density. However, the essentially repulsive nature of the pseudopotential itself tends to favor the ions sitting at the positions of low electron density. The latter effect wins out in aluminum and the ions tend to sit at positions of low electron density.

We may directly compute the asymptotic form of the effective interaction between ions by partial integration, just as we computed the electron-density fluctuations. The asymptotic form is found to be

$$\mathscr{V}(r) \sim \frac{9\pi Z^{*2}\langle -k_F|w|+k_F\rangle^2}{E_F} \quad \frac{\cos 2k_F r}{(2k_F r)^3} \tag{4.68}$$

for a local pseudopotential. Note that the sign of the oscillations is independent of the sign of the pseudopotential that enters squared. In contrast we found that the sign of the charge-density fluctuations depended on the sign of the pseudopotential. These two effects together cause the tendency for aluminum ions to sit at positions of low electron density.

Some properties may be immediately deduced from the effective interaction between ions which would require lengthy calculations with the energy-wavenumber characteristic. For example, we may consider the question raised earlier as to the relaxation of neighbors to a vacancy. We see that near neighbors exert an attractive force on each other; thus when an ion is removed the nearest neighbors will relax outward. This may not be the case in other materials. If the nearest-neighbor minimum shifts with the asymptotic form then we would find that in sodium the nearest neighbors interact with a repulsive force and the neighbors to a vacancy should relax inward.

It is also rather easy to show that the interaction between two vacancies in a metal is of precisely the same form as the effective interaction between ions. Thus we may estimate the binding of two vacancies to neighboring sites (in the absence of distortion) by simply reading off the value of the effective interaction between ions at the near-neighbor distances.

6.3 Atomic properties of insulators and semiconductors The separation of small configuration-dependent terms in the energy from the very large contributions determining the cohesion was an extremely important aspect of the theory we have given for metals. It is difficult to imagine proceeding with what we have called atomic properties without that separation. Although the theory of the atomic properties of insulators and semiconductors is not in nearly such a tidy form, it has been possible to make an analogous separation and methods have been developed for treating atomic properties. We will describe these approaches in somewhat general terms; the manner in which they may be applied to specific properties is at once obvious and is quite similar to the use of the effective interaction between ions in metals.

In Chap. II we noted that the cohesive energy of ionic crystals was given approximately by the electrostatic energy gained in bringing the ions together in the crystal from infinity (minus the energy required to form the separated ions from atoms). Similarly, we have indicated that the weak binding in molecular crystals arose primarily from van der Waals forces, the latter arising from the correlated electronic motion in interacting atoms or molecules and a resulting dipole-dipole attraction. In both cases these attractive forces are much longer range than repulsive forces that must also come into play. (Without repulsive forces the lattice would collapse under these attractive forces.) We may imagine bringing the ions or atoms together, gaining an energy equal to the binding energy before the repulsive forces come into play. Then the ions are essentially in contact and no further contraction occurs. All of the binding energy is associated with the long-range forces, but once the crystal is constructed the much more abrupt repulsive forces completely dominate, for example, the elastic behavior. For the configuration-dependent atomic properties the repulsive forces are all-important.

There have been a large number of approaches, most of them phenomenological, for treating this repulsive interaction; Born-Mayer and Lennard-Jones are the names associated with the two most familiar ones.[1] In all cases, the total potential used is of the general form shown in Fig. 4.27.

Two important distinctions should be noted immediately between the interaction shown in Fig. 4.27 and the effective interaction between ions in a metal shown in Fig. 4.26. First is that the curve purports to describe the entire interaction energy; there are no additional volume-dependent terms as in the metal. Second, the potential rises monotonically beyond the minimum; there are no Friedel oscillations. These features have a number of consequences that will apply to any theory based upon such a model. First are the Cauchy relations between the elastic constants[2] derivable for a system of particles in equilibrium under the action of spherically symmetric potentials alone. These are satisfied to within about 10 percent in sodium chloride but deviate much more in many other insulators. We may also note that under such potentials the interaction force between all neighbors but the nearest will inevitably be attractive for a system such as a rare gas in which all atoms are identical, if the lattice is to be in equilibrium. It follows that if one atom is removed to form a vacancy the nearest neighbors will be displaced inward. Finally, we may note that for both the ionic and molecular insulators the lattice will tend to expand with increased temperature. Because of the asymmetry of the potential near the minimum, vibrations will lead to an average outward displacement.

Because it has not been possible to reliably calculate the repulsive interaction, a form has been assumed, ordinarily exponential, and parameters adjusted to fit such properties as the equilibrium displacement and the compressibility. In the absence of theoretical calculations of the interaction this has been a reasonable approach but it can have only limited success. Experimental deviations from the Cauchy relations imply the existence of contributions to the energy which are comparable to the difference in energy between different structures and therefore the importance of contributions

[1] For extensive discussion of these approaches, see F. Seitz, "Modern Theory of Solids," McGraw-Hill, New York, 1940.

[2] *Ibid.*, p. 95.

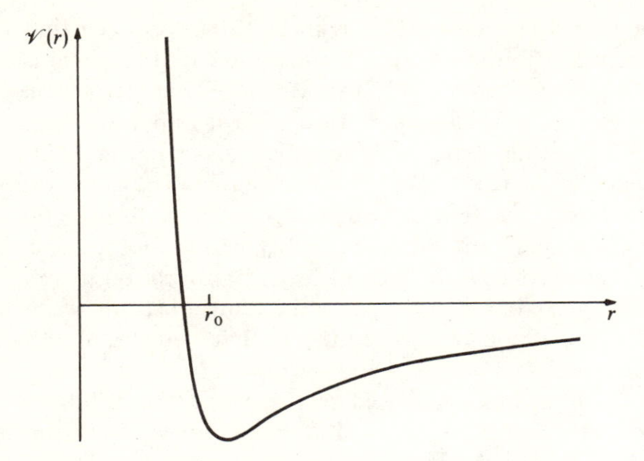

Fig. 4.27 The general form of the interaction between ions in an insulator. The attractive interaction used here is coulombic and therefore appropriate to ions of opposite charge in an ionic insulator. For molecular crystals the attractive interaction goes to zero much more rapidly at large distances, being proportional to r^{-6}. The distance r_0 represents the equilibrium nearest-neighbor spacing.

to the energy which cannot be included in the spherically symmetric models.

The theory of the atomic properties of semiconductors is on even weaker footing. Again, it is not the cohesive energy that is the problem. Even if we ignore the semiconducting nature of silicon, and treat it as a simple metal in the Wigner-Seitz approximation, we obtain approximately the correct cohesive energy and even the equilibrium atomic volume.[1] This sheds no light on the configuration dependence of the energy that arises entirely from the small changes in energy as the electrons are rearranged from their metallic states to lie in the bonds. This success with the cohesive energy, however, might suggest the use of the pseudopotential approach as we have done for simple metals.[2] Such an approach clearly has no relevance to the electronic properties where the vanishing of the Fermi surface is the central point. In addition, the treatment of screening is conspicuously in error at long wavelengths since it leads to a dielectric function which diverges at long wavelengths rather than approaching a constant as it should. However, to the extent that the properties are determined by Fourier components of the potential with wavelengths of the order of the lattice distance,

[1] H. Brooks, *Trans. AIME*, **227**:546 (1963).

[2] W. A. Harrison, *Physica*, **31**:1692 (1965).

the approach may not be unreasonable. This is in fact the approach that led to the electron-density distribution in silicon shown in Fig. 1.6, which is at least semiquantitatively in accord with experiment. That approach, however, did not yield the correct stable structure[1] and it apparently has not yet been applied to the calculation of the vibration spectrum. Unquestionably, such a calculation would lead to Kohn anomalies which do not exist in reality, and the question of whether it gave a *meaningful* description of the vibration spectrum would probably be a nebulous one. Finally, there is reason to believe that the second-order perturbation theory envisaged in the metallic approximation may miss the central feature of the configuration-dependent interaction. We will return to that evidence after a discussion of Jones zones.

The traditional view that has been used in discussing the bands in semiconductors begins very much as in the metallic model (for a discussion of this approach, see Anderson).[2] We imagine again the valence electrons as a free-electron gas with a spherical Fermi surface. We then introduce the Bragg reflection planes as in an extended-zone scheme. We then imagine that a particular group of these planes, forming the *Jones zone*, dominates the band structure and the Fermi surface disappears into them. These planes should have large structure factors (when the approach was developed there was no knowledge of the relative magnitude of pseudopotential form factors); they should form a rather spherical zone; and the zone should contain a volume just sufficient to accommodate the appropriate number of valence electrons per primitive cell. In the diamond structure the choice was natural; the Jones zone is made up of planes bisecting lattice wavenumbers of the type $[220]2\pi/a$. The structure factors are unity and the zone is of exactly the shape of the Brillouin zone for body-centered cubic structures shown in Fig. 2.2 which is indeed rather spherical and has the appropriate volume. However, we now have estimates of the pseudopotential form factors which also determine the relative importance of different lattice wavenumbers. In silicon, we find[3] that the form factor goes to zero very close to these wavenumbers, casting doubt on the entire picture.

When the matrix elements are accidentally very small, however, we cannot neglect higher-order corrections to those matrix elements. The form of such corrections was given in Eq. (2.69). In this case, the second-order corrections are larger than the first-order form factor,[4] the main contribution coming from second-order terms in which both matrix elements

[1] *Ibid.*

[2] P. W. Anderson, "Concepts in Solids," pp. 28ff., Benjamin, New York, 1964.

[3] W. A. Harrison, *Physica*, **31**:1692 (1965).

[4] L. Kleinman and J. C. Phillips, *Phys. Rev.*, **125**:819 (1962); J. C. Phillips, *Phys. Rev.*, **166**:832 (1968).

derive from [111]$2\pi/a$ lattice wavenumbers. Thus it is not unreasonable to assert that the band gaps on the Jones zone faces are the essential ones. In fact, Heine and Jones[1] have recently described the energy bands of diamond-type semiconductors in the Jones zone rather than in the reduced zone, finding a gap that is roughly constant over the entire zone. This allowed a simple description of the band, the dielectric constant, and the optical absorption.

This does not yet give a tractable approach for the treatment of the atomic properties of semiconductors. It does, however, point to an essential feature of a semiconductor which is omitted in the simple-metal theory, a point also emphasized by Heine and Jones. The inclusion of second-order terms in the matrix elements would give fourth-order terms in the energy. Such terms are neglected in simple metals but seem essential here. The presence of such terms precludes the definition of an energy-wavenumber characteristic independent of the configuration and therefore precludes an expansion of the energy in two-body interactions. It would seem possible nevertheless to proceed with this calculation of the total energy keeping these fourth-order terms, though it would probably be necessary to discard others. Such an analysis has not been carried out but it would seem a promising one. It would be a rather direct extension of the simple-metal pseudopotential approach to valence crystals. It has become customary to define *covalency* to be the presence of these higher-order corrections which are omitted in the theory of simple metals. Thus such an approach would be adding covalency effects to the theory of simple metals.

There is an alternative approach which is currently being investigated and which would appear to be applicable to insulators as well as to valence crystals. This approach arose naturally out of the transition-metal pseudopotential treatment that was described earlier. There we constructed conduction-band states by perturbation theory based upon single OPWs. However, we found that in summing these through resonance we omitted a single state for each resonance present. We then returned to the omitted states, taking a linear combination of atomic orbitals (tight-binding states) as zero-order approximations and admixing plane waves by perturbation theory. Precisely the same approach that was used in the determination of these *d*-like states appears to be directly applicable to the valence states in insulators and semiconductors. It in fact becomes very much simpler since the plane waves that are admixed have zero-order energies far removed from the valence-band energies. Thus there are no resonant energy denominators and the terms in the perturbation expansion are all well behaved. There is a compensating complication in that the overlap of adjacent

[1] V. Heine and R. O. Jones, *J. Phys. C.* (*Solid St. Phys.*), [2], **2**:719 (1969).

atomic states cannot be neglected, particularly in the case of semiconductors. However, the overlap of these atomic states enters as an expansion parameter and this method appears to provide a systematic method of computing the total energy to any desired order in that parameter. As in the case of simple metals we would carry the expansion sufficiently far to obtain the leading term in the interaction of interest. In analogy with the simple metals we would hope to obtain directly the configuration-dependent energy but would very likely need to determine the atomic volume from experiment or from a separate calculation.

Although the theory of the atomic properties of insulators and semiconductors is still in a rudimentary state, it seems unlikely that this will remain true.

6.4 Dislocations[1] We will conclude our discussion of atomic properties with some description of a type of defect which is very important in all classes of solids and which can be discussed without reference to the microscopic theory of the interatomic interactions.

There are two properties of solids that are totally incomprehensible in terms of the perfect crystalline structures that we have discussed. The first is the plastic deformation of solids under high stresses. We may imagine a perfect crystalline solid with an applied shear stress tending to make atomic planes slide over each other. We would expect the elastic energy to increase until some critical value is reached, after which the planes would slide one atom spacing, as shown in Fig. 4.28. If we make an estimate of the stress required to cause such a slip, we find it to be extremely large in comparison to the observed stresses at which slip occurs in real crystals.

[1] For a complete discussion, see J. Friedel, "Dislocations," Addison-Wesley, Reading, Mass., 1964.

Fig. 4.28 The shearing of a perfect lattice. For this case the lattice would support immense shear stresses before snapping to a 45° distortion.

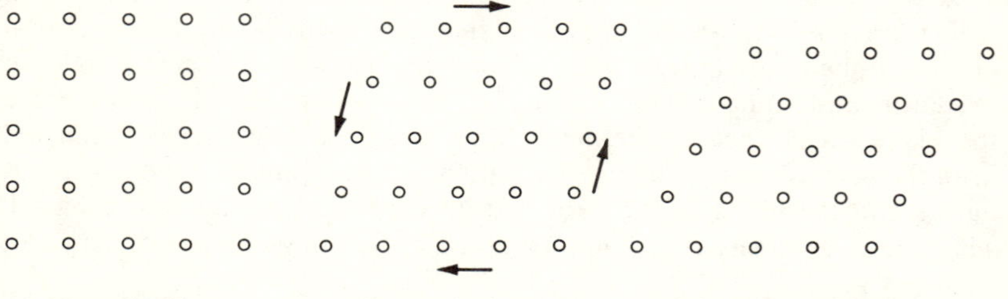

The second anomalous property concerns the rate of growth of crystals from the vapor phase. An atom from the vapor impinging on an atomically flat crystalline face is only very weakly bound to the surface and tends very quickly to boil back into the vapor. If, on the other hand, there is a partially completed atomic plane on the original plane, atoms from the vapor will be strongly bound at the edge of the new plane. We would expect therefore that any partial plane would be rapidly completed but that no further growth would occur until the nucleation of a sufficiently large section of a new plane. Estimates of the rate at which such planes would be nucleated yield a rate of growth from the vapor many orders of magnitude smaller than that observed.

Both of these difficulties were resolved by the suggestion that crystals in general are not perfect but contain defects called *dislocations*. Two types of dislocations may be readily visualized in terms of which the resolution of these two puzzles may be immediately seen. An *edge dislocation* may be envisaged as an extra half-plane of atoms inserted in the middle of a perfect crystal as shown in Fig. 4.29. The horizontal plane in which the extra plane of atoms terminates is called the slip plane. Consider two atoms situated vertically above and below each other across the slip plane at the left of the figure. We successively consider such pairs, moving to the right across the figure, and find that the upper atom is displaced successively greater distances

Fig. 4.29 *An edge dislocation in a square lattice. The dislocation line is perpendicular to the paper and forms the edge of the inserted half-plane of atoms. The dislocation may move in the slip plane by local displacement of the atoms.*

Slip plane

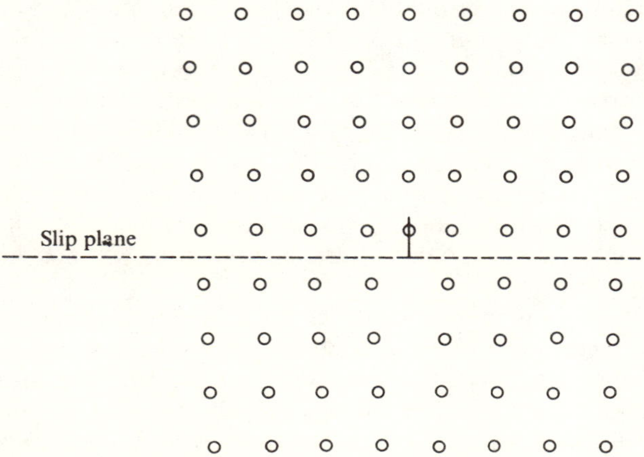

to the left from the lower atom until as we move beyond the dislocation the upper atoms are displaced a full lattice distance to the left.

We may make a deformation of the crystal which corresponds to the motion of the dislocation to the right or to the left and a negligible activation energy is required for this motion. Thus the edge dislocation is extremely mobile in the glide plane. We may further note that the motion of a dislocation across the entire slipped plane will give rise to a displacement of the upper half of the crystal one atom distance with respect to the lower. In particular, if a shear stress is applied which tends to shift the top half of the crystal to the right, energy will be lowered if the dislocation also moves to the right. Thus if a crystal contains edge dislocations, as indicated in the figure, they will move rapidly and easily such as to relieve the stress on the crystal.

The force exerted on a dislocation line due to an applied stress is easily calculated, particularly for the geometry shown in Fig. 4.29. Let the crystal be a cube of edge L with lattice distance a, and let the applied stress (force per unit area) be σ. Then if f is the force per unit length on the dislocation line (the quantity we wish to compute), the work done on the line as it moves across the entire crystal is fL^2. This must equal the work done by the applied stress as the upper surface is displaced by one lattice distance, $\sigma L^2 a$. Thus f is equal to σa and is independent of the crystal size. The force is normal to the dislocation line and lies in the slip plane. Similarly, a dynamic mass per unit length can be defined in terms of the total atomic kinetic energy for a given dislocation velocity and the dynamics of the dislocation discussed.

The deformation of a crystal by the motion of edge dislocations is closely analogous to the movement of a rug across the floor when a ripple is introduced at one edge and pushed across the rug. If the rug is large it may be very difficult to move it by pulling the edge of the rug, but it may be moved in small increments by pushing ripples across.

Distortion by the motion of edge dislocations may be observed directly in a raft of soap bubbles. If small uniform bubbles are blown in a soapy solution they will form close-packed rafts on the surface. If we then apply a shear strain to such a raft, dislocations will be seen to form at the edge of the raft and dart rapidly across, relieving the strain on the raft. It may seem surprising that dislocations were not postulated early on the basis of accidental observation of bubble rafts; however, this was not the order in which things occurred.

The *screw dislocation* may be constructed by conceptually slicing a crystal half-way through and then displacing the two sides one lattice distance parallel to the slice. This is illustrated in Fig. 4.30. As in the edge dislocation the crystal remains perfect (though slightly distorted) except along the line of the dislocation core. If such a dislocation threads the surface

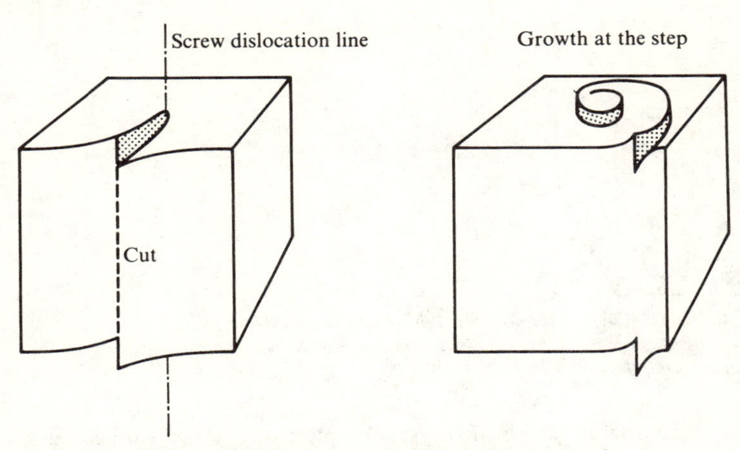

Fig. 4.30 The construction of a screw dislocation may be imagined as a cut partway through the crystal and a displacement of the two cut surfaces parallel to the edge of the cut plane. (Note that a displacement perpendicular to this but again in the cut plane would generate an edge dislocation.) A screw dislocation threading a surface leaves an atomic step on that surface. If atoms are added from a vapor, the step will advance.

of a crystal we may see that it provides a step upon which the crystal may grow from the vapor. As atoms are added to this step it moves around the crystal but never disappears. It eliminates the need for a nucleation center for growth of a plane. Thus it resolves the difficulty we raised in that connection.

In such growth we would not of course expect the step on the surface to remain a straight line. Atoms should be added near the dislocation just as rapidly as they are added far away. It is easy to see that the step will wind around the dislocation core and produce a spiral structure as shown in the right side of Fig. 4.30. Such configurations were proposed early in the study of dislocations and were soon observed by microscopic studies of organic crystals, providing striking support for the theory.

We have of course greatly oversimplified the form of the dislocations by postulating simple edge and simple screw dislocations with cores forming straight lines. We may generalize this very simply to *dislocation loops* by constructing four planes forming a prism within a cubic crystal as shown in Fig. 4.31. We may displace the atoms within that prism and in the front half of the crystal backward by one lattice distance. This will form a rectangular line of edge dislocations in the center of the crystal as indicated. The four planes making up the sides of the prism are the glide planes in this case.

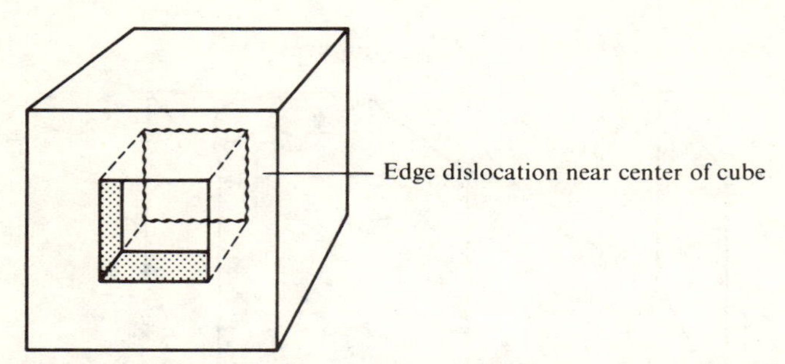

Fig. 4.31 A dislocation loop is formed by making four partial cuts, as in Fig. 4.30, and displacing the atoms enclosed by the four cut surfaces one lattice distance toward the cut edges.

If we were now to strain the crystal by pushing the top surface back and by pulling the bottom surface of the crystal forward, the dislocation forming the upper edge of the rectangle would tend to move forward; that forming the lower edge would move backward as shown in Fig. 4.32. This also tends to tilt and extend the dislocations making up the lateral edges of the rectangle. There is, however, elastic energy associated with the dislocation lines and energy associated with the dislocation core. Increasing the length of a dislocation line will increase its energy, and therefore under

Fig. 4.32 An applied stress to a crystal containing a dislocation loop will cause that loop to deform. The deformation under small stresses is shown on the left. If sufficiently large stresses are applied, such that the loop reaches the surface, a permanent deformation will occur and will remain after the stress is removed. This condition is illustrated on the right. A prism extends beyond the back surface which is of the same shape as the prismatic cavity on the front surface.

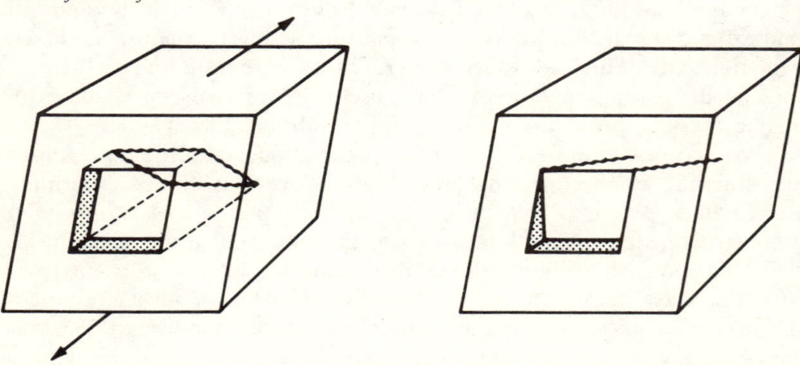

small strains the loop will stretch like a rubber band and come to equilibrium at some configuration. At larger stresses we would expect the lines to move to the surface leaving a configuration as shown to the right in Fig. 4.32. The combined elastic stresses remaining within and outside the prism would lead to a net distortion of the crystal even after the applied stress had been removed. Such permanent deformation is simply *plastic flow*. We may also note that the remaining dislocation lines threading the crystal in this case are not simple edge dislocations as we have defined them above. This can be more clearly discussed, however, with a different geometry.

We consider now the glide plane of an edge dislocation. The dislocation forms a line in this plane and at some point may leave the plane as it did in the case above. For the situation described above the dislocation moved easily under applied stresses. In real crystals, however, it frequently happens that at some point on the dislocation line it may be *pinned*. For example, a large impurity atom may have its energy significantly lowered by being localized in the dilated side of an edge-dislocation core and thereby immobilize the dislocation. Let us consider a section of dislocation line that is pinned at two end points as shown in Fig. 4.33. We will assume that beyond those pinning points the dislocation moves out of the glide plane. Then an applied stress will bow out the dislocation line between the pinning points but it will remain fixed at its ends. It again behaves as a rubber band with an increase in energy proportional roughly to the length of the line. Under stresses there will also be a decrease in energy as the line moves upward, which is of course responsible for the force bowing the dislocation. At the center of the arc this remains a pure edge dislocation. Near the pinning points, however, the displacements are parallel to the line of the dislocation and this may readily be seen to correspond to a pure screw dislocation.

Fig. 4.33 A glide plane containing an edge dislocation pinned at the points P. The arrows indicate the displacement (by a lattice distance) of atoms in this region above the plane of the figure.

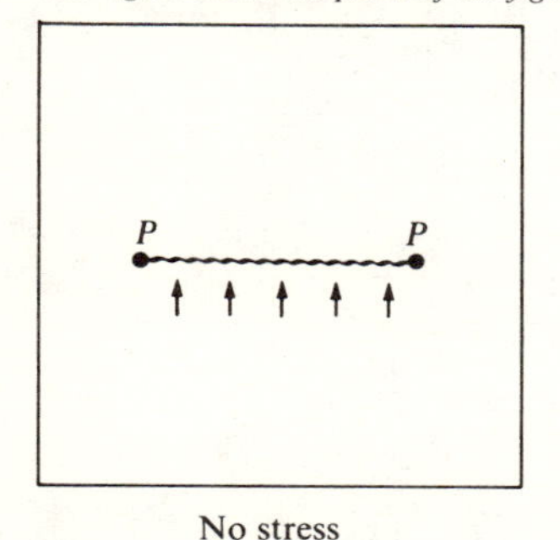

No stress

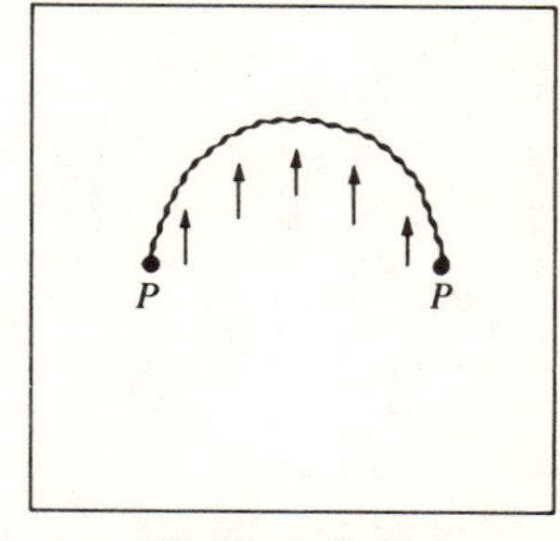

Stress applied

In the region between, the dislocation is mixed. As the stress is increased further the area under the curve tends to increase still further as shown on the left in Fig. 4.34. The line tends always to move away from the region of upward displacement since such motion decreases the elastic energy of the applied stress. Eventually the lines from the two sides will meet, forming a new dislocation ring which can expand to the surface of the crystal and the process is repeated.

We see that the pinned dislocation line that we have constructed provides a source of dislocations. Such a source of dislocations was proposed by Frank and Read[1] and subsequently demonstrated experimentally.

Two techniques have been particularly successful for direct study of dislocation behavior. One is the diffusion into a crystal of impurities that selectively condense upon dislocation lines, just as the large impurities discussed above reside favorably upon dislocation lines. These decorated dislocation lines then can be observed directly. It is in this way that Frank-Read sources have been observed. The decoration of a dislocation line of course pins that line along its entire length and prevents further motion. It has also been possible to detect the emergence of a dislocation line at a crystalline surface. This has been done with etchants that selectively attack the dislocation core and produce pits on the surface at the points of emergence. This does not pin the line and subsequent stress may move it. Its new location may be detected by subsequent etching.

It has been possible to obtain large sections of silicon crystals that are free of dislocations. This is a very special situation, however, and in almost all crystals there is a significant concentration, typically 10^{15} dislocation

Fig. 4.34 Increased stress applied to the system shown in Fig. 4.33. On the left the line has bowed out below the initial line. On the right a further stress has brought the two arcs together to form a new pinned line and a loop. This is a Frank-Read source of dislocations.

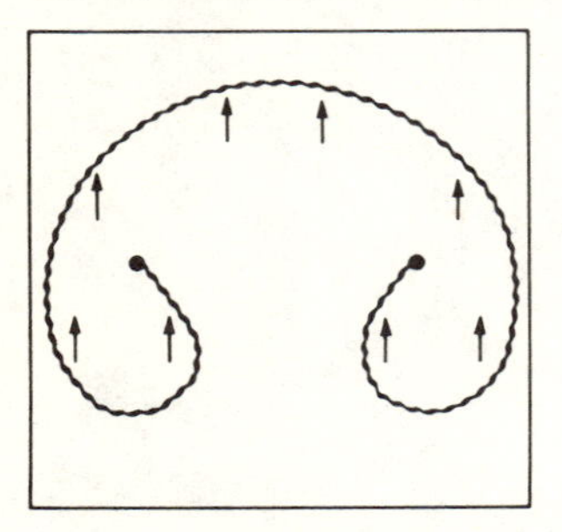

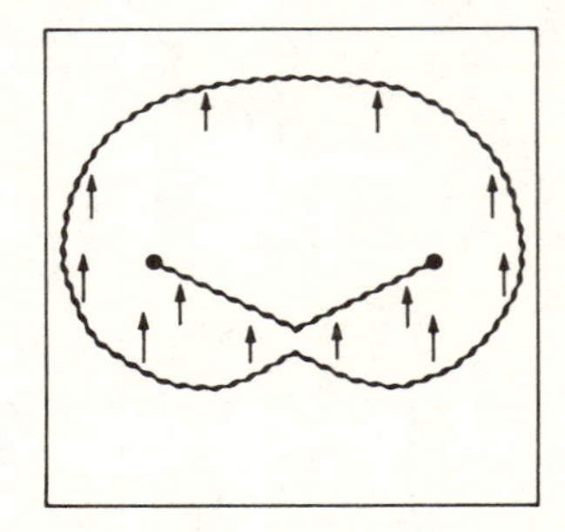

[1] F. C. Frank and W. T. Read, *Phys. Rev.*, **79**:722 (1950).

lines per square centimeter. They are clearly of primary importance in the plastic properties of solids. They may also have appreciable effects in reducing the elastic constants. They also provide channels for easy diffusion in solids and sinks for interstitials and vacancies. (Note that a vacancy may be absorbed by an edge dislocation by producing a slight *jog* in the dislocation.) They may even become important in the electrical properties in heavily cold-worked materials. There the high density of dislocations produced creates important electron scattering and the corresponding increase in electrical resistivity.

The study of dislocations and the effects of dislocations is a large field in itself and one that we have only introduced here.

Problems

4.1 Consider a simple cubic crystal with three types of force constants. Each atom is coupled to each of its six near neighbors by a spring with spring constant κ_1; that is, $\delta E = \frac{1}{2}\kappa_1(\delta r)^2$ where δr is the change in separation from equilibrium. Each next-nearest-neighbor pair is similarly coupled by a spring constant κ_2. Finally, a change in angle $\delta\theta$ between any adjacent right-angle pair of near-neighbor bonds gives $\delta E = \frac{1}{2}\kappa_3 a^2(\delta\theta)^2$.

a. Evaluate the composite spring constant for longitudinal waves propagating along [100]. Also find the maximum frequency in a [100] direction.

b. Obtain the frequency with which a given atom would vibrate in a [100] direction if all other atoms were fixed. This is the manner in which the Einstein frequency is envisaged.

4.2 Take as a model of NaCl point charges of $\pm e$ coupled by springs as well as by electrostatic interaction. (Thus we neglect the electronic polarizability of individual ions, which in real NaCl is appreciable.) Take as the static dielectric constant $\epsilon_s = 4.4$ (approximately equal to

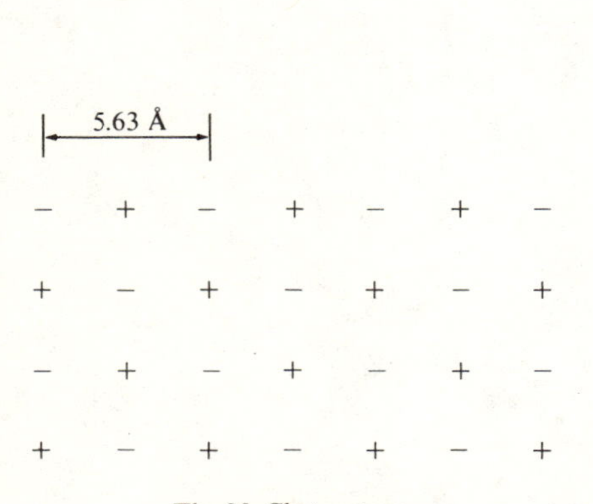

The NaCl structure

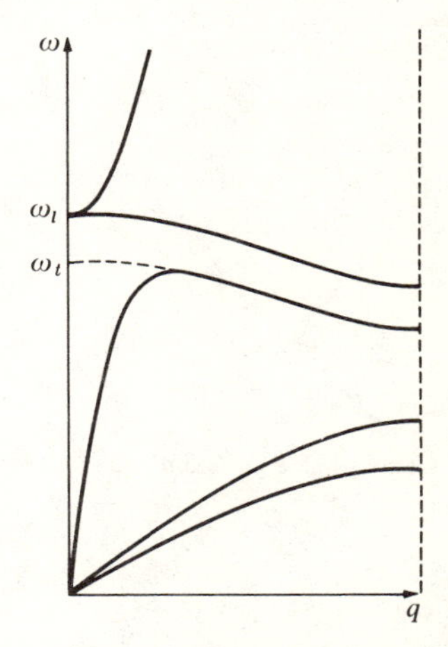

Frequencies ω_l and ω_t being estimated

the true value minus effects of electronic polarizability). The [100] spacing of sodium ions is 5.63 Å. The object of the problem is to estimate the longitudinal and transverse optical frequencies indicated on the preceding page.

a. As a first step deduce a composite spring constant for uniform displacement of the Na^+ with respect to the Cl^- ions by deducing the static dielectric constant in terms of it. The dielectric constant may be written, for example, as the ratio of the voltage across a capacitor with vacuum between the plates to that with NaCl between plates, all with the same charge density on the plates.

b. The transverse frequency may be computed by assuming no charge accumulation or surface charges. The longitudinal frequency may be computed as an infinite wavelength mode in a plate with displacements perpendicular to the surfaces, and therefore with surface charge accumulated.

4.3 Imagine a copper bar standing in a gravitational field. There is, of course, a gravitational force on any electron given by mg. Assuming that the copper structure is not distorted by the gravitational field, an electronic "screening field" will arise.

a. What is this electric force? (The answer may be clearer if you replace the mg force by an equivalent applied electric field.)

b. Compute the additional force arising because the structure is distorted and there is a dilatation varying with height. For simplicity assume all ion displacements are vertical. (Noting that the corresponding elastic constant c_{11} and the copper density ρ are related to the speed of longitudinal sound by $v_s^2 = c_{11}/\rho$, the answer may be written in simple form.)

4.4 Consider a gas of N free electrons (all in the same spin state) in a volume Ω. Let each pair have a contact interaction energy

$$V(\mathbf{r}_i - \mathbf{r}_j) = \beta\,\delta(\mathbf{r}_i - \mathbf{r}_j) \qquad \text{for } i \neq j$$

where β is a constant. The ground state of the system is again composed of plane-wave states

$$\Psi = c_N^+ c_{N-1}^+ \ldots c_2^+ c_1^+ \,|0\rangle$$

By computing the expectation value of the electron-electron interaction, obtain the direct interaction energy and the exchange energy. (It may help in the counting to imagine N and Ω to be finite.)

4.5 Consider a gas of free spinless electrons. The Hamiltonian is

$$H_0 = \sum_k \epsilon_k c_k^+ c_k$$

and the ground state is

$$\Psi_0 = \prod_{k<k_F} c_k^+ \,|0\rangle$$

We introduce a perturbation,

$$H_1 = \sum_k V_q c_{k+q}^+ c_k$$

with a single $q \neq 0$ and seek the ground-state energy to second order in H_1.

$$E = E_0 + \langle\Psi_0|H_1|\Psi_0\rangle + \sum_n \frac{|\langle\Psi_n|H_1|\Psi_0\rangle|^2}{E_0 - E_n}$$

The zero-order energy is

$$E_0 = \langle\Psi_0|H_0|\Psi_0\rangle = \langle 0|c_1 c_2 \ldots c_N \sum_k \epsilon_k c_k^+ c_k c_N^+ c_{N-1}^+ \ldots c_1^+|0\rangle$$

$$= \sum_{k<k_F} \epsilon_k \langle \Psi_0 | \Psi_0 \rangle = \sum_{k<k_F} \epsilon_k$$

Similarly, evaluate the first- and second-order energies in terms of V_q and ϵ_k. The second-order energy will contain a sum over a region R of **k**. Sketch that region for a two-dimensional gas.

4.6 Let the vibrations of a crystal be describable by an Einstein model,

$$H_\varphi = \hbar\omega_E \sum_q a_q^+ a_q$$

with q taking N values. Let the electron Hamiltonian be

$$H_e = \sum_k \epsilon_k c_k^+ c_k \qquad \text{with } \epsilon_k = \frac{\hbar^2 k^2}{2m}$$

Finally let the electron-phonon interaction be

$$H_{e\varphi} = \frac{V}{N^{1/2}} \sum_{k,q} c_k^+ c_k (a_q + a_q^+)$$

Consider a zero-order, no-phonon state $c_{k_0}^+ |0\rangle$ with zero-order energy ϵ_{k_0}.

a. Compute the energy to second order in V.

b. What is the expectation value of the total number n of phonons in the corresponding first-order state? This number is a dimensionless *coupling constant*. If $\langle n \rangle \ll 1$ we have weak coupling and perturbation theory is appropriate; $\langle n \rangle \gg 1$ represents strong coupling.

4.7 Compute the polaronlike shift in electron energy for a conduction band with $m^* = m$ in a semiconductor. Assume only longitudinal waves, and the Debye approximation to the dispersion curve. Start with the expression

$$\delta E = \sum_{q,k} D^2 q^2 \frac{\hbar}{2MN\omega_q} \frac{\langle 0 | c_k^+ c_{k+q} c_{k+q}^+ c_k | 0 \rangle}{\epsilon_k - \epsilon_{k+q} - \hbar\omega_q}$$

Use as the zero-order state $|0\rangle = C_k^+ |\text{vacuum}\rangle$ with $k = 0$. Compute only to lowest order in $mv_s/\hbar q$ or $mv_s^2/\hbar\omega_D$. Evaluate the expression for $D = 5$ ev, $M = 6 \times 10^4$ electron masses, and $\hbar\omega_D = 0.025$ ev.

4.8 The interaction potential for scattering of an electron by an impurity fixed at the position $\mathbf{r}_0$ is $w(|\mathbf{r} - \mathbf{r}_0|)$. A free impurity would recoil, but a real impurity is coupled to its neighbors. Using the approximations of Sec. 4.4, estimate the probability that it will not recoil (at zero temperature) when a scattering event takes place. Obtain a rough estimate for scattering from a zinc impurity in copper.

4.9 Consider a one-dimensional crystal, with *total energy* (at constant length) given by

$$E = \sum_q{}' F(q) S(q) S(q)$$

with $F(q) = A/q^2$ rydbergs per ion, A is constant.

Let it have a structure with lattice distance a, but with two ions per cell; the second a distance δ from the first: $x_i = 0, \delta, a, a + \delta, 2a, 2a + \delta, \ldots$.

Find the total energy as a function of δ. What is the stable structure (with respect to variation in δ). Note:

$$\sum_{n=1}^{\infty} \frac{1}{n^2} = \frac{\pi^2}{6}$$

$$\sum_{n=1}^{\infty} \frac{\cos nx}{n^2} = \frac{(x - \pi)^2}{4} - \frac{\pi^2}{12} \qquad \text{for } 0 \le x \le 2\pi$$

4.10 Set up the calculation of the activation energy for diffusion of vacancies in a metal with one ion per primitive cell:

Assume that diffusion occurs by the motion of a near neighbor (displaced by $\mathbf{r}_0$ from the vacancy) to the vacant site with no other distortion occurring.

Obtain expressions for S^*S for the system as a function of the position of this ion, and write the difference between S^*S with and without displacement of the vacancy.

The band-structure energy, $\sum_q \delta[S^*(\mathbf{q})S(\mathbf{q})]F(q)$, can be written as an integral plus a sum. (The electrostatic energy is also of this form, with suitable choice of $F(q)$, but we will not treat it.) Write out the total change in band-structure energy, simplifying the integral by performing the angular integration. This is the end of the problem.

Note that the energy is symmetric around the midpoint of the displacement, so the maximum is expected there. At that point $\mathbf{q} \cdot \mathbf{r}$ is an integral multiple of π so the lattice sum becomes simple.

4.11 Consider a simple-cubic metallic lattice with one electron per ion and lattice distance a. Assume that when the energy is divided into a volume-dependent contribution and an effective interaction between ions, the former is just the electron kinetic energy

$$E_{\mathrm{fe}} = \frac{3}{5}\frac{\hbar^2 k_F{}^2}{2m} \qquad \text{per ion}$$

(and, of course, depends on volume through k_F) and the latter is

$$\frac{1}{2N}\sum_{i,j}{}' V(|\mathbf{r}_i - \mathbf{r}_j|) \qquad \text{per ion}$$

with

$$V(r) = V_0\left(\frac{a_0{}^2}{r^2} - \frac{a_0}{r}\right)$$

for r near the nearest-neighbor distances and zero beyond. The a_0 and V_0 are constants, independent of volume. These are the only terms in the energy.

Obtain a formula for the equilibrium lattice distance a. Would nearest neighbors to a vacancy in this lattice relax inward or outward?

4.12 Consider the pinned edge dislocation shown in Fig. 4.33. Let the lattice spacing be a and the distance between pinning points be L. If the formation energy per unit length $\delta E/\delta l$ of the dislocation line is independent of orientation, a uniform applied stress σ will bow the line into a circle. Obtain an expression for the radius of the circle.

V

COOPERATIVE PHENOMENA

Studies of magnetism and of superconductivity in solids form two very large and active areas. Both are cooperative phenomena since they involve directly the interaction between the electrons. One electron alone cannot become ferromagnetic or superconducting; a cooperative condensation of many electrons is required. We consider first magnetism for which the treatment is somewhat simpler and will develop the ideas that are used in its study. As in our earlier discussions we will emphasize the self-consistent-field description of properties though in some cases recent many-body theories have given more accurate descriptions.

A. MAGNETISM[1]

Soon after the advent of quantum mechanics it was recognized that the origin of ferromagnetism was the exchange interaction. However, exchange is a difficult effect to treat mathematically, and phenomenological represen-

[1] For a current and definitive treatment, see C. Herring, in G. T. Rado and H. Suhl (eds.), "Magnetism," Academic, New York, 1966.

tations of it were introduced. Almost all of the study of ferromagnetism is based upon these simple phenomenological forms. Within recent years it has been possible to calculate ferromagnetic properties more directly based upon the exchange interaction itself. We will begin by discussing one such recent effort before going on to the phenomenologies upon which most of the theory of magnetism is based. In almost all of this discussion we will direct our attention at cooperative magnetic phenomena and not at the magnetic properties of simple materials, some of which have been discussed earlier.

1. Exchange

In Sec. 4.2 of Chap. IV we wrote the interaction between electrons in second-quantized form.

$$V(\mathbf{r}_1 - \mathbf{r}_2) \rightarrow \tfrac{1}{2} \sum_{\substack{k_1,k_2 \\ k_3,k_4}} \langle \mathbf{k}_4,\mathbf{k}_3 | V | \mathbf{k}_2,\mathbf{k}_1 \rangle \, c_{k_4}{}^+ c_{k_3}{}^+ c_{k_2} c_{k_1} \tag{5.1}$$

The $\mathbf{k}_i$ are quantum numbers for any complete set of one-electron states which may or may not be plane waves, and the $\mathbf{k}_i$ again include the spin quantum number.

We will now consider the solution of the many-electron problem more carefully. If we *approximate* the true state of the system as we have before by a single Slater determinant $|\Psi\rangle = \prod_{k_i} c_{k_i}{}^+|0\rangle$ we obtain, as described in Sec. 3 of Chap. II, the Hartree-Fock approximation. Once we have obtained the ψ_{k_i} from the Hartree-Fock equation the electron-electron interaction contributes to the energy only through

$$\begin{aligned} \tfrac{1}{2}\langle\Psi|V(\mathbf{r}_1 - \mathbf{r}_2)|\Psi\rangle = \tfrac{1}{2} \prod_{k_i} \langle 0| c_{k_i} \sum_{k,k'}{}' (\langle \mathbf{k}',\mathbf{k}|V|\mathbf{k}',\mathbf{k}\rangle c_{k'}{}^+ c_k{}^+ c_{k'} c_k \\ + \langle \mathbf{k}',\mathbf{k}|V|\mathbf{k},\mathbf{k}'\rangle c_{k'}{}^+ c_k{}^+ c_k c_{k'}) c_{k_i}{}^+ |0\rangle \end{aligned} \tag{5.2}$$

which is a direct generalization of Eq. (4.30) to many electrons. The product is over all occupied states $\mathbf{k}_i$ and the sum is over all $\mathbf{k}'$ different from $\mathbf{k}$. The term in $\langle \mathbf{k}',\mathbf{k}|V|\mathbf{k},\mathbf{k}'\rangle$ is the direct interaction and can be included with the external potential. The term in $\langle \mathbf{k}',\mathbf{k}|V|\mathbf{k}',\mathbf{k}\rangle$ is the exchange energy and contributes only if $\mathbf{k}$ and $\mathbf{k}'$ have the same spin. The matrix element itself is given by

$$\langle \mathbf{k}',\mathbf{k}|V|\mathbf{k}',\mathbf{k}\rangle = \int \psi_{k'}^*(\mathbf{r}_1)\psi_k^*(\mathbf{r}_2)V(\mathbf{r}_1 - \mathbf{r}_2)\psi_{k'}(\mathbf{r}_2)\psi_k(\mathbf{r}_1)\, d^3r_1\, d^3r_2 \tag{5.3}$$

We may see immediately how the exchange interaction may lead to magnetism. In its absence the Slater determinant with lowest energy will have the lowest-lying one-electron states occupied in pairs, spin up and spin down. However, the exchange term lowers the energy (if $\langle \mathbf{k}',\mathbf{k}|V|\mathbf{k}',\mathbf{k}\rangle$ is positive since $c_{k'}{}^{+}c_{k}{}^{+}c_{k'}c_{k} = -n_{k'}n_{k}$ is negative) for configurations of predominantly parallel spin. Thus, for example, in a tight-binding crystal with one electron per atom, the exchange interaction between neighbors may favor the alignment of their spins and favor ferromagnetism.

The state $|\Psi\rangle = \prod_{k_i} c_{k_i}{}^{+}|0\rangle$ cannot, however, be the ground state of the system since it is not an eigenstate of the Hamiltonian. Operation of Eq. (5.1) upon $|\Psi\rangle$ will yield new Slater determinants when $\mathbf{k}_4$, $\mathbf{k}_3$ are not equal to $\mathbf{k}_2$, $\mathbf{k}_1$ as well as the original $|\Psi\rangle$ when $\mathbf{k}_4$, $\mathbf{k}_3$ *are* equal to $\mathbf{k}_2$, $\mathbf{k}_1$. The electron-electron interaction couples many different Slater determinants and the exact solution will be a linear combination of all of them. As with the Hartree-Fock terms of Eq. (5.2), there will be only one matrix element coupling $|\Psi\rangle$ to each new Slater determinant generated by a term in Eq. (5.1) in which the spin of $|\mathbf{k}_1\rangle$ and $|\mathbf{k}_2\rangle$ is different, but two when the spin is the same. The extra interaction term for parallel spins is again called exchange interaction. If all of these terms were included we could obtain an exact solution of the many-electron problem. Field-theoretic techniques have been used in an attempt to obtain the important matrix elements in the free-electron gas. In the studies of magnetism two distinct approaches have been used.

The first is to replace the electron-electron interaction in the Hamiltonian by a self-consistent potential and a fictitious spin-dependent term. The Heisenberg approach is of this type. It is then hoped that the spin-dependent term in the Hamiltonian will represent the effects of all of the spin-dependent matrix elements of the electron-electron interaction. It has been an extremely successful approach but is not one that is directly derivable from the fundamental equations.

The second approach is to discard all off-diagonal matrix elements and include only the direct and exchange matrix elements included in Eq. (5.2). This is the Hartree-Fock approximation. This approach is more in line with the discussions of electronic structure that we have made and we will begin in this way, first making the additional simplification of free-electron exchange which we discussed in connection with screening in Sec. 3.2 of Chap. II and Sec. 4.5 of Chap. III.

2. *Band Ferromagnetism*

We may evaluate the exchange interaction based upon plane-wave electronic states. The matrix element for exchange, Eq. (5.3), becomes

$$\langle \mathbf{k}',\mathbf{k}|V|\mathbf{k}',\mathbf{k}\rangle = \frac{1}{\Omega^2}\int e^{-i(k'-k)\cdot(r_2-r_1)}\,V(\mathbf{r}_2 - \mathbf{r}_1)\,d^3r_1\,d^3r_2$$

for parallel spin states. The $V(\mathbf{r}_1 - \mathbf{r}_2)$ is simply the coulomb potential and the integration over $\mathbf{r}_2 - \mathbf{r}_1$ simply gives the Fourier transform of V. The remaining integral gives a factor of the normalization volume, Ω.

$$\langle \mathbf{k}',\mathbf{k}|V|\mathbf{k}',\mathbf{k}\rangle = \frac{4\pi e^2}{\Omega|\mathbf{k}' - \mathbf{k}|^2}$$

Furthermore, $c_{k'}{}^{+}c_k{}^{+}c_{k'}c_k = -c_{k'}{}^{+}c_{k'}c_k{}^{+}c_k = -n_{k'}n_k$ if $\mathbf{k}'$ is different from $\mathbf{k}$. For an electron gas with all states occupied for $k < k_F$ we may directly compute the total exchange energy with the factor $\frac{1}{2}$ appropriate to electron-electron interactions.

$$E_{\text{ex}} = -\frac{1}{2}\frac{4\pi e^2}{\Omega}\sum_{k' \neq k,k}\frac{n_{k'}n_k}{|\mathbf{k}' - \mathbf{k}|^2} \tag{5.4}$$

Since this expression is colinear in $n_{k'}$ and n_k we might instead sum only over $\mathbf{k}'$ to evaluate exchange energy for an electron of wavenumber $\mathbf{k}$. The result will depend on the wavenumber $\mathbf{k}$, in contrast to the energy arising from the direct interaction. Thus we cannot replace it by a self-consistent field in the sense that we could for the direct interaction. Even if we chose to include this wavenumber dependence and add a wavenumber-dependent exchange potential, we would be led into difficulties. The exchange energy is found to have a logarithmic singularity at $k = k_F$ [obtained by integrating Eq. (5.4)]. This in turn leads to a vanishing density of states at the Fermi surface contrary to experiment. This error arises from our assumption of a single Slater determinant as the state of the system. In many-body theory, coupling to different Slater determinants is included as a perturbation and the resulting terms are found to remove the singularity at $k = k_F$. These additional corrections beyond the direct and exchange interactions are the correlation energy discussed in Sec. 4.5 of Chap. III.

We may, however, compute the total exchange energy per electron as a function of electron density using Eq. (5.4) and define an exchange potential in terms of it as we did in Sec. 4.5 of Chap. III. The total exchange energy per electron is given by

$$\frac{E_{\text{ex}}}{N} = -\frac{4\pi e^2}{2N\Omega}\sum_{k,k'}\frac{n_{k'}n_k}{|\mathbf{k}' - \mathbf{k}|^2} = -\frac{4\pi e^2}{2N}\frac{\Omega}{(2\pi)^6}\int d^3k'\,d^3k\,\frac{n_{k'}n_k}{|\mathbf{k}' - \mathbf{k}|^2}$$

This may be evaluated directly for the Fermi sea with all states occupied for $k < k_F$; however, the evaluation is somewhat tricky.[1] The interesting aspects of the result, however, may be seen directly. The integrals can be performed within the Fermi sphere where n_k and $n_{k'}$ are unity. These variables of integration may be changed to k/k_F and the value of the integral is seen

[1] C. Kittel, "Quantum Theory of Solids," p. 91, Wiley, New York, 1963.

to be proportional to k_F^4. Furthermore the volume per electron, Ω/N, is proportional to k_F^{-3}. Thus the exchange energy per electron is equal to a numerical constant times $e^2 k_F$. It is more convenient to write the result in terms of the electron density which is of course proportional to k_F^3. We obtain

$$\frac{E_{\text{ex}}}{N} = -\frac{3}{4}e^2\left(\frac{3\rho}{\pi}\right)^{1/3}$$

where ρ is the total density of electrons (of both spins). This is the exchange energy per electron for the uniform electron gas of density ρ.

Finally we may, following Kohn and Sham, define an exchange potential equal to dE_{ex}/dN or, following Slater, equal to $2E_{\text{ex}}/N$. In either case we have obtained a self-consistent potential that is proportional to the one-third power of the density. (It is frequently called $\rho^{1/3}$ exchange.) It is only an approximate treatment of exchange but in that approximation it has eliminated the singularity at the Fermi surface which arises from true exchange. In addition, in using this potential we are ignoring correlation effects.

The addition of free-electron exchange to the calculation of energy bands need not make any fundamental change. It may simply modify the dielectric function determining the potential as we found earlier. It is not even clear that this improves the result.

It is possible, however, for the exchange introduced to cause spontaneous magnetization of the system. This is illustrated in Fig. 5.1. In a normal metal we would assume equal population of electrons with spin up and with spin down. Then the exchange potential would be the same for electrons of either spin. The energy bands would be identical and we would have found a self-consistent solution. On the other hand, if we were to populate the spin-up states to a greater extent than the spin-down states we would find a larger exchange potential for the electrons of spin up. Since this potential is attractive these bands would be lowered in energy in comparison to the spin-down bands. After completing the calculation we could see if the Fermi levels that we postulated for the two spins in fact came at the same energy. This would ordinarily not be the case but there could be some choice of net spin that did lead to a self-consistent solution. For the free-electron gas there is only one self-consistent solution, the normal gas, except for very low electron density. Thus one does not expect ferromagnetism in the simple metals.

In a transition metal, however, with its very irregular density of states, it is possible that there is a second self-consistent solution of nonzero total spin. (We will develop a criterion for such spontaneous magnetization for a simple system when we discuss local moments in Sec. 7.) This has

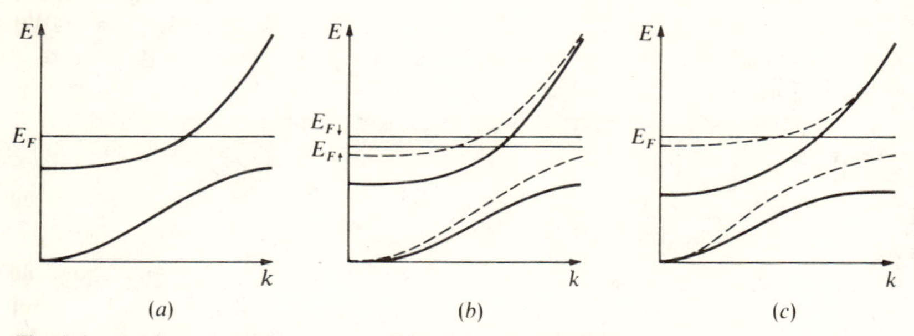

Fig. 5.1 *A schematic representation of band ferromagnetism. In (a) we assume equal occupation of up-spin and down-spin states; the exchange energy is the same for both, the bands are the same, and a single Fermi energy is obtained. In (b) we assume slightly more spin-up electrons and slightly fewer spin-down electrons. The change in exchange energy lowers the spin-up bands (solid curves) and raises the spin-down bands (dashed curves). Then the Fermi energies in the two bands, defined by the assumed occupation of the two, may differ. If the spin-up Fermi energy is lower, the nonmagnetic state (a) was unstable. A sufficient further transfer to spin-up states will bring the two Fermi energies together to form the self-consistent ferromagnetic state (c).*

turned out to be the case for nickel and iron in work by Wakoh and Yamashita[1] and work by Connolly[2] on nickel. Band ferromagnetism had been discussed in a more phenomenological way earlier by Stoner.[3] Using this self-consistent solution one may immediately obtain the total number of unbalanced spins and multiply by the magnetic moment per spin to obtain the equilibrium magnetization of the material. This turns out to be in reasonable agreement with experiment.

Though the treatment of exchange here has been approximate it nevertheless represents an attempt at a first-principles calculation of the ferromagnetism in these materials. The approximation made has been first that the states are describable as a single Slater determinant, and second in the use of free-electron exchange. We will see that phenomenological Heisenberg exchange does not entail these approximations.

We have also made the assumption in this treatment that the electronic states are itinerant, i.e., describable as Bloch states. As we have indicated earlier there is some question as to whether something in the nature of a Mott transition has occurred and a localized description might be more appropriate.

[1] S. Wakoh and J. Yamashita, *J. Phys. Soc. Japan*, **21**:1712 (1966).

[2] J. W. D. Connolly, *Phys. Rev.*, **159**:415 (1967).

[3] E. C. Stoner, *Proc. Roy. Soc.*, **169A**:339 (1939).

3. *Spin Operators*

In the remainder of our discussion of magnetism it will be convenient to write the Hamiltonian and other operators in terms of spin operators. Since we have not used that formalism before, we should summarize it before proceeding. At this point we are simply defining the notation. When we use it later, we will replace results we have derived using second quantization by equivalent expressions in spin operators. Equivalence may be verified by carrying out the operations defined here. We begin with the states of a single electron and generalize afterward to atoms with total spin greater than $\frac{1}{2}$.

The spin *state* of an electron can be represented by a normalized vector $\binom{a}{b}$; a spin-up electron corresponds to $\binom{1}{0}$, a spin-down electron to $\binom{0}{1}$. Thus a state of the ith electron might, e.g., be $\psi_k(\mathbf{r}_i)\binom{1}{0}_i$. The spin *operators*, which appear in the Hamiltonian or other dynamical variables, can be written in terms of Pauli spin matrices; the three components of the spin operator $\mathbf{S}$ are

$$S_i^x = \tfrac{1}{2}\sigma_x = \tfrac{1}{2}\begin{pmatrix}0 & 1\\ 1 & 0\end{pmatrix}_i \qquad S_i^y = \tfrac{1}{2}\sigma_y = \tfrac{1}{2}\begin{pmatrix}0 & -i\\ i & 0\end{pmatrix}_i$$

$$S_i^z = \tfrac{1}{2}\sigma_z = \tfrac{1}{2}\begin{pmatrix}1 & 0\\ 0 & -1\end{pmatrix}_i \tag{5.5}$$

The subscript i indicates that the matrix operates on the ith spin state. The results of operation of each component upon a given spin state can be directly obtained. For example, $S_i^x = \binom{1}{0}_i = \frac{1}{2}\binom{0}{1}_i$.

In expressions containing spin operators, the three components are treated as components of vectors. Thus, for example, we may evaluate the dot product for two states,

$$\mathbf{S}_i \cdot \mathbf{S}_j = S_i^x S_j^x + S_i^y S_j^y + S_i^z S_j^z \tag{5.6}$$

Components of $\mathbf{S}$ for different electrons operate of course on different coordinates and therefore commute. Commutation relations for components on the same electron may be obtained from Eq. (5.5).

$$S_i^x S_i^y - S_i^y S_i^x = iS_i^z \tag{5.7}$$

$$S_i^y S_i^z - S_i^z S_i^y = iS_i^x \tag{5.8}$$

$$S_i^z S_i^x - S_i^x S_i^z = iS_i^y \tag{5.9}$$

In most cases it is convenient to write expressions in terms of the raising and lowering operators, S_i^+ and S_i^-, defined by

$$S_i^+ = S_i^x + iS_i^y \left[= \begin{pmatrix}0 & 1\\ 0 & 0\end{pmatrix}_i \text{ for electrons}\right] \tag{5.10}$$

$$S_i^- = S_i^x - iS_i^y \left[= \begin{pmatrix}0 & 0\\ 1 & 0\end{pmatrix}_i \text{ for electrons}\right] \tag{5.11}$$

We see immediately that

$$S_i^+\binom{1}{0}_i = 0 \qquad S_i^+\binom{0}{1}_i = \binom{1}{0}_i \tag{5.12}$$

$$S_i^-\binom{1}{0}_i = \binom{0}{1}_i \qquad S_i^-\binom{0}{1}_i = 0 \tag{5.13}$$

$$S_i^z\binom{1}{0}_i = \tfrac{1}{2}\binom{1}{0}_i \qquad S_i^z\binom{0}{1}_i = -\tfrac{1}{2}\binom{0}{1}_i \tag{5.14}$$

Thus, in analogy with phonon creation and annihilation operators, S_i^+ raises the z component of the spin by 1, and S_1^- lowers it. The S_i^z is the operator giving that component. In terms of these we may readily verify that a dot product is given by

$$\mathbf{S}_i \cdot \mathbf{S}_j = \tfrac{1}{2}(S_i^+ S_j^- + S_i^- S_j^+) + S_i^z S_j^z \tag{5.15}$$

From Eqs. (5.7), (5.8), and (5.9) we obtain the commutation relation

$$S_i^+ S_i^- - S_i^- S_i^+ = 2S_i^z \tag{5.16}$$

$$S_i^- S_i^z - S_i^z S_i^- = S_i^- \tag{5.17}$$

$$S_i^z S_i^+ - S_i^+ S_i^z = S_i^+ \tag{5.18}$$

All of these results have been developed for electrons, with spin $\frac{1}{2}$. Equations (5.5), (5.12) to (5.14) and the square brackets in (5.10) are valid only for that case. However the dot-product expressions and the commutation relations are those for general angular-momentum operators and relations Eqs. (5.6) to (5.9) and (5.15) to (5.18) are applicable to ions or atoms with arbitrary total spin.[1] We may verify immediately, using Eqs. (5.6) and (5.7) to (5.9) that S_i^z commutes with $\mathbf{S}_i \cdot \mathbf{S}_i$ so states may be simultaneously eigenstates of both. We may also show immediately that S_i^+ operating on an eigenstate of S_i^z raises the eigenvalue by 1 and S_i^- lowers it by 1 without changing the eigenvalue of $\mathbf{S}_i \cdot \mathbf{S}_i$.

It is clear from Eq. (5.6) or (5.15) that $\langle S_i^z \rangle^2$ is limited for a fixed total spin so successive operation on a given state by S_i^+ must lead to a state upon which operation by S_i^+ gives zero. Calling this maximum eigenvalue of S_i^z the number S, we see from Eq. (5.15) [using also Eq. (5.16)] that the eigenvalue of $\mathbf{S}_i \cdot \mathbf{S}_i$ is $S + S^2 = S(S + 1)$. Calling S the total spin on the atom we find, as is well known, the eigenvalue of $\mathbf{S}_i \cdot \mathbf{S}_i$ is $S(S + 1)$ and S_i^z can take on eigenvalues running from $-S$ to S by integral steps. The S may be an integer or half-odd integer.

We may note also that, like the phonon raising and lowering operators, S_i^+ and S_i^- do not conserve the normalization. For given total spin S, we specify a state by the eigenvalue S_z of the operator S^z. Then if $|S_z\rangle$ is normalized the normalization integral for the state $S^+|S_z\rangle$ is

$$\langle S_z | S^- S^+ | S_z \rangle = \langle S_z | S^+ S^- | S_z \rangle - \langle S_z | 2S^z | S_z \rangle$$

[1] L. I. Schiff, "Quantum Mechanics," pp. 140ff., McGraw-Hill, New York, 1949.

Now consider the state $S_z = -S$. Then $\langle S_z|S^+S^-|S_z\rangle = 0$ so $\langle S_z|S^-S^+|S_z\rangle = 2S$ and $S^+|S_z\rangle$ is not normalized except for spin $\frac{1}{2}$ particles such as electrons.

With this brief introduction of spin operators we may proceed with the phenomenological treatment of exchange.

4. *Heisenberg Exchange*

We begin with the description of the spin-dependent interaction between single electrons, which can be identified with the exchange interaction given earlier. The formalism and results, however, are directly applicable to ions and atoms. We postulate a spin-dependent interaction represented by a term in the Hamiltonian,

$$\mathscr{H}_{\text{ex}} = -\sum_{i>j} J_{ij}\,\mathbf{S}_i \cdot \mathbf{S}_j \tag{5.19}$$

This is called *Heisenberg exchange* and the sum is over all *pairs* of electrons. The coefficients J_{ij} are called exchange integrals and will be identified later with matrix elements in Hartree-Fock theory.

If the two states of interest are two electronic states in the free atom, J tends to be positive. The spins tend to align as indicated by Hund's rule. If the interaction is between two states on different atoms, J tends to be negative. It corresponds to the fact that bonding states of the electron have antiparallel spin. Within solids the sign may be either way.

It is of interest first to evaluate the expectation value of the exchange Hamiltonian of Eq. (5.19) for a single Slater determinant as we did in the preceding sections. We consider the expectation value for a two-electron state with both spins up and the value for a two-electron state with one up and one down. In order to do this we write the spin state of the ith electron $\binom{1}{0}_i$ and then note that the operator S_i operates only upon this state. Thus operation on the two-electron spin states using Eq. (5.6) gives us

$$\begin{aligned} \mathbf{S}_i \cdot \mathbf{S}_j \tbinom{1}{0}_i \tbinom{1}{0}_j &= \tfrac{1}{4}\tbinom{0}{1}_i\tbinom{0}{1}_j - \tfrac{1}{4}\tbinom{0}{1}_i\tbinom{0}{1}_j + \tfrac{1}{4}\tbinom{1}{0}_i\tbinom{1}{0}_j = \tfrac{1}{4}\tbinom{1}{0}_i\tbinom{1}{0}_j \\ \mathbf{S}_i \cdot \mathbf{S}_j \tbinom{1}{0}_i \tbinom{0}{1}_j &= \tfrac{1}{4}\tbinom{0}{1}_i\tbinom{1}{0}_j + \tfrac{1}{4}\tbinom{0}{1}_i\tbinom{1}{0}_j - \tfrac{1}{4}\tbinom{1}{0}_i\tbinom{0}{1}_j \end{aligned} \tag{5.20}$$

We see that the x and y components of the dot product have flipped both spins and given terms orthogonal to the initial state. Thus only the z component contributes to the expectation values which are given by

$$\langle \tbinom{1}{0}_j \tbinom{1}{0}_i | \mathscr{H}_{\text{ex}} | \tbinom{1}{0}_i \tbinom{1}{0}_j \rangle = -\tfrac{1}{4} J_{ij}$$

$$\langle \tbinom{0}{1}_j \tbinom{1}{0}_i | \mathscr{H}_{\text{ex}} | \tbinom{1}{0}_i \tbinom{0}{1}_j \rangle = +\tfrac{1}{4} J_{ij}$$

If we wish to make an identification of these matrix elements with exchange in the Hartree-Fock approximation we must identify the difference between the two matrix elements as the exchange integral $\langle ij|V|ij\rangle$ and Heisenberg exchange has given us an additional direct term. With sufficiently detailed definition of exchange integrals and of additional direct terms it would be possible to reproduce all of the matrix elements of the many-electron problem. This would have gained us nothing, of course, and it is not the way in which Heisenberg exchange is used. In the first place the interaction is most usually used to describe the interactions between total spins on different atoms. In addition the exchange integrals are taken to be of very simple form. In spite of these simplifications the Heisenberg exchange which is ordinarily used goes beyond the Hartree-Fock approximation in other respects. In particular, off-diagonal matrix elements are usually included. This may be seen most clearly by returning to the two-electron problem described above.

Only the first of the two-electron states that we wrote down is an eigenstate of the total Hamiltonian including Heisenberg exchange. This follows from the fact that the operation of $\mathbf{S}_i \cdot \mathbf{S}_j$ upon the second one led to states of different spin. We may, however, find the eigenstates for the exchange operator using Eq. (5.20) and seeking linear combinations of coupled states. We obtain

$$\psi_1 = \frac{1}{\sqrt{2}}\left[\binom{1}{0}_i\binom{0}{1}_j - \binom{0}{1}_i\binom{1}{0}_j\right] \qquad \mathscr{H}_{\text{ex}}\psi_1 = \tfrac{3}{4}J_{ij}\psi_1$$

$$\left.\begin{aligned}\psi_2 &= \frac{1}{\sqrt{2}}\left[\binom{1}{0}_i\binom{0}{1}_j + \binom{0}{1}_i\binom{1}{0}_j\right]\\ \psi_3 &= \frac{1}{\sqrt{2}}\left[\binom{1}{0}_i\binom{1}{0}_j + \binom{0}{1}_i\binom{0}{1}_j\right]\\ \psi_4 &= \frac{1}{\sqrt{2}}\left[\binom{1}{0}_i\binom{1}{0}_j - \binom{0}{1}_i\binom{0}{1}_j\right]\end{aligned}\right] \qquad \begin{aligned}&\mathscr{H}_{\text{ex}}\psi_n = -\tfrac{1}{4}J_{ij}\psi_n\\ &\text{for } n = 2, 3, 4\end{aligned}$$

These are the familiar singlet and triplet states for two coupled electron spins. Note that any orthogonal linear combinations of the last three would have been acceptable.

The eigenvalues themselves could have been obtained more easily using the vector model for the spin operators.

$$(\mathbf{S}_i + \mathbf{S}_j)^2 = \mathbf{S}_i^2 + \mathbf{S}_j^2 + 2\mathbf{S}_i \cdot \mathbf{S}_j$$

or

$$\mathbf{S}_i \cdot \mathbf{S}_j = \tfrac{1}{2}[(\mathbf{S}_i + \mathbf{S}_j)^2 - \mathbf{S}_i^2 - \mathbf{S}_j^2] = \tfrac{1}{2}S(S+1) - S_i(S_i+1)$$

where S is the total-spin quantum number, 1 or 0, and S_i is the individual electron-spin quantum number of $\tfrac{1}{2}$. This leads immediately to the eigenvalues given above. The corresponding results could be obtained for this simple case with the electron-electron interaction though it is not nearly so simple.

Let us now turn our attention to atomic spins. If we imagine a lattice of localized moments interacting through Heisenberg exchange we may intuitively see immediately the nature of the states. If the exchange integrals are positive the Hamiltonian favors the alignment of spins. We saw that for the two-electron state a parallel alignment of the spins was an eigenstate of the Heisenberg exchange [see Eq. (5.20)]. Thus we may schematically represent the ground state of the system by its classical analog shown in Fig. 5.2*a*. This of course corresponds to a ferromagnetic state. In this frame of reference, where the state of each atom is represented only by its spin, we could also find that this corresponds to a quantum-mechanical eigenstate of the electron-electron interaction, which connects only states of the same total spin.

If the exchange interaction were negative and coupled only nearest-neighbor spins, we may also make conjecture as to the nature of the state. That shown in Fig. 5.2*b* would be favored by Heisenberg exchange and corresponds to the usual conception of the antiferromagnetic state. We should note, however, that such a state is not an eigenstate of the Heisenberg exchange, which may be seen from Eq. (5.20). The operation on such a state with the Heisenberg exchange operator would lead to states with neighboring spins flipped with respect to the postulated state. The true antiferromagnetic ground state then is quite complicated but in that state neighboring spins are predominantly antiparallel as in the classical state of Fig. 5.2*b*. Antiferromagnetism is common in oxides of the transition metals. In many such

Fig. 5.2 *The states of a system of classical spins coupled by an interaction* $-\sum_{i>j} J_{ij}\, \mathbf{S}_i \cdot \mathbf{S}_j$. *These give a schematic representation of the states of coupled quantum-mechanical spins. In all cases we imagine that the nearest neighbors dominate the interaction. In* (*a*) *the corresponding* J_{ij} *are positive; in* (*b*) *and* (*c*) *they are negative. In* (*c*) *alternate spins have differing magnetic moments.*

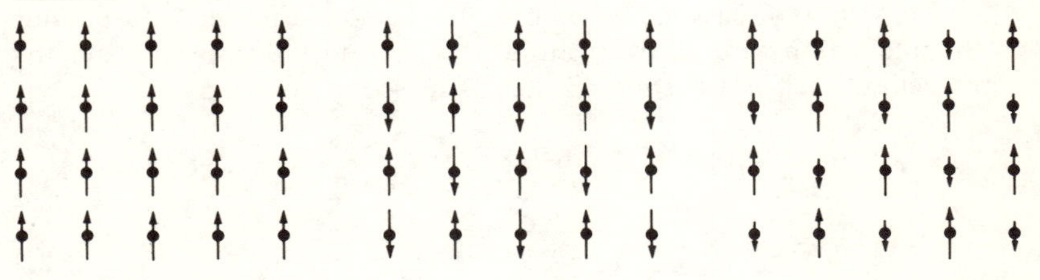

(*a*) Ferromagnetic (*b*) Antiferromagnetic (*c*) Ferrimagnetic

instances the origin of the exchange interaction between the moments on the transition-metal ions is described by *superexchange*. This is an indirect interaction between the moments on the transition metal ions through the intermediary of the oxides. The spin on one transition-metal ion polarizes a neighboring oxygen which in turn interacts with a neighboring transition-metal ion. In many such cases the spins on the two sublattices are better described as being *canted* with respect to each other and not simply antiparallel. Another situation occurs in ferrites in which there is an antiferromagnetic configuration, but the moments on the two sublattices are unequal. There is then a net magnetization. Such states are called *ferrimagnetic* and are illustrated in Fig. 5.2*c*.

In the treatment of the true antiferromagnetic ground state, in the treatment of spin waves, and in the treatment of many scattering processes involving magnetic ions, Heisenberg exchange provides a very convenient formalism and one that has been very effective. For some simpler properties it is possible to approximate Heisenberg exchange by the Ising model or a molecular-field approximation.

The state of a collection of spins may be specified by giving the spin component of each along a particular axis. In the *Ising model* the Heisenberg exchange is replaced by $-\sum_{i>j} J_{ij} S_z(i) S_z(j)$, that is, only the final terms in Eq. (5.20). The antiferromagnetic state shown in Fig. 5.2*b* is an eigenstate in the Ising model. In addition the Ising model is useful in many statistical studies of magnetism. However, it is an approximation to Heisenberg exchange.

5. *The Molecular-Field Approximation and the Ferromagnetic Transition*

One of the earliest treatments of ferromagnetism was that given by Weiss in 1907 which is called the molecular-field approximation. At that time of course it was strictly phenomenological and predates the recognition of exchange as the mechanism for spin alignment. It may be desirable, however, to see how this approximation arises from Heisenberg exchange.

We wish to discuss the behavior of the spin on a particular atom due to the interaction with all other atoms. We can do this approximately by constructing a self-consistent field; just as in Sec. 4.2 of Chap. IV

$$\mathscr{H}_{ex} = -\sum_{i>j,j} J_{ij}\mathbf{S}_i \cdot \mathbf{S}_j \approx -\sum_k \left(\sum_{i>k} J_{ik} \langle \mathbf{S}_i \rangle \cdot \mathbf{S}_k + \sum_{j<k} J_{kj} \mathbf{S}_k \cdot \langle \mathbf{S}_j \rangle \right)$$

$$= -\sum_k \mathbf{S}_k \cdot \left(\sum_i{}' J_{ik} \langle \mathbf{S}_i \rangle \right) \tag{5.21}$$

where $\langle \mathbf{S}_j \rangle$ is the expectation value of the spin $\mathbf{S}_j$. In the first form J_{ij} appears only for $i > j$. In the final form we define $J_{ij} = J_{ji}$ for $j > i$. In a ferromagnet $\langle \mathbf{S}_j \rangle$ will be a vector parallel to the total spin of the system; in an antiferromagnet it will be parallel or antiparallel to a sublattice spin. By making this approximation we have replaced the two-electron operator by a one-electron operator. Let us say now that there is a net alignment (ferromagnetic) of spins in the system, which we take to lie in the z direction. Then if any given spin $\mathbf{S}_i$ interacts with a sufficient number of neighbors to be statistically significant (even 8 to 12 nearest neighbors may be sufficient), the interaction associated with the ith ion is simply given by the product of its z component and a field proportional to this net spin polarization. We have obtained an expression very much like that in the Ising model but with all equivalent ions seeing the same field.

It is interesting to note that this term has the same form as the interaction between the magnetic moment of the ion and a magnetic field, $-\boldsymbol{\mu} \cdot \mathbf{H}$. The magnetic moment of an ion is given by the gyromagnetic ratio (the g value, equal to 2 for a free electron) times the Bohr magneton, $\mu_0 = e\hbar/2mc$, times the spin, $\mathbf{S}$. It is convenient to write the interaction of Eq. (5.21) in terms of an effective magnetic field H_I, sometimes called the *molecular field* and sometimes the internal field. Taking the z axis along $\langle \mathbf{S}_j \rangle$

$$\mathscr{H}_{\text{ex}} = -\sum_i (g\mu_0 S_i^z) H_I$$

with

$$H_I = \frac{1}{g\mu_0} \sum_j{}' J_{ij} \langle S_j^z \rangle$$

Note that H_I is proportional and parallel to the magnetization per unit volume, $M_z = g\langle S_j^z \rangle \mu_0/\Omega_0$. We then have a phenomenological parameter relating these two.

$$\mathbf{H}_I = \lambda \mathbf{M} \tag{5.22}$$

The molecular field may be added to any applied field to obtain the Hamiltonian with exchange. For isotropic systems both may be taken in the same direction, the z direction, and our Hamiltonian including interaction with the magnetic field is given by

$$\mathscr{H} = -\sum_i \mu_i^z (H + H_I) \tag{5.23}$$

with $\mu_i^z = g\mu_0 S_i^z$.

The Hamiltonian has become sufficiently simple that we may very easily compute the state of the system with or without applied fields.

We begin by computing the magnetization, and therefore the magnetic susceptibility, as a function of applied field and of temperature. The magnetization is obtained by summing over a unit volume,

$$M = \sum_i \mu_i^z = N\langle \mu^z \rangle$$

where N is the number of ions per unit volume. Using Eq. (5.23) and ordinary statistical mechanics

$$\langle \mu^z \rangle = \frac{\sum_{\mu_z} \mu_z e^{(H_I + H)\mu_z/KT}}{\sum_{\mu_z} e^{(H_I + H)\mu_z/KT}} \tag{5.24}$$

where the sum over μ^z is over possible spin orientations. In the classical theory this is an average over all angles. In quantum theory it is a sum over the $2S + 1$ components. For simplicity we evaluate Eq. (5.24) for $S = \frac{1}{2}$ and $g = 2$. Then $\mu^z = \pm \mu_0$ and

$$\langle \mu^z \rangle = \mu_0 \tanh \frac{(H + H_I)\mu_0}{KT} \tag{5.25}$$

Before continuing the calculation of the molecular field it is of some interest to examine the simple case in which there is no exchange field, $H_I = 0$. Then the total magnetization per unit volume is given by

$$M = N\mu_0 \tanh \frac{H\mu_0}{KT}$$

At small fields this becomes

$$M = \frac{N\mu_0{}^2}{KT} H$$

The proportionality factor $N\mu_0{}^2/KT$ is the *susceptibility*. When written in terms of $\langle \mu^2 \rangle = (2\mu_0)^2 S(S + 1) = 3\mu_0{}^2$ this becomes the *Curie law* for magnetic susceptibility.

$$\chi = \frac{N\langle \mu^2 \rangle}{3KT} \tag{5.26}$$

It gives a good account of the paramagnetic susceptibility of many solids with unbalanced moments on each ion. This corresponds ordinarily to a very weak magnetic response. The net polarization is of the order of 2×10^{-4} for 1 kilogauss at room temperatures.

Let us now again include the molecular field, writing it in terms of the magnetization as in Eq. (5.22). Then using Eq. (5.25) the magnetization becomes

$$M = N\mu_0 \tanh \frac{(H + \lambda M)\mu_0}{KT} \tag{5.27}$$

which is to be solved for M.

If the magnetization is again small we may expand the hyperbolic tangent and solve directly. We obtain a susceptibility of $N\mu_0{}^2/(KT - N\mu_0{}^2\lambda)$. Written again in terms of $\langle\mu^2\rangle$, this becomes the *Curie-Weiss law*

$$\chi = \frac{N\langle\mu^2\rangle}{3K(T - \theta)}$$

where the Curie temperature is $\theta = N\langle\mu^2\rangle\lambda/3K$. This accounts well for the magnetic susceptibility of ferromagnetic materials above the Curie temperature where the magnetization is small. However, we note that it predicts a divergent magnetization as the temperature approaches the Curie temperature. At that temperature a phase transition occurs and we must improve our solution of Eq. (5.27). This can readily be done graphically.

In Fig. 5.3 we have plotted the right- and left-hand sides of Eq. (5.27). Here we have made the plot for a vanishing applied magnetic field. At high temperatures there remains only the single solution, no magnetization with no applied field. The application of a field simply shifts the tanh curve to one side and a paramagnetism corresponding to the Curie-Weiss law. At low temperatures, however, there will be three solutions, the paramagnetic solution and two solutions with nonvanishing magnetism. Those with nonvanishing magnetism are the ferromagnetic solutions and correspond

Fig. 5.3 *Graphical solution for the magnetization M when no magnetic field is applied. Intersections of the curve M and $N\mu_0$ tanh$(\lambda M\mu_0/KT)$ give self-consistent solutions for the magnetization.*

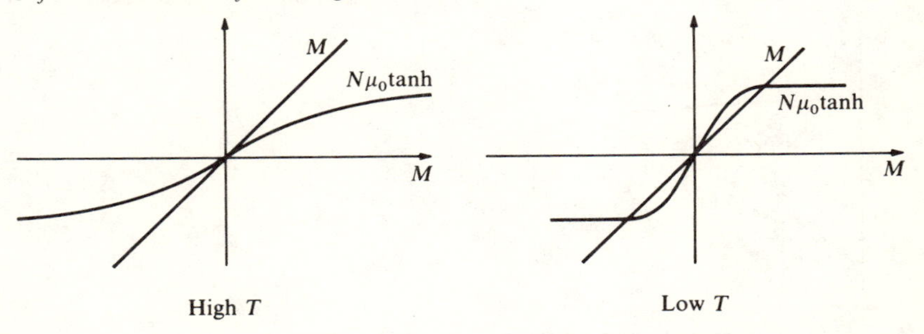

to lower energy. Thus when the temperature drops below the Curie temperature, spontaneous ferromagnetism will occur corresponding to a cooperative transition.

The spontaneous magnetization will be of the order of that corresponding to all spins aligned. (They may of course be aligned parallel or antiparallel to the z axis.) By carrying out the graphical solution as a function of temperature we may also obtain the spontaneous magnetization as a function of temperature. The result is as indicated in Fig. 5.4.

Clearly an analogous treatment may be given of antiferromagnetism. The lattice is divided into sublattices, one with spin up and one with spin down. The spin on any given atom is then assumed to respond to a molecular field due to the other sublattice. We obtain two coupled equations in the magnetizations of the sublattices. Again a critical temperature is obtained which in antiferromagnetism is called the *Néel temperature*. The magnetization may be computed as a function of applied magnetic field and of temperature. Clearly, also, this formalism can be used to describe the thermodynamic properties of ferromagnets and antiferromagnets. In both cases it is an approximate treatment but one that includes much of the physics of the problem.

In recent years improved approximations have been used and particular attention has been focused on the transition itself. Near the transition temperature statistical fluctuations of the magnetic order occur and strongly influence the properties. The treatment of these is formally the same as the treatment of the corresponding fluctuations in superconductors which we discuss in Sec. 10.5. We have simply chosen to discuss them in that context rather than here. A study which has been particularly active but which we

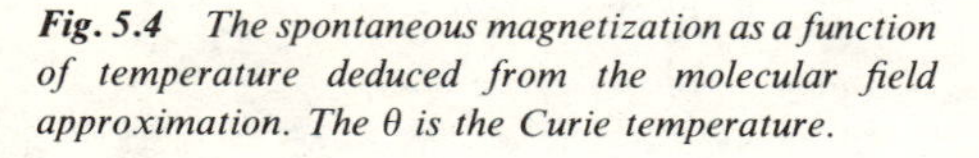

Fig. 5.4 *The spontaneous magnetization as a function of temperature deduced from the molecular field approximation. The θ is the Curie temperature.*

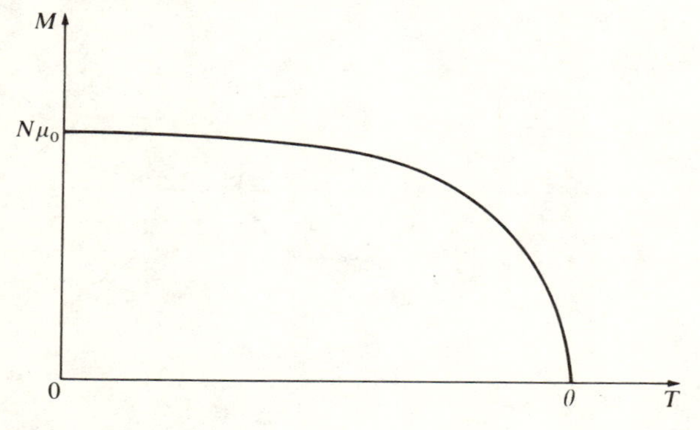

will not discuss in detail in either context is that of *critical-point exponents.* One such exponent may be defined in terms of Fig. 5.4. We see that as we approach the critical temperature from below the magnetization drops to zero as $(\theta - T)^{\beta}$. By expanding the tanh in Eq. (5.27) with $H = 0$ we may readily see that we predict that the critical-point exponent β is equal to $\frac{1}{2}$. Experimentally it is found instead to be very close to $\frac{1}{3}$. Calculations using the Ising model have also given values close to $\frac{1}{3}$. These have been exceedingly difficult numerical computations and seem not to have shed great light on the problem. Intensive activity persists, however, in the treatment of such exponents which describe the behavior of the magnetic transition and many other phase transitions.

6. Inhomogeneities

The study of magnetism also includes the properties of nonuniform ferromagnets. Two such situations which are easy to visualize classically are Bloch walls and spin waves.

6.1 Bloch walls A ferromagnetic crystal will ordinarily be divided by *Bloch walls* into *domains*, each with a different orientation of the magnetization. This is in fact favored energetically since it can remove the external magnetic fields arising from a single direction of magnetization. Such a configuration is shown in Fig. 5.5*a*. It is arranged such that there are no

Fig. 5.5 A simple domain structure in a ferromagnet. In (a) there is no net magnetic moment. In (b) a magnetic field is applied and the domain walls move, giving a net magnetization.

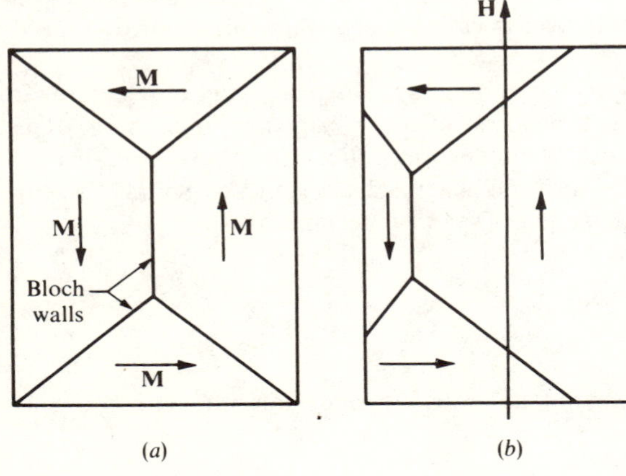

Fig. 5.6 The arrows represent individual atomic magnetic moments. The transition region between the two oppositely magnetized domains is called the Bloch wall.

fields outside the crystal. When magnetic fields are applied, the Bloch walls shift so as to produce a net magnetization as shown in Fig. 5.5*b*. The figure is appropriate for a highly anisotropic material where the magnetization within any domain tends to lie in a symmetry direction.

At the Bloch wall boundaries we may imagine the spins turning over as shown in Fig. 5.6. There is an increase in energy due to such a configuration, which can be described as a surface energy for Bloch walls. However, with no applied fields this increase in energy will ordinarily be more than compensated for by the decrease in the magnetic-field energy.

6.2 Spin waves The excited states of a magnetic system also correspond to nonuniformities. In a ferromagnet we might imagine that the lowest-lying excited state corresponds to the reversal of a single spin requiring an energy $2\mu H_I$. We may readily see, however, that this is not an eigenstate of Heisenberg exchange. For a system with nearest-neighbor interactions, for example, the Heisenberg operator on a state with a particular spin flipped will lead to states with adjacent spins flipped. We may, however, obtain excited states which are very nearly eigenstates of the Heisenberg exchange by taking a linear combination of ferromagnetic states, each with a single spin flipped, and modulating these by a phase factor $e^{i\mathbf{q}\cdot\mathbf{r}_j}$, where $\mathbf{r}_j$ is the position of the flipped spin. Such an excitation is called a *spin wave* or, when quantized, a *magnon.*

We may describe spin waves more systematically and quite simply in terms of a linear chain with nearest-neighbor exchange interactions. This model, like the force-constant model for lattice vibrations, contains the physics of the system. (For a more general discussion, see Kittel.[1]) In this model the Heisenberg exchange of Eq. (5.19) becomes

$$\begin{aligned} H &= -J \sum_i \mathbf{S}_i \cdot \mathbf{S}_{i+1} \\ &= -J \sum_i \left[S_i^z S_{i+1}^z + \tfrac{1}{2}(S_i^+ S_{i+1}^- + S_i^- S_{i+1}^+)\right] \end{aligned} \tag{5.28}$$

We first write the ground state for ferromagnetic interactions, $J > 0$.

[1] Kittel, *op. cit.*, p. 49.

This is obtained by letting each ion be in a state of $S_z = S$.

$$\Psi_0 = \prod_i |S\rangle_i$$

To verify that this is an eigenstate of the system we operate on Ψ_0 with H from Eq. (5.28), noting that any S_i^+ operating on Ψ_0 gives zero and any S_i^z operating on Ψ_0 gives $S\Psi_0$. We obtain immediately

$$H\Psi_0 = -NJS^2\Psi_0$$

where N is the number of ions in the chain. The ground-state energy is $E_0 = -NJS^2$.

We could then seek an excited state in which a single spin component is reduced by 1. (For spin ½ the spin is "flipped.") Letting the nth spin component be reduced we obtain a state $\Psi_n = S_n^-\Psi_0 = S_n^- \prod_i |S\rangle_i$, within a normalization constant. However, this is not an eigenstate of H as we indicated earlier and can verify here in detail.

$$H\Psi_n = -J(NS^2 - 2S)\Psi_n - JS(S_{n+1}^-\Psi_0 + S_{n-1}^-\Psi_0) \tag{5.29}$$

We noted in this derivation, for example, that

$$\begin{aligned} S_{n+1}^- S_n^+(S_n^-\Psi_0) &= S_{n+1}^- S_n^- S_n^+\Psi_0 + S_{n+1}^- 2S_n^z\Psi_0 \\ &= 0 + 2S(S_{n+1}^-\Psi_0) \end{aligned}$$

where we have used Eq. (5.16). Thus from Eq. (5.29) we see that $H\Psi_n$ contains not only a constant multiple of Ψ_n, but additional terms with neighboring spins reduced.

As can be deduced from the translational symmetry of the system, the excited state must be of the form

$$\Psi = \sum_n \Psi_n e^{iqan} = \sum_n S_n^-\Psi_0 e^{iqan} \tag{5.30}$$

where a is the spacing between ions. The normalization constant is ignored. We can verify that this is an eigenstate of H. By using Eqs. (5.29) and (5.30) we obtain immediately

$$\begin{aligned} H\Psi &= -J(NS^2 - 2S)\Psi - JS\sum_n (S_{n+1}^-\Psi_0 e^{iqan} + S_{n-1}^-\Psi_0 e^{iqan}) \\ &= -J(NS^2 - 2S)\Psi - JS(e^{-iqa} + e^{+iqa})\Psi \end{aligned}$$

In performing the final two sums we changed dummy indices to $n' = n + 1$ and $n' = n - 1$, respectively. Thus Ψ is an eigenstate with energy depending upon q and given by

$$E_q = E_0 + 2JS(1 - \cos qa) = E_0 + 4JS\sin^2\frac{qa}{2}$$

At small q the excitation energies go to zero as JSq^2a^2, in contrast to the energy required to flip a single spin which is of order $2JS$. At larger q they approach a constant at the Brillouin zone face. The existence of excitations at arbitrarily low energies can be of importance in many problems, just as is the difference between an Einstein and a Debye approximation to the phonon spectrum.

Because the excitation energy is quadratic in q at small q, we may think of the excitations as particles, *magnons*, with an effective mass m^* given by $\hbar^2q^2/2m^* = JSa^2q^2$. This mass tends to be of the order of 10 electron masses.

It is interesting to carry the magnon concept further. We see as in Eq. (5.30) that the operator

$$a_q{}^+ = A^+ \sum_j S_j{}^- e^{i\mathbf{q}\cdot\mathbf{r}_j}$$

creates a magnon of wavenumber $\mathbf{q}$ since the state $\Psi = a_q{}^+\Psi_0$ contains such an excitation. Here we have included a constant A^+ such that Ψ is now normalized (A^+ will depend upon S and N), and have made a generalization to a three-dimensional system. The operator $a_q{}^+$ and its inverse $a_q{}^-$ are called magnon creation and annihilation operators respectively.

We can see that a magnon very nearly obeys Bose statistics. Each term in the state

$$a_q{}^+\Psi_0 = A^+ \sum_j S_j{}^- \prod_i |S_i\rangle$$

differs from the ground state only in one of the N spin states. Thus the operating a second time with a_q to obtain $a_q{}^+ a_q{}^+ \Psi_0$ again gives an eigenstate except for an error of one part in N and the energy of the new state differs from the ground state by twice the single excitation energy, to one part in N. Thus in each magnon mode (i.e., for each $\mathbf{q}$) we may have a number n_q magnons and the excitation energy will be n_qE_q, as long as we keep $n_q \ll N$. The limitation on n_q comes ultimately of course from the limitation in the number of reductions allowed for the spin component of each ion. The magnon modes, like the phonon modes, form a set of excitations of the ferromagnetic system. Thus they provide the basis for a statistical treatment of the thermodynamic properties of ferromagnets. Similarly, magnons may be constructed for antiferromagnetic crystals[1] and provide the basis for the theory of their thermodynamic properties. In both cases these excitations contribute to the specific heat. In the case of ferromagnets, a thermal distribution of magnons clearly decreases the magnetization of the system.

The magnon modes interact with neutrons through the magnetic moment of the neutron, and can be detected in neutron diffraction. They also will interact with phonons through spin-dependent forces between ions.

[1] *Ibid.*

Since the dispersion curves cross, as shown in Fig. 5.7, there will be ranges of wavenumber where they are strongly mixed.

It is easy to see that with the application of a magnetic field along the z axis, the magnon energies are shifted by $2\mu_0 H$, where μ_0 is the Bohr magneton, since the expectation value of the total spin component for the system is reduced by 1 for each magnon excitation. Thus by applying magnetic fields we may shift the curves of Fig. 5.7.

7. *Local Moments*

The study of the properties of magnetic impurities, such as manganese and iron, dissolved in normal metals, such as copper, has been extremely active in recent years. Many free transition-metal ions, with partially filled d shells, have an unbalanced electronic spin and therefore a net magnetic moment in the ground state. Our discussion of transition-metal bands and resonant states would suggest that if one such ion were dissolved in a simple metal it would retain much of its atomic character and therefore might be expected to produce a localized magnetic moment in the simple metal, localized near the impurity site. This turns out frequently to be the case and such moments contribute to the paramagnetic susceptibility of the alloy and to many other properties.

We should recognize here that we are going beyond the one-electron band picture in the sense that we have been using it; we are specifically interested in cases where Koopmans' theorem is not applicable. Even

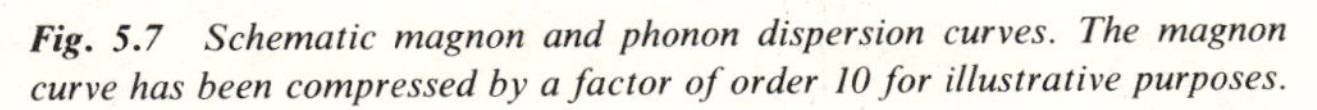

Fig. 5.7 *Schematic magnon and phonon dispersion curves. The magnon curve has been compressed by a factor of order 10 for illustrative purposes.*

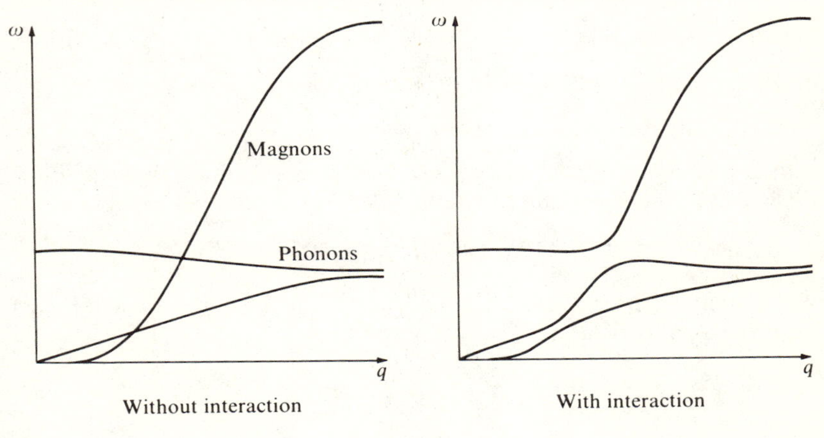

though the states may be described as resonant rather than as localized, the occupation of a resonant state of spin up may shift the energy of the corresponding spin-down resonance above the Fermi energy.

We will treat the resonant d states in the spirit of the transition-metal pseudopotentials discussed earlier, but will now add an explicit electron-electron interaction term to the Hamiltonian. In addition, we will improve upon perturbation theory in the treatment of the hybridization term. By treating initially a single resonant center and by sufficiently simplifying the Hamiltonian it will be possible to treat the resonant interaction more accurately than we did before. The results are equivalent when the resonances are far from the Fermi energy; here the results will be meaningful even when the resonance lies close to the Fermi energy.

7.1 The formation of local moments The model we use is due to Anderson.[1] It attempts to describe the behavior of a transition-metal impurity in a simple metal. Here we will develop it as an approximation to the transition-metal pseudopotential equation, which came much later and which was in fact partly inspired by the Anderson model. This approach will have the advantage of providing a precise meaning to the parameters that enter Anderson's model and a means for calculating them.

It is desirable here to make a slightly different formulation of the pseudopotential equation. In particular, we wish to make the pseudowavefunction orthogonal to the free-ion d states; this was not done before, but it is always possible. In addition we will absorb the contribution of the core states into the potential since it is not of direct interest here. We then expand the wavefunction (exactly) in the form

$$|\psi\rangle = \sum_k a_k|\mathbf{k}\rangle + \sum_d a_d|d\rangle \tag{5.31}$$

but now the function $\varphi = \sum_k a_k|\mathbf{k}\rangle$ is orthogonal to the $|d\rangle$. As before, operation of the Hamiltonian on the d state gives

$$(T + V)|d\rangle = E_d|d\rangle - \Delta|d\rangle$$

with Δ and E_d defined as before. We may then substitute the expansion, Eq. (5.31), in the eigenvalue equation, $(T + V - E)|\psi\rangle = 0$, to obtain

$$\sum_k a_k(T + V - E)|\mathbf{k}\rangle + \sum_d a_d(E_d - E - \Delta)|d\rangle = 0 \tag{5.32}$$

We multiply on the left with $\langle d|$, letting $T + V$ operate to the left, and noting the orthogonality of $|d\rangle$ to $\sum_k a_k|\mathbf{k}\rangle$. We obtain

[1] P. W. Anderson, *Phys. Rev.*, **124**:41 (1961).

$$-\sum_k a_k \langle d|\Delta|\mathbf{k}\rangle + a_d(E_d - E) = 0 \tag{5.33}$$

Similarly, we multiply Eq. (5.32) on the left by a particular $\langle \mathbf{k}|$, using Eq. (5.33) to rewrite a term $\sum_d a_d(E_d - E)\langle \mathbf{k}|d\rangle$. First changing $\mathbf{k}$ to $\mathbf{k}'$ in Eq. (5.32) this leads to

$$a_k\left(\frac{\hbar^2 k^2}{2m} + \langle \mathbf{k}|V|\mathbf{k}\rangle + \sum_d \langle \mathbf{k}|d\rangle\langle d|\Delta|\mathbf{k}\rangle - E\right)$$
$$+ \sum_{k'} a_{k'}(\langle \mathbf{k}|V|\mathbf{k}'\rangle + \sum_d \langle \mathbf{k}|d\rangle\langle d|\Delta|\mathbf{k}'\rangle) - \sum_d a_d\langle \mathbf{k}|\Delta|d\rangle = 0 \tag{5.34}$$

Thus if we define a pseudopotential by

$$\langle \mathbf{k}|\tilde{W}|\mathbf{k}'\rangle = \langle \mathbf{k}|V|\mathbf{k}'\rangle + \sum_d \langle \mathbf{k}|d\rangle\langle d|\Delta|\mathbf{k}'\rangle$$

which differs slightly from our earlier form due to the orthogonality requirement used here, and let $E_k = \hbar^2k^2/2m + \langle \mathbf{k}|\tilde{W}|\mathbf{k}\rangle$, Eq. (5.34) becomes

$$(E_k - E)a_k + \sum_{k'} \langle \mathbf{k}|\tilde{W}|\mathbf{k}'\rangle a_{k'} - \sum_d \langle \mathbf{k}|\Delta|d\rangle a_d = 0 \tag{5.35}$$

Up to this point no approximations have been made. Eqs. (5.33) and (5.35) provide a set of simultaneous algebraic equations which, if solved exactly, lead to the same eigenvalues as the original Schroedinger equation.

At this point we associate the matrix elements of $\tilde{W}$ with band-structure effects which are not of interest and drop them. This is plausible, but not completely justified, since it removes all nonhermiticity from the problem and therefore all effects of nonorthogonality of the overcomplete set. However, we saw in the discussion of transition-metal pseudopotentials that there was a one-to-one correspondence between the basis states we retain and the solutions of the problem so we may expect that the approximation is a meaningful one. In addition, for simplicity, we neglect the dependence of the matrix elements of Δ upon $\mathbf{k}$ and write $\langle \mathbf{k}|\Delta|d\rangle$ simply as a parameter Δ. This is an inessential approximation. Then Eqs. (5.33) and (5.35) become precisely the equations derivable from a Hamiltonian

$$H = \sum_k E_k c_k^+ c_k + \sum_d E_d c_d^+ c_d - \sum_{k,d} (\mathring{\Delta} c_k^+ c_d + \Delta^* c_d^+ c_k)$$

Use of second-quantized notation at this point is irrelevant, but will be useful later. We have, as an approximation, derived a simple equation containing hybridization; all parameters are well defined and derivable from the atomic structure of the atom in question.

We now wish to add an explicit electron-electron interaction between electrons in d-like states on the atom. In particular, we discuss only a single atomic d state and assert that there is an additional coulomb interaction energy U if both the spin-up and spin-down states are occupied in a free atom; such terms are known to exist and have been discussed before. Then in the metal such an interaction energy will arise if both the states $|d+\rangle$ and $|d-\rangle$ are occupied, where the sign represents the spin. But in our model the operators corresponding to such occupation are $n_{d+} = c_{d+}{}^{+}c_{d+}$ and $n_{d-} = c_{d-}{}^{+}c_{d-}$. Note that this has been made more precise by our orthogonality requirement. Thus an extra term in the Hamiltonian, $Un_{d+}n_{d-}$ is added. As with the other terms in the Hamiltonian the coefficient is derivable from the free-atom wavefunctions. We are led then to the *Anderson Hamiltonian*, which was postulated by Anderson initially on more intuitive grounds

$$H = \sum_{k,\sigma} E_k n_{k\sigma} + E_d(n_{d+} + n_{d-}) + Un_{d+}n_{d-} - \Delta \sum_{k,\sigma} (c_{k\sigma}{}^{+}c_{d\sigma} + c_{d\sigma}{}^{+}c_{k\sigma}) \quad (5.36)$$

We have taken Δ real and have written $c_{k\sigma}{}^{+}c_{k\sigma} = n_{k\sigma}$, with σ the spin index.

Let us first neglect the hybridizing term by setting $\Delta = 0$. Further, let us assume that $E + U$ lies above the Fermi energy of the simple metal while E lies below. Then clearly the ground state of the system will have each of the states $\mathbf{k}$ occupied for E_k less than the Fermi energy and a single localized d state occupied. The ground state is degenerate since either the state $d+$ or $d-$ may be occupied. Thus the local moment of the d state can contribute to the paramagnetism and will obey a Curie law as in Eq. (5.26).

When the hybridization is introduced the problem becomes considerably more difficult. The occupied d orbital, say $d+$, becomes hybridized with the conduction band and the expectation value of the corresponding n_{d+} becomes slightly less than 1. Similarly the unoccupied d state becomes hybridized and the corresponding n_{d-} becomes greater than zero. In pure transition metals the corresponding states are those of band ferromagnetism discussed in Sec. 2. Here we have a single impurity and may describe the states in terms of phase shifts. The spin-up resonance is below the Fermi energy; the spin-down resonance is above.

Before the introduction of hybridization the operators n_d were either 1 or 0 and the spin-up and the spin-down problem became essentially independent-particle problems. With hybridization the coulomb interaction term makes the problem intrinsically a many-body one, but one that may be solved in a self-consistent-field approximation. In treating spin-up states we replace n_{d-} by its expectation value in the coulomb term. This is completely analogous to replacing the coulomb energy in the band-structure problem by a potential arising from the expectation value of the charge density due to other electrons. The Hamiltonian, Eq. (5.36), for spin-up states becomes simply

$$H_+ = \sum_k E_k n_{k+} + (E_d + U\langle n_{d-}\rangle)n_{d+} - \Delta \sum_k [c_{k+}{}^+ c_{d+} + c_{d+}{}^+ c_{k+})$$

This Hamiltonian has one-electron eigenstates of the form

$$|\psi_k\rangle = (ac_k{}^+ + bc_d{}^+)|0\rangle \tag{5.37}$$

where a and b are functions only of $\mathbf{k}$. The problem for each spin is exactly that of solving the transition-metal pseudopotential equation discussed in Sec. 9 of Chap. II, but now we wish to perform sums over the occupied states to determine, for example, $\langle n_{d-}\rangle$.

At this point we will proceed in an intuitive way. This analysis can be done with more rigor[1] using the one-electron Green's functions described in Sec. 10.3 of Chap. II; however, the answers are the same.

We first seek the eigenstates, Eq. (5.37), treating the hybridizing term as a perturbation. The states, which in zero order are conduction-band states, in first order take values of $a = 1$ and $b_\pm = -\Delta/(E_k - E_{d\pm})$. The subscript on the b denotes the spin on the state in question and $E_{d\mp} = E_d + U\langle n_{d\mp}\rangle$ (actually the energy of the state $d\pm$).

To be specific we evaluate $\langle n_{d+}\rangle$ by summing the squared amplitude of the d component, $b_+{}^2$ for occupied states. Our first-order expression for b is appropriate to lowest order in Δ except where the energy denominator approaches zero. In contrast to the summations giving electron density and total energy we see that the summation of $b_+{}^2$ gives a divergent result. We remove the divergence by adding a small positive constant ϵ^2 to the denominator.

$$b_+{}^2 = \frac{\Delta^2}{(E_k - E_{d-})^2 + \epsilon^2}$$

ϵ^2 must go to zero as Δ goes to zero if the form is to be correct to lowest order in Δ away from resonance. We may evaluate ϵ explicitly by requiring that the contribution from the integration across the resonance be simply that from the additional state in the limit as Δ approaches zero. By requiring this limit we obtain the contribution of d-like states as well as k-like states when we sum over wavenumber. Thus

$$\lim_{\epsilon,\Delta\to 0} \int_-^+ \frac{\Delta^2 n(E)\,dE}{(E - E_{d-})^2 + \epsilon^2} = \lim \frac{\Delta^2 n(E_{d-})}{\epsilon} \arctan \frac{E - E_{d-}}{\epsilon}\bigg|_-^+$$

$$= \frac{\Delta^2 \pi n(E_{d-})}{\epsilon} = 1$$

[1] *Ibid.*

where the limits of integration are taken slightly above and below E_{d-}. We can in fact neglect variations in the density of states over the energies of interest and obtain for both spin up and spin down, $\epsilon = \Delta^2 \pi n(E_F)$. In terms of ϵ we may now immediately evaluate $\langle n_{d+} \rangle$

$$\langle n_{d+} \rangle = \int_{-\infty}^{E_F} b_+{}^2 n(E)\, dE = \frac{1}{\pi} \int_{-\infty}^{E_F} \frac{\epsilon\, dE}{(E - E_{d-})^2 + \epsilon^2}$$

$$= \frac{1}{\pi} \operatorname{arccot} \frac{E_{d-} - E_F}{\epsilon} \tag{5.38}$$

We of course have the comparable expression for $\langle n_{d-} \rangle$. The equations that must be solved self-consistently are

$$\langle n_{d+} \rangle = \frac{1}{\pi} \operatorname{arccot} \frac{E_d - E_F + \langle n_{d-} \rangle U}{\epsilon}$$

$$\langle n_{d-} \rangle = \frac{1}{\pi} \operatorname{arccot} \frac{E_d - E_F + \langle n_{d+} \rangle U}{\epsilon} \tag{5.39}$$

The perturbation approach that we used earlier in treating transition-metal pseudopotentials corresponds to an expansion of the arccot in these expressions to lowest order in ϵ. The additional terms that make up the expansion of the arcot in Eq. (5.39) remove the divergence when the resonance is near the Fermi energy and the results are meaningful throughout the energy range.

These equations can most easily be solved graphically by plotting $\langle n_{d-} \rangle$ as a function of $\langle n_{d+} \rangle$ for each and seeking intersections. In Fig. 5.8 we have shown such a plot for two choices of parameters. In part *a* the *d*-state splitting is large and the unsplit *d* state lies well below the Fermi energy. It can readily be seen that the lowest energy occurs at the intersections with unequal moments; the other intersection corresponds to a local maximum in the energy. In the lowest-energy states the local moment is given by $\langle n_{d+} \rangle - \langle n_{d-} \rangle = \pm 0.644$ Bohr magnetons.

In part *b* the splitting parameter U is small and the unperturbed state lies close to the Fermi energy. Only one solution occurs and it corresponds to a vanishing magnetic moment.

We see that the formation of a local moment is a cooperative effect requiring an appropriate range of parameters if it is to be formed. It arose from the interaction term $U n_{d+} n_{d-}$ in the Hamiltonian. It is also interesting to note that though we began with a localized *d* state we have obtained nonintegral values for the local moment. Nonintegral moments are usual not only for magnetic impurities but also for pure magnetic elements such as

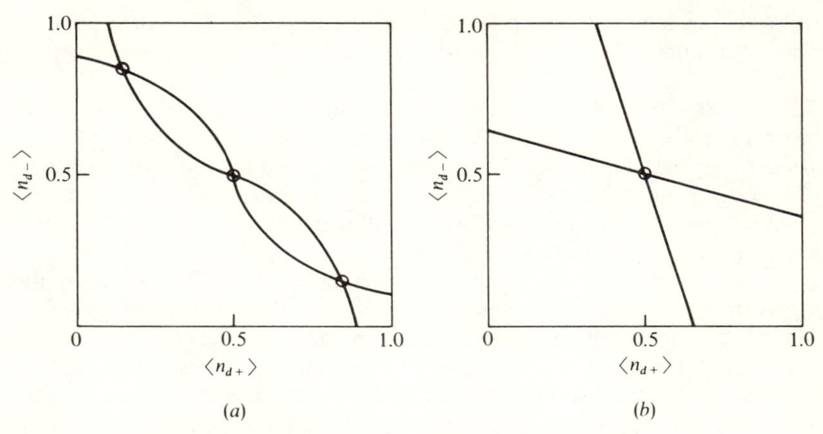

Fig. 5.8 *The self-consistency plot of* $\langle n_{d-} \rangle$ *against* $\langle n_{d+} \rangle$. (*a*) $U = 5\epsilon$ *and* $E_F - E_d = 2.5\epsilon$. *There are three solutions, with two corresponding to a local moment.* (*b*) $U = \epsilon$ *and* $E_F - E_d = 0.5\epsilon$. *The only solution is for* $\langle n_{d-} \rangle = \langle n_{d+} \rangle$ *and no local moment is formed.* [*After P. W. Anderson,* Phys. Rev., ***124****:41 (1961)*.]

iron. We could see how such nonintegral moments formed in pure materials when we discussed band ferromagnetism. We could alternatively describe the pure material by first imagining the formation of local moments on each atom and then the broadening of these moments into bands.

The latter point of view would have prevented confusion in the interpretation of optical properties and photoelectric emission in ferromagnetic metals at the Curie temperature. It was initially imagined that the formation of ferromagnetism at the Curie temperature was accompanied by a change in band structure from a nonmagnetic state, illustrated in Fig. 5.1*a*, to a magnetic-band structure, illustrated in Fig. 5.1*c*. Such a large change would be apparent in the optical properties but was not observed. It now seems clear that when the material was taken above the Curie temperature the local moments on each atom were retained but became disordered rather than ferromagnetically aligned. Thus even above the Curie temperature the density of states would correspond roughly to the band structure of Fig. 5.1*c* rather than 5.1*a*; the disorder would complicate the bands, as in liquids, but would not greatly influence their position in energy, their breadth, or the number of states they contain. Shifts in the density of states may be expected to be on the scale of $K\theta$ (of the order of 0.01 to 0.1 ev) rather than on the scale of U (of the order of 1 ev).

Finally we might mention that when a local moment is almost formed but U is slightly too weak, the system is very "soft" against the formation of local spin fluctuations; i.e., little energy is required. The corresponding zero-point fluctuations in spin density are large and are called *paramagnons.*

7.2 The Ruderman-Kittel interaction We have found above that under favorable circumstances an isolated impurity in a simple metal may form a local moment. This was a cooperative effect arising from the interaction of two electrons on a single ion. We may also imagine that with many local moments present there might be interaction between these and a cooperative alignment of these moments might occur.

There will of course be a direct magnetic interaction between the magnetic moments on individual impurities but this is always very small. There is also an interaction between moments through the electron-electron interaction. Physically this interaction between moments arises from a conduction electron scattering from one impurity, sensing the corresponding local spin, propagating with the spin information to the second impurity, and scattering from the second in a way dependent upon its local spin. To understand this effect in more detail we first consider the interaction of an electron with an individual impurity.

The s-d interaction. We look first at terms arising from the electron-electron interaction. We use the form Eq. (4.28) of Sec. 4.2 of Chap. IV.

$$V(\mathbf{r}_1,\mathbf{r}_2) = \tfrac{1}{2} \sum_{k_1k_2k_3k_4} \langle \mathbf{k}_4,\mathbf{k}_3|V|\mathbf{k}_2,\mathbf{k}_1\rangle c_{k_4}{}^+ c_{k_3}{}^+ c_{k_2} c_{k_1} \tag{5.40}$$

with

$$\langle \mathbf{k}_4,\mathbf{k}_3|V|\mathbf{k}_2,\mathbf{k}_1\rangle = \int d^3r_1\, d^3r_2\, \psi_{k_4}^*(\mathbf{r}_1)\psi_{k_3}^*(\mathbf{r}_2)V(\mathbf{r}_1 - \mathbf{r}_2)\psi_{k_2}(\mathbf{r}_2)\psi_{k_1}(\mathbf{r}_1) \tag{5.41}$$

Our interest is in processes in which a conduction electron scatters and a d state is occupied before and after. Thus one of the states $\mathbf{k}_4,\mathbf{k}_3$ must be a d state and also one of the states $\mathbf{k}_2$, $\mathbf{k}_1$. Thus we may take $\mathbf{k}_2$ to be a d state and $\mathbf{k}_1$ a conduction-band state and eliminate the factor $\tfrac{1}{2}$ in front. Since $V(\mathbf{r}_1 - \mathbf{r}_2)$ is independent of spin, the spin of $\mathbf{k}_4$ and $\mathbf{k}_1$ must be the same as must that of $\mathbf{k}_3$ and $\mathbf{k}_2$. There are eight types of terms, four of which are indicated in the accompanying table. The remainder are obtained by reversing all spins. Note that two different spatial matrix elements occur, though they would be equal for a contact interaction $V(\mathbf{r}_1 - \mathbf{r}_2) = V_0\,\delta(\mathbf{r}_1 - \mathbf{r}_2)$. It may be readily shown, using Eq. (5.20) of Sec. 4, that the operator

$$V_{sd} = \sum_{\substack{k,k'\\ \sigma,\sigma'}} c_{k'\sigma'}^{+}\left[-2V_1(\mathbf{k}',\mathbf{k})\mathbf{S}_d\cdot\mathbf{S}_c - \frac{V_1(\mathbf{k}',\mathbf{k})}{2}\delta_{\sigma\sigma'} + V_2(\mathbf{k}',\mathbf{k})\,\delta_{\sigma\sigma'}\right]c_{k\sigma} \tag{5.42}$$

Contributing matrix elements of the s-d interaction

$\langle \mathbf{k}_4,\mathbf{k}_3\lvert V\rvert \mathbf{k}_2,\mathbf{k}_1\rangle c_{k_4}{}^+c_{k_3}{}^+c_{k_2}c_{k_1}$	Times $c_{d+}{}^+c_{k+}{}^+\lvert 0\rangle$	Times $c_{d-}{}^+c_{k+}{}^+\lvert 0\rangle$
$V_1(\mathbf{k}',\mathbf{k})c_{d+}{}^+c_{k'+}{}^+c_{d+}c_{k+}$	$-V_1(\mathbf{k}',\mathbf{k})c_{d+}{}^+c_{k'+}{}^+\lvert 0\rangle$	0
$V_2(\mathbf{k}',\mathbf{k})c_{k'+}{}^+c_{d+}{}^+c_{d+}c_{k+}$	$+V_2(\mathbf{k}',\mathbf{k})c_{d+}{}^+c_{k'+}{}^+\lvert 0\rangle$	0
$V_1(\mathbf{k}',\mathbf{k})c_{d+}{}^+c_{k'-}{}^+c_{d-}c_{k+}$	0	$-V_1(\mathbf{k}',\mathbf{k})c_{d+}{}^+c_{k'-}{}^+\lvert 0\rangle$
$V_2(\mathbf{k}',\mathbf{k})c_{k'+}{}^+c_{d-}{}^+c_{d-}c_{k+}$	0	$V_2(\mathbf{k}',\mathbf{k})c_{d-}{}^+c_{k'+}\lvert 0\rangle$

Plus equivalent entries with all spins reversed.

yields exactly the same results as Eq. (5.40) when operating upon the states $c_{d+}{}^+c_{k+}{}^+\lvert 0\rangle$ and $c_{d-}{}^+c_{k+}{}^+\lvert 0\rangle$, as well as upon the two states obtained by reversing all spins. In order to show this we must clarify the operator algebra. The spin operator $\mathbf{S}_d$ operates on the spin coordinates of the d states and commutes with the c, c^+ and $\mathbf{S}_c$ which operate on the conduction-electron coordinates. The $\mathbf{S}_c$ may be expanded in components, $\mathbf{S}_c{}^+$, $\mathbf{S}_c{}^-$, and $\mathbf{S}_c{}^z$. Then for a term such as $c_{k'\sigma'}^+\mathbf{S}_c{}^+c_{k\sigma}$, the $\mathbf{S}_c{}^+$ raises the spin index σ. Using Eq. (5.12) we obtain

$$\sum_{\sigma,\sigma'} c_{k'\sigma'}^+\mathbf{S}_c{}^+c_{k\sigma} = c_{k'+}{}^+c_{k-} \tag{5.43}$$

Similarly,

$$\sum_{\sigma,\sigma'} c_{k'\sigma'}^+\mathbf{S}_c{}^-c_{k\sigma} = c_{k'-}{}^+c_{k+} \tag{5.44}$$

$$\sum_{\sigma,\sigma'} c_{k'\sigma'}^+\mathbf{S}_c{}^zc_{k\sigma} = \tfrac{1}{2}(c_{k'+}{}^+c_{k+} + c_{k'-}{}^+c_{k-}) \tag{5.45}$$

The final two terms in Eq. (5.42) are independent of the spin of either state, and conserve the spin of each. They have the same form as terms arising from a simple potential. The first term is of the form of Heisenberg exchange

$$V_{sd} = -\frac{1}{N}\sum_{\substack{k,k'\\ \sigma,\sigma'}} J(\mathbf{k}',\mathbf{k})c_{k'\sigma'}^+\mathbf{S}_d\cdot\mathbf{S}_c c_{k\sigma} \tag{5.46}$$

the so-called *s-d exchange interaction*, with J arising from the electron-electron interaction. We have inserted a factor of the reciprocal of the number of electrons present so that J will be independent of the size of the system. The J has the units of energy and should be of the order of electron volts. We will use this form in treating the interaction between moments on different ions and drop the scattering term in $V_2(\mathbf{k}',\mathbf{k})$ which does not contribute to the coupling.

Interaction between moments The original derivation of the interaction we wish to consider was directed at the coupling of nuclear moments[1] but used an interaction of the form Eq. (5.46) and is therefore entirely equivalent.

In the analysis to this point we have implicitly taken the local state at the origin. With two moments we must now note the position dependence of the matrix elements. We return to the matrix element of Eq. (5.41) taking the conduction-band states as plane waves and the d states as centered at $\mathbf{R}$. From the preceding table of matrix elements, we identify $V_1(\mathbf{k}',\mathbf{k})$ and write

$$V_1(\mathbf{k}',\mathbf{k}) = \frac{1}{\Omega}\int d^3r_1\, d^3r_2\, \psi_d^*(\mathbf{r}_1 - \mathbf{R})e^{-ik'\cdot r_2}\, V(\mathbf{r}_1 - \mathbf{r}_2)\psi_d(\mathbf{r}_2 - \mathbf{R})e^{ik\cdot r_1}$$

$$= e^{-i(k'-k)\cdot R}\,\frac{1}{\Omega}\int d^3r_1\, d^3r_2\, \psi_d^*(\mathbf{r}_1 - \mathbf{R})e^{-ik'\cdot(r_2 - R)}\, V(\mathbf{r}_1 - \mathbf{r}_2)$$

$$\psi_d(\mathbf{r}_2 - \mathbf{R})e^{ik\cdot(r_1 - R)}$$

The integral is now independent of $\mathbf{R}$ and the spatial dependence comes from the leading phase factor. Thus we take matrix elements

$$J(\mathbf{k}',\mathbf{k};\mathbf{R}) = e^{-i(k'-k)\cdot R}\, J(\mathbf{k}',\mathbf{k})$$

which are the same as before for $\mathbf{R} = 0$.

We now seek the shift in energy of the system to the lowest order in V_{sd} which gives an interaction. This will clearly be in second order, with one matrix element from each impurity, and will be of the form

$$\delta E = \sum_i \frac{\langle 0|V_{sd}(\mathbf{R}_1)|i\rangle\langle i|V_{sd}(\mathbf{R}_2)|0\rangle}{E_0 - E_i} + \sum_i \frac{\langle 0|V_{sd}(\mathbf{R}_2)|i\rangle\langle i|V_{sd}(\mathbf{R}_1)|0\rangle}{E_0 - E_i} J$$

where $\mathbf{R}_1$ and $\mathbf{R}_2$ are the positions of the two impurities, $|0\rangle$ is the initial ground state, and the $|i\rangle$ are intermediate states.

The two sums give identical contributions so we consider only the first and multiply by 2. We obtain, using Eq. (5.46),

$$\delta E = \frac{2}{N^2}\sum_{\substack{k,k'\\ \sigma,\sigma'}} \frac{J(\mathbf{k},\mathbf{k}')J(\mathbf{k}',\mathbf{k})}{\epsilon_k - \epsilon_{k'}}\, e^{-i(k'-k)\cdot(R_2 - R_1)}$$

$$\langle 0|c_{k\sigma}{}^{+}(\mathbf{S}_{d_1}\cdot\mathbf{S}_c)c_{k'\sigma}c_{k'\sigma'}^{+}(\mathbf{S}_{d_2}\cdot\mathbf{S}_c)c_{k\sigma}|0\rangle$$

where the $\mathbf{S}_{d_1}$ and $\mathbf{S}_{d_2}$ are the spin operators for the local states at $\mathbf{R}_1$ and $\mathbf{R}_2$.

The product of annihilation and creation operators restricts the sum over $\mathbf{k}$ to states below the Fermi energy and $\mathbf{k}'$ to states above. However,

[1] M. A. Ruderman and C. Kittel, *Phys. Rev.*, **96**:99 (1954).

we may let $\mathbf{k}'$ run over all wavenumbers since for $\mathbf{k}'$ less than the Fermi energy each pair of states $\mathbf{k}_1$, $\mathbf{k}_2$ will appear once as $\mathbf{k}$, $\mathbf{k}'$ and once as $\mathbf{k}'$, $\mathbf{k}$ and, because of the energy denominator, will cancel. (Thus the Pauli principle does not enter the results in second order; this will not be true in the scattering calculation in the next section.)

The product of spin operators is best evaluated by expanding in the form

$$(\mathbf{S}_d \cdot \mathbf{S}_c) = [S_d^z S_c^z + \tfrac{1}{2}(S_d^+ S_c^- + S_d^- S_c^+)]$$

as in Eq. (5.15). The conduction-electron states in $|0\rangle$ have well-defined spin. In operating, e.g., on a spin-up conduction-electron state the S_c^+ in the second factor $(\mathbf{S}_{d_1} \cdot \mathbf{S}_c)(\mathbf{S}_{d_2} \cdot \mathbf{S}_c)$ gives zero and the S_c^- contributes only with the S_c^+ in the first factor giving a contribution $\tfrac{1}{4} S_{d_1}^- S_{d_2}^+ (S_c^+ S_c^-) = \tfrac{1}{4} S_{d_1}^- S_{d_2}^+$. Similarly, the S_c^z factors contribute together to give $S_{d_1}^z S_{d_2}^z S_c^z S_c^z = \tfrac{1}{4} S_{d_1}^z S_{d_2}^z$. Adding the contribution for spin-down conduction-electron states we obtain

$$\begin{aligned}&\langle 0|(\mathbf{S}_{d_1} \cdot \mathbf{S}_c)(\mathbf{S}_{d_2} \cdot \mathbf{S}_c)|0\rangle \\ &\quad = \langle 0|\tfrac{1}{4} S_{d_1}^- S_{d_2}^+ + \tfrac{1}{4} S_{d_1}^+ S_{d_2}^- + \tfrac{1}{2} S_{d_1}^z S_{d_2}^z|0\rangle = \tfrac{1}{2}\langle 0|\mathbf{S}_{d_1} \cdot \mathbf{S}_{d_2}|0\rangle\end{aligned}$$

This factor depends only upon the spins of the d states and may be taken from under the sum over $\mathbf{k}$ and $\mathbf{k}'$. We replace $\mathbf{k}'$ by $\mathbf{k} + \mathbf{q}$ and obtain

$$\delta E = \langle 0|\mathbf{S}_{d_1} \cdot \mathbf{S}_{d_2}|0\rangle \sum_q \frac{e^{-iq\cdot(R_2 - R_1)}}{N^2} \sum_{k<k_F} \frac{J(\mathbf{k},\mathbf{k}+\mathbf{q})J(\mathbf{k}+\mathbf{q},\mathbf{k})}{\epsilon_k - \epsilon_{k+q}} \tag{5.47}$$

The sum over spins on $\mathbf{k}$ has been performed.

Except for the spin-dependent factor in front, Eq. (5.47) is the same form as the indirect interaction between ions in the pseudopotential method given by Eqs. (4.63) and (4.64) in Sec. 5 of Chap. IV. To make the identification we imagine a monovalent metal so that the number of electrons N is also the number of ions. The structure factor is evaluated by summing over the two impurities, $S^*(\mathbf{q})S(\mathbf{q}) = (1/N^2)(2 + e^{-iq\cdot(R_2 - R_1)} + e^{-iq\cdot(R_1 - R_2)})$, giving the same structure-dependent factors as in Eq. (5.47). Then if $J(\mathbf{k}+\mathbf{q},\mathbf{k})/2$ is identified with $\langle \mathbf{k}+\mathbf{q}|w|\mathbf{k}\rangle$ the equations are equivalent.

Having made this identification we may immediately describe the coupling between moments as an *indirect exchange interaction* between moments depending upon separation R. The asymptotic form at large R is obtained immediately from Eq. (4.68) of Sec. 6.2 of Chap. IV.

$$V_{\text{ind}}(R) \sim \frac{9\pi}{4E_F} J(-k_F,k_F)^2 \frac{\cos 2k_F R}{(2k_F R)^3} \langle 0|\mathbf{S}_{d_1} \cdot \mathbf{S}_{d_2}|0\rangle$$

or a term in the ion Hamiltonian of the form $\tfrac{1}{2}\sum_{ij} V_{\text{ind}}(\mathbf{R}_i - \mathbf{R}_j)$ with

$\langle 0|\mathbf{S}_{d_1} \cdot \mathbf{S}_{d_2}|0\rangle$ replaced by $\mathbf{S}_{d_1} \cdot \mathbf{S}_{d_2}$. In the corresponding result given by Kittel,[1] a wavenumber-independent coupling parameter J was introduced, related to ours by $J = \Omega J(\mathbf{k}',\mathbf{k})/N$.

We have found an oscillatory exchange coupling between local moments which, comparing with the pseudopotential interaction, we see will typically favor parallel spins at near-neighbor distances. Because of the oscillations, however, this interaction can also give rise to complicated antiferromagnetic structures.[2]

7.3 The Kondo effect We saw in Sec. 7.2 how the interaction between conduction electrons and local moments could be represented by a Heisenberg exchange interaction. We saw further how this interaction shifted the energies of the electrons and gave rise to an indirect interaction between two local moments in the metal. It is of course also true that this interaction will cause a scattering of conduction electrons and a contribution to the resistivity. This contribution to the electron scattering is distinctive in that, in contrast to scattering by an ordinary potential, the electron spin may be flipped in the process. Calculated in lowest order, however, this simply appears as an additional contribution to the resistivity.

Kondo[3] noted that the situation was not so simple if the scattering were carried to higher order. We have noted explicitly in our derivation of the Ruderman-Kittel interaction that the application of the Pauli principle to the intermediate states had no effect on our second-order energy. This is a common feature of solid state calculations. However, in the calculation of scattering by local moments the corresponding effects of the Pauli principle do not cancel out. Kondo, in fact, found that they led in second order to a divergence of the scattering rate for electrons very near the Fermi energy which in turn leads to a divergent resistivity as the temperature approaches zero. This gave a mechanism for the existence of a long-known and puzzling minimum in the resistivity as a function of temperature (see Fig. 5.9) in dilute alloys containing local moments.

Of course the appearance of a divergence in the second order of perturbation theory suggests the general failure of perturbation theory for the corresponding phenomena and brings into question the meaning of the calculation itself. Subsequent to Kondo's work a sizable effort has been expended, using sophisticated techniques, to correctly evaluate the scattering in such situations. However, the presence of the divergence, like the instability against Cooper pairs which we will discuss in the following section on superconductivity, reflects the essence of the problem and should be discussed briefly.

[1] Kittel, *op. cit.*, p. 364.

[2] D. Mattis and W. E. Donath, *Phys. Rev.*, **128**:1618 (1962).

[3] J. Kondo, *J. Appl. Phys.*, **37**:1177 (1966).

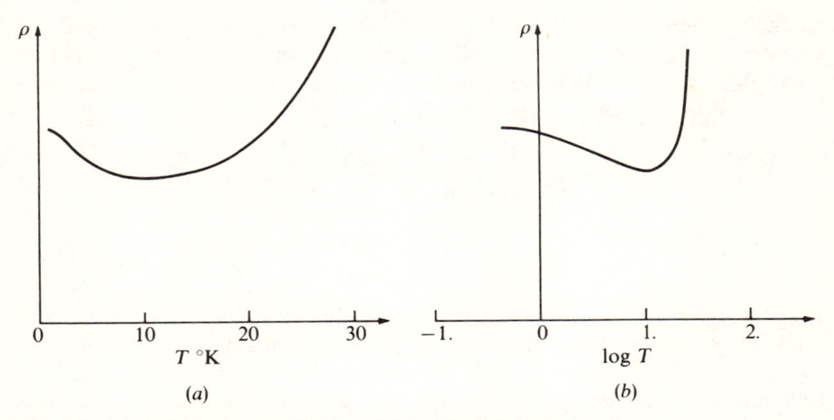

Fig. 5.9 *(a) A schematic representation of the minimum in resistance ρ as a function of temperature now attributed to the Kondo effect. In (b) the resistance is plotted as a function of the logarithm of temperature, showing the linearity in log T. At the lowest temperatures the resistance drops below the log T line.*

We again use the *s-d* interaction given in Eq. (5.46). For simplicity, however, we neglect the dependence of J upon $\mathbf{k}'$ and $\mathbf{k}$. In addition we write out the product $\mathbf{S}_d \cdot \mathbf{S}_c$ to obtain

$$V_{sd} = -\frac{J}{N} \sum_{\substack{k,k' \\ \sigma,\sigma'}} c_{k'\sigma'}^{+} \left[\tfrac{1}{2}(S_d^{+} S_c^{-} + S_d^{-} S_c^{+}) + S_d^{z} S_c^{z} \right] c_{k\sigma}$$

We wish to write out the matrix element, to second order, which couples two states $\mathbf{k}$, σ and $\mathbf{k}'\sigma'$. This matrix element may be written

$$\langle \mathbf{k}',\sigma' | M | \mathbf{k},\sigma \rangle = \langle \mathbf{k}',\sigma' | \left(V_{sd} + \sum_i \frac{V_{sd}|i\rangle\langle i| V_{sd}}{E_0 - E_i} \right) | \mathbf{k},\sigma \rangle \tag{5.48}$$

where the sum is over intermediate states. The algebra very quickly becomes complicated but will be simplified somewhat if in both initial and final states the conduction-electron spin is up.

We note first that there is a first-order contribution to the matrix element given by $-JS_d^z/2N$. In this term and in others we will retain the d-state spin operators which must later be evaluated in terms of the spin state of the d electron during the scattering event. We eliminate the spin operator for the *conduction* electron in each term using Eqs. (5.43) to (5.45).

Each second-order term may be labeled by the intermediate conduction-band state $\mathbf{k}''$, σ'' which enters the intermediate state. Such a state may enter in two ways; if the state $\mathbf{k}''\sigma''$ is initially unoccupied it may become occupied in the first step and emptied in the second; if it is initially occupied it may be emptied in the first step and refilled in the second. These two possibilities correspond to two different intermediate states. It is convenient to write these two contributions separately.

$$\sum_i \frac{V_{sd}|i\rangle\langle i|V_{sd}}{E_0 - E_i} = \frac{J^2}{N^2} \sum_{k'',\sigma''} \left(\frac{c_{k'+}^{+} \mathbf{S}_d \cdot \mathbf{S}_c c_{k''\sigma''} c_{k''\sigma''}^{+} \mathbf{S}_d \cdot \mathbf{S}_c c_{k+}}{\epsilon_k - \epsilon_{k''}} + \frac{c_{k''+}^{+} \mathbf{S}_d \cdot \mathbf{S}_c c_{k+} c_{k'+}^{+} \mathbf{S}_d \cdot \mathbf{S}_c c_{k''\sigma''}}{\epsilon_{k''} - \epsilon_{k'}} \right) \tag{5.49}$$

Both types of second-order terms take us between the same initial and final states and we have included all terms transferring an electron from $\mathbf{k}+$ to $\mathbf{k}'+$. We have noted that the relative energies of the intermediate and initial states depend only upon the wavenumbers of the occupied conduction-electron states in the two. In the scattering calculation we will be interested only in matrix elements between initial and final states of the same energy and may therefore write $\epsilon_{k''} - \epsilon_{k'} = -(\epsilon_k - \epsilon_{k''})$.

Let us look first at intermediate states of spin up. In all cases the only contributing term in $\mathbf{S}_d \cdot \mathbf{S}_c$ is $S_d^z S_c^z$ [see, for example, Eqs. (5.43) to (5.45)]. The evaluation of the corresponding terms in Eq. (5.49) is immediate.

$$\frac{1}{4} \frac{J^2}{N^2} \sum_{k''} \frac{S_d^z S_d^z}{\epsilon_k - \epsilon_{k''}} (c_{k''+} c_{k''+}^{+} + c_{k''+}^{+} c_{k''+}) c_{k'+}^{+} c_{k+} \tag{5.50}$$

We have again used Eqs. (5.43) to (5.45) to eliminate the conduction-electron spin operators and have commuted annihilation and creation operators, noting that $\mathbf{k}$, $\mathbf{k}'$, and $\mathbf{k}''$ are all different. A change in sign due to the difference in energy denominators was cancelled by a change in sign in commuting $c_{k'+}^{+}$ and c_{k+}. Just as in the calculation of energy in second order and scattering by potentials in second order, the dependence on the occupation of the intermediate state disappears since $c_{k''+} c_{k''+}^{+} + c_{k''+}^{+} c_{k''+} = 1$.

We must next evaluate the contribution to Eq. (5.49) for intermediate states with spin down. Then the only contribution from each product $\mathbf{S}_d \cdot \mathbf{S}_c$ is from either $\frac{1}{2} S_d^+ S_c^-$ or $\frac{1}{2} S_d^- S_c^+$. For this case we obtain

$$\frac{1}{4} \frac{J^2}{N^2} \sum_{k''} \frac{1}{\epsilon_k - \epsilon_{k''}} (S_d^- S_d^+ c_{k''-} c_{k''-}^{+} + S_d^+ S_d^- c_{k''-}^{+} c_{k''-}) c_{k'+}^{+} c_{k+} \tag{5.51}$$

Here, because S_d^- and S_d^+ do not commute, we cannot eliminate the dependence upon the occupation of the intermediate state. The difficulty occurs only in scattering events which involve a spin flip in going to the intermediate step.

If we considered instead matrix elements between initial and final states with opposite spins, a similar effect would have arisen from the fact that S_d^+ and S_d^z do not commute or that S_d^- and S_d^z do not commute.

Equation (5.51) may be written in more convenient form by noting, for example, that $S_d^- S_d^+ = \frac{1}{2}(S_d^- S_d^+ + S_d^+ S_d^-) - S_d^z$. We obtain

$$\frac{1}{4}\frac{J^2}{N^2}\sum_{k''}\frac{1}{\epsilon_k - \epsilon_{k''}}\left[\frac{S_d^+ S_d^- + S_d^- S_d^+}{2} + S_d^z(c_{k''-}^+ c_{k''-} - c_{k''-} c_{k''-}^+)\right]c_{k'+}^+ c_{k+} \tag{5.52}$$

The contributions from Eqs. (5.50) and (5.52) may be combined and the first-order term added to give the matrix element of Eq. (5.48),

$$\langle \mathbf{k}'+|M|\mathbf{k}+\rangle = -\frac{JS_d^z}{2N} + \frac{1}{4}\frac{J^2}{N^2}\sum_{k''}\frac{\mathbf{S}_d\cdot\mathbf{S}_d + S_d^z[2f(\mathbf{k}'') - 1]}{\epsilon_k - \epsilon_{k''}}$$

where we have replaced the number operator $c_{k''-}^+ c_{k''-}$ by its expectation value, the Fermi function, $f(\mathbf{k}'')$.

There are of course other contributions to the matrix element between $\mathbf{k}+$ and $\mathbf{k}'+$ such as those from the scattering potential. The term in $\mathbf{S}_d \cdot \mathbf{S}_d$ enters in precisely the same way as do the potential terms and we discard these together. For the terms we retain, the scattering rate is proportional to

$$|\langle \mathbf{k}'+|M|\mathbf{k}+\rangle|^2 = \left(\frac{J}{2N}\right)^2 \langle S_{dz}\rangle^2\left[1 - \frac{J}{2N}\sum_{k''}\frac{2f(\mathbf{k}'') - 1}{\epsilon_k - \epsilon_{k''}}\right]^2$$

to third order in J.

We may now see how this result leads to a divergence. If there were no dependence on $f(\mathbf{k}'')$ a convergent result for the sum over $\mathbf{k}''$ could be obtained by taking principal values across the singularity at $\epsilon_{k''} = \epsilon_k$. However, the numerator in that sum changes sign across the Fermi energy, and if the initial energy ϵ_k lies at the Fermi energy the sum diverges at $T = 0$. This would correspond to an infinite scattering rate as the temperature approached zero.

We may see the form of the divergence by replacing the sum by an integral. We measure all energies from the Fermi energy and write the density per unit energy of spin-down electrons as $N(0)/2$. We let the initial state energy differ from the Fermi energy by $\delta\epsilon$. Evaluating the Fermi function at $T = 0$ the term in $\sum_{k''}$ becomes

$$-\frac{JN(0)}{4N}\int\frac{d\epsilon''[2f(\epsilon'') - 1]}{\epsilon_k - \epsilon''} \approx -\frac{JN(0)}{4N}\left(\int^0 \frac{d\epsilon''}{\delta\epsilon - \epsilon''} - \int_0 \frac{d\epsilon''}{\delta\epsilon - \epsilon''}\right)$$

$$\approx \frac{JN(0)}{2N}\ln\left|\frac{\delta\epsilon}{E_F}\right| \tag{5.53}$$

where we have taken the distant limits to be of the order of the Fermi energy. Finally we note that the electrons of interest will lie within an energy of the order of KT of the Fermi energy at finite temperature and that $N(0)/N$ is of the order of E_F^{-1}. Thus we may expect a divergence in the resistivity of the form

$$R \approx R_0\left[1 + \frac{J}{E_F}\ln\left(\frac{KT}{E_F}\right)\right]$$

This is the divergent resistivity at low temperatures which is called the Kondo effect. It is difficult to state the physical origin in a simple way. We have seen that it is a divergence that originates from the Pauli principle when the scattering center contains a degree of freedom, which in this case is the spin orientation.

We should note also before leaving this effect that we may expect a strong dependence of the divergence upon magnetic fields, though the point is a tricky one. Let us imagine a sufficiently strong field that the moments on all d states are aligned. The spin-flip scattering can occur only with a reduction in the spin component of the d state as the conduction-electron spin flips from down to up. If the gyromagnetic ratio is the same for the local state as for the conduction electron, the two changes in magnetic energy cancel and the energy denominators remain the same. However, the *conduction-electron energy* is lowered while the Fermi energy is the same for spin-up and spin-down electrons. Thus for an electron initially at the Fermi energy the occupation of the intermediate state enters as if the initial state were displaced from the Fermi energy by the Zeeman energy, $\delta\epsilon = -\mu_0 H$. We see from Eq. (5.53) that this reduces the divergence just as does a change in temperature. This dependence of the singularity upon the magnetic field is observed experimentally.

There is some mathematical similarity between the treatment of the Kondo effect and the treatment of superconductivity. For some time it was thought that a cooperative phase transition must occur at a sufficiently low temperature in a system showing the Kondo effect. It is now believed that the similarity is only formal and that no phase transition occurs.

B. SUPERCONDUCTIVITY

Three general references on the theory of superconductivity might be mentioned. Rickaysen[1] is most closely at the level we use here. Schrieffer[2]

[1] G. Rickaysen, "The Theory of Superconductivity," Interscience-Wiley, New York, 1965.

[2] J. R. Schrieffer, "Theory of Superconductivity," Benjamin, New York, 1964.

tends to be more formal and is directed at discussion of the fundamentals of the microscopic theory. DeGennes[1] includes a wide range of applications with extensive use of the phenomenological theory.

We will focus our attention first on the microscopic theory of superconductivity due to Bardeen, Cooper, and Schrieffer.[2] It is now well known that the superconducting state arises from the interaction between the electrons and the lattice vibrations in metals. This was not obvious, however, from the early experiments in superconductivity. The superconducting state was first discovered in 1911 but it was not until 1950 that Froehlich recognized that the electron-phonon interaction was involved. At about the same time the isotope effect, a dependence of the superconducting temperature upon the isotope mass of the nuclei in the metal, was observed experimentally, confirming Froehlich's speculation. Early attempts by Bardeen and Froehlich to account for superconductivity on the basis of this interaction did not succeed. Both were attempts to obtain the superconducting state from the normal ground state by perturbation theory. It is now widely recognized that the superconducting state is not obtainable in this way.

In analogy we may note that the ferromagnetic state is not obtainable from the normal paramagnetic state by perturbation theory. We found in particular that the unmagnetized state was also an eigenstate of the system in our discussion of band ferromagnetism. This, however, was an unstable state and the energy could be lowered by successively aligning the electron spins. The corresponding instability in the formation of superconductors was noted by Cooper[3] and we will discuss that instability first.

8. Cooper Pairs

In our discussion of the electron-phonon interaction in second-quantized form we found that this interaction gave rise to an effective interaction between the electrons themselves. We now note that such an effective interaction between electrons can cause an instability of the normal (nonsuperconducting) state of the electrons. We will consider two electrons in the system which interact with each other and will neglect all other interactions between electrons. We can then construct the state for these two electrons in terms of their coordinates alone. All other electrons are assumed to form a state $\prod_{k<k_F} c_k^+|0\rangle$.

The interaction between electrons is assumed to conserve total

[1] P. G. deGennes, "Superconductivity of Metals and Alloys," Benjamin, New York, 1966.

[2] J. Bardeen, L. N. Cooper, and J. R. Schrieffer, *Phys. Rev.*, **108**:1175 (1957).

[3] L. N. Cooper, *Phys. Rev.*, **104**:1189 (1956).

momentum and total spin. This is true of the effective interaction between electrons that we discussed earlier. The total spin of the system must be either 0 or 1. If it is 0, the spin states are antisymmetric and the spatial wavefunctions must therefore be symmetric. If the total spin is 1, then the spatial wavefunction must be antisymmetric. It turns out that a lower-energy state is obtained with a symmetric spatial wavefunction; therefore in the lowest-energy state the spins of the two electrons will be antiparallel. In addition it is clear that the lowest-energy state of the two electrons will have vanishing total momentum. Thus if the electron coordinates are written in terms of the relative coordinate, $\rho = \mathbf{r}_1 - \mathbf{r}_2$ and the center of mass coordinate, the wavefunction will depend only upon the former. We may then write the spatial wavefunction as a single function

$$\psi(\mathbf{r}_1,\mathbf{r}_2) \rightarrow \psi(\boldsymbol{\rho})$$

Now in the *normal* (not superconducting) ground state of the noninteracting system this wavefunction will be $\Phi = \Omega^{-1} e^{i(k \cdot r_1 - k \cdot r_2)}$. Since the other electrons occupy all of the plane-wave states with $k < k_F$, the lowest energy for the pair will be $2E_F$.

To lower the energy of the *interacting* system, we try writing the pair of interest as a superposition of pairs with opposite momentum and opposite spin, $\mathbf{k}\uparrow$, $-\mathbf{k}\downarrow$, all of which lie outside the Fermi surface. This will leave all other electronic states the same. In the absence of an electron-electron interaction the expectation value of the energy of this state will clearly be higher because of the extra kinetic energy associated with these states outside the Fermi surface. However, in the presence of interactions we may obtain a state of lower energy than the normal state.

We show this by writing a general expansion of the wavefunction as a sum of pairs.

$$\psi = \sum_{k' > k_F} a_{k'} e^{ik' \cdot \rho} = \sum_{k' > k_F} a_{k'} e^{ik' \cdot r_1} e^{-ik' \cdot r_2}$$

We substitute this into the energy eigenvalue equation, which we write

$$(H_0 + V_{\text{elel}})\psi = E\psi$$

or

$$(E - H_0)\psi = V\psi$$

where H_0 is the Hamiltonian without interaction, which may be just the kinetic energy, and $V = V_{\text{elel}}$ is the effective interaction between electrons. Substituting for ψ we multiply on the left by $(e^{ik \cdot r_1} e^{-ik \cdot r_2})^*/\Omega$ and integrate over $\mathbf{r}_1$ and $\mathbf{r}_2$. We obtain

$$(E - 2\epsilon_k)a_k = \sum_{k'} \langle \mathbf{k}, -\mathbf{k} | V | \mathbf{k}', -\mathbf{k}' \rangle a_{k'} \tag{5.54}$$

where

$$\langle \mathbf{k}, -\mathbf{k}|V|\mathbf{k}', -\mathbf{k}'\rangle = \frac{1}{\Omega^2}\int\int d^3r_1\, d^3r_2 e^{i(k'-k)\cdot(r_1-r_2)}\, V(\mathbf{r}_1 - \mathbf{r}_2)$$

and the ϵ_k are the one-electron eigenvalues of H_0. This is to be solved for the a_k and the state of lowest energy sought.

This equation cannot be solved for a general interaction, nor in fact for an interaction of the form we have found previously. However, we may study the properties of the equation by assuming a very simple form for the interaction. In particular we assume a separable form,

$$\langle \mathbf{k}, -\mathbf{k}|V|\mathbf{k}', -\mathbf{k}'\rangle = \lambda w_{k'} w_k$$

The λ will be negative for an attractive interaction and positive for a repulsive interaction. For this simple interaction we may see the form of the solution. Equation (5.54) now takes the form

$$(E - 2\epsilon_k)a_k = \lambda w_k \sum_{k'} w_{k'} a_{k'} \tag{5.55}$$

The result of the summation over $\mathbf{k}'$ is independent of $\mathbf{k}$ so we may write it as a constant

$$C = \sum_{k'} w_{k'} a_{k'}$$

In terms of this constant we may solve immediately for a_k.

$$a_k = \frac{\lambda w_k C}{E - 2\epsilon_k}$$

We may now evaluate C in terms of these a_k.

$$C = \sum_k w_k a_k = C\lambda \sum_k \frac{w_k^2}{E - 2\epsilon_k}$$

The C's cancel on the two sides and we have obtained a condition for the solutions of Eq. (5.55) analogous to the secular equation for the solutions of simultaneous equations.

$$1 = \lambda \sum_k \frac{w_k^2}{E - 2\epsilon_k} \tag{5.56}$$

Again the sum is over $k > k_F$.

In order to solve Eq. (5.56) explicitly we need to know w_k. However, we may see the form of the solutions graphically by simply assuming that w_k is a slowly varying function of k and plotting the right- and left-hand sides. We note that Eq. (5.56) is of essentially the same form as Eq. (4.21) in Sec.

3 of Chap. IV, giving the frequency of localized vibrational modes in crystals containing defects, and that the plot based upon Eq. (5.56) is very similar to Fig. 4.12 given there. We will not repeat the graphical argument but will obtain the corresponding conclusions directly from Eq. (5.56).

If the interaction is repulsive corresponding to $\lambda > 0$, there can be no solution with E less than the lowest energy $2E_F$ for the noninteracting system since then every term on the right in Eq. (5.56) would be negative. As for the vibrational frequencies with a heavy impurity, *every* excitation energy is slightly shifted, in this case raised. If, on the other hand, the interaction is attractive, $\lambda < 0$, a value of E less than $2E_F$ makes every term on the right in Eq. (5.56) positive and a solution exists well below the excitation energy of the noninteracting system. In addition, the energy of each of the higher excited states is lowered slightly. The one state that is pulled off corresponds to a bound electronic pair, called a *Cooper pair*. We have gained more than enough energy from the interaction to counterbalance the increase in kinetic energy. This represents an instability of the system. For the model problem we have found a state of lower energy than that of the normal state.

We may carry this somewhat further and evaluate the binding energy for the Cooper pair. If w_k were independent of k the integrals would diverge. We therefore assume it to be a constant over a range of states near the Fermi surface but to drop discontinuously to zero at a cutoff value $\epsilon_k - E_F = E_c$. This cutoff energy is taken to be very small in comparison to the Fermi energy. Thus the sum over wavenumber may be replaced by an integral over electron energy (which is legitimate since the E of interest is outside the range of integration), using a density of states evaluated at the Fermi surface. Measuring our energies from the Fermi energy we obtain

$$1 = \lambda \sum_k \frac{{w_k}^2}{E - 2\epsilon_k} = \lambda {w_k}^2 \frac{n(E_F)}{2} \int_0^{E_c} \frac{d\epsilon}{E - 2\epsilon}$$

$$= -\lambda {w_k}^2 \frac{n(E_F)}{4} \ln \left| \frac{E - 2E_c}{E} \right|$$

Exponentiating both sides we obtain

$$\left| \frac{E - 2E_c}{E} \right| = e^{-4/\lambda {w_k}^2 n(E_F)} \tag{5.57}$$

We have written the density of coupled states as $n(E_F)/2$ since the matrix element couples only pairs of the same spin configuration and therefore only half of the total number of states. Since E is negative for the bound pair, the left-hand side of Eq. (5.57) may be written $(|E| + 2E_c)/|E|$. Again $\lambda {w_k}^2$

is a measure of the strength of the electron-electron interaction and we take it to be negative. We can expect this interaction to be very weak; thus the exponent will be a very large positive number and therefore $E_c \gg |E|$. We can write the binding energy,

$$|E| = 2E_c e^{-4/|\lambda w_k^2| n(E_F)}$$

Note that this binding energy cannot be written as an expansion in powers of the interaction λw_k^2. Thus the state we have found is not obtainable by perturbation theory. We may also note that a_k has its largest value at the Fermi surface and drops off at higher energies. In our model it drops exactly to zero for $\epsilon_k > E_c$, the cutoff energy. (Again, we measure ϵ_k from the Fermi energy.) We can also show that the energy E is less negative for states with net momentum. Thus the stationary pair represents the strongest instability.

In demonstrating an instability in the electron gas it suffices to find any state of the system which has lower energy than the ordinary normal state. These Cooper pairs represent such an instability of our model system. We wish now to find the ground state of the full system. This is accomplished approximately in the microscopic theory of superconductivity.

9. Bardeen-Cooper-Schrieffer (BCS) Theory

In the discussion above we considered only the interaction between a single pair of electrons, though of course all electrons interact. We might at first attempt to put every electron pair in a Cooper pair state. However, this clearly violates the exclusion principle. We therefore must proceed more carefully and construct the many-electron state from the beginning.

We will give the BCS variational solution of this problem. Another method is the spin-analog method due to Anderson.[1]

It is necessary first to rewrite the electron-electron interaction in the form of the BCS reduced Hamiltonian, including only interaction between the two electrons of a pair, these electrons having opposite momentum and spin. We begin with the electron-electron interaction that we obtained in Eq. (4.58) in Sec. 4.3 of Chap. IV, and include only terms coupling such pairs. It is convenient in doing this to add to each term the conjugate-transpose term. This includes each term twice so we must divide by 2.

$$V_{\text{elel}} = \frac{1}{2N} \sum_{k,q} \left(\frac{V_q^* V_q}{\epsilon_k - \epsilon_{k+q} - \hbar\omega_q} c_{k+q}^+ c_{-k-q}^+ c_{-k} c_k + \frac{V_q^* V_q}{\epsilon_k - \epsilon_{k+q} - \hbar\omega_q} c_k^+ c_{-k}^+ c_{-k-q} c_{k+q} \right)$$

[1] Kittel, *op. cit.*, p. 157.

We allow the summation over $\mathbf{k}$ to include all wavenumbers by selecting always the state $\mathbf{k}$ to have spin up and the state $-\mathbf{k}$ to have spin down, then we have not counted any pair twice. These two terms can be combined if we write $\mathbf{k} + \mathbf{q} = \boldsymbol{\kappa}$ in the second term and change the sign of the dummy variable $\mathbf{q}$. That sum becomes

$$\frac{1}{2N} \sum_{\kappa,q} \frac{V_q^* V_q}{\epsilon_{\kappa+q} - \epsilon_\kappa - \hbar\omega_q} c_{\kappa+q}^+ c_{-\kappa-q}^+ c_{-\kappa} c_\kappa$$

Changing the variable $\boldsymbol{\kappa}$ to $\mathbf{k}$ we may combine the two terms directly.

$$V_{\text{elel}} = \sum_{k,q} \frac{V_q^* V_q \hbar\omega_q / N}{(\epsilon_k - \epsilon_{k+q})^2 - (\hbar\omega_q)^2} (c_{-k-q} c_{k+q})^+ (c_{-k} c_k) \tag{5.58}$$

We have collected our creation and annihilation operators together to give *pair-annihilation* and *-creation* operators. The operator

$$b_k = c_{-k} c_k$$

annihilates a pair of electrons and the operator

$$b_k^+ = c_k^+ c_{-k}^+ = (c_{-k} c_k)^+$$

creates the corresponding pair. In this form we see that the electron-electron interaction is explicitly attractive if the two coupled states are sufficiently near each other in energy; i.e., if both are sufficiently close to the Fermi energy. It is the attractive interaction in this energy range that leads to the instability. The energy range over which the interaction may be attractive is restricted to something of the order of the Debye energy $\hbar\omega_D$. In the simplest approximation the coefficient of the pair-annihilation and -creation operators in Eq. (5.58) is taken to be a constant over this energy range and zero beyond it in analogy with our approximate treatment of the interaction giving rise to Cooper pairs.

Here we add the effect of the coulomb electron-electron interaction. Note that the corresponding matrix element is $4\pi e^2/q^2\Omega$, also inversely proportional to N. It will simplify notation to absorb the $1/N$ in the net coefficient. We write

$$V_{k+q,k} = \frac{V_q^* V_q \hbar\omega_q / N}{(\epsilon_k - \epsilon_{k+q})^2 - (\hbar\omega_q)^2} + V_{k+q,k}(\text{coulomb})$$

and add the kinetic energy of the electrons to obtain the *BCS reduced Hamiltonian.*

$$\mathscr{H} = \sum_k \epsilon_k c_k^+ c_k + \sum_{k,k'} V_{k'k} b_{k'}^+ b_k$$

We note that if we assume a state that is completely paired, $|\psi\rangle = \prod_k b_k^+|0\rangle$, the BCS reduced Hamiltonian will give us matrix elements only with other states that are completely paired. In the ground state of the system only such states will be necessary and the BCS reduced Hamiltonian may be rewritten

$$\mathscr{H} = \sum_k 2\epsilon_k b_k^+ b_k + \sum_{k,k'} V_{k'k} b_{k'}^+ b_k$$

9.1 The ground state We seek a variational eigenstate of the BCS reduced Hamiltonian. This many-electron state is written

$$|\psi\rangle = \prod_k (u_k + v_k b_k^+)|0\rangle \tag{5.59}$$

The state $|0\rangle$ is again the vacuum state. The u_k and v_k are variational parameters which are taken to be real; two variational parameters have been introduced for each wavenumber. This state has the property that electrons occur only as pairs; however, this is a rather peculiar state. By writing out the product we see that there is a term with no pairs present, a large number of terms with one pair present, and so forth. The number of pairs or the number of electrons in this state is ill defined. The uncertainty in the number of pairs, however, turns out to be of the order of $N^{1/2}$ which is so small in comparison to the number of electrons N that it is not a serious problem.

We wish to vary the parameters u_k and v_k in order to obtain a minimum energy subject to the subsidiary condition fixing the expectation value of the number of electrons present at N. This is to be applied with the use of Lagrange multipliers.

Let us first examine the normalization integral.

$$\begin{aligned}\langle\psi|\psi\rangle &= \langle 0|\prod_k (u_k + v_k b_k)(u_k + v_k b_k^+)|0\rangle \\ &= \prod_k \langle 0|u_k^2 + v_k^2 b_k b_k^+|0\rangle = \prod_k (u_k^2 + v_k^2)\end{aligned}$$

where in the second step we have noted that $\langle 0|b_k|0\rangle = \langle 0|b_k^+|0\rangle = 0$. This wavefunction would be normalized if $u_k^2 + v_k^2 = 1$ for all $\mathbf{k}$, but for the moment we do not require normalization since we seek to minimize $\langle\psi|H|\psi\rangle/\langle\psi|\psi\rangle$.

We fix the number of particles by requiring

$$\frac{\left\langle\psi\left|\sum_k c_k^+ c_k\right|\psi\right\rangle}{\langle\psi|\psi\rangle} = 2\frac{\left\langle\psi\left|\sum_k b_k^+ b_k\right|\psi\right\rangle}{\langle\psi|\psi\rangle} = N$$

In the method of Lagrange multipliers we minimize a function $f(x)$ subject to the condition $A(x) = 0$ by minimizing the expression $f(x) - \lambda A(x)$ without restriction and then solving for $f(x)$ and λ using this minimum condition and the condition $A(x) = 0$. Thus in this case we wish to minimize

$$W = \frac{\langle\psi|\mathscr{H}|\psi\rangle}{\langle\psi|\psi\rangle} - 2\mu \frac{\left\langle\psi\left|\sum_k b_k{}^+ b_k\right|\psi\right\rangle}{\langle\psi|\psi\rangle} + \mu N$$

where μ is a Lagrange multiplier. The third term is independent of the variational parameters u_k and v_k and need not be considered. The first and second terms may be combined by writing out explicitly the form of the BCS reduced Hamiltonian.

$$W = \frac{\left\langle\psi\left|\sum_k 2(\epsilon_k - \mu) b_k{}^+ b_k + \sum_{k,k'} V_{k'k} b_{k'}{}^+ b_k\right|\psi\right\rangle}{\langle\psi|\psi\rangle}$$

This must be evaluated in terms of the variational parameters in order that it may be minimized. Terms in $b_k{}^+ b_k$ may be evaluated immediately by commuting the b_k to the right until it reaches the $(u_k + v_k b_k{}^+)$ factor in $|\psi\rangle$ and they may be commuted together to the right to $|0\rangle$. We note

$$b_k(u_k + v_k b_k{}^+)|0\rangle = v_k|0\rangle$$

Similarly, the $b_k{}^+$ may be commuted to the left, giving another factor of v_k. The remaining factors may be contracted as in the evaluation of $\langle\psi|\psi\rangle$ to give $\prod_{k' \neq k} (u_{k'}{}^2 + v_{k'}{}^2)$. Thus the contribution to W from a single term in $b_k{}^+ b_k$ is

$$\delta W = \frac{2(\epsilon_k - \mu) v_k{}^2}{u_k{}^2 + v_k{}^2} \tag{5.60}$$

after cancellation of the $(u_{k'}{}^2 + v_{k'}{}^2)/(u_{k'}{}^2 + v_{k'}{}^2)$ factors. Similarly, we may evaluate the expression $b_{k'}{}^+ b_k$ moving b_k to the right. It selects from the corresponding factor only the term v_k. This term contributes only with a factor u_k from the left-hand wavefunction. In a like fashion the $b_{k'}{}^+$ extracts a factor $u_{k'} v_{k'}$. Our expression for W above becomes

$$W = \sum_k 2(\epsilon_k - \mu) \frac{v_k{}^2}{u_k{}^2 + v_k{}^2} + \sum_{k',k} \frac{V_{k'k} u_{k'} v_{k'} u_k v_k}{(u_{k'}{}^2 + v_{k'}{}^2)(u_k{}^2 + v_k{}^2)} \tag{5.61}$$

Rather than minimize this directly we will drop the factors $u_k{}^2 + v_k{}^2$ and $u_{k'}{}^2 + v_{k'}{}^2$ in the denominator and minimize the result subject to the subsidiary condition that these are unity. This introduces an additional Lagrange multiplier, λ_k, for each value of k. The quantity to be minimized then is given by

$$W' = \sum_k [2(\epsilon_k - \mu)v_k^2 + \lambda_k(u_k^2 + v_k^2)] + \sum_{k,k'} V_{k'k}u_{k'}v_{k'}u_kv_k$$

We are to simultaneously solve equations

$$\frac{\partial W'}{\partial u_k} = 0 \qquad \frac{\partial W'}{\partial v_k} = 0 \qquad u_k^2 + v_k^2 = 1 \qquad N = \sum_k 2v_k^2$$

which will determine the u_k, the v_k, the λ_k, and μ. In terms of these we will obtain the energy of the transition and other relevant quantities.

The first two of these variational equations become

$$\frac{\partial W'}{\partial u_k} = 2\lambda_k u_k - 2v_k\Delta_k = 0$$

$$\frac{\partial W'}{\partial v_k} = [4(\epsilon_k - \mu) + 2\lambda_k]v_k - 2u_k\Delta_k = 0$$

where Δ_k is called the *energy-gap parameter* and is given by

$$\Delta_k = -\sum_{k'} V_{k'k}u_{k'}v_{k'} \tag{5.62}$$

These equations can be written in more convenient form by defining an energy,

$$E_k = \epsilon_k - \mu + \lambda_k$$

which will turn out to be the energy of an excited state in the superconductor. This definition simply replaces the Lagrange multiplier λ_k by a more useful parameter. Substituting for the $2\lambda_k$ we obtain

$$\begin{aligned} [E_k - (\epsilon_k - \mu)]u_k - \Delta_k v_k &= 0 \\ [E_k + (\epsilon_k - \mu)]v_k - \Delta_k u_k &= 0 \end{aligned} \tag{5.63}$$

It will be convenient to obtain two other equations from these. We multiply the first above by v_k, the second by u_k and add, noting that $u_k^2 + v_k^2 = 1$. This leads to

$$u_k v_k = \frac{\Delta_k}{2E_k} \tag{5.64}$$

Substituting $u_k = \Delta_k/2E_k v_k$ in the first of the above equations and solving for v_k we obtain

$$v_k^2 = \frac{E_k - (\epsilon_k - \mu)}{2E_k} \tag{5.65}$$

Now we may solve the first of the Eqs. (5.63) for v_k and substitute in the second. Cancelling the u_k we obtain

$$E_k^2 - (\epsilon_k - \mu)^2 - \Delta_k^2 = 0$$

or

$$E_k = \pm\sqrt{(\epsilon_k - \mu)^2 + \Delta_k^2} \tag{5.66}$$

Only the + gives a form for v_k approaching zero at large energies ϵ_k, and thus only the + corresponds to a physical solution. We note that v_k^2 is the probability of finding an electron in the kth state. From Eqs. (5.65) and (5.66) we see that this probability varies with electron energy ϵ_k as shown in Fig. 5.10*a*. The corresponding expression in the normal state at absolute zero is simply a step function as shown in Fig. 5.10*b*. These figures would suggest that the Lagrange multiplier μ is equal to the Fermi energy and this may be shown to be the case. We will continue to use the symbol μ for the thermodynamic Fermi energy rather than the ξ which we used in normal metals. This is conventional and in superconductivity the symbol ξ is reserved for the coherence length to be defined presently.

The variational solution has given us an occupation of states that drops off smoothly near the Fermi energy. We have yet, however, to determine the energy-gap parameter Δ_k. We may obtain an equation for Δ_k by writing its definition and substituting for $u_k v_k$ from Eq. (5.64). We obtain

$$\Delta_k = -\sum_{k'} V_{k'k} u_{k'} v_{k'} = -\tfrac{1}{2} \sum_{k'} \frac{V_{k'k}\,\Delta_{k'}}{E_{k'}} \tag{5.67}$$

which is called the *energy-gap equation*. Note that this is an integral equation for Δ_k which may be solved if we know the form of $V_{k'k}$.

We make this evaluation only for the simple model in which $V_{k'k}$ is a

Fig. 5.10 *The probability v_k^2 of occupation of one-electron states as a function of one-electron energy ϵ_k at $T = 0$. (a) A superconductor; Δ has been taken to be one-eighth of the Fermi energy (rather than of the order of one-hundredth) for illustrative purposes. (b) A normal electron gas.*

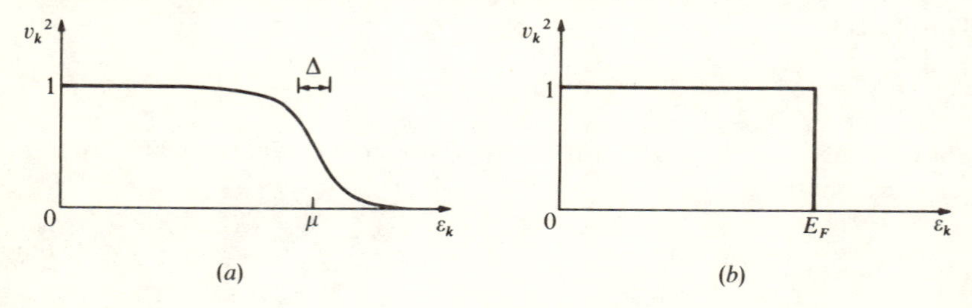

constant, $-V$, for $|\epsilon_k|$ and $|\epsilon_{k'}| < \hbar\omega_D$ and zero otherwise. We have chosen a cutoff equal to the Debye energy. Then from Eq. (5.67) we see that Δ_k is equal to zero for $|\epsilon_k| > \hbar\omega_D$ since $V_{k'k}$ is equal to zero over the entire range of integration. For $\epsilon_k < \hbar\omega_D$ Eq. (5.67) leads to a constant value independent of ϵ_k. We call this value Δ_0. This simple form for Δ_k arises from our simple model of the interaction. We may evaluate the energy-gap parameter Δ_0 by performing the appropriate integration. We have

$$\Delta_0 = \frac{V}{2}\Delta_0 \sum_k \frac{1}{\sqrt{(\epsilon_k - \mu)^2 + \Delta_0{}^2}} = V\frac{\Delta_0 n(E_F)}{4}\int_{-\hbar\omega_D}^{\hbar\omega_D} \frac{d\epsilon}{\sqrt{\epsilon^2 + \Delta_0{}^2}}$$

In the integration we have again measured energy from the Fermi energy and noted that the density of coupled states is $n(\epsilon)/2$ which we evaluate at the Fermi energy.

We cancel Δ_0 and perform the integration,

$$1 = \frac{V}{4} n(E_F) ln \left| \frac{\sqrt{\hbar^2\omega_D{}^2 + \Delta_0{}^2} + \hbar\omega_D}{\sqrt{\hbar^2\omega_D{}^2 + \Delta_0{}^2} - \hbar\omega_D} \right|$$

In the weak-coupling limit, which we expect to obtain for superconductivity, we may expand the logarithm in $\Delta_0/\hbar\omega_D$ to obtain finally

$$\Delta_0 = 2\hbar\omega_D e^{-2/Vn(E_F)} \tag{5.68}$$

As in the case of binding of pairs we find a result that is not obtainable by perturbation theory in the electron-electron interaction.

Since $Vn(E_F)$ is much less than 1 we expect Δ_0 to be very small compared to the Debye energy $\hbar\omega_D$. In practice the energy-gap parameter Δ_0 is of the order of KT_C where T_C is the critical temperature for superconductivity. This turns out to be only a few degrees (the highest known values are slightly over 20°K) in comparison to Debye temperatures of several hundred degrees. We note also that the energy-gap parameter shows an isotope effect because of our choice of a cutoff parameter $\hbar\omega_D$ which depends upon the isotopic mass.

We may obtain the condensation energy by evaluating the expectation value of the energy W in the superconducting state and subtracting that for the normal state W_N. From Eqs. (5.61), (5.64), and (5.65) we have

$$W - W_N = \sum_k (2\epsilon_k v_k{}^2 - u_k v_k \Delta_k) - W_N$$

$$= \sum_k \left[\frac{\epsilon_k(E_k - \epsilon_k)}{E_k} - \frac{\Delta_k{}^2}{2E_k}\right] - \sum_{\epsilon_k < 0} 2\epsilon_k$$

where energies are now measured from the Fermi energy μ. We note the E_k is independent of the sign of ϵ_k so these expressions may be combined

$$W - W_N = \sum_{\epsilon_k < 0} \left[\frac{\epsilon_k(-E_k - \epsilon_k)}{E_k} - \frac{\Delta_k^2}{2E_k} \right] + \sum_{\epsilon_k > 0} \left[\frac{\epsilon_k(E_k - \epsilon_k)}{E_k} - \frac{\Delta_k^2}{2E_k} \right]$$
$$= 2 \sum_{\epsilon_k > 0} \left(\epsilon_k - \frac{\epsilon_k^2}{E_k} - \frac{\Delta_k^2}{2E_k} \right)$$

We may now write E_k using Eq. (5.66) and set $\Delta_k = \Delta_0$ over the small range $0 < \epsilon_k < \hbar\omega_D$. Taking the density of wavenumber states as half the total density of states at the Fermi energy, that is, $\sum_{\epsilon_k} \to \frac{1}{2} n(0) \int d\epsilon$, we may perform the integrations to obtain

$$W - W_N = \frac{n(0)}{2} \left[(\hbar\omega_D)^2 - \hbar\omega_D \sqrt{(\hbar\omega_D)^2 + \Delta_0^2} \right] \approx \frac{n(0)}{4} \Delta_0^2$$

since $\Delta_0 \ll \hbar\omega_D$. The critical magnetic field that suppresses superconductivity also may be evaluated in terms of the condensation energy from thermodynamic arguments.

9.2 Excited states Excited states may be obtained by adding a single electron in the state of wavenumber **k**. Then in constructing the superconducting state as above we must omit the wavenumber **k** from our product wavefunction. This approach is a little tricky because the ground state does not have a well-defined number of particles as we have indicated before. We postulate an excited state, or a *quasiparticle excitation*,

$$\psi = c_k^+ \prod_{k' \neq k} (u_{k'} + v_{k'} b_{k'}^+)|0\rangle \tag{5.69}$$

We may keep the total number of particles fixed by exchanging electrons with a system in equilibrium with the superconductor. Any transferred electron is removed at the Fermi energy μ so Eq. (5.69) contains an increase in energy over the ground state of $\epsilon_k - \mu$. In addition, by omitting the pair $\mathbf{k}\uparrow - \mathbf{k}\downarrow$ from the ground state we modify the energy given in Eq. (5.61) by

$$\delta W = -2(\epsilon_k - \mu)v_k^2 - \sum_{k'} (V_{k'k} + V_{kk'}) u_{k'} v_{k'} u_k v_k$$
$$= -2(\epsilon_k - \mu)v_k^2 + 2u_k v_k \Delta_k$$

where we have included both contributions in the double sum of the final term. Eliminating v_k^2 and $u_k v_k$ with Eqs. (5.64) and (5.65), we obtain a total change in energy of

$$\epsilon_k - \mu + \delta W = \frac{(\epsilon_k - \mu)^2 + \Delta_k^2}{E_k} = E_k$$

and we are to take the positive sign for E_k in Eq. (5.66). As we indicated earlier, the energy of the excitation (or the energy required to introduce an electron from a normal system in equilibrium with the superconductor) is $E_k = \sqrt{\epsilon_k^2 + \Delta_0^2}$.

Similarly, we may obtain the energy to remove an electron from the superconductor, producing a "hole" excitation and find again an energy E_k. We see immediately that the minimum energy for an excited state is Δ_0. There are no excited states separated from the ground state by an infinitesimal energy as there are in a normal metal. In this regard the superconducting state resembles the state of an intrinsic semiconductor with a band gap of $2\Delta_0$, the energy required to make an electron-hole pair (break a Cooper pair). The first direct measure of this superconducting energy gap was a measure of the corresponding absorption edge in the infrared.[1]

9.3 Experimental consequences Of course the finding of an electronic phase transition is a consequence in agreement with experiment. The specification of the ground state and the excitation spectrum allows a treatment of a wide range of properties for homogeneous superconductors. Many of these were given in the original BCS paper. Here we will consider only a few consequences that follow immediately from the basic theory described above.

Persistent currents The most striking property of a superconductor is the vanishing of the electrical resistance. This can be directly understood in terms of the microscopic theory. We have constructed the ground state by pairing electrons of wavenumber $\mathbf{k}$ and $-\mathbf{k}$. We could instead have constructed a state by pairing states of wavenumber $\mathbf{k} + \mathbf{q}$ and $-\mathbf{k} + \mathbf{q}$. The resulting state is entirely equivalent to the original ground state viewed from a coordinate system moving with a velocity $-\hbar\mathbf{q}/m$. The center of gravity of every pair is drifting with the velocity $\hbar\mathbf{q}/m$, and the current density is $-Ne\hbar\mathbf{q}/m\Omega$, with N/Ω the electron density. Also the total energy of the system is increased by the kinetic energy $N\hbar^2q^2/2m$. We could similarly construct a drifting state of a normal electron gas; the difference is in the decay of the current. In the normal metal an impurity or defect can scatter an electron from the "leading edge" of the Fermi surface to the trailing edge, decreasing the current as in Fig. 5.11*a*. The scattering potential couples the two states of the system with a term in the Hamiltonian $\langle \mathbf{k}'|V_{\text{scat}}|\mathbf{k}\rangle c_{k'}^+ c_k$ and the energy of the system is conserved if $|\mathbf{k}'| = |\mathbf{k}|$. Such scattering processes cause the decay of the current as we have described in our discussion of transport properties.

In the superconductor the same scattering term couples the drifting

[1] R. E. Glover and M. Tinkham, *Phys. Rev.*, **108**:243 (1957).

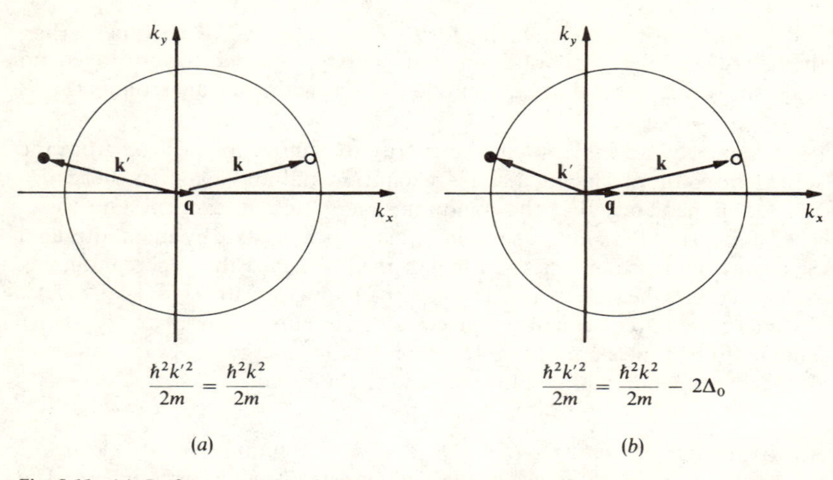

Fig. 5.11 *(a) Drifting current $-e\hbar\mathbf{q}/m$ per electron in a normal electron gas showing a scattering event ($\mathbf{k}$ to $\mathbf{k}'$) which reduces the current. (b) A sufficiently large current density in a superconductor to allow the spontaneous creation of an electron-hole pair with a reduction in current.*

BCS state to the state of reduced current, but if the transition is to cause a change in current it requires the creation of an excited electron and an excited hole with the expense of an energy $2\Delta_0$ in addition to any difference in kinetic energy. Thus the energy delta function prevents the transitions unless the drift is so large that enough kinetic energy is picked up in back-scattering to make up the energy $2\Delta_0$; that is, unless $(\hbar^2 k_F/m)(2q) \geq 2\Delta_0$. Such an event is illustrated in Fig. 5.11*b*. This corresponds to a current of $-Ne\Delta_0/\hbar k_F\Omega$. Taking Δ_0 as 1 millielectronvolt and reasonable choices of the other parameters, we obtain a very large limiting current of the order of 10^7 amp/cm^2.

We see that, within limits, a current can be set up and ordinary scattering processes cannot cause its decay. One-electron perturbations do not couple states of slightly different current (and the same energy) in superconductors as they do in normal metals.

There are at least three defects in the argument given here which require mention. The drifting current we have constructed will produce magnetic fields, but the theory is based on an electron gas in the absence of magnetic fields. In fact, it has long been known that superconductors exclude static magnetic fields from their interior: the *Meissner effect*, which follows from the BCS theory but is not easy to deduce. Thus the drifting solution cannot be correct in the bulk of a superconductor. For thin films or fine wires, however, the fields become negligible and the drifting state is relevant. In thin films or fine wires there is danger of inhomogeneities that may cause one spot to become normal, to heat up and to cause a normal transition of the entire specimen. We will return to this geometry in Sec. 10.3. In addition the electron-electron interaction is not invariant under the translation to a moving coordinate system since it arises through the transfer of lattice vibrations in the lattice at rest. Finally, persistent currents exist even in "gapless superconductors" which have excited states at all energies. It is therefore not surprising that experimentally, though persistent currents are limited, the maximum value is well below that we have deduced and is frequently of the order of 10^4 amp/cm^2. We will return to a discussion of the critical current in Sec. 10.3.

Giaever tunneling[1] The most direct observation of excited states in superconductors came several years after the BCS theory in the tunneling of electrons from normal metals, through a thin oxide, into a superconductor. Such a situation is very close to that we envisaged in our discussion of excited states in Sec. 9.2. The lowest single-particle energy in the superconductor lies at an energy Δ_0 above the Fermi energy which will be the same in the normal and superconducting metals with no applied voltage. Thus we would expect no tunneling current to flow (at zero temperature) until a voltage of Δ_0/e, of the order of millivolts, is applied.

This argument is not without flaws. One could note, as I did prior to the experiment, that this argument can be maintained as the oxide becomes arbitrarily thin; yet one knows that with no oxide the current flows between normal and superconducting metals with no difficulty. The failure of this objection became clear with the Josephson effect (to which we will return) and the additional mechanism for current transport through very thin films. The fact is that Giaever's experiments did show negligible tunneling currents until a sufficient voltage was applied and gave thereby a direct voltmeter measure of the energy-gap parameter Δ_0.

The experiments were easier to interpret than to predict. We imagine a tunneling term in the Hamiltonian, such as discussed in Sec. 2.5 of Chap. III, which couples one-electron states on the two sides of the oxide; it transfers single electrons. In the normal metal (on the right in Fig. 5.12) all one-electron states are occupied for $\epsilon < \mu$; the density of states per unit energy is nearly constant over the few millivolts of interest. In the superconductor (on the left in Fig. 5.12) there are no one-electron states below the energy $\Delta_0 + \mu$. Above $\Delta_0 + \mu$ there are excited states, all empty at $T = 0$, with a density

[1] Pronounced Gā'ver. I. Giaever, *Phys. Rev. Letters*, **5**: 147, 464 (1960).

which we may evaluate from the usual periodic boundary conditions and excitation energy of Eq. (5.66).

$$n_s(E_k)\,dE_k = \frac{2}{(2\pi)^3}\,4\pi k_F^{\,2}\,\frac{dE_k}{(dE_k/dk)}$$

$$= \frac{mk_F}{\pi^2\hbar^2}\,\frac{E_k}{\epsilon_k - \mu}\,dE_k$$

This density of excited states obtains for $\epsilon_k \geq \mu$ and therefore $E_k \geq \Delta_0$. These states may be represented as unoccupied quasiparticle states as in Fig. 5.12. Similarly, the possibility of excited hole states may be represented by filled quasiparticle states below the Fermi energy as in Fig. 5.12.

Now taking the tunneling matrix element T as independent of energy near the Fermi energy (the onset of superconductivity does not modify

Fig. 5.12 *A single-particle energy-level diagram for Giaever tunneling. With no voltage applied the Fermi energy μ is the same on both sides.*

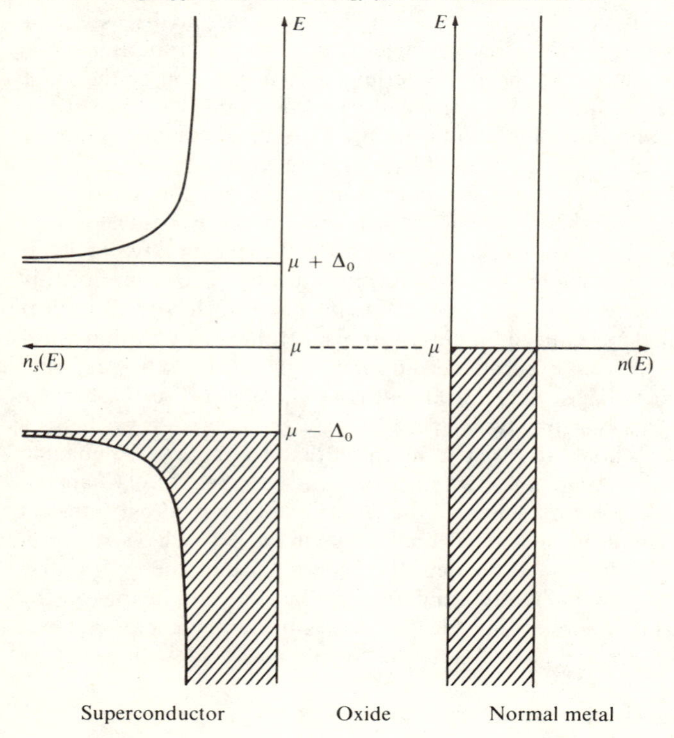

these one-electron matrix elements), we obtain the tunneling current from the superconductor with an applied voltage V to be

$$I = \frac{2\pi}{\hbar}(-e)T^2 \int dE_k n_s(E_k) n(E_k + eV) f_0(E_k)[1 - f_0(E_k + eV)] \quad (5.70)$$

This follows immediately from the Golden Rule; n_s is the density of quasiparticle states in the superconductor (as shown in Fig. 5.12) and n is the density of states in the normal metal. The f_0 is the Fermi distribution function. We now neglect the variation of $n(\mathrm{E})$ with energy in the normal metal and evaluate the current at zero temperature; then the f_0 factors become unity in an energy range of width eV. A particularly useful quantity is the dynamic conductivity, dI/dV, obtained by differentiating Eq. (5.70) with these approximations.

$$\frac{dI}{dV} = \left[\frac{2\pi}{\hbar} e^2 T^2 n(\mu)\right] n_s(-eV)$$

which is directly proportional to the density of quasiparticle states in the superconductor. The corresponding results are shown in Fig. 5.13 and give a remarkably accurate description of the experiment.

Of course at finite temperature the f_0 in Eq. (5.70) are not step functions and these curves are smoothed exactly as expected on the basis of the Fig. 5.12 model. In addition there is weak structure in the experimental curves which is interpreted as variation of the energy-gap parameter Δ_k with energy. Specifically, peaks in the density of phonon states as a function energy are expected to produce fluctuations in Δ_k at energies above the gap by the corresponding phonon energies; these expectations are in good accord

Fig. 5.13 *(a) Current versus voltage for a Giaever junction. (b) Dynamic conductivity for the same junction. In both cases the dashed curve gives the corresponding result if the superconductor is made normal.*

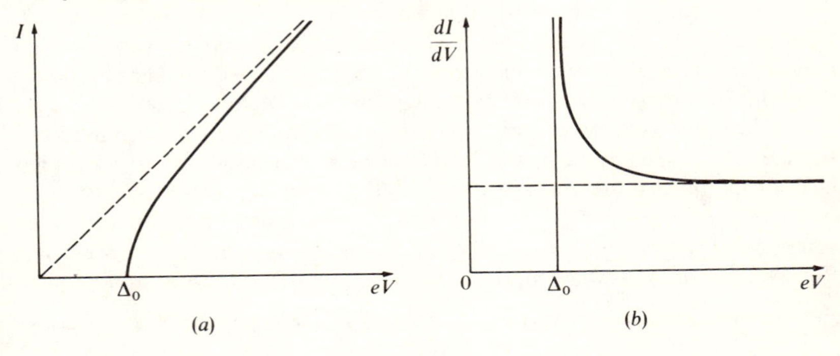

with observed fluctuations. Indeed, it is possible to work back from careful tunneling experiments, using the energy-gap equation, and deduce a detailed distribution of phonon frequencies and an estimate of the electron-phonon interaction.[1] This provides a remarkably detailed confirmation of the BCS theory.

Giaever tunneling experiments can also be carried out with both metals superconducting. A model of the system can be constructed as a direct generalization of that shown in Fig. 5.12, either for the case where both metal films are the same or for the case where they are different. Equation (5.70) for the current remains valid with $n(E_k + eV)$ replaced by the $n_s(E_k + eV)$ for the second superconductor. The current threshold appears at the sum of the Δ_0 values for the two metals, and at finite temperatures a peak occurs also at the difference of the two due to the thermal excitation of quasiparticle states.

Thermodynamic properties In our calculation of the quasiparticle energy we found a contribution from the reduction of the ground-state energy due to the omission of one Cooper pair. When we have a thermal distribution of quasiparticles, it is clear that we must redo our calculation of Δ_0 with the reduced number of factors in the wavefunction of Eq. (5.59). Clearly also this will reduce the computed Δ_0 by reducing the "useful" density of states $n(E_F)$ entering Eq. (5.68). This in turn lowers the energy of quasiparticle excitations and increases their number at a given temperature. Thus we see explicitly that the superconducting transition is cooperative and will have a transition temperature T_c above which superconductivity is suppressed. The calculation of the critical temperature is straightforward in the framework we have given and leads to a value $KT_c = 2\Delta_0/3.52$, where Δ_0 is evaluated at $T = 0$. This is confirmed reasonably well in the "weak coupling" superconductors such as aluminum. Below the critical temperature Δ_0 is a function of temperature such as shown in Fig. 5.14. The curve drops to zero with infinite slope at $T = T_c$. The behavior at T_c can be described by a critical-point exponent just as was the behavior of the magnetization at the Curie temperature. As in that case the simple theory (here the BCS theory) does not give the correct exponent.

The presence of an energy gap in the excitation spectrum also modifies the electronic specific heat and leads, as in semiconductors, to a specific heat dropping exponentially to zero as T approaches zero, the exponent being of the form $e^{-2\Delta_0/KT}$. At any finite temperature there are of course some quasiparticles excited and these influence a number of properties. For example, a long-wavelength sound wave (with $\hbar\omega < 2\Delta_0$) cannot cause

[1] J. M. Rowell and W. L. McMillan, *Phys. Rev. Letters*, **14**: 108 (1965) and to be published.

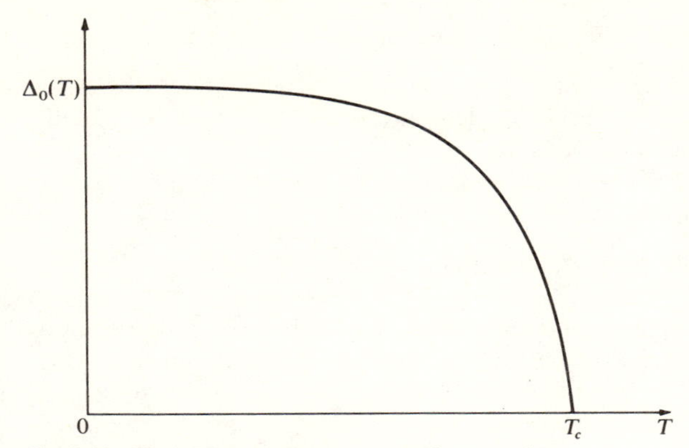

Fig. 5.14 The energy-gap parameter as a function of temperature from the BCS theory.

excitations from the ground state but can scatter any quasiparticles present. Thus the electronic contribution to ultrasonic attenuation will approach zero exponentially at low temperature but may approach the normal-metal value as T approaches T_c. (This is not true for transverse waves where the electron-phonon coupling is modified by the presence of superconductivity.)

9.4 The superconducting wavefunction or order parameter We may note the remarkable property of the BCS ground state, that it contains a non-vanishing expectation value of pair-annihilation and pair-creation operators. As we indicated in our discussion of coherent phonon states, this is closely associated with the fact that the number of particles in the BCS state is ill-defined. We consider in particular the operator

$$\mathbf{B} = \frac{1}{\Omega} \sum_k b_k$$

The expectation value of **B** may be obtained directly for the BCS state using steps of just the type that led to Eq. (5.60). We obtain

$$\langle \mathbf{B} \rangle = \frac{1}{\Omega} \sum_k u_k v_k \tag{5.71}$$

By comparison with Eq. (5.62), and using again the approximation of constant matrix elements $V_{k'k} = -V$, we obtain

$$\langle \mathbf{B} \rangle = \langle \mathbf{B}^+ \rangle = \frac{\Delta}{\Omega V} \tag{5.72}$$

where of course $\mathbf{B}^+ = \sum_k b_k^+$. This constant value represents the long-range order in the superconducting ground state. If we weaken the superconductivity by reducing Δ, the order parameter $\langle \mathbf{B} \rangle$ becomes smaller and equal to zero in the normal state.

We may define a more general order parameter in terms of the electron-field operators developed in Sec. 4.2 of Chap. IV.

$$\mathbf{B}(\mathbf{r}) = \psi_\uparrow(\mathbf{r})\psi_\downarrow(\mathbf{r}) = \sum_{k,k'} \psi_{k'}(\mathbf{r})\psi_k(\mathbf{r})c_{k'\uparrow}c_{k\downarrow}$$

The arrows indicate spin up and spin down. We again seek the expectation value of $\mathbf{B}(\mathbf{r})$ with respect to the BCS ground state. We obtain in analogy with Eq. (5.71)

$$\langle \mathbf{B}(\mathbf{r}) \rangle = \sum_k u_k v_k \psi_k(\mathbf{r})\psi_{-k}(\mathbf{r})$$

which for plane-wave states is again $\Delta/V\Omega$; for the BCS ground state $\langle \mathbf{B}(\mathbf{r}) \rangle$ is independent of position. However, if we seek the expectation value of $\mathbf{B}(\mathbf{r})$ with respect to the drifting BCS state described in Sec. 9.3, we obtain

$$\langle \mathbf{B}(\mathbf{r}) \rangle = \frac{\Delta}{\Omega V} e^{2iq \cdot r}$$

Thus the current density is obtainable from $\langle \mathbf{B}(\mathbf{r}) \rangle$ as

$$\mathbf{j}(\mathbf{r}) = \frac{\hbar N \Omega}{4mi} e \left(\frac{V}{\Delta} \right)^2 [\langle \mathbf{B}(\mathbf{r}) \rangle^* \nabla \langle \mathbf{B}(\mathbf{r}) \rangle - \langle \mathbf{B}(\mathbf{r}) \rangle \nabla \langle \mathbf{B}(\mathbf{r}) \rangle^*] \tag{5.73}$$

Properties of the system may be computed directly from this function $\langle \mathbf{B}(\mathbf{r}) \rangle$ of a single spatial variable which we call the *superconducting wavefunction*, or the *order parameter*. It has become a macroscopic variable describing ODLRO in the superconductor in precisely the same sense as did the nonvanishing value of the displacement amplitude in the phonon system.

In order to compute properties of superconductors we need only a method for computing $\langle \mathbf{B}(\mathbf{r}) \rangle$ itself; we need the counterpart of the Schroedinger equation for the usual wavefunction. We can obtain this equation immediately for simple systems and will return afterward to the more complete method based upon self-consistent fields.

We look first at our evaluation of $\langle \mathbf{B}(\mathbf{r}) \rangle$ for the BCS ground state. In constructing the BCS state with u_k and v_k real, we have of course chosen states with no time dependence, and therefore a $\langle \mathbf{B}(\mathbf{r}) \rangle$ independent of time. Equivalently, we may say that we have chosen our zero of energy at the energy for the eigenstate. This can be done only for an eigenstate or for a system that is in equilibrium. When we discuss the Josephson effect in the following section we will wish to consider two superconductors, each of which is in equilibrium by itself but between which there is an applied potential difference. In this simple case the time dependence of $\langle \mathbf{B}(\mathbf{r}) \rangle$ can be seen immediately.

Shifting the electron potential energy of a superconductor by δV produces a phase factor $e^{-i\delta Vt/\hbar}$ in the wavefunction for each electron present, or equivalently a factor of $e^{-2i\delta Vt/\hbar}$ in each factor v_k. From Eq. (5.71) we see that this is also precisely the time dependence given to $\langle \mathbf{B} \rangle$. Thus if we define φ to be the difference in the phase of $\langle \mathbf{B}(\mathbf{r}) \rangle$ between two superconductors, the introduction of an applied potential between the two causes φ to change with time according to

$$\hbar \frac{\partial \varphi}{\partial t} = -2\,\delta V \tag{5.74}$$

This phase difference is of no importance if the two superconductors are indeed separate but becomes important when they are coupled. Equations (5.73) and (5.74) together will provide a basis for the understanding of the Josephson effect which arises when there is weak coupling between two superconductors.

We can also see from Eqs. (5.73) and (5.74) some aspects of the behavior of $\langle \mathbf{B}(\mathbf{r}) \rangle$ under more general circumstances. If we imagine a superconductor in the ground state, $\langle \mathbf{B}(\mathbf{r}) \rangle$ constant, and then apply a nonuniform potential, the phase will advance more rapidly in the regions of lower potential. Then from Eq. (5.73) we see that a supercurrent will arise. The flow of electrons will be into the region of low potential, causing a restoring electrostatic potential. A steady state of the system can exist only if $\langle \mathbf{B}(\mathbf{r}) \rangle$ is a constant in time (or changing phase uniformly over the sample) and this will occur only in the absence of potential gradients.

To obtain a more general method for the determination of $\langle \mathbf{B}(\mathbf{r}) \rangle$, we must return to the BCS Hamiltonian. The equations may then be solved by a self-consistent-field method. This provides an alternative to the BCS solution (the method of Bogoliubov[1]) but one that is equivalent. We will give only the first steps in such a formulation (for a more complete treatment see deGennes[2]). We will then proceed to the Ginsburg-Landau theory which accomplishes the same task approximately.

We recall in our discussion of field operators in Sec. 4.2 of Chap. IV that electron-electron interactions were included in the Hamiltonian through a term of the form

$$V_{\text{elel}} = \tfrac{1}{2} \int d^3r\, d^3r'\, \psi^+(\mathbf{r})\psi^+(\mathbf{r}')V(\mathbf{r} - \mathbf{r}')\psi(\mathbf{r}')\psi(\mathbf{r}) \tag{5.75}$$

[1] N. Bogoliubov, *Nuovo Cimento*, **7**:6, 794 (1958).

[2] P. G. deGennes, "Superconductivity of Metals and Alloys," p. 137, Benjamin, New York, 1966.

The self-consistent-field approximation was made by replacing pairs of operators, such as $\psi^+(\mathbf{r}')\psi(\mathbf{r}')$, by their expectation value which reduced the interaction to the form of interaction with a simple potential. We have seen that in the superconducting state the products $\psi^+(\mathbf{r})\psi^+(\mathbf{r})$ and $\psi(\mathbf{r})\psi(\mathbf{r})$ also have nonvanishing expectation values so that additional terms occur in the self-consistent-field approximation. In our approximate treatment in the BCS theory we took the potential to be independent of wavenumber (at least over the range of interest) corresponding to a contact interaction in real space $V(\mathbf{r} - \mathbf{r}') = -V\Omega\,\delta(\mathbf{r} - \mathbf{r}')$ and to couple states of opposite spin. Thus the interaction given in Eq. (5.75) becomes

$$V_{\text{elel}} \approx -V\Omega \int d^3r \psi_\uparrow{}^+(\mathbf{r})\psi_\downarrow{}^+(\mathbf{r})\psi_\downarrow(\mathbf{r})\psi_\uparrow(\mathbf{r})$$

where we have added terms with spins interchanged to cancel the factor of $\frac{1}{2}$. Using Eq. (5.72) and giving Δ the same phase as $\langle \mathbf{B} \rangle$, this becomes in the self-consistent-field approximation,

$$V_{\text{elel}} \approx -\int d^3r [\Delta(\mathbf{r})\mathbf{B}^+(\mathbf{r}) + \Delta^*(\mathbf{r})\mathbf{B}(\mathbf{r})]$$

The term $\Delta(\mathbf{r})$ is then sometimes called a *pair potential* but it must be noted that this term in the Hamiltonian does not conserve particles as does an ordinary potential.

The entire Hamiltonian may again be written in terms of electron-field variables and the solution is sought by making a canonical transformation to field operators which are linear combinations of, for example, $\psi_\uparrow(\mathbf{r})$ and $\psi_\downarrow{}^+(\mathbf{r})$, leading to the so-called Bogoliubov equations. The results when applied to the ground state are equivalent to the BCS treatment that we have given before. They may also be solved in principle for nonuniform systems but the solution is difficult. The problem is greatly simplified for temperatures near the superconducting temperature where $\langle \mathbf{B} \rangle$ becomes very small and can be used as an expansion parameter. This is the approach used in the Ginsburg-Landau theory, which in fact predates the microscopic theory. We will return to that approach in Sec. 10.3.

9.5 The Josephson effect In Giaever tunneling, current is carried through the insulating barrier as single electrons, and quasiparticle excitations are left on the two sides of the junction. In 1962 Josephson[1] predicted that there may also be tunneling of supercurrent in which the number of quasiparticle excitations in the superconductors remains unchanged. In such a situation a lossless current is carried through the tunneling barrier with no

[1] B. D. Josephson, *Phys. Letters (Netherlands)*, **1**:25 (1962).

applied voltage. This remarkable effect can be treated in two distinct ways, illustrating the two approaches in the theory of superconductivity which we have discussed. The first is based directly on the BCS theory and can be carried out by an ingenious technique due to Ferrell.[1] This gives explicitly the possible magnitudes of supercurrent flow though the evaluation is complicated. Second, we can describe the effect in terms of the superconducting wavefunction. We will only use the second approach here. The results will contain an undetermined parameter but will illustrate clearly how the phase of the superconducting wavefunction can be treated as a macroscopic variable of the problem, like the voltage difference across the junction, allowing direct analysis of circuits containing superconducting elements. It seems likely that this approach will become familiar to electrical engineers as superconductors become a more and more important part of technology.

We imagine first two isolated superconductors which are described by superconducting wavefunctions which do not overlap each other (since the superconducting wavefunction must go to zero where the electron density is zero). Let each be an eigenstate so each has a single well-defined phase and the two superconducting wavefunctions may be written respectively $\mathbf{B}_1(\mathbf{r})e^{i\varphi_1}$ and $\mathbf{B}_2(\mathbf{r})e^{i\varphi_2}$ with $\mathbf{B}_1$ and $\mathbf{B}_2$ taken to be real. Initially we take the Fermi energies to be the same in both superconductors and therefore can take both φ_1 and φ_2 to be independent of time, though they may be different.

Now let us bring the two superconductors sufficiently close that the electron wavefunctions, and therefore the superconducting wavefunctions, in the two overlap each other. We assume in the spirit of the tight-binding approximation that the superconducting wavefunction for the system may be written as a sum of the two separate superconducting wavefunctions.

$$\mathbf{B}(\mathbf{r}) = \mathbf{B}_1(\mathbf{r})e^{i\varphi_1} + \mathbf{B}_2(\mathbf{r})e^{i\varphi_2} \tag{5.76}$$

Since the phase of the superconducting wavefunction now varies with position in the region of overlap there may be current flowing which can be evaluated using Eq. (5.73). We see immediately that any current flow is proportional to $\sin(\varphi_1 - \varphi_2)$. Thus we may write the total superconducting current from 1 to 2 as

$$J = -J_1 \sin \varphi \tag{5.77}$$

where $\varphi = \varphi_2 - \varphi_1$ is the phase difference between the two superconductors and J_1 is a positive constant depending upon the detail of the overlaps. The sign was determined by noting that for small phase differences electrons

[1]See R. A. Ferrell and R. E. Prange, *Phys. Rev. Letters*, **10**:479 (1963), or a more complete discussion by deGennes, *op. cit.*, p. 118.

will flow into the superconductor with larger phase, and current will flow in the opposite direction.

In the tunneling region, since the superconducting wavefunction is quadratic in the electronic wavefunction, J_1 is of fourth order in the magnitude of the electronic wavefunctions. A similar treatment of normal electron tunneling would indicate that the tunneling matrix element is quadratic in the electronic wavefunctions in the barrier. Thus the supercurrent flow is quadratic in the tunneling matrix element just as is the tunneling of normal electrons. We may therefore expect this supercurrent, the Josephson current, to be comparable to currents for normal metals in the same configuration. This is by no means obvious since we may expect the superconducting wavefunction to decay more rapidly than the electron density, but the conclusion turns out to a good approximation to be true. A detailed calculation indicates that J_1 is equal to the current that would flow if both superconductors were made normal and a voltage equal to $\pi/2$ times the energy-gap parameter Δ were applied. The dc supercurrent described by Eq. (5.77) is the *dc Josephson effect*.

If we now apply a voltage difference between the two superconductors but retain the assumption that the total superconducting wavefunction may be written as in Eq. (5.76), we see that the phase difference changes with time according to Eq. (5.74). To keep the signs consistent with Eq. (5.77) and to put $V = V_2 - V_1$ in units of the voltage difference, we may write

$$\hbar \frac{d\varphi}{dt} = 2eV \tag{5.78}$$

with e the magnitude of the electronic charge. If the voltage difference is independent of time we see from Eq. (5.78) that the phase advances with a constant rate and we find an alternating current with a frequency $2eV/\hbar$. This is the *ac Josephson effect*.

Equations (5.77) and (5.78) characterize the tunnel junction and allow for the ultimate elimination of the phase from the final results. A more complete description of the junction and the circuit is necessary if a realistic study of its behavior is to be made. We note in particular that a Josephson junction always contains an appreciable capacitance which allows additional current flow through the junction. The junction itself must be drawn with a parallel capacitance.

Let us then construct a circuit and examine the behavior. Such a circuit is shown in Fig. 5.15. We let the phase difference φ and the potential V across the junction be given by the values at (2) minus the values at (1). Then the current J flowing out of the junction is given by

$$J = J_1 \sin\varphi - C\frac{dV}{dt}$$

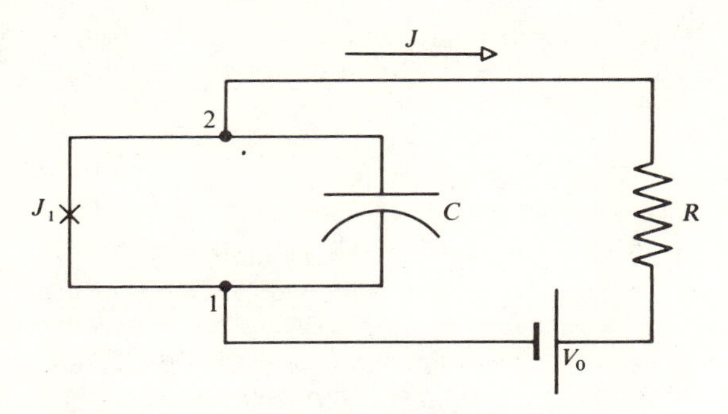

Fig. 5.15 A simple circuit containing a Josephson junction with parameter J_1 and capacitance C. The tunnel junction is designated by a cross, reminiscent of the configuration of a deposited junction; the capacitance of the junction is drawn in parallel with the tunnel path.

Furthermore, from the external circuit we see that

$$V = V_0 + RJ$$

We need one additional equation if we are to evaluate J, V, and φ, and this is of course Eq. (5.78).

This set of equations is most conveniently solved by eliminating V and J. We obtain

$$\frac{d^2\varphi}{dt^2} = -\frac{2eJ_1}{\hbar C}\sin\varphi + \frac{2eV_0}{\hbar RC} - \frac{1}{RC}\frac{d\varphi}{dt} \tag{5.79}$$

Because of the term in sin φ this equation is difficult to handle mathematically. Fortunately, however, the solutions are intuitively obvious since this is precisely the equation describing the motion of a pendulum which may have arbitrarily large angular displacements φ. The left side of Eq. (5.79) represents the angular acceleration. On the right side the first term represents the angular component of the gravitational force when the deflection is through an angle φ. The second term represents an applied torque on the pendulum proportional to V_0 and the third term represents a viscous damping due to the external resistance. Thus the solutions of Eq. (5.79) correspond to the oscillations of the pendulum which may be imagined or calculated. We may then relate the voltage across the junction directly to the instantaneous velocity of the pendulum through Eq. (5.78) and the supercurrent flowing through the junction to the lateral displacement of the pendulum.

We may immediately understand the dc and ac Josephson effects in these terms. The dc effect corresponds to $d\varphi/dt = 0$ and Eq. (5.79) may be solved immediately to give

$$J = -J_1 \sin\varphi = \frac{-V_0}{R}$$

This corresponds to a steady displacement of the pendulum under an appropriate applied torque. The dc supercurrent is limited to the magnitude J_1. If we seek to increase the current further by increasing the voltage we may see that an oscillatory current arises by again imagining the pendulum. With a sufficiently large torque the pendulum will flip over and will begin to run away. The angular velocity will ultimately be limited by the viscous damping term, and the phase increases approximately linearly with time. The final two terms in Eq. (5.79) become equal and opposite giving a frequency of $\omega = 2eV_0/\hbar$, the ac Josephson effect.

It is interesting to consider two other situations which can easily arise. Let us first imagine that $V_0 = 0$ and the solution of Eq. (5.79) becomes $\varphi = 0$. If we instantaneously apply a small V_0, for example, by closing a switch, we can again imagine the behavior by visualizing the corresponding pendulum. When the instantaneous torque is applied the pendulum will swing out and oscillate around its final value, slowly damping to a steady value. Correspondingly the current rises as shown in Fig. 5.16.

We may also imagine a behavior in the ac Josephson effect as the applied voltage is reduced. We imagine that the external resistance is sufficiently high that the damping term is small. Considering again the pendulum as the applied torque is reduced, the angular velocity of the pendulum decreases. If the applied torque becomes sufficiently small it will be

Fig. 5.16 The instantaneous Josephson supercurrent as a function of time after a small dc voltage is applied, and a sketch of the corresponding pendulum.

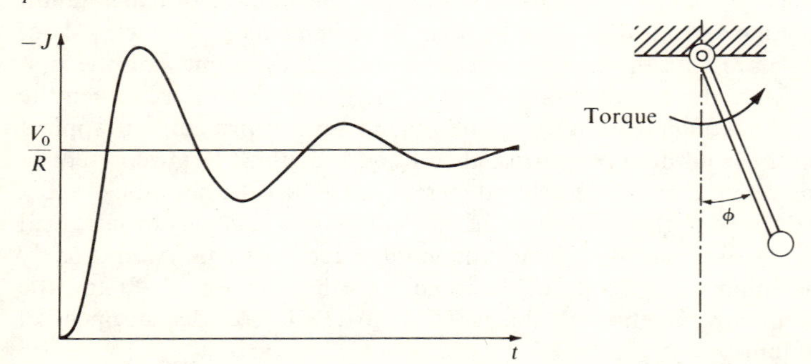

just sufficient to take the pendulum over the top on each pass. Thus the angular velocity will become very small when the pendulum is at the top but will become rapid at the bottom. The corresponding current and voltage are sketched in Fig. 5.17. We note that the ac Josephson effect becomes severely distorted at low voltages. From a study of the electromagnetic fields within the junction it may be seen that each of these pulses can be interpreted as the passage of a flux quantum, to be described in Sec. 10.4, through the junction.

We have not discussed the behavior of the superconducting wavefunction in the presence of magnetic fields. This will be done through the Ginsburg-Landau equations. We could see from the analysis there that a magnetic field through the tunneling barrier would cause a variation in the phase of the superconducting wavefunction over the junction and a corresponding cancellation of the currents flowing in different portions of the junction. Thus the behavior of the junction is extremely sensitive to the presence of magnetic fields. (See Prob. 5.10.)

10. The Ginsburg-Landau Theory

Some seven years prior to the appearance of the BCS theory, Ginsburg and Landau proposed a phenomenological theory of superconductivity.[1] It was little known in the West and in any case did not provide an under-

[1] V. L. Ginsburg and L. D. Landau, *JETP (USSR)*, **20**:1064 (1950).

Fig. 5.17 The voltage and supercurrent as a function of time in the ac Josephson effect when the applied voltage is just sufficient to support the oscillatory state. Also shown is a sketch of the corresponding pendulum.

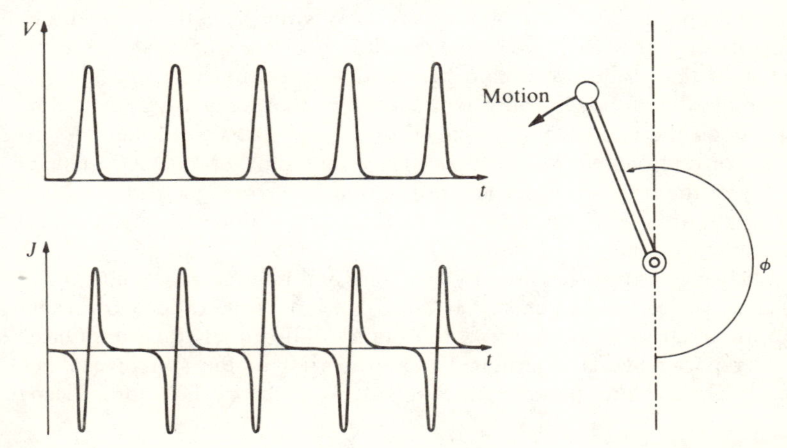

standing of the microscopic mechanism of superconductivity which was being widely sought. During the first few years after the appearance of the BCS theory almost all theoretical activity in the West used the BCS theory as the starting point and most experiments were directed at checking the various predictions of the microscopic theory. When the essential correctness of the microscopic theory seemed established, attention was turned to superconductors for which the microscopic theory in its simplest form was not applicable, and to inhomogeneous systems. At this stage much of the theoretical effort and virtually all of the interpretation of experiments used the older phenomenological theory as the starting point.

The conceptual basis of the theory is an older two-fluid model of superconductors. In this model electrons are thought to exist either in a normal state, corresponding to the quasiparticle excitations of the subsequent microscopic theory, or in superconducting or condensed states. The superconducting electrons are capable of carrying a lossless current but the normal electrons can, for example, take up thermal energy. We write the *fraction* of electrons which are superconducting as n_s; it is proportional to the *density* of superconducting electrons. The fraction n_s depends upon temperature and in fact drops to zero at the critical temperature. Ginsburg and Landau sought a theory of superconductors near the critical temperature; then the density of superconducting electrons is sufficiently small that it could be used as an expansion parameter. More specifically they sought to describe the superconductor by a wavefunction $\psi(\mathbf{r})$ in terms of which the fraction of electrons that are superconducting could be computed.

$$n_s(\mathbf{r}) = \psi^*(\mathbf{r})\psi(\mathbf{r}) \tag{5.80}$$

It is now recognized that $\psi(\mathbf{r})$ is proportional to the superconducting wavefunction or order parameter $\langle \mathbf{B}(\mathbf{r}) \rangle$ that we defined in Sec. 9.4 and is therefore proportional also to $\Delta(\mathbf{r})$. This is not obvious from the formulation we have given but we will see at the conclusion of Sec. 10.2 that it is consistent with the results we obtain here and with the microscopic theory. The formulation of the Ginsburg-Landau theory that we give here does not depend upon the microscopic origin of $\psi(\mathbf{r})$, for which we will henceforth use the conventional term *order parameter*, but only upon its existence. A theory of the order-disorder transformation in ferromagnetism can also be given in terms of such an order parameter, which in that case becomes the local magnetization of the system.

The general approach used by Ginsburg and Landau was to construct an expression for the free energy as an expansion in the order parameter and, using a variational argument, to obtain a differential equation. Once the order parameter is determined the properties of the system can be deduced. The various parameters that enter the theory have now been related to parameters in the microscopic theory; however, we can perhaps best display the intuitive achievement of the theory by postponing that identification.

10.1 Evaluation of the free energy We first write the free energy for a uniform system at any temperature as an expansion in the density of superconducting electrons.

$$F(T) = F_n(T) + a(T)n_s + \frac{b(T)}{2} n_s^2 + \cdots$$

The F_n is the free energy of the normal metal for which of course n_s is zero.

We can obtain some of the properties of the coefficients a and b by minimizing the free energy with respect to n_s, dropping all terms in the free energy of higher order than n_s^2. We obtain

$$n_s = \frac{-a(T)}{b(T)} \tag{5.81}$$

which must be greater than zero if there is to be a superconducting solution. When there is, the equilibrium free energy is given by

$$F(T) = F_n(T) - \frac{a(T)^2}{2b(T)} \tag{5.82}$$

The density of superconducting electrons must go to zero at the critical temperature so we expect the leading term in an expansion of $a(T)$ in $T - T_c$ to be linear. Furthermore, the solution must have lower energy than that of the normal state for temperatures less than T_c and higher energy for temperatures above T_c. Therefore the proportionality constant must be positive. In addition, we expect the constant term in the corresponding expansion of $b(T)$ to be nonzero. Thus for temperatures very close to T_c we may write

$$\begin{aligned} a &= \frac{\alpha(T - T_c)}{T_c} \\ b &= \beta \end{aligned} \tag{5.83}$$

where α and β are positive constants. Furthermore, substituting Eq. (5.83) back into the expression for the free energy, we obtain

$$F = F_n(T) - \frac{\alpha^2(T - T_c)^2}{2\beta T_c^2} \tag{5.84}$$

This result obtains only for temperatures below T_c since otherwise the superconducting density is found to be negative. This corresponds to a second-order phase transition at the temperature T_c. At that temperature the free energy and its derivative with respect to temperature are continuous but the second derivative is discontinuous. Correspondingly there is a discontinuity in the specific heat as is observed in the superconducting transition.

We now wish to consider a system that is inhomogeneous and therefore in which the density of superconducting electrons and the order parameter vary with position. Since the state of minimum free energy in a uniform system corresponds to a uniform order parameter we can expect an increase in the free energy proportional to $|\nabla\psi|^2$. At this stage we rewrite all expressions in terms of the order parameter. The total free energy is written $F = \int f(\mathbf{r})\,d\tau$ and if the a and b given above are evaluated for a system of unit volume we have

$$f = f_n + a(T)|\psi(\mathbf{r})|^2 + \tfrac{1}{2}b(T)|\psi(\mathbf{r})|^4 + c(T)|\nabla\psi(\mathbf{r})|^2 \tag{5.85}$$

We expect c to be nonzero and positive at the critical temperature so we take it to be positive constant near the critical temperature.

Finally, we wish to add an interaction with magnetic fields by including the vector potential $\mathbf{A}$, which gives rise to a magnetic field $\mathbf{H} = \nabla \times \mathbf{A}$. The free energy must depend only on the field; it must not change if the gradient of some function of position is added to the vector potential (since its curl would be zero). This is simply the statement that the free energy must be *gauge invariant*. This is accomplished in the usual Schroedinger equation by introducing the vector potential as a term $ie\mathbf{A}/\hbar c$ added to the gradient, where e is again the magnitude of the charge on the electron. Ginsburg and Landau made that choice here, though clearly gauge invariance is maintained even if the magnitude of the linear coefficient, $e/\hbar c$, is modified. This can be done in general by replacing $-e$ by q^*, an effective charge. It is now recognized that the proper choice is $q^* = -2e$ where the 2 arises from the pairing condition in the microscopic theory. For the present we will retain the general form with q^*. Thus we have for the free-energy density

$$f = f_n + a|\psi(\mathbf{r})|^2 + \tfrac{1}{2}b|\psi(\mathbf{r})|^4 + c\left|\left(\nabla - \frac{iq^*\mathbf{A}}{\hbar c}\right)\psi(\mathbf{r})\right|^2 + \frac{1}{8\pi}\mathbf{A}\cdot(\nabla \times \nabla \times \mathbf{A}) \tag{5.86}$$

where of course the final term is the energy density due to the magnetic field.

10.2 The Ginsburg-Landau equations We now seek to minimize the total free energy with respect to variations in the order parameter and in the vector potential; i.e., we minimize F with respect to arbitary variations $\delta\psi$, $\delta\psi^*$ (regarded as independent), and $\delta\mathbf{A}$. This variation may be written out explicitly.

$$\delta F = \delta \int f\, d\tau = {}' \int \left\{ a\psi + b\psi^*\psi\psi + c\left[\left(\boldsymbol{\nabla} - \frac{iq^*\mathbf{A}}{\hbar c}\right)\psi\right] \right.$$

$$\left. \left[\boldsymbol{\nabla} + \frac{iq^*\mathbf{A}}{\hbar c}\right]\right\} \delta\psi^*\, d\tau + \text{c.c.} + c \int \left[\frac{iq^*\psi^*}{\hbar c}\left(\boldsymbol{\nabla} - \frac{iq^*\mathbf{A}}{\hbar c}\right)\psi + \text{c.c.}\right]$$

$$\cdot\, \delta\mathbf{A}\, d\tau + \frac{1}{4\pi}\int (\boldsymbol{\nabla} \times \boldsymbol{\nabla} \times \mathbf{A}) \cdot \delta\mathbf{A}\, d\tau$$

which is to be set equal to zero. A single partial integration on $\boldsymbol{\nabla}\,\delta\psi^*$ may be performed in the first integral to obtain an integrand in which $\delta\psi^*$ appears as a simple multiplicative factor and the coefficient must be equal to zero.

$$a\psi + b\psi^*\psi\psi - c\left(\boldsymbol{\nabla} - \frac{iq^*\mathbf{A}}{\hbar c}\right)^2 \psi = 0 \tag{5.87}$$

The complex-conjugate term will of course lead simply to the complex conjugate of this equation. In performing the partial integration we dropped a surface term which may be shown to vanish if the normal component of current vanishes at the surface.

Similarly, the final two integrals lead to an equation for $\mathbf{A}$. However, since $\boldsymbol{\nabla} \times \boldsymbol{\nabla} \times \mathbf{A}$ is equal to $4\pi\mathbf{j}/c$, where $\mathbf{j}$ is the current density, we may use this equation to obtain an expression for that current density.

$$\mathbf{j} = -\frac{iq^*c}{\hbar}\psi^*\left(\boldsymbol{\nabla} - \frac{iq^*\mathbf{A}}{\hbar c}\right)\psi + \text{c.c.} \tag{5.88}$$

The coefficient c may now be rewritten. The current density should be proportional to the density of superconducting electrons, while our normalization of ψ is such that $\psi^*\psi$ is the *fraction* of electrons which are superconducting; thus we need a factor n^* of the order of the density of electrons. Further, we need an inverse factor of mass to obtain proper units. This is taken to be m^*, of the order of the electron mass. Thus c is written $\hbar^2 n^*/2m^*$ and Eqs. (5.87) and (5.88) become:

$$-\frac{\hbar^2}{2m}\left(\boldsymbol{\nabla} - \frac{iq^*\mathbf{A}}{\hbar c}\right)^2 \psi + \frac{m^*a}{mn^*}\psi + \frac{m^*b}{mn^*}\psi^*\psi\psi = 0 \tag{5.89}$$

$$\mathbf{j} = \frac{n^*\hbar q^*}{2im^*}\left[\psi^*\left(\nabla - \frac{iq^*\mathbf{A}}{\hbar c}\right)\psi - \text{c.c.}\right] \tag{5.90}$$

Thus we have obtained a Schroedinger-like equation and the usual expression for current density for a particle of charge q^* and mass m^*. The nonlinearity of Eq. (5.89) is an important feature and has a profound effect upon the results of the theory. If q^* is replaced by $-e$, the electronic charge, m^* is replaced by m, the electron mass, and n^* is replaced by the density of electrons, these become the Ginsburg-Landau equations of 1950. If q^* is replaced by $-2e$, the charge on a Cooper pair, and m^* is replaced by $2m$, the mass of a Cooper pair, and n^* is replaced by half the density of electrons, these become the Ginsburg-Landau equations as presently used.

We can now see the power of this phenomenology. Given a choice of m^*, q^*, and n^*, there are only two parameters that enter the theory, α and β, from which a and b are given at any temperature near the critical temperature by Eq. (5.83). These two parameters can be determined by two experiments and from them the wide variety of properties of the superconductor may be computed.

One such experimental number is of course the difference in free energy between the normal and superconducting states which was given in Eq. (5.82). Such a measurement gives the ratio a^2/b, or equivalently α^2/β.

This quantity is in fact most conveniently measured indirectly by measuring the *critical field*. The two quantities may be related by consideration of Eq. (5.86). In the absence of a magnetic field the final term vanishes and we have a well-defined expression for the difference in free energy in terms of the order parameter. If a magnetic field is applied to a system it is well known that it will not penetrate the superconductor appreciably. This is again the Meissner effect. Thus the main portion of the condensation energy associated with the superconductor will remain unchanged; i.e., the order parameter remains the same within the bulk of the superconductor. However, the magnetic-field energy represented by the last term in Eq. (5.86) will be greater due to the presence of a superconductor that deforms the field lines around it. This extra field energy is in fact given by $H^2/8\pi$ times the volume of the superconductor, a result obtainable from thermodynamic arguments.[1] When the field is increased sufficiently this extra field energy will exceed the energy gain associated with the superconducting transition. At that point the free energy is lowered if the metal becomes normal and the field uniform. Thus the difference in free-energy density

[1] M. W. Zemansky, "Heat and Thermodynamics," p. 283, McGraw-Hill, New York, 1943.

between the normal and superconducting states is given by $H_c^2/8\pi$. Using also Eq. (5.82) we obtain

$$\frac{a^2}{b} = \frac{H_c^2}{4\pi} \tag{5.91}$$

We will see that there is in fact some penetration of the magnetic field into the superconductor. A weak field will decay into the superconductor as $H_0 e^{-z/\lambda(T)}$, where the *penetration depth* λ is related to a and b by

$$\frac{a}{b} = -\frac{m^* c^2}{4\pi n^* q^* \lambda(T)^2} \tag{5.92}$$

Equations (5.91) and (5.92) together give us a and b in terms of experimental quantities.

The Ginsburg-Landau equations have now been derived by Gor'kov.[1] (For an outline of the derivation see Schrieffer.[2]) This derivation leads to the effective charge, the effective mass, and n^* given above. It indicates also, as we stated earlier, that the order parameter ψ is proportional to the local value of the energy-gap parameter Δ of the microscopic theory. This is consistent with Eqs. (5.80), (5.81) and (5.83) which together imply that $\psi^*\psi$ is proportional to $T_c - T$, with Eq. (5.82) giving a condensation energy proportional to $(T - T_c)^2$ and the condensation energy from the microscopic theory of $n(E_F)\Delta_0^2/4$.

10.3 Applications of the Ginsburg-Landau theory Although the Ginsburg-Landau equations are derivable from the microscopic theory, they predate it and should be thought of as a theoretical achievement in their own right. Furthermore, in studying complex situations with the Ginsburg-Landau theory, results are most usually obtained in terms of parameters of the theory which ultimately come from experiment rather than from the microscopic theory. Thus in practice it is frequently utilized as an independent theory and the applications here will be discussed from that point of view.

The existence of a relation between the microscopic theory and the phenomenology does give a unity to the understanding of superconductivity and in a practical vein gives methods for estimating the parameters of the theory from first principles. It will be helpful therefore to make contact with the results of the microscopic theory at several points.

[1] L. P. Gor'kov, *JETP (USSR)*, **36**: 1918 (1959), translated in *Soviet Phys. JETP*, **9**: 1364 (1959).

[2] J. R. Schrieffer, "Theory of Superconductivity," p. 248, Benjamin, New York, 1964.

Zero-field solutions We see that there are trivial solutions of the first Ginsburg-Landau equation, Eq. (5.89), when there are no magnetic fields present. Then **A** may be taken equal to zero and ψ = constant becomes a solution, with $\psi^*\psi = -a/b$. From Eq. (5.90) it is then seen that the current is equal to zero. The value of the order parameter corresponds to that obtained through Eqs. (5.80) and (5.81). We note that as we have defined the order parameter it will not be normalized. In fact, the expansion parameter for the theory is $\int \psi^*\psi\, d\tau$, for a system with unit volume.

There are also solutions of Eq. (5.89) with $\mathbf{A} = 0$ in which ψ is a plane wave, $\psi = \psi_0 e^{i\mathbf{k}\cdot\mathbf{r}}$. If this is substituted in Eq. (5.90) it is found immediately that it corresponds to a uniform current density, $\mathbf{j} = (n^*\hbar q^*\mathbf{k}/m^*)\psi^*\psi$. This corresponds to the drifting state that we discussed in Sec. 9.3; a current generated in a superconducting ring will persist indefinitely without apparent decay. It should perhaps be mentioned that this is not the ground state of the system and there are processes by which the system may return to the ground state. However, these require a macroscopic activation energy and such processes are therefore exceedingly unlikely.

It is important to recognize that such a current will generate a nonzero vector potential so that this is not a self-consistent solution. The vector potential for a uniform current **j** lying in the z direction can be written $\mathbf{A} = -2\pi\mathbf{j}x^2/c$. The solution, however, can be very close to self-consistent if the current is in a very fine wire or a thin film, since then x^2 will always be small within the superconductor. These are interesting geometries and the physics is particularly clear in the simple case where the field can be neglected, so we will pursue it further.

In this state the magnitude of the order parameter is independent of position and therefore the energy-gap parameter and the density of superconducting electrons is the same everywhere. However, the phase of the order parameter (or the energy-gap parameter) varies with position and this variation of phase gives rise to the current.

We may obtain the magnitude of the current density as a function of the wavenumber **k** by substituting the plane-wave order parameter into Eq. (5.89), again dropping the term in the vector potential. We may solve for the order parameter and obtain

$$\psi_0^*\psi_0 = \frac{-a}{b} - \frac{n^*\hbar^2k^2}{2m^*b}$$

Recalling that a is negative, we see that the order parameter decreases as k increases. The result may be combined with Eq. (5.90) to obtain a current density of

$$\mathbf{j} = \frac{\hbar q^*\mathbf{k}n^*\psi_0^*\psi_0}{m^*} = \frac{\hbar q^*\mathbf{k}n^*}{m^*b}\left(-a - \frac{n^*\hbar^2k^2}{2m^*}\right)$$

As k increases from zero the current increases, reaches a maximum, and drops again to zero. The maximum value which the current may reach is readily evaluated and is given by

$$j_c = \frac{q^*\sqrt{n^*}}{\sqrt{m^*b}}\left(-\frac{2a}{3}\right)^{3/2}$$

The system cannot sustain a larger supercurrent and will become normal at current densities that exceed this *critical-current density.* The existence of a critical-current density was discussed in connection with the BCS theory also. There we noted that its measurement is tricky. It is important to notice that in both discussions we dropped the self-generated magnetic field and therefore were restricted to a particular geometry.

Nonuniform systems The above description has been restricted to homogeneous superconductors. The Ginsburg-Landau theory is of particular interest when applied to systems in which $|\psi|$ varies with position. This is a difficult problem to do exactly because of the nonlinearity of Eq. (5.89). However, the equation may be linearized for small deviations from uniformity. We will seek such solutions for a real-order parameter and will see that this is consistent with our results. We will also drop the vector potential, but since our solutions will correspond to vanishing current, they will be consistent in that sense also.

We write the order parameter as a constant term ψ_0 with an additional small term ψ_1 which may vary with position. By substituting $\psi = \psi_0 + \psi_1$ into Eq. (5.89) and separating the terms of zero and first order in ψ_1, we obtain

$$a\psi_0 + b\psi_0{}^3 = 0$$

$$-\frac{\hbar^2}{2m}\nabla^2\psi_1 + \frac{m^*a}{mn^*}\psi_1 + \frac{3m^*b}{mn^*}\psi_0{}^2\psi_1 = 0$$

Solving the first for $\psi_0{}^2$ and substituting in the second, we obtain

$$-\frac{\hbar^2}{2m}\nabla^2\psi_1 - \frac{2m^*a}{mn^*}\psi_1 = 0$$

We see that there is a natural unit of length for spatial variations in the order parameter. It is called the *coherence distance* $\xi(T)$ which is defined by

$$\xi(T) = \sqrt{-\frac{\hbar^2 n^*}{2m^*a}} \tag{5.93}$$

We see there are solutions which decay in one dimension as $\exp(-\sqrt{2}\, z/\xi)$ or spherically symmetric decaying solutions of the form $\exp(-\sqrt{2}\, r/\xi)/r$. There is a rigidity of the superconducting wavefunctions which inhibits variations of the order parameter over distances less than the coherence length.

The coherence length enters in a very natural way also in the microscopic theory. The superconducting state was constructed by making suitable linear combinations of wavefunctions over an energy range of the order of Δ. The degree to which such a coherent state of the system can be localized is limited by the possible localization of packets using states in this energy range. But the possible localization is to distances of the order of $\partial E/\partial k|_{k_F}/\Delta = \hbar v_F/\Delta$ which, except possibly for a numerical constant, defines a natural coherence length in the superconductor. Specifically the coherence length is more usually defined from the microscopic theory as $\xi_0 = \hbar v_F/\pi\Delta$. Taking again an energy-gap parameter of 1 mev and a Fermi velocity of 10^8 cm/sec we obtain a coherence length of 2000 Å.

Applied magnetic fields A second characteristic length in superconductors is the penetration depth in weak magnetic fields, which we mentioned earlier. We consider a semi-infinite superconductor occupying the region $z > 0$ and apply a weak magnetic field in the y direction. It will be a constant H_0 for $z < 0$. We take the curl of both sides of the equation for the current density in the superconductor [Eq. (5.90)] and may note that only the term in $\nabla \times \mathbf{A}$ appears on the right to lowest order in the magnetic field. This may be seen by again writing the order parameter as a constant term plus a small spatially dependent term which will be first order in H. The expression $\nabla \times (\psi^* \nabla \psi)$ contains terms such as $\dfrac{\partial \psi^*}{\partial y} \dfrac{\partial \psi}{\partial z}\, \hat{x}$ but terms of the form $\dfrac{\psi^*\, \partial^2 \psi}{\partial y\, \partial z}\, \hat{x}$ cancel. The contributing terms are second order in ψ_1 and negligible in comparison to the terms in $\nabla \times \mathbf{A}$. Thus we obtain by taking the curl of Eq. (5.90)

$$\nabla \times \mathbf{j} = -\frac{n^* q^{*2} \psi_0^* \psi_0}{m^* c} \nabla \times \mathbf{A} = -\frac{n^* q^{*2} \psi_0^* \psi_0}{m^* c} \mathbf{H} \tag{5.94}$$

where again ψ_0 is the zero-field value of the order parameter. But from Maxwell's equations, $\nabla \times \mathbf{H} = 4\pi \mathbf{j}/c$, these may be combined to give

$$\nabla \times \nabla \times \mathbf{H} = -\frac{4\pi q^{*2} n^*}{m^* c^2} \psi_0^* \psi_0\, \mathbf{H}$$

This has a solution for fields in the y direction of $H = H_0 e^{-z/\lambda}$, with the *penetration depth* λ given by

$$\left[\frac{1}{\lambda(T)}\right]^2 = \frac{4\pi n^* q^{*2} \psi_0^* \psi_0}{m^* c^2} = -\frac{4\pi n^* q^{*2} a}{m^* c^2 b}$$

We see as we indicated in Sec. 10.2 that the penetration depth depends upon the ratio a/b.

We note that both the penetration depth defined in Eq. (5.94) and the coherence distance defined in Eq. (5.93) vary with temperature as $a^{-1/2}$ or $(T_c - T)^{-1/2}$ near the critical temperature. Thus their ratio is a dimensionless constant which is characteristic of the superconductor and not of the temperature. This ratio,

$$\kappa = \frac{\lambda}{\xi}$$

is called the *Landau-Ginsburg parameter* for the material.

In simple metals which are reasonably pure the Landau-Ginsburg parameter tends to be less than 1. These are called superconductors of the first kind or *type I superconductors*. In simple metals which are quite impure and in transition-metal superconductors, the Landau-Ginsburg parameter is greater than 1. These are called superconductors of the second kind or *type II superconductors*. It is perhaps not surprising that the response of these two types of superconductors to applied fields is quite different.

10.4 Flux quantization We consider next a multiply-connected superconductor which might be, for example, a superconducting ring as shown in Fig. 5.18. There may be magnetic fields threading the ring but the fields will be zero within the body of the superconductor. Furthermore, since the fields within the superconductor are zero, the current density, which is proportional to the curl of the magnetic field, must also be zero. Currents are restricted to the penetration depth. We may therefore define a path of integration around the ring within the interior of the superconductor such that the current is everywhere zero along that path. We wish to consider formally the integral $\oint \mathbf{j} \cdot d\mathbf{l}$ around such a path using Eq. (5.90).

$$\oint \mathbf{j} \cdot d\mathbf{l} = \frac{n^* \hbar q^*}{2im^*}\left[\oint \psi^* \nabla \psi \cdot d\mathbf{l} - \frac{iq^*}{\hbar c}\oint \psi^* \psi \mathbf{A} \cdot d\mathbf{l}\right] + \text{c.c.}$$

The magnitude of the order parameter is taken to be the same over the entire path but its phase may vary. Thus we write the order parameter along the path $\mathbf{l}$ as $\psi = \psi_0 e^{i\theta(l)}$. Then $\psi^*\psi$ may be taken out of the second

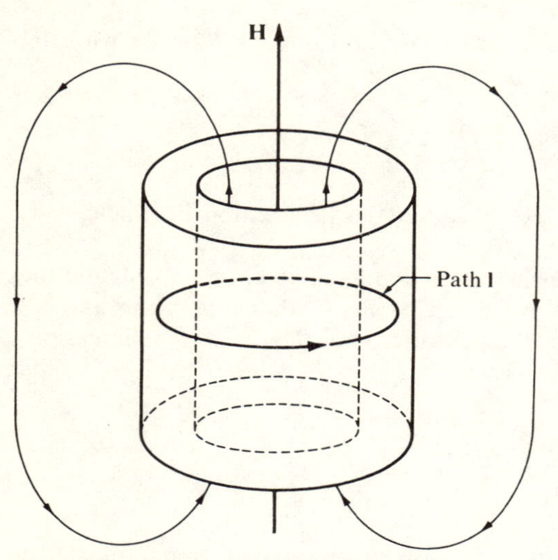

Fig. 5.18 A superconducting ring threaded by a magnetic field **H**. *Along the path* **l** *in the interior the current density* **j** *is identically zero and therefore the integral* $\oint \mathbf{j} \cdot d\mathbf{l} = 0$.

integral and $\oint \mathbf{A} \cdot d\mathbf{l}$ is simply the flux contained within the ring, Φ. The integral.

$$\oint \psi^* \nabla \psi \cdot d\mathbf{l} = i\psi_0^* \psi_0 \oint \nabla \theta(\mathbf{l}) \cdot d\mathbf{l}$$

is simply $i\psi_0^*\psi_0$ times the change in the value of θ in completing the cycle. The order parameter itself must be single-valued but of course θ may increase by an integral multiple of 2π. The terms in $\mathbf{A}$ and in $\nabla\psi$ must cancel identically since $\oint \mathbf{j} \cdot d\mathbf{l} = 0$, which leads to a condition on the flux contained within the ring,

$$\Phi = \frac{2\pi \hbar c}{q^*} n \tag{5.95}$$

where n is an integer. Flux can occur within the superconducting ring only in integral multiples of the *flux quantum* hc/q^*. This is a very small flux indeed. It corresponds to the flux associated with the magnetic moment of a single electron. This quantization of flux has been confirmed experimentally.[1]

[1] B. S. Deaver, Jr., and W. M. Fairbank, *Phys. Rev. Letters*, **7**:43 (1961); R. Doll and M. Näbauer, *Phys. Rev. Letters*, **7**:51 (1961).

It was found that the appropriate effective charge is in fact twice the electronic charge. The derivation of Eq. (5.95) depended upon the fact that the superconductor is sufficiently thick that there are current-free regions within the superconductor and modifications must be made if it is not.

The argument is also of interest in superconductors that are not multiply connected. Clearly if a moderate magnetic field is applied normal to a thin superconducting sheet, the lowest-energy configuration will not be that in which the flux is excluded from the entire sheet since this would require a very large field energy. Furthermore, it is clear that the minimum free energy will not be obtained with the entire system normal since then all of the condensation energy is lost. Instead we may expect a minimum energy configuration when the flux lines penetrate the superconducting sheet in small bundles distributed over the sheet. This will retain most of the condensation energy, although perhaps small normal regions will be required, and at the same time will greatly lower the field energy. This state of the system is known as the *intermediate state*.

We may apply the quantization argument given above to each of those bundles and find they also must be quantized. It turns out that the minimum energy ordinarily corresponds to single flux quanta threading the superconducting sheet. These are the so-called *vortices* of superconductors. Each consists of a single flux quantum and associated supercurrents surrounding it. In type II superconductors such vortices may occur readily even without the geometry of a thin plate.

10.5 Fluctuations in superconductors In the framework of the Ginsburg-Landau theory we describe the state of the superconducting electrons in terms of a well-defined order parameter ψ. Viewing the order parameter as a superconducting wavefunction we may imagine that there are neighboring states of similar energy, and at a temperature near the transition temperature, where the Ginsburg-Landau theory is applicable, the system should be describable by a statistical distribution of such states. Thus the order parameter as we have used it really represents some average and we may anticipate thermal fluctuations around that average. Such fluctuations near the transition temperature have been the center of extensive research in recent years not only in superconductivity but in other systems showing phase transitions. Here we will see how they can be studied within the framework of the Ginsburg-Landau theory.

We consider a homogeneous superconductor with no fields present. We write ψ as a constant term ψ_0 plus fluctuations which are expanded in a Fourier series,

$$\psi = \psi_0 + \sum_k a_k e^{i\mathbf{k}\cdot\mathbf{r}} \tag{5.96}$$

We imagine such a state to have a well-defined energy $E(\{a_k\})$ which depends upon all of the parameters $\{a_k\}$. The statistical probability of finding the system with each of the parameters in some range $d\{a_k\} = da_{k_1} da_{k_2} \ldots$ is given by

$$dP = d\{a_k\}\, P(\{a_k\}) e^{-E(\{a_k\})/KT} \tag{5.97}$$

where $P(\{a_k\})$ is the appropriate statistical weight depending upon the number of microscopic states corresponding to a given $\{a_k\}$. We do not know that function but fortunately if we write

$$P(\{a_k\}) = \exp[(KT/KT) \ln|P(\{a_k\})|] = e^{TS/KT}$$

where S is the entropy of the system, $K \ln|P(\{a_k\})|$, to within a constant term, we see that the probability given in Eq. (5.97) becomes simply $d\{a_k\} e^{-F(\{a_k\})/KT}$, to within a constant factor, where F is again the total *free energy*. Thus we may estimate the probability of given fluctuations using Eq. (5.85) for the free-energy density.

Substituting Eq. (5.96) for ψ in Eq. (5.85) and integrating over the volume Ω of the system, we obtain

$$F = F_n + \Omega \left[a\psi_0^* \psi_0 + \frac{b}{2} (\psi_0^* \psi_0)^2 + \sum_k a_k^* a_k (a + 2b\psi_0^* \psi_0 + k^2 c) \right] \tag{5.98}$$

to lowest order in the $\{a_k\}$. We have dropped terms in $a_k a_{-k}$ and $a_k^* a_{-k}^*$ since the phases of different fluctuations are taken to be independent and such terms will vanish in a statistical average.

We now fix ψ_0 by minimizing Eq. (5.98) with respect to ψ_0 to zero order in $a_k^* a_k$. This minimization leads to

$$a + b\psi_0^* \psi_0 = 0$$

from which

$$F = F_n + \Omega \left[-\frac{b\psi_0^* \psi_0}{2} + \sum_k a_k^* a_k (-a + k^2 c) \right]$$

The first two terms may be absorbed in the proportionality constant in the statistical probability and we may evaluate the mean-square fluctuation amplitude for a particular $\mathbf{k}$ from

$$\langle a_k^* a_k \rangle = \frac{\int e^{-a_k^2(-a + k^2 c)\Omega/KT}\, a_k^2\, da_k}{\int e^{-a_k^2(-a + k^2 c)\Omega/KT}\, da_k}$$

$$= \frac{KT}{2(-a + k^2 c)\Omega}$$

This should give a good estimate of the amplitude of fluctuations in the realm where the Ginsburg-Landau theory obtains, i.e., at temperatures near the critical temperature and for fluctuations of wavelength long compared to the coherence length. For shorter wavelengths the rigidity of the superconducting wavefunction discussed in Sec. 10.3 suppresses the fluctuations. We note that the total volume appears in the denominator. However, when we compute any property in terms of the fluctuations, we will need to sum over $\mathbf{k}$ and this sum will be written in terms of an integral over wavenumber space with a density factor $\Omega/(2\pi)^3$ and the total volume will drop out. In particular the expectation value for the amplitude of the squared fluctuation at any point in real space will be given by a direct sum of $\langle a_k^* a_k \rangle$ over wavenumber. Such a sum diverges if carried out over all wavenumbers and should be restricted to wavenumbers less than of the order of $2\pi/\xi$ (where ξ is the coherence length), beyond which the Ginsburg-Landau theory breaks down. This has the effect, in order of magnitude, of replacing Ω by the *coherence volume*, ξ^3. We have

$$\sum_k \langle a_k^* a_k \rangle \approx \frac{KT}{2(-a)\xi^3} \tag{5.99}$$

where we have neglected ck^2 in the denominator in comparison to $-a$.

This result can be made more meaningful by comparing it to the mean-squared order parameter, $\psi_0^*\psi_0$, given by $-a/b$ in Eq. (5.81). Finally we note that the difference in free energy Δf per unit volume between the normal and superconducting state is given in Eq. (5.82) by $a^2/2b$. Then Eq. (5.99) becomes (dropping the numerical factor)

$$\frac{\sum_k \langle a_k^* a_k \rangle}{\psi_0^* \psi_0} \approx \frac{KT}{\Delta f\, \xi^3}$$

The relative mean-square deviation at any given temperature is given by KT divided by the energy required to take a coherence volume of the material from the superconducting to the normal state at that temperature. Roughly speaking, because of the rigidity of the superconducting wavefunction, one point in the superconductor cannot go normal without taking a coherence volume of superconductor with it. The fluctuations tend to be very small but become important when we study the properties very close to the transition temperature.

Problems

5.1 In a degenerate electron gas with equal population of spin-up and spin-down electrons, the kinetic energy increases when a small number of electron spins are flipped (by raising and lowering the two Fermi wavenumbers). However, the exchange energy drops when this is done.

The latter contribution is enhanced with respect to the former when the electron density is decreased or the effective mass is increased. Find the effective mass at which the nonmagnetic state becomes unstable under the influence of these two energies alone, and evaluate it for an electron gas with the electron density of aluminum ($k_F = 0.927$ atomic unit^{-1}).

5.2 Consider a system of N spins (each spin $\frac{1}{2}$) coupled equally by Heisenberg exchange. (Take N even.)

$$\mathscr{H} = J \sum_{i>j,j} \mathbf{S}_i \cdot \mathbf{S}_j = \frac{J}{2} \sum_{i,j}{}' \mathbf{S}_i \cdot \mathbf{S}_j$$

with $J > 0$. The prime indicates the terms with $i = j$ are to be omitted.

a. What is the exact ground-state energy? (Hint: It will also be an eigenstate of $\mathbf{S}^2$ where $\mathbf{S} = \sum_i \mathbf{S}_i$.) What is the total spin of the ground state?

b. What is the expectation value of

$$\mathscr{H}_I = J \sum_{i>j,j} S_i^z S_j^z = \frac{J}{2} \sum_{i,j}{}' S_i^z S_j^z$$

in the antiferromagnetic state of Fig. 5.2 and in the ground state of part *a*.

5.3 Consider three magnetic ions, each of spin $\frac{1}{2}$, each coupled to the other by Heisenberg exchange. We apply a field $\mathbf{H}$, so the Hamiltonian is

$$\mathscr{H} = -\sum_{i,j}{}' J \mathbf{S}_i \cdot \mathbf{S}_j - 2\mu_0 \sum_i \mathbf{H} \cdot \mathbf{S}_i$$

In the Ising model states can be specified by assigning $+\frac{1}{2}$ or $-\frac{1}{2}$ to the individual ions ($\mathbf{S}_i \cdot \mathbf{S}_j = \pm\frac{1}{4}$). By taking a thermal average over the eight states, obtain the magnetization

$$\langle \mathbf{M} \rangle = 2\mu_0 \langle \mathbf{S}_1 + \mathbf{S}_2 + \mathbf{S}_3 \rangle$$

and the contribution to the susceptibility,

$$\left. \frac{\partial \langle M \rangle}{\partial H} \right|_{H=0}$$

(Leave μ_0^2 in the answer rather than attempting to evaluate $\langle \mu^2 \rangle$.)

5.4 We have constructed magnons as linear combinations of single-spin flips. We might instead attempt to construct excited states in one-dimensional ferromagnets using double-spin flips,

$$\Psi = \mathbf{A} \sum_n S_n^- S_{n+1}^- \Psi_0 e^{iqan}$$

which are approximate eigenstates of the system. The $\mathbf{A}$ is a normalization constant.

a. Demonstrate that this is not an eigenstate of the Hamiltonian we used to construct magnons. What is the form of the additional terms needed? (Not the coefficients, just the combinations of spin operators.)

b. If Heisenberg exchange is replaced by the Ising model with nearest neighbor for this system, the magnon energy becomes $2JS$ and is independent of q. What is the energy of the state given above in this model? Viewed as a pair of interacting magnons bound to each other, what is the binding energy? Bound magnons have been discussed by Torrance and Tinkham.[1]

[1] J. B. Torrance, Jr., and M. Tinkham, *J. Appl. Phys.*, **39**:822 (1968), and J. B. Torrance, Jr., to be published.

5.5 Consider a simple metal with a local moment (spin ½) on a dissolved transition-metal impurity. Let the Hamiltonian be approximated by

$$\mathscr{H} = \sum_k E_k(n_{k+} + n_{k-}) + E_d(n_{d+} + n_{d-}) + Un_{d+}n_{d-} - \frac{J}{N}\sum_{\substack{k,k'\\ \sigma,\sigma'}} c_{k'\sigma'}^{+}\,\mathbf{S}_d \cdot \mathbf{S}_\sigma c_{k\sigma}$$

The S_σ operates on the spin state indexed σ. Find an expression for the total scattering of a state

$$c_{k_0\sigma_0}^{+} \prod_{k<k_F} c_{k+}{}^{+} c_{k-}{}^{+} c_{d+}{}^{+}|0\rangle$$

(where $k_0 > k_F$) due to the final term in the Hamiltonian. Include only processes in which the state $k_0\sigma_0$ is emptied and compute transition probabilities only to order J^2, but treat both cases, σ_0 plus and minus. No operators should appear in the result.

5.6 Consider a metal described by a BCS reduced Hamiltonian

$$\mathscr{H} = \sum_k 2\epsilon_k b_k{}^{+} b_k + \sum_{k,k'} V_{kk'} b_{k'}{}^{+} b_k$$

and let

$$V_{kk'} = \lambda|\epsilon_k||\epsilon_{k'}|$$

for

$$|\epsilon_k| < \hbar\omega_D \qquad \text{and} \qquad |\epsilon_{k'}| < \hbar\omega_D$$

and zero otherwise. Here ϵ_k is measured from the Fermi energy.

a. By solving the energy-gap equation, find Δ_k, and also write a criterion for the existence of superconductivity. [This will be a condition on $N(0)\lambda(\hbar\omega_D)^2$ required for there to be a solution with real Δ_k.]

b. Sketch the density of excited states as a function of energy (out to and beyond $\hbar\omega_D$). This would be called a "gapless superconductor."

5.7 Estimate the superconducting critical temperature for a simple model. We obtained an approximate form for the energy-gap equation at $T = 0$,

$$1 = \frac{Vn(E_F)}{4}\int_{-\hbar\omega_D}^{\hbar\omega_D} \frac{dE}{\sqrt{\epsilon^2 + \Delta^2}}$$

assuming that all single-electron states were available for construction of the superconducting state.

Because of thermal excitation of quasiparticles, fewer pair states are available and the integrand above is reduced by a factor tanh $(\sqrt{\epsilon^2 + \Delta^2}/2KT)$.[1] This form differs from what would be obtained from a simple-minded approach, but approaches 0 for excitation energies much less than KT and 1 for large excitation energies, as expected.

a. Obtain the critical temperature T_c as the temperature at which Δ goes to zero in the temperature-dependent energy-gap equation. For simplicity, use the approximation

$$\tanh x \to x \qquad \text{for } 0 < x \leq 1$$
$$\qquad 1 \qquad \text{for } 1 < x < \infty$$

in the integration.

b. Noting $e^{2/n(E_F)V} \gg 1$, compare KT_c and the value of Δ at $T = 0$.

[1] deGennes, *op. cit.*, p. 123.

5.8 Consider the circuit of Fig. 5.15 with a large voltage V_0 applied. If C is large, almost no current will flow through R, and the current through J and C will be $J_1 \sin(2eVt/\hbar)$. No energy is lost in the circuit and no work is done by the battery.

a. Determine the current through R to lowest order in $1/C$ and by conservation of energy find what other current must be flowing in the circuit. [It will be of order $(1/C)^2$].

b. Viewed from the junction, what causes it to pass this additional current?

5.9 In the Ginsburg-Landau theory, the free energy of the superconducting state is given relative to that of the normal state by

$$f = f_n + a|\psi|^2 + \tfrac{1}{2}b|\psi|^4 + c|\nabla\psi|^2$$

with $a(T) = \alpha(T - T_c)/T_c$ negative below the critical temperature. It is reasonable to extend the expression, with α, $b = \beta$, and c constant above the critical temperature, though the stable state is normal, $n_s = 0$.

a. Construct the *linearized* Ginsburg-Landau equation for the material above T_c and obtain solutions for variation in one dimension.

b. Using this description for a normal metal with $\alpha_1, \beta_1, c_1, T_{c1} < T$, find the order parameter as a function of position near a plane boundary separating the normal metal from a superconductor with α_0, β_0, c_0, $T_{c0} > T$. (Linearize the equations for the superconductor around a constant ψ_0.) Take the order parameter everywhere continuous and obtain the boundary condition for the gradient by requiring continuous current, Eq. (5.88), across the boundary.

c. Sketch the order parameter for two identical superconductors separated by a normal metal film. The film may provide a weak link between superconductors just as in the Josephson effect.

5.10 Consider a Josephson junction with a uniform magnetic field in the oxide, as shown below, in terms of the Ginsburg-Landau order parameter ψ. In the upper superconductor (beyond the penetration depth which we take here equal to zero), the ψ may be written $\psi_+(x) = \psi_0 e^{i(\alpha x + \delta\varphi)}$ with ψ_0 and $\delta\varphi$ constant. In the lower superconductor, $\psi_-(x) = \psi_0 e^{-i\alpha x}$.

a. Determine α in terms of the magnetic field H. In the line integral you may neglect the contribution of the path parallel to z at the ends.

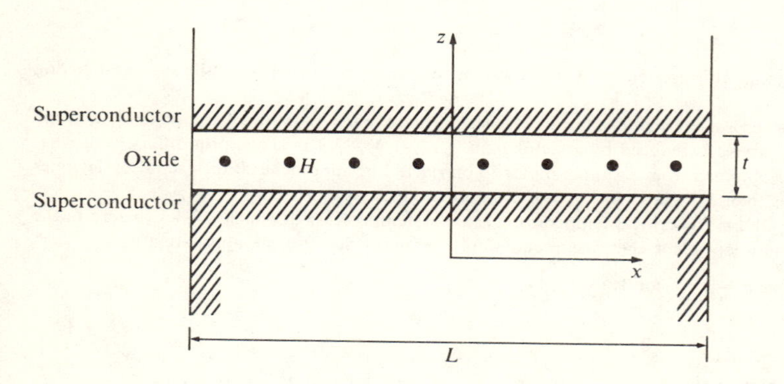

b. Determine the maximum dc supercurrent $J_1(\mathrm{H})$ which the junction can pass. You will need to define a Josephson parameter per unit length j_1 such that

$$J_1(0) = \int_{-L/2}^{L/2} j_1 \, dx = L j_1$$

and the current *density* at any point in the junction is proportional to j_1 times the sine of the phase difference between ψ_+ and ψ_- at that point.

c. Write J_1 as a function of the number (which need not be an integer) of flux quanta in the junction and sketch the result.

(Note: In a real junction, the penetration depth will ordinarily be much greater than the oxide thickness and t should be replaced by a distance of the order twice the penetration depth.)

INDEX